Current Topics in Microbiology and Immunology

Volume 335

Current Topics in Microbiology and Immunology

Previously published volumes

Further volumes can be found at springer.com

Vol. 310: **Doerfler, Walter; Böhm, Petra (Eds.):** DNA Methylation: Development, Genetic Disease and Cancer. 2006. 25 figs. X, 284 pp. ISBN 3-540-31180-7

Vol. 311: **Pulendran, Bali; Ahmed, Rafi (Eds.):** From Innate Immunity to Immunological Memory. 2006. 13 figs. X, 177 pp. ISBN 3-540-32635-9

Vol. 312: **Boshoff, Chris; Weiss, Robin A. (Eds.):** Kaposi Sarcoma Herpesvirus: New Perspectives. 2006. 29 figs. XVI, 330 pp. ISBN 3-540-34343-1

Vol. 313: **Pandolfi, Pier P.; Vogt, Peter K.(Eds.):** Acute Promyelocytic Leukemia. 2007. 16 figs. VIII, 273 pp. ISBN 3-540-34592-2

Vol. 314: **Moody, Branch D. (Ed.):** T Cell Activation by CD1 and Lipid Antigens, 2007, 25 figs. VIII, 348 pp. ISBN 978-3-540-69510-3

Vol. 315: **Childs, James, E.; Mackenzie, John S.; Richt, Jürgen A. (Eds.):** Wildlife and Emerging Zoonotic Diseases: The Biology, Circumstances and Consequences of Cross-Species Transmission. 2007. 49 figs. VII, 524 pp. ISBN 978-3-540-70961-9

Vol. 316: **Pitha, Paula M. (Ed.):** Interferon: The 50th Anniversary. 2007. VII, 391 pp. ISBN 978-3-540-71328-9

Vol. 317: **Dessain, Scott K. (Ed.):** Human Antibody Therapeutics for Viral Disease. 2007. XI, 202 pp. ISBN 978-3-540-72144-4

Vol. 318: **Rodriguez, Moses (Ed.):** Advances in Multiple Sclerosis and Experimental Demyelinating Diseases. 2008. XIV, 376 pp. ISBN 978-3-540-73679-9

Vol. 319: **Manser, Tim (Ed.):** Specialization and Complementation of Humoral Immune Responses to Infection. 2008. XII, 174 pp. ISBN 978-3-540-73899-2

Vol. 320: **Paddison, Patrick J.; Vogt, Peter K.(Eds.):** RNA Interference. 2008. VIII, 273 pp. ISBN 978-3-540-75156-4

Vol. 321: **Beutler, Bruce (Ed.):** Immunology, Phenotype First: How Mutations Have Established New Principles and Pathways in Immunology. 2008. XIV, 221 pp. ISBN 978-3-540-75202-8

Vol. 322: **Romeo, Tony (Ed.):** Bacterial Biofilms. 2008. XII, 299. ISBN 978-3-540-75417-6

Vol. 323: **Tracy, Steven; Oberste, M. Steven; Drescher, Kristen M. (Eds.):** Group B Coxsackieviruses. 2008. ISBN 978-3-540-75545-6

Vol. 324: **Nomura, Tatsuji; Watanabe, Takeshi; Habu, Sonoko (Eds.):** Humanized Mice. 2008. ISBN 978-3-540-75646-0

Vol. 325: **Shenk, Thomas E.; Stinski, Mark F. (Eds.):** Human Cytomegalovirus. 2008. ISBN 978-3-540-77348-1

Vol. 326: **Reddy, Anireddy S.N; Golovkin, Maxim (Eds.):** Nuclear pre-mRNA processing in plants. 2008. ISBN 978-3-540-76775-6

Vol. 327: **Manchester, Marianne; Steinmetz, Nicole F. (Eds.):** Viruses and Nanotechnology. 2008. ISBN 978-3-540-69376-5

Vol. 328: **van Etten, (Ed.):** Lesser Known Large dsDNA Viruses. 2008. ISBN 978-3-540-68617-0

Vol. 329: **Diane E. Griffin; Michael B.A. Oldstone (Eds.):** Measles 2009. ISBN 978-3-540-70522-2

Vol. 330: **Diane E. Griffin; Michael B.A. Oldstone (Eds.):** Measles 2009. ISBN 978-3-540-70616-8

Vol. 331 **Villiers, E. M. de (Eds.):** TT Viruses. 2009. ISBN 978-3-540-70917-8

Vol. 332 **Karasev A.(Ed.):** Plant produced Microbial Vaccines. 2009. ISBN 978-3-540- 70857-5

Vol. 333 **Richard W. Compans; Walter A. Orenstein (Eds.):** Vaccines for Pandemic Influenza. 2009. ISBN 978-3-540-92164-6

Vol. 334 **Dorian McGavern; Michael Dustin (Eds.):** Visualizing Immunity. 2009. ISBN 978-3-540-93862-0

Beth Levine · Tamotsu Yoshimori · Vojo Deretic
Editors

Autophagy in Infection and Immunity

Springer

Editors
Beth Levine
Howard Hughes Medical Institute
Department of Internal Medicine
and Microbiology
University of Texas
Southwestern Medical Center
5323 Harry Hines Blvd.
Dallas TX 75390-9113,
USA
beth.levine@utsouthwestern.edu

Tamotsu Yoshimori
Department of Cell Regulation
Research Insitute for Microbial Diseases
Osaka University
3-1, Yamadaoka, Suita, Osaka 565-0871
Japan
tamyoshi@biken.osaka-u.ac.jp

Vojo Deretic
Department of Molecular
Genetics and Microbiology
Health Sciences Center
University of New Mexico
915 Camino de Salud NE.,
Albuquerque NM 87131
USA
vderetic@salud.unm.edu

ISBN 978-3-642-00301-1 e-ISBN 978-3-642-00302-8
DOI 10.1007/978-3-642-00302-8
Springer Dordrecht Heidelberg London New York

Current Topics in Microbiology and Immunology ISSN 0070-217x

Library of Congress Catalog Number: 2009926065

Cover legend: The cover image illustrates different stages of the autophagic degradation of bacteria, including an isolation membrane surrounding bacteria, an autophagosome containing intact bacteria, and an autolyososome containing partially degraded bacteria. The discovery of autophagic degradation of bacteria was a seminal observation that opened up the field of autophagy in infection and immunity.

Cover illustration by Angela Diehl and Anthony Orvedahl, University of Texas Southwestern Medical Center, Dallas, Texas, USA

Cover design: WMX Design Gmbh, Heidelberg

Printed on acid-free paper

Springer is part of Springer Science+Business Media (www.springer.com)

Preface

Autophagy is a fundamental biological process that enables cells to adjust cytoplasmic mass, quality and organization through capture and autodigestion of their cytoplasmic components. Autophagic targets range in size and complexity from individual long-lived macromolecules to whole organelles and microbial invaders. The principal role of this ubiquitous eukaryotic homeostatic mechanism is to ensure cell survival under adverse conditions, including nutrient absence, growth factor withdrawal, accumulation of toxic protein aggregates, and faulty organelles (e.g., leaky mitochondria), or infection by intracellular pathogens. The physiologic and pathophysiologic roles of autophagy (and defects in autophagy) are vast, encompassing cancer, neurodegeneration, metabolic diseases, aging, and (as of more recently) immunity. The immunological roles of autophagy fall into two broad categories, including: (1) effects on the control of general homeostasis in immune cells that parallel its roles in other cell types in the body, and; (2) effects on specialized functions of immune cells or other cellular targets of infection that enable the host to effectively deal with microbes or microbial products.

In this volume, experts in autophagy provide overviews and more detailed dissections of the basic molecular and cellular mechanisms of autophagy (chapter by Yang and Klionsky), the signaling cascades that control these processes (chapter by Codogno and colleagues), and the fundamental and applied physiological roles of autophagy (chapter by Mizushima). The role of autophagy in cellular homeostasis as it relates to immunity is covered in two chapters; Pua and He describe the role of autophagy in lymphocyte homeostasis and Espert and Biard-Piechaczyk describe the effects of HIV on lymphocyte cell death through autophagy. These chapters provide contrasting examples of how autophagy can be used to normally adjust lymphocyte populations or be misdirected by a potent virus to deplete certain types of immune cells. A similar theme of "pros and co s" is seen in the chapters that comprise the bulk of this volume, covering what is likely to be the most ancient specialized immune function of autophagy: the direct elimination of intracellular microbes. In the chapters by Orvedahl and Levine, Deretic and colleagues, Huang and Brumell, Yoshimori and Amano, and Sabauste, autophagy is revealed in its primordial immune form of an innate, cell-autonomous defense against the microbes that manage to erode into or specifically invade the interior of the eukaryotic cell. The chapters by Kirkegaard, Orvedahl and Levine, and Campoy and Colombo reveal the flip side of these relationships, addressing specialized adaptations that successful intracellular pathogens have evolved to deal with autophagy as an antimicrobial mechanism. Moreover, the broad role of autophagy as a cell-autonomous innate immunity mechanism that is deeply engrained in the eukaryotic cell is evidenced by its role in plants, as discussed in the chapter by Dinesh-Kumar and colleagues. Lastly, the chapters by Tal and Iwasaki, Gannagé and Münz, and Virgin and colleagues cover some of the most intriguing immunological applications of autophagy, including: (1) the ability of autophagosomes to capture cytosolic microbial products for presentation to the lumenally-oriented innate immunity receptors such as Toll-like receptors (Tal and Iwasaki), thus activating innate immunity; (2) a similar topological inversion function of autophagy in antigen presentation, whereby microbial antigens expressed in the cytosol of the host cell can be processed and loaded onto lumen-facing MHC II molecules for proper presentation to T cells (Gannagé and Münz), thereby functioning in adaptive immunity as well as in thymic T-cell selection; and (3) the role of autophagy genes as their function relates to specialized aspects of highly differentiated cells in the context of understanding inflammatory disorders such as Crohn's disease (Virgin and colleagues).

The chapters in the book describe an increasingly complex set of interactions that are beginning to be unraveled between the autophagy pathway, infectious diseases, and immunity. In its most primal form, autophagy may be a universal defense mechanism by which individual eukaryotic cells (or unicellular organisms) protect themselves in a cell-autonomous fashion by "eating" the microbes that attack them. With the evolution of metazoan organisms, the autophagy pathway has likely been utilized to enable such organisms to develop a more complex immune system. Not only is autophagy a primordial defense mechanism, it is also a pathway that shapes the dynamics of immune cell populations, that contributes to immunological tolerance and the control of inflammation, and that links pathogen recognition to the activation of innate and adaptive immunity. And—based upon the pace of new discoveries in this area—this list may just be the tip of the iceberg. Hopefully, this volume of *Current Topics in Microbiology and Immunology* will stimulate investigators to delve even deeper into this exciting field.

Dallas, TX, USA — Beth Levine
Osaka, Japan — Tamotsu Yoshimori
Albuquerque, NM, USA — Vojo Deretic

Contents

An Overview of the Molecular Mechanism of Autophagy 1
Zhifen Yang and Daniel J. Klionsky

Macroautophagy Signaling and Regulation 33
Audrey Esclatine, Magali Chaumorcel, and Patrice Codogno

Physiological Functions of Autophagy 71
Noboru Mizushima

Autophagy and Lymphocyte Homeostasis 85
Heather H. Pua and You-Wen He

Autophagy and Innate Recognition Systems 107
Michal Caspi Tal and Akiko Iwasaki

Autophagy in MHC Class II Presentation of Endogenous Antigens 123
Monique Gannagé and Christian Münz

Role of Autophagy and Autophagy Genes in Inflammatory Bowel Disease 141
Ken Cadwell, Thaddeus S. Stappenbeck, and Herbert W. Virgin

Autophagy in Immunity Against *Mycobacterium tuberculosis*: a Model System to Dissect Immunological Roles of Autophagy 169
Vojo Deretic, Monica Delgado, Isabelle Vergne, Sharon Master, Sergio De Haro, Marisa Ponpuak, and Sudha Singh

Autophagy in Immunity Against Intracellular Bacteria 189
Ju Huang and John H. Brumell

Group A *Streptococcus*: A Loser in the Battle with Autophagy 217
Tamotsu Yoshimori and Atsuo Amano

Autophagy Subversion by Bacteria 227
Emanuel Campoy and María I. Colombo

Autophagy in Immunity Against *Toxoplasma gondii* 251
Carlos S. Subauste

Autophagy in Mammalian Antiviral Immunity 267
Anthony Orvedahl and Beth Levine

Something Old, Something New: Plant Innate Immunity and Autophagy 287
Montrell Seay, Andrew P. Hayward, Jeffrey Tsao, and S.P. Dinesh-Kumar

Autophagy in HIV-Induced T Cell Death 307
Lucile Espert and Martine Biard-Piechaczyk

Subversion of the Cellular Autophagy Pathway by Viruses 323
Karla Kirkegaard

Index 335

Contributors

Atsuo Amano
Department of Oral Frontier Biology,
Osaka University Graduate School of Dentistry,
Suita-Osaka 565-0871, Japan

Martine Biard-Piechaczyk
CPBS, CNRS UMR5236/University of Montpellier,
Institut de Biologie, 4, Bd Henri IV, CS 69033,
34965 Montpellier Cedex 2, France
martine.biard@univ-montp1.fr

John H. Brumell
Cell Biology Program, Hospital for Sick Children, Toronto,
ON, M5G 1X8, Canada;
Department of Molecular Genetics and Institute of Medical Science,
University of Toronto, Toronto, ON, Canada M5S 1A8
john.brumell@sickkids.ca

Ken Cadwell
Department of Pathology and Immunology,
Washington University School of Medicine,
660 South Euclid Avenue, St. Louis, MO 63110, USA

Emanuel Campoy
Laboratorio de Biología Celular y Molecular,
Instituto de Histología y Embriología (IHEM)-CONICET,
Facultad de Ciencias Médicas, Universidad Nacional de Cuyo,
Mendoza, Argentina

Magali Chaumorcel
INSERM U756 Faculté de Pharmacie, 5 rue Jean-Baptiste Clément
92290 Châtenay-Malabry, France;
Université Paris-Sud 11, Faculté de Pharmacie, 5 rue Jean-Baptiste Clément
92290 Châtenay-Malabry, France

Patrice Codogno
INSERM U756 Faculté de Pharmacie, 5 rue Jean-Baptiste Clément
92290 Châtenay-Malabry, France
Université Paris-Sud 11, Faculté de Pharmacie, 5 rue Jean-Baptiste Clément
92290 Châtenay-Malabry, France
Patrice.codogno@u-psud.fr

María I. Colombo
Laboratorio de Biología Celular y Molecular,
Instituto de Histología y Embriología (IHEM)-CONICET,
Facultad de Ciencias Médicas, Universidad Nacional de Cuyo,
Casilla de Correo 56, Centro Universitario,
Parque General San Martín, (5500) Mendoza, Argentina
mcolombo@fcm.uncu.edu.ar

Sergio De Haro
Department of Molecular Genetics and Microbiology,
University of New Mexico Health Sciences Center,
915 Camino de Salud NE, Albuquerque, NM 87131, USA

Monica Delgado
Department of Molecular Genetics and Microbiology,
University of New Mexico Health Sciences Center,
915 Camino de Salud NE, Albuquerque, NM 87131, USA

Vojo Deretic
Department of Molecular Genetics and Microbiology,
University of New Mexico Health Sciences Center,
915 Camino de Salud, NE, Albuquerque, NM 87131, USA
vderetic@salud.unm.edu

S.P. Dinesh-Kumar
Department of Molecular, Cellular, and Developmental Biology,
Yale University, New Haven, CT 06520-8103, USA
Savithramma.dinesh-kumar@yale.edu

Audrey Esclatine
INSERM U756 Faculté de Pharmacie,
5 rue Jean-Baptiste Clément 92290 Châtenay-Malabry, France;
Université Paris-Sud 11, Faculté de Pharmacie,
5 rue Jean-Baptiste Clément 92290 Châtenay-Malabry, France

Lucile Espert
CPBS, CNRS UMR5236/University of Montpellier,
Institut de Biologie, 4, Bd Henri IV, CS 69033,
34965 Montpellier Cedex 2, France
lucile.espert@univ-montp1.fr

Monique Gannagé
Viral Immunobiology, Institute of Experimental Immunology,
University Hospital of Zürich, Winterthurerstrasse 190,
8057 Zürich, Switzerland

Andrew Hayward
Department of Molecular, Cellular, and Developmental Biology,
Yale University, New Haven, CT 06520-8103, USA

You-Wen He
Box 3010, Department of Immunology, DUMC, Durham,
NC 27710, USA
he000004@mc.duke.edu

Ju Huang
Cell Biology Program, Hospital for Sick Children, Toronto,
ON, Canada M5G 1X8

Akiko Iwasaki
Department of Immunobiology, Yale University School of Medicine,
300 Cedar Street, New Haven, CT 06520, USA
akiko.iwasaki@yale.edu

Karla Kirkegaard
Department of Microbiology and Immunology,
Stanford University School of Medicine,
Stanford CA 94305, USA
karlak@stanford.edu

Daniel J. Klionsky
Life Sciences Institute and Departments of Molecular,
Cellular and Developmental Biology and Biological Chemistry,
University of Michigan, Ann Arbor,
MI 48109-2216, USA
klionsky@umich.edu

Beth Levine
Howard Hughes Medical Institute,
Departments of Internal Medicine and Microbiology,
University of Texas Southwestern Medical Center, Dallas,
TX 75390, USA
beth.levine@utsouthwestern.edu

Sharon Master
Department of Molecular Genetics and Microbiology,
University of New Mexico Health Sciences Center,
915 Camino de Salud NE, Albuquerque, NM 87131, USA

Noboru Mizushima
Department of Physiology and Cell Biology, Tokyo Medical and Dental University, Tokyo 113-8519, Japan
nmizu.phy2@tmd.ac.jp

Christian Münz
Viral Immunobiology, Institute of Experimental Immunology, University Hospital of Zürich, Winterthurerstrasse 190, 8057 Zürich, Switzerland
christian.muenz@usz.ch

Anthony Orvedahl
Departments of Internal Medicine and Microbiology, University of Texas Southwestern Medical Center, Dallas, TX 75390, USA

Marisa Ponpuak
Department of Molecular Genetics and Microbiology, University of New Mexico Health Sciences Center, 915 Camino de Salud NE, Albuquerque, NM 87131, USA

Heather H. Pua
Department of Immunology, Duke University Medical Center, Durham, NC 27710, USA

Montrell Seay
Department of Molecular, Cellular, and Developmental Biology, Yale University, New Haven, CT 06520-8103, USA

Sudha Singh
Department of Molecular Genetics and Microbiology, University of New Mexico Health Sciences Center, 915 Camino de Salud NE, Albuquerque, NM 87131, USA

Thaddeus S. Stappenbeck
Department of Pathology and Immunology, Washington University School of Medicine, 660 South Euclid Avenue, St. Louis, MO 63110, USA

Carlos S. Subauste
Department of Ophthalmology and Visual Sciences, and Division of Infectious Diseases and HIV Medicine, Department of Medicine, Department of Pathology, Case Western Reserve University School of Medicine, Cleveland, OH 44106, USA
carlos.subauste@case.edu

Michal Caspi Tal
Department of Immunobiology, Yale University School of Medicine,
300 Cedar Street, New Haven, CT 06520, USA

Jeffrey Tsao
Department of Molecular, Cellular, and Developmental Biology,
Yale University, New Haven, CT 06520-8103, USA

Isabelle Vergne
Department of Molecular Genetics and Microbiology,
University of New Mexico Health Sciences Center,
915 Camino de Salud NE, Albuquerque, NM 87131, USA

Herbert W. Virgin
Department of Pathology and Immunology,
Washington University School of Medicine, 660 South Euclid Avenue,
St. Louis, MO 63110, USA
virgin@wustl.edu

Zhifen Yang
Life Sciences Institute and Departments of Molecular,
Cellular and Developmental Biology and Biological Chemistry,
University of Michigan, Ann Arbor, MI 48109-2216, USA

Tamotsu Yoshimori
Department of Cell Regulation, Research Institute for Microbial Diseases,
Osaka University, Suita-Osaka 565-0871, Japan
tamyoshi@biken.osaka-u.ac.jp

An Overview of the Molecular Mechanism of Autophagy

Zhifen Yang and Daniel J. Klionsky

Contents

1 Introduction 2
2 Molecular Mechanism of Autophagy 5
2.1 Induction and Regulation of Autophagy 5
2.2 The Cvt Pathway and Other Selective Types of Autophagy 10
2.3 Phosphatidylinositol 3-Kinase Complex 13
2.4 Two Ubiquitin-Like Protein Conjugation Systems 15
2.5 Atg9 and Its Cycling Systems 19
2.6 De Novo Vesicle Formation 21
2.7 Vesicle Docking and Fusion with the Vacuole 23
2.8 Vesicle Breakdown and Recycling of the Resulting Macromolecules 24
3 Conclusion 25
References 25

Abstract Autophagy is a highly conserved cellular degradation process in which portions of cytosol and organelles are sequestered into a double-membrane vesicle, an autophagosome, and delivered into a degradative organelle, the vacuole/lysosome, for breakdown and eventual recycling of the resulting macromolecules. This process relieves the cell from various stress conditions. Autophagy plays a critical role during cellular development and differentiation, functions in tumor suppression, and may be linked to life span extension. Autophagy also has diverse roles in innate and adaptive immunity, such as resistance to pathogen invasion. Substantial progress has been made in the identification of many autophagy-related (*ATG*) genes that are essential to drive this cellular process, including both selective and nonselective types of autophagy. Identification of the *ATG* genes in yeast, and the finding of orthologs in other organisms, reveals the conservation of the autophagic machinery in all eukaryotes. Here, we summarize our current knowledge about the machinery and molecular mechanism of autophagy.

Z. Yang and D.J. Klionsky (✉)
Life Sciences Institute and Departments of Molecular, Cellular and Developmental Biology and Biological Chemistry, University of Michigan,
Ann Arbor, MI, 48109-2216, USA
e-mail: klionsky@umich.edu

B. Levine et al. (eds.), *Autophagy in Infection and Immunity*,
Current Topics in Microbiology and Immunology 335,
DOI 10.1007/ 978-3-642-00302-8_1,

1 Introduction

Autophagy, "self-eating" at the subcellular level, has gained tremendous attention in the past few years, and our knowledge concerning the mechanism of autophagy has expanded dramatically (Yorimitsu and Klionsky 2005b). There are three major types of autophagy in eukaryotic cells—macroautophagy, microautophagy, and chaperone-mediated autophagy (CMA)—and they are mechanistically different from each other (Klionsky 2005; Massey et al. 2004). Both macro- and microautophagy involve dynamic membrane rearrangement to engulf portions of the cytoplasm, and they have the capacity for the sequestration of large structures, such as entire organelles. Microautophagy involves the direct engulfment of cytoplasm at the lysosome surface by invagination, protrusion, and septation of the lysosome membrane. In contrast, during macroautophagy, portions of cytoplasm are sequestered into a de novo-formed double-membrane vesicle termed an autophagosome. Subsequently, the completed autophagosome fuses with the lysosome/vacuole and the inner single-membrane vesicle is released into the lumen. In either case, the membrane of the resulting autophagic body is lysed to allow the contents to be broken down, and the resulting macromolecules are transported back into the cytosol through membrane permeases for reuse. In contrast, CMA does not involve a similar type of membrane rearrangement; instead, it translocates unfolded, soluble proteins directly across the limiting membrane of the lysosome.

In this chapter, we will focus on macroautophagy, hereafter referred to as autophagy. Autophagy is an evolutionarily conserved process that occurs ubiquitously in all eukaryotic cells (Reggiori and Klionsky 2002) and has many physiological roles. Autophagy is active at a basal level for the turnover of long-lived proteins and also for the removal of superfluous or damaged organelles. This latter function might provide a connection to autophagy's proposed role in life span extension (Levine and Klionsky 2004). On the other hand, autophagy is induced as a cellular response to various stress conditions, such as nutrient limitation, heat, and oxidative stress. Autophagy also plays a role in cellular development and differentiation (Levine and Klionsky 2004). Moreover, autophagy is implicated in a wide range of diseases (Huang and Klionsky 2007; Mizushima et al. 2008; Shintani and Klionsky 2004a), including cancer and neurodegenerative disorders such as Alzheimer', Parkinson' and Huntington' diseases. In addition, autophagy has diverse roles in innate and adaptive immunity (Levine and Deretic 2007). For example, autophagy can eliminate invasive pathogens, including viruses, parasites and bacteria; autophagy also promotes MHC class II presentation of microbial (and self) antigens. Finally, in the absence of apoptosis, autophagy may participate in a type of programmed cell death (type II) that is distinct from apoptosis, although the physiological relevance of the former is not clear (Levine and Yuan 2005).

The morphology of autophagy was first identified in mammalian cells in the 1950s, and extensive morphological and pharmacological studies defined the basic steps of this process. Subsequent work in various fungi starting in the 1990s allowed the identification of individual molecular components that participate in autophagy. To date, there are 31 autophagy-related (*ATG*) genes (Huang and Klionsky 2007;

Klionsky et al. 2003). The *ATG* genes were discovered from genetic screens for mutants that affected protein turnover (nonspecific autophagy), peroxisome degradation (pexophagy) and delivery of a resident vacuolar hydrolase (the cytoplasm to vacuole targeting (Cvt) pathway). Although the Cvt, pexophagy, and autophagy pathways are morphologically and mechanistically similar and share most of the Atg components, they differ in several aspects (Fig. 1). Autophagy and pexophagy are degradative, whereas the Cvt pathway is biosynthetic. Autophagy is generally considered nonselective, whereas pexophagy and the Cvt pathway are highly selective. The Cvt pathway is used to deliver two resident vacuolar hydrolases, aminopeptidase

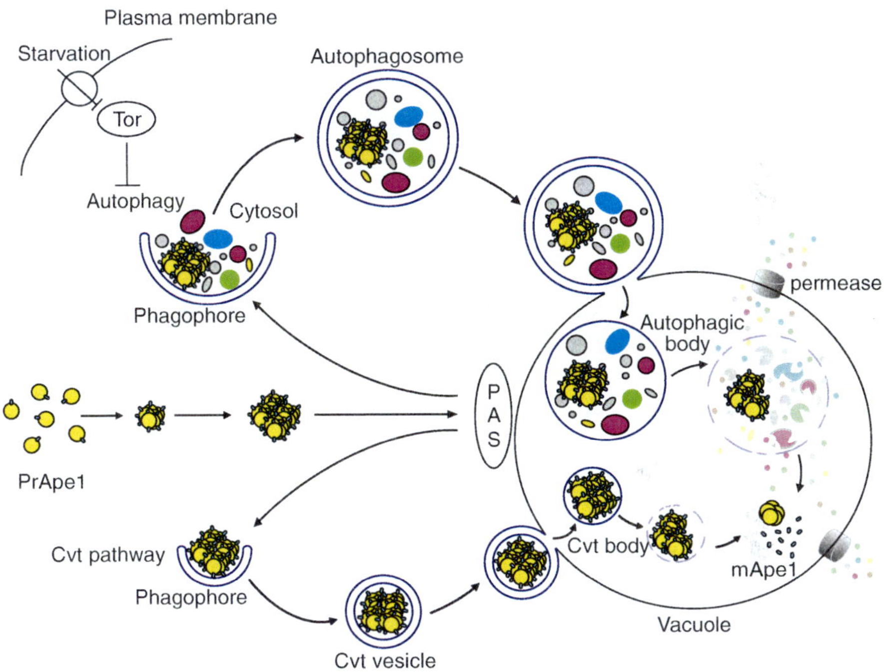

Fig. 1 Schematic overview of autophagy and the Cvt pathway in yeast. Both pathways involve the engulfment of cargos within distinct double-membrane vesicles, which are thought to originate from the phagophore assembly site (PAS). The Cvt pathway is biosynthetic and is used for the delivery of two resident vacuolar hydrolases, aminopeptidase I (Ape1), and α-mannosidase (Ams1), and it occurs under vegetative conditions. The Cvt vesicle is approximately 140–160 nm in diameter and appears to closely enwrap the specific cargo, the Cvt complex (consisting of prApe1 and the Atg19 receptor), and exclude bulk cytoplasm. In contrast, autophagy is degradative and is induced by inactivation of Tor kinase upon nutrient starvation. The autophagosome, which is 300–900 nm in diameter, sequesters cytoplasm, including organelles, and can also specifically sequester the Cvt complex. Once completed, the double-membrane vesicles dock and fuse with the vacuole, and release the inner single-membrane vesicles (autophagic or Cvt body) into the lumen. Subsequently, these vesicles are broken down, allowing the maturation of prApe1 and the degradation of cytoplasm, with recycling of the resulting macromolecules through vacuolar permeases. This figure is modified from Fig. 1 of Yorimitsu and Klionsky (2005b)

I (Ape1) and α-mannosidase (Ams1) (Hutchins and Klionsky 2001; Scott et al. 1997). A double-membrane vesicle that sequesters these two proteins is termed a Cvt vesicle; this is relatively consistent in size but significantly smaller than the autophagosome, being 140–160 nm in diameter compared to 300–900 nm for the autophagosome (Baba et al. 1997). Similarly, the vesicle formed during pexophagy, the pexophagosome, is also larger than the Cvt vesicle in order to accommodate its specific cargos, peroxisomes (Hutchins et al. 1999). In contrast to the autophagosome, both the Cvt vesicle and pexophagosome appear to closely enwrap the cargo and exclude bulk cytoplasm.

These dynamic pathways can be broken down into a series of steps (Fig. 2), including induction, cargo recognition and packaging, vesicle nucleation, vesicle expansion and completion, Atg protein cycling, vesicle fusion with the vacuole/

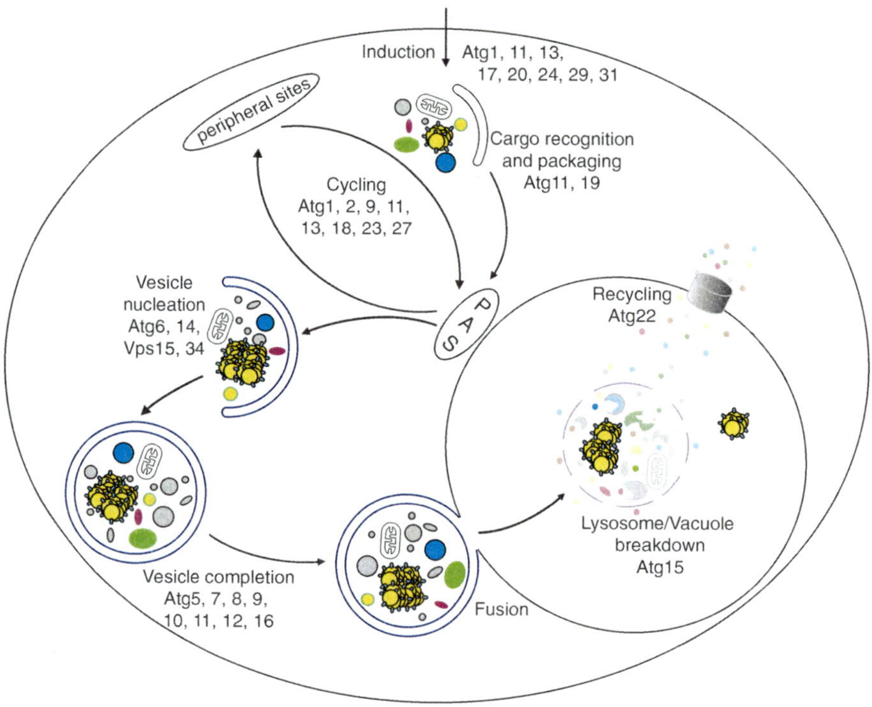

Fig. 2 Schematic representation of autophagy and autophagy-related pathways. These dynamic pathways can be broken down into a series of steps including induction, cargo recognition and packaging, vesicle nucleation, vesicle expansion and completion, Atg protein cycling, vesicle fusion with the vacuole/lysosome, vesicle breakdown and recycling of the resulting macromolecules. The Atg proteins can be classified into several different groups according to their functions at the different steps of the pathway. The Atg1 complex may act at multiple steps of the pathway, including induction and Atg protein cycling. During the vesicle formation process, several Atg proteins are involved in cycling between the peripheral sites and the PAS. *PAS*, phagophore assembly site; thought to be the organizing site for phagophore formation. This figure is modified from Fig. 2 of Huang and Klionsky (2007)

lysosome, vesicle breakdown, and recycling of the resulting macromolecules (Huang and Klionsky 2007). Thus, the Atg proteins can be classified into several different groups according to their functions at the different steps of the pathway. Many orthologs of the *ATG* gene products have also been identified and studied in higher eukaryotes, such as worms, insects, plants and mammals, and they have essentially similar roles as those in yeast (Xie and Klionsky 2007; Yorimitsu and Klionsky 2005b). Continued investigation of functions of the *ATG* gene products in yeast will greatly expand our understanding of autophagy. In this chapter, we will mainly discuss the molecular machinery of autophagy, with an emphasis on yeast.

2 Molecular Mechanism of Autophagy

2.1 Induction and Regulation of Autophagy

Insufficient autophagy can be deleterious (Komatsu et al. 2007a; Kuma et al. 2004), but excessive levels may also be harmful. Accordingly, autophagy is a tightly regulated process in all eukaryotes. The induction and regulation of autophagy have been studied extensively in yeast, mammalian cells and *Drosophila*. Several signaling pathways, as summarized in the following, are involved in the control of autophagy.

TORC1. The protein target of rapamycin, Tor, plays a major regulatory role in autophagy induction (Carrera 2004). Tor forms two functionally distinct protein complexes, Tor complex 1 and 2 (TORC1 and TORC2) (Loewith et al. 2002), and TORC1 has the primary role in regulating autophagy. Under nutrient-rich conditions, TORC1 is active and inhibits autophagy, whereas TORC1 is inhibited upon nutrient deprivation, allowing an increase in autophagic activity (Noda and Ohsumi 1998).

In yeast, TORC1 acts on autophagy in two ways (Klionsky 2005). First, TORC1 regulates the Atg1–Atg13–Atg17 kinase complex (Fig. 3a). The formation of this ternary complex correlates with an increase in autophagic activity. Atg1, a serine/threonine kinase, is one of the key Atg proteins required for both autophagy and the Cvt pathway (Matsuura et al. 1997). Based on yeast two-hybrid data and affinity isolation, Atg1 is found to be in a complex with Atg13 and Atg17 (Kamada et al. 2000; Kabeya et al. 2005). The observation that Atg17 interacts with Atg13 in the absence of Atg1 but not vice versa suggests that Atg13 mediates the interaction between Atg1 and Atg17. TORC1 regulates (directly or indirectly) the Atg13 phosphorylation state. Under nutrient-rich conditions, Atg13 is highly phosphorylated, and has a lower affinity for Atg1 and Atg17. Upon the inactivation of TORC1 by rapamycin or nutrient deprivation, Atg13 is rapidly and partially dephosphorylated, leading to a higher affinity for Atg1 and Atg17. The identities of the phosphatase(s) that control Atg13 phosphorylation are currently unknown. The interaction of Atg1 with hypophosphorylated Atg13 and Atg17 allows the activation of Atg1 kinase activity. Loss of interaction between Atg1 and Atg13 or between Atg13 and Atg17 leads to a decrease in Atg1 kinase activity and decreased autophagy. The kinase

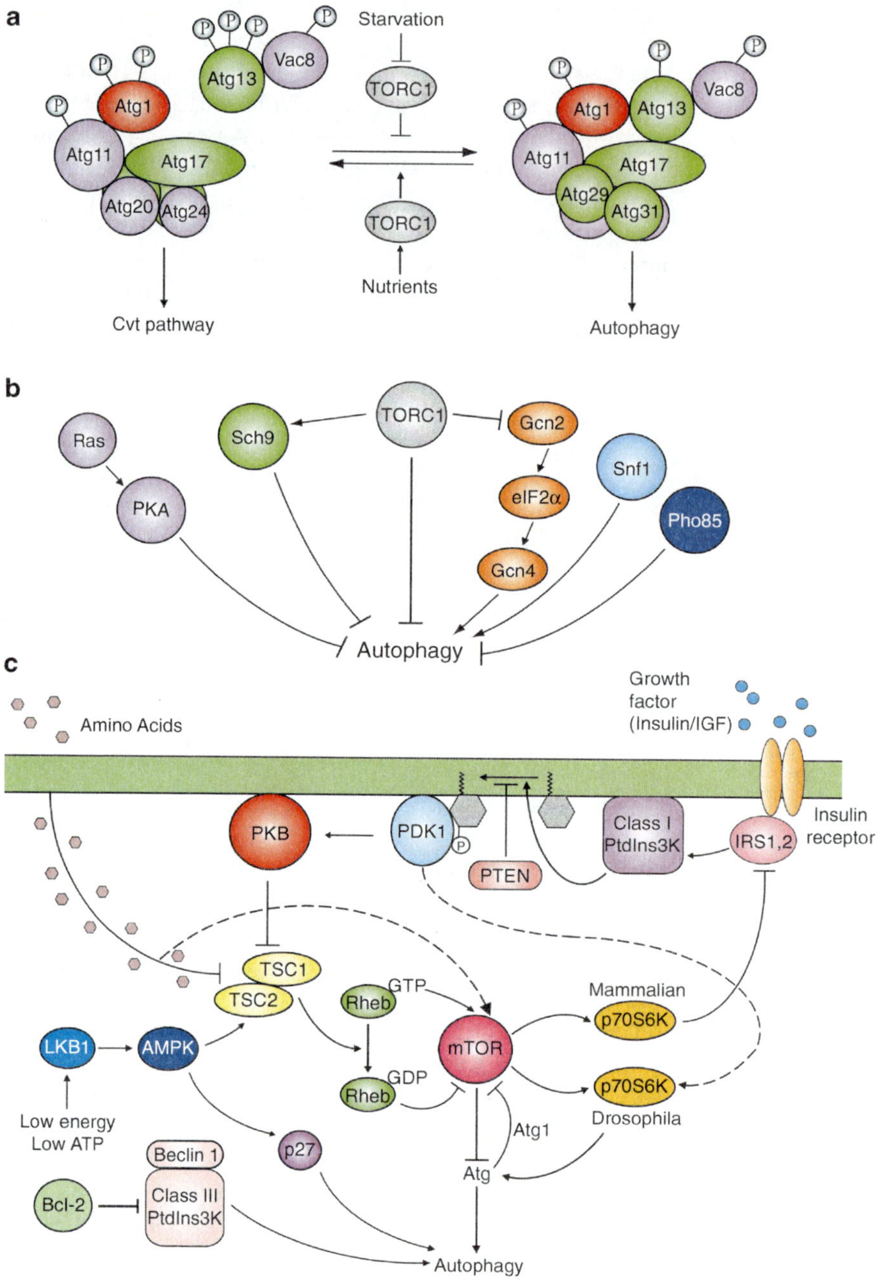

Fig. 3a–c Regulation of autophagy induction in yeast and mammalian cells. **a** Regulatory complex for autophagy induction in yeast. In yeast, autophagy is mainly a starvation response, and Tor kinase complex 1 (TORC1) regulates the induction of autophagy upon sensing the nutrient conditions. Atg1 kinase, which is essential for both autophagy and the Cvt pathway, forms a putative complex with several Atg proteins that are primarily required for autophagy (*in green*) or the Cvt pathway (*in purple*). Under nutrient-rich conditions, TORC1 is active and Atg13 is highly phosphorylated, and this hyperphosphorylated Atg13 has a lower affinity for Atg1 and Atg17.

activity of Atg1 is essential for both autophagy and the Cvt pathway, although a higher level of kinase activity appears to be needed for the Cvt pathway (Kamada et al. 2000; Kabeya et al. 2005; Cheong et al. 2008; Abeliovich et al. 2003). It is possible that the kinase activity of Atg1 is critical to the magnitude of autophagy but not its initiation (Nair and Klionsky 2005). The downstream substrate of Atg1 kinase is unclear, and it is still a matter of debate as to whether Atg1 primarily acts on autophagy through its kinase activity or through a structural role during autophagic complex formation. However, one role of the Atg1–Atg13–Atg17 ternary complex is thought to be that of regulating the switch between autophagy and the Cvt pathway in response to environmental changes.

Homologs of Atg1 are involved in autophagy in various multicellular organisms such as *Dictyostelium discoideum* (Otto et al. 2004), *Drosophila melanogaster* (Scott et al. 2004), *C. elegans* (Melendez et al. 2003), *Arabidopsis thaliana* (Hanaoka et al. 2002), and mammals (Yan et al. 1998, 1999). In *Drosophila*, Atg1 activity is

Fig. 3 (continued) Upon inactivation of TORC1 by nutrient starvation, Atg13 is rapidly and partially dephosphorylated, leading to a higher affinity for Atg1 and Atg17. The formation of the Atg1–Atg13–Atg17 ternary complex allows the activation of Atg1 kinase activity, which may regulate the switch between autophagy and the Cvt pathway in response to environmental changes. The function of additional components of the putative complex depicted here, including Atg20, Atg24, Atg29, Atg31 and Vac8, are not known. Atg11 may function in part as a scaffold protein. This figure is modified from Fig. 2 of Yorimitsu and Klionsky (2005b) **b** Multiple nutrient-sensing kinase signaling pathways converge on autophagy in yeast. TORC1 plays a major role in the regulation of autophagy. Ras is active under nutrient-rich conditions and allows the activation of PKA, which inhibits autophagy. The PKA and Sch9 signaling pathways cooperatively regulate the induction of autophagy in parallel with Tor, although Sch9 is also a direct substrate of TORC1. The eIF2α kinase signaling pathway positively regulates autophagy, and Gcn2 might be another target of TORC1. Snf1 and Pho85 are additional positive and negative regulatory components, respectively, of autophagy in yeast. **c** Regulation of autophagy in mammalian cells. mTor activation depends on several inputs, including nutrients (amino acids), energy (ATP) and growth factor (insulin/IGF). In response to insulin receptor stimulation, a class I phosphoinositide 3-kinase (PtdIns3K) is activated and generates PtdIns(3,4)P_2 and PtdIns(3,4,5)P_3 at the plasma membrane, and the latter two activate 3-phosphoinositide-dependent protein kinase 1 (PDK1) and protein kinase B (PKB)/Akt. PKB phosphorylates and inhibits the GTPase-activating protein complex TSC1–TSC2, leading to the stabilization of Rheb-GTP, which stimulates mTor, causing inhibition of autophagy. PTEN, a 3'-phosphoinositide phosphatase, antagonizes PKB and has a stimulatory effect on autophagy. Both mTor and PDK1 stimulate p70S6 kinase (p70S6K). In one model, under nutrient-rich conditions, activation of S6K directly stimulates autophagy, or it is stimulated indirectly through inhibition of PtdIns3K, allowing a basal level of autophagy for homeostatic purposes. Under starvation conditions, inhibition of mTor prevents further activation of S6K, which limits and prevents excessive autophagy. Both ATP and amino acid deprivation result in mTor inactivation independent of the insulin signaling pathway. Amino acids activate mTor via inhibition of the TSC1–TSC2 complex or are sensed by mTor directly. Energy stress causes activation of the LKB1–AMPK pathway, which inhibits mTor by activating TSC1–TSC2. AMPK phosphorylates and stabilizes p27, a cyclin-kinase inhibitor, leading to activation of autophagy. An antiapoptotic protein, Bcl-2, associates with Beclin 1, the mammalian homolog of Atg6, and inhibits a class III PtdIns3K complex, whereas the latter serves a stimulatory role in autophagy. Also shown is the notion that Atg1 overexpression negatively feeds back on Tor activity in *Drosophila*

modulated by TORC1 as in yeast, because the induction of autophagy that results from the overexpression of Atg1 is suppressed when the TORC1 signaling pathway is impaired (Scott et al. 2007). Normally, a feedback mechanism may occur in which Atg1 downregulates Tor activity, resulting in a further activation of Atg1 and a further increase in autophagy (Fig. 3c). Because these studies are based on overexpressed Atg1, however, the physiological significance is not clear at present. In mammals, the uncoordinated 51-like kinases 1 and 2 (Ulk1 and Ulk2) appear to be the functionally equivalent mammalian homologs of yeast Atg1. Knockdown of *ULK1* inhibits the induction of autophagy by rapamycin treatment, indicating that Ulk1 functions downstream of mTOR in autophagy regulation (Chan et al. 2007). In contrast to the result in *Drosophila*, overexpression of *ULK1* or *ULK2* suppresses autophagy. Furthermore, moderate expression of kinase-dead *ULK* mutants also efficiently suppresses autophagy, indicating that kinase activity of the Ulk proteins is critical during this process (Hara et al. 2008). FIP200 is a recently identified Ulk-interacting protein that is required for autophagy (Hara et al. 2008). Ulk and FIP200 function together and form a complex that is essential during an early step in autophagosome formation; FIP200 is thus thought to be a counterpart of yeast Atg17. Further identification and analysis of a functional homolog of mammalian Atg13 might help to clarify the functional relationship between the yeast and mammalian Atg1 complex.

Second, TORC1 acts through its downstream effectors to control autophagy. Several, but not all, TORC1 readouts, including autophagy, are regulated through protein phosphatase type 2A (PP2A) and/or 2A-related protein phosphatase (Sit4) (De Virgilio and Loewith 2006). PP2A and Sit4 are in distinct complexes containing Tap42. Under nutrient-rich conditions, Tap42 is phosphorylated and tightly associates with PP2A and Sit4. Starvation or rapamycin treatment causes dephosphorylation and dissociation of Tap42 or a change in conformation, resulting in the activation of Sit4. TORC1 may directly phosphorylate Tap42, or it may indirectly regulate Tap42 via Tip41. Upon the inactivation of TORC1, Tip41 is dephosphorylated and has a high affinity for Tap42, resulting in the inhibition of the latter. One report suggests that Tap42 does not transmit the signal from TORC1 to regulate autophagy (Kamada et al. 2000). However, more recent data indicate a role for Tap42 in the negative regulation of this process (Yorimitsu et al. 2009).

The conserved Tor protein in mammalian cells (mTor) also senses nutrient status and modulates autophagy, but the mechanism of regulation is more complex than in fungi, which are not responsive to hormones. As shown in Fig. 3c, the regulatory cascade upstream of mTor includes an insulin receptor, insulin-receptor substrates 1 and 2, class I phosphoinositide 3-kinase (PtdIns3K), 3-phosphoinositide-dependent protein kinase 1 (PDK1), and protein kinase B (PKB)/Akt (Meijer and Codogno 2006). mTor activity is controlled by the heterodimer TSC1–TSC2, which acts as a GTPase-activating protein (GAP) for the GTPase Rheb. The GDP-bound form of Rheb inhibits mTor, whereas the GTP-bound form stimulates the enzyme. PKB phosphorylates and inhibits the TSC1–TSC2 complex, leading to the activation of mTor signaling. PTEN, a 3'-phosphoinositide phosphatase, antagonizes PKB, and has a stimulatory effect on autophagy (Arico et al. 2001). The best characterized signaling pathway, located downstream of mTor, includes components such as ribosomal subunit S6 kinase (p70S6K). In one model, S6K exerts a negative feedback on mTor signaling by phosphorylating IRS1 to downregulate insulin signaling, leading to a decline in PtdIns(3,4,5)P_3, an inhibitor of autophagy; this feedback regulation may ensure a basal level of autophagy even under nutrient-rich conditions (Klionsky et al. 2005).

Ras/cAMP-dependent protein kinase A (PKA). In addition to TORC1, the Ras/PKA signaling pathway also regulates autophagy from yeast to mammals (Budovskaya et al. 2004; Furuta et al. 2004; Mavrakis et al. 2006; Schmelzle et al. 2004; Yorimitsu et al. 2007). Under nutrient-rich conditions in yeast, two redundant small GTPases (Ras1 and Ras2) are activated, and stimulate adenylyl cyclase to produce cAMP. cAMP binds to the PKA regulatory subunit (Bcy1) and allows its dissociation from the three PKA catalytic subunits (Tpk1, Tpk2, and Tpk3), resulting in the activation of PKA (Thevelein and de Winde 1999). Constitutive activation of PKA through a dominant hyperactive allele of *RAS2*, *RAS2*^{G19}*V*, or deletion of *BCY1* prevents the induction of autophagy by nutrient starvation or rapamycin, whereas inactivation of PKA by a dominant negative allele of *RAS2*, *RAS2*^{G22A}, induces autophagy under rich conditions without rapamycin (Budovskaya et al. 2004; Schmelzle et al. 2004). Thus, in addition to TORC1, Ras/PKA is another negative regulator of autophagy (Fig. 3b). Among the Atg proteins, Atg1, Atg13, Atg18 and Atg21 contain PKA phosphorylation sites. However, it is still unclear whether the phosphorylation of these Atg proteins by PKA has any functional link to autophagy (Budovskaya et al. 2005).

Sch9 is a homolog of mammalian PKB or p70S6 kinase (Urban et al. 2007). A recent report shows that PKA and Sch9 signaling pathways cooperatively regulate the induction of autophagy (Yorimitsu et al. 2007). Simultaneous inactivation of PKA and Sch9 triggers the induction of autophagy under rich conditions independent of effects on TORC1, whereas further inactivation of TORC1 causes an additive effect. These observations suggest a model wherein PKA, Sch9, and TORC1, at least in part, regulate autophagy in parallel (Fig. 3b). This model is supported by the finding that TORC1 and Ras/PKA function as two parallel pathways that independently act in regulating cell growth (Zurita-Martinez and Cardenas 2005). However, Sch9 is a direct substrate of TORC1 (Urban et al. 2007); furthermore, it is also suggested that TORC1 transmits signals through the Ras/PKA pathway to its downstream targets (Schmelzle et al. 2004). Therefore, the connection between PKA, Sch9, and TORC1 with regard to their effects in autophagy regulation is still not clear.

eIF2α kinase signaling and GCN4 general control. In response to amino acid starvation, budding yeast initiates a general amino acid control to induce the transcription of numerous genes. Central to this response is Gcn4, a master transcriptional activator of gene expression (Hinnebusch 2005). Gcn4 synthesis is mainly regulated at the translational level. Derepression of *GCN4* mRNA translation requires a protein kinase, Gcn2, whose only known substrate is the α subunit of translation initiation factor 2 (eIF2). The eIF2α kinase signaling pathway is also involved in the regulation of autophagy from yeast to mammals (Fig. 3b) (Talloczy et al. 2002).

Upon the loss of Gcn2 or Gcn4, or in the presence of the eIF2α nonphosphorylatable mutant *SUI2-S51A*, autophagic activity is impaired. Intriguingly, TORC1 is implicated in the eIF2α kinase signaling pathway because rapamycin activates Gcn2, at least in part, through dephosphorylation of Ser577 (Kubota et al. 2003). Thus, Gcn2 might be another target of TORC1.

Other signaling pathways controlling autophagy. Snf1, the closest yeast homolog of the mammalian AMP-activated protein kinase, and Pho85, a cyclin-dependent kinase (CDK), antagonistically control autophagy in yeast (Fig. 3b) (Wang et al. 2001b). Snf1, which is activated upon glucose depletion to allow transcription of glucose-repressed genes, is required for starvation-induced autophagy. Pho85, which has multiple functions through associations with its ten different cyclins (Pcls), is a negative regulator of autophagy, although the functions of the various Pcl proteins and the pathways that they regulate are currently unknown (Carroll and O'Shea 2002).

In mammalian cells, AMPK is also required for autophagy (Meley et al. 2006). During energy stress, AMP accumulation causes activation of the LKB1-AMPK pathway, which inhibits mTor by activating TSC1/TSC2 (Hoyer-Hansen and Jaattela 2007). Furthermore, AMPK phosphorylates p27, a cyclin-kinase inhibitor, thereby stabilizing p27, whereas ectopic expression of wild-type or a stabilized p27 mutant induces autophagy (Liang et al. 2007).

2.2 The Cvt Pathway and Other Selective Types of Autophagy

Although autophagy is generally considered to be a nonselective pathway for the degradation of bulk cytoplasmic components, recent findings indicate that there are many types of selective autophagy in both yeast and higher eukaryotes. In yeast, even bulk autophagy can be selective; cytosolic acetaldehyde dehydrogenase, Ald6, is preferentially sequestered into autophagosomes relative to other cytosolic proteins (Onodera and Ohsumi 2004). Several organelles are selectively degraded through autophagy. For example, the selective degradation of mitochondria is termed mitophagy (Kim et al. 2007). This type of selective process is thought to play a crucial role in mitochondrial homeostasis; however, the mechanism underlying mitophagy remains unclear. The use of electron microscopy to observe mitochondrial degradation indicates that mitophagy occurs both selectively and nonselectively. A recent report demonstrates that mature ribosomes are rapidly degraded by autophagy in yeast through a process termed ribophagy. This degradation involves a type of selective autophagy in that it specifically requires catalytic activity of the Ubp3/Bre5 ubiquitin protease (Kraft et al. 2008).

In fungi such as *Saccharomyces cerevisiae*, *Pichia pastoris*, *Hansenula polymorpha*, and *Yarrowia lipolytica*, peroxisomes are selectively engulfed and degraded through two morphologically distinct autophagic degradation processes, micro- and macropexophagy (Gunkel et al. 1999; Hutchins and Klionsky 2001; Sakai et al. 2006; Tuttle et al. 1993; Veenhuis et al. 1983). When fungi grow on specific carbon sources, such as oleic acid or methanol, peroxisome proliferation is induced to adapt to the new physiological conditions that require peroxisome metabolism. When peroxisome proliferation becomes unnecessary and peroxisomes become superfluous (as occurs after shifting to a preferred carbon source such as glucose), peroxisomes are rapidly and specifically degraded. The two main modes of pexophagy, micro- and macropexophagy share most of the molecular components with nonspecific autophagy. However, the presence of Pex14 at the peroxisomal membrane is necessary for the specific recognition of the organelle by the macropexophagy machinery (Bellu et al. 2001). A specificity factor, Atg11, which is required for the Cvt pathway, is also essential for the selective transport of peroxisomes to the vacuole (Kim et al. 2001). A recently identified pexophagy-specific protein, PpAtg30, functions as a peroxisome receptor through interactions with PpPex3, PpPex14, PpAtg11, and PpAtg17 to deliver peroxisomes to the site for pexophagosome formation (Farre et al. 2008). Furthermore, a fully functional actin cytoskeleton is required for selective autophagy, including the Cvt pathway and pexophagy, but not for nonselective autophagy (Reggiori et al. 2005a).

The Cvt pathway is a unique type of specific autophagy. The mechanism of the selective recognition and packaging of prApe1 has been relatively well clarified (Fig. 4). The Ape1 protein is synthesized in the cytoplasm as a precursor form (prApe1) (Klionsky et al. 1992). After synthesis, prApe1 assembles into a dodecamer (Kim et al. 1997), which is further packaged into a larger oligomeric structure called the Ape1 complex (Shintani et al. 2002). The prApe1 propeptide contains vacuolar targeting information (Martinez et al. 1997; Oda et al. 1996). In addition, the propeptide also mediates the interaction between prApe1 and its receptor protein, Atg19, to form the Cvt complex in the cytosol (Scott et al. 2001). Another Cvt cargo, Ams1, also binds Atg19 via a site that is distinct from the prApe1 binding site and is concentrated at the Cvt complex (Shintani et al. 2002). The Cvt complex is subsequently enwrapped by a double membrane that forms a Cvt vesicle. The Cvt complex can be also sequestered within autophagosomes, depending on the nutrient conditions (Baba et al. 1997), but this still occurs through a selective process that involves Atg19.

Atg11 subsequently associates with Atg19, acting like an adapter or tethering protein to bring the Cvt complex to the phagophore assembly site (PAS), a potential site for the formation of the Cvt vesicle and autophagosome. Several lines of evidence support the idea that Atg11 assembles with the Cvt complex before targeting to the PAS (Yorimitsu and Klionsky 2005a). However, how Atg11 guides the Cvt complex to the PAS is still unclear. A C-terminal coiled-coil domain of Atg11 is critical for interaction with the C terminus of Atg19, whereas the N-terminal and/or central coiled-coil domains contain information necessary for the Cvt complex to be targeted to the PAS (Yorimitsu and Klionsky 2005a). Besides Atg19, Atg11 has several other interacting partners, including Atg1, Atg9, Atg17, Atg20, and itself (Chang and Huang 2007; He et al. 2006; Yorimitsu and Klionsky 2005a). Atg9, the only characterized transmembrane protein that is required for sequestering vesicle formation, interacts with Atg11 independent of Atg1 or Atg19, suggesting that there are distinct and multiple populations of Atg11 within the cell. Atg11 homo-oligomerization may allow various

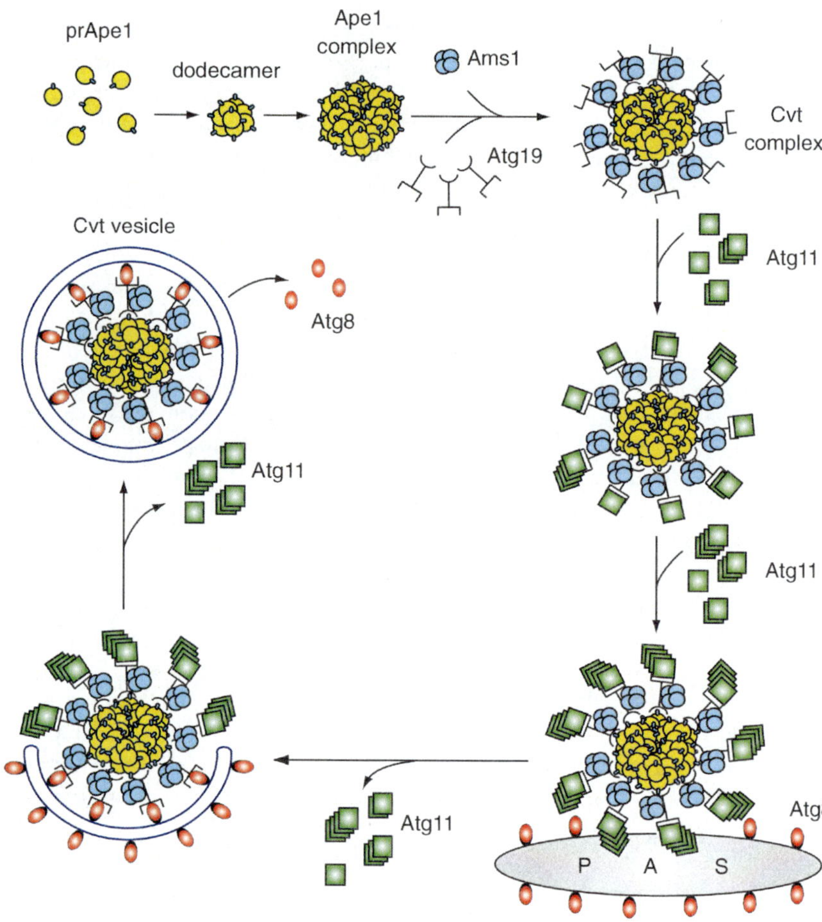

Fig. 4 Temporal order of action of cargo recognition, packaging and sequestration in the Cvt pathway. A selective type of autophagy, the Cvt pathway, specifically transports the vacuolar hydrolases precursor Ape1 and Ams1 into the vacuole. Precursor Ape1 is synthesized in the cytosol, assembled into a dodecamer, and then further packaged into a larger oligomeric structure, called the Ape1 complex. Atg19 binds to the propeptide of prApe1 to form the Cvt complex in the cytosol; Ams1 is also incorporated into this complex via binding to Atg19. Atg11 subsequently associates with Atg19, acting as an adapter to bring the Cvt complex to the phagophore assembly site or PAS, a potential site for Cvt vesicle formation. The PAS may organize the formation of the sequestering vesicle, or it may literally become the sequestering vesicle as shown. Atg11 assembles with the Cvt complex before targeting to the PAS, and it forms a homodimer or homo-oligomer at the PAS, although it is not clear whether this self-interaction occurs before or after the arrival at this site. Several Atg components, including Atg8, are recruited to the PAS independent of Atg11. Atg8 is conjugated into Atg8—PE for subsequent vesicle formation. Atg8—PE interacts with Atg19, and allows the correct incorporation of the Cvt complex into the forming vesicle. Atg19 is delivered into the vacuole together with the cargo proteins and degraded there. The scaffold protein Atg11, however, dissociates from the Cvt complex before vesicle completion, although the exact timing and mechanism of its release remain to be resolved. This figure is modified from Fig. 3 of Yorimitsu and Klionsky (2005b)

Atg11 populations, along with its various interacting partners, to be delivered to the PAS (Yorimitsu and Klionsky 2005a). A point mutation (H192L) in Atg9 disrupts the interaction with Atg11, preventing movement of Atg9 to the PAS and blocking the Cvt pathway, but not bulk autophagy (He et al. 2006), in agreement with the finding that Atg11 is not required for nonspecific autophagy (Kim et al. 2001).

After the arrival of the Atg11–Atg19-cargo complex at the PAS, Atg19 interacts with Atg8—PE to allow the transfer of the Atg19-cargo complex to the forming Cvt vesicle (or autophagosome); interaction between these two proteins may ensure the incorporation of the Cvt complex into the Cvt vesicle (Shintani et al. 2002). Unlike most receptors that recycle between donor and acceptor membranes, Atg19 is delivered into the vacuole together with the cargo proteins and degraded there. The scaffold protein Atg11, however, does not appear to remain associated with the Cvt complex; rather, it is thought to be released from Atg19 after delivery to the PAS and to dissociate from the complex before vesicle formation (Kim et al. 2001). It remains unknown whether there is a role for Atg11 during the process of Cvt vesicle completion, and the exact timing and mechanism of its release remain to be resolved. However, disassembly of the homo-oligomerized Atg11 requires the Atg1–Atg13–Atg17 kinase complex (Yorimitsu and Klionsky 2005a).

Increasing evidence indicates that selective autophagy also occurs in mammals. For example, the p62/SQSTM1/sequestosome protein preferentially recognizes polyubiquitinated protein aggregates and connects these with the autophagic machinery through interaction with the Atg8 mammalian homolog, LC3 (Bjørkøy et al. 2005; Komatsu et al. 2007b). Thus, p62 could function as a receptor protein similar to Atg19 to link polyubiquitinated proteins to autophagosomes. Another recent example of selective autophagy is seen with the clearance of mitochondria and ribosomes during reticulocyte maturation (Kundu et al. 2008). In this case, Ulk1 plays a critical role in selective autophagy, but is not essential for the induction of starvation-induced bulk autophagy. Selectivity is also seen with the degradation of peroxisomes in mammalian cells (Iwata et al. 2006). Finally, some pathogens are selectively targeted by autophagy, such as *Mycobacterium tuberculosis* and *Streptococcus pyogenes* (Gutierrez et al. 2004; Nakagawa et al. 2004). It is important to note, however, that other microbes (including bacteria and viruses) regulate autophagy for their own survival (Nakagawa et al. 2004; Ogawa et al. 2005; Orvedahl and Levine 2008). *Shigella*, an invasive bacteria, is able to escape autophagy by secreting IcsB on the bacterial surface. The IcsB protein interacts with VirG, which prevents the latter from binding Atg5 and triggering specific autophagic sequestration (Ogawa et al. 2005).

2.3 Phosphatidylinositol 3-Kinase Complex

The class III phosphatidylinositol 3-kinase (PtdIns3K) is known to participate in various membrane trafficking events. Vps34 is the only PtdIns3K in yeast, and it forms at least two distinct complexes, complex I and II (Fig. 5). Each complex contains three common components, Vps34, Vps15, and Vps30/Atg6 (Kihara et al.

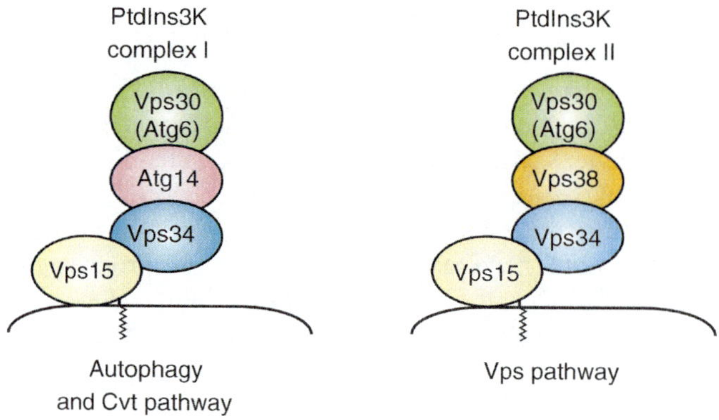

Fig. 5 Two phosphatidylinositol 3-kinase (PtdIns3K) complexes in yeast. Each complex contains three common components, Vps15, Vps34, and Vps30/Atg6. Vps34 is the PtdIns3K enzyme, and Vps15 is thought to be a regulatory component; the function of Vps30/Atg6 is not known. In addition, each complex contains another specific component, Atg14 (complex I) or Vps38 (complex II), which is thought to act as a connector between Vps30 and Vps15–Vps34. Complex I functions in autophagy and the Cvt pathway, whereas complex II acts in the Vps pathway, including the CPY and MVB pathways. This figure is modified from Fig. 5a of Yorimitsu and Klionsky (2005b)

2001). The function of Vps34 is dependent on a serine/threonine kinase, Vps15, which is required for Vps34 membrane association and activity (Stack et al. 1995). The role of Vps30/Atg6 within these PtdIns3K complexes is not well understood. These three common proteins are involved in autophagy, the Cvt pathway and the sorting of carboxypeptidase Y (CPY), which is normally transported from the late Golgi to the vacuole through the CPY pathway. In addition, each complex contains another specific component, Atg14 (complex I) or Vps38 (complex II), which is thought to act as a connector between Vps30 and Vps15–Vps34. The region containing the coiled-coil domains I and II within the N-terminal half of Atg14 is responsible for the interaction between Vps34 and Vps30/Atg6. Loss of Atg14 disrupts complex I and causes a defect only in autophagy and the Cvt pathway, whereas Vps38 deletion disrupts complex II and blocks only the CPY pathway. The association of Atg14 or Vps38 confers functional specificity on the two PtdIns3K complexes by targeting Vps34 to distinct compartments, thus regulating different protein trafficking events. Vps15–Vps34 complexed with Vps30 and Atg14 localizes to the PAS, and functions in autophagy and the Cvt pathway; Vps15–Vps34 complexed with Vps30 and Vps38 localizes to endosomes, and functions in the CPY pathway (Obara et al. 2006).

PtdIns3K is a lipid kinase and the kinase activity of Vps34 is essential for autophagy and the Cvt pathway. One possible role of PtdIns3K is to produce PtdIns(3)P at the PAS to recruit PtdIns(3)P-binding proteins, which in turn recruit additional downstream effectors to the PAS. PtdIns(3)P is bound by proteins that have specific binding sites, such as the PX (phox homology) domain and the FYVE (for conserved in Fab1, YOTB, Vac1, and EEA1) zinc finger domain (Ellson et al. 2002; Stenmark et al. 2002). Two PX domain-containing proteins, Atg20 and Atg24, bind to PtdIns(3)P (Nice et al. 2002). These proteins are essential only for the Cvt pathway, not bulk autophagy. Their functional PX domains are necessary for membrane localization to the PAS and the endosome, which in turn depend on PtdIns3K complexes I and II, respectively. The role of endosomal localization is unknown since the CPY pathway is normal in the absence of Atg20 or Atg24; however, the endosomal localization is not necessary for Cvt transport. Atg20 and Atg24 interact with each other, and Atg24 and possibly Atg20 interact with Atg17 (Nice et al. 2002). In addition, Atg20 interacts with Atg11 (Yorimitsu and Klionsky 2005a). Thus, the Atg20–Atg24 complex might be part of the Atg1 kinase complex. Atg18 and Atg21 are also PtdIns(3)P-binding proteins, although neither of them contain known phosphoinositide-binding domains. Both proteins are recruited to the PAS in a manner dependent on PtdIns3K complex I (Guan et al. 2001; Stromhaug et al. 2004). Atg18 is needed for the correct movement of Atg9, but the function of Atg21 is not known.

In contrast to yeast, there are two types of PtdIns3K in mammalian cells: class I and class III PtdIns3K. Mammalian class III PtdIns3K, hVps34—similar to yeast Vps34—generates PtdIns(3)P, and plays a stimulatory role in autophagy (Fig. 3c). It forms a complex with its regulator, p150, the homolog of Vps15, and its accessory protein Beclin 1, the homolog of Vps30/Atg6 (Liang et al. 1999; Panaretou et al. 1997). Class I PtdIns3K uses PtdIns(4,5)P_2 as substrate to yield PtdIns(3,4,5)P_3. It functions at the plasma membrane and acts through an insulin signaling cascade to activate mTOR and PKB; hence it has an inhibitory effect on autophagy (Jacinto and Hall 2003). A major pathway by which amino acids control mTor is not mediated through class I PtdIns3K but instead through activation of the class III PtdIns3K, hVps34 (Nobukuni et al. 2005). Thus, hVps34 might also have an inhibitory effect on autophagy in mammalian cells. The specific function of PtdIns(3)P in mammalian cells has not yet been clarified, but it could function similar to that in yeast. Moreover, the effectors of PtdIns(3)P are also not clear. Atg20 and Atg24 do not have mammalian homologs. Atg18 has a human homolog and binds to PtdIns(3)P, but its role in autophagy has not yet been elucidated (Jeffries et al. 2004).

2.4 Two Ubiquitin-Like Protein Conjugation Systems

There are two protein conjugation systems that function in selective and nonselective autophagy, and they include the ubiquitin-like proteins Atg12 and Atg8 (Fig. 6) (Ohsumi 2001). Both conjugation systems are evolutionarily conserved from yeast to humans. Although Atg12 and Atg8 do not have apparent sequence homology with ubiquitin, each of them contains a ubiquitin fold at the C terminus, based on the crystal structures of Atg12 and Atg8 homologs from plants and mammals, respectively (Paz et al. 2000; Suzuki et al. 2005).

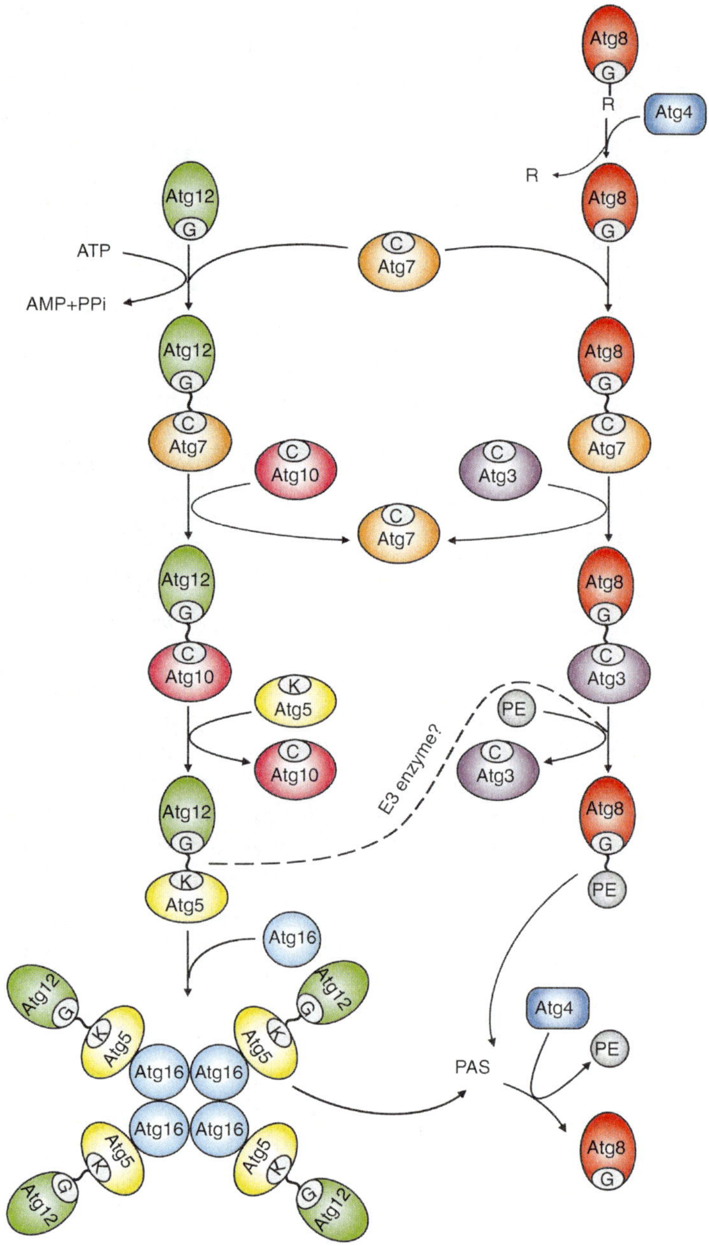

Fig. 6 Two ubiquitin-like protein conjugation systems. The conjugation of Atg12 to Atg5 starts with activation by Atg7, which is homologous to the E1 ubiquitin-activating enzyme. Atg7 hydrolyzes ATP, resulting in the activation of Atg12 via the formation of a thioester bond between the C-terminal glycine of Atg12 and the active site cysteine of Atg7; subsequently, the activated Atg12 is transferred to the active site cysteine of Atg10, an E2-like enzyme, which catalyzes the conjugation of Atg12 to Atg5 through the formation of an isopeptide bond between the activated

Atg12 is covalently attached to Atg5 through an isopeptide bond between a C-terminal glycine of Atg12 and an internal lysine residue of Atg5. The conjugation reaction is catalyzed by two additional proteins, Atg7 and Atg10 (Mizushima et al. 1998a). Atg7 is homologous to the E1 ubiquitin-activating enzyme, Uba1, in the ATP-binding region and the active cysteine residue, but not in terms of its overall structure (Tanida et al. 1999). Atg10 functions as an E2 ubiquitin-conjugating enzyme, although Atg10 shows no homology to the E2 enzymes that participate in the ubiquitin system (Shintani et al. 1999). As occurs during ubiquitination, Atg7 hydrolyzes ATP, resulting in the activation of Atg12 via the formation of a high-energy thioester bond between the C-terminal glycine of Atg12 and the active cysteine 507 of Atg7; subsequently, the activated Atg12 is directly transferred to the active cysteine 133 of Atg10 to form an Atg12—Atg10 thioester; finally, Atg12 is transferred to the target protein Atg5 to form the final conjugate. Atg5 is further bound noncovalently to another coiled-coil protein, Atg16, to form an Atg12—Atg5—Atg16 multimeric structure through homo-oligomerization of Atg16. This multimer has a molecular mass of approximately 350 kDa in yeast, and is predicted to represent a tetramer of the Atg12—Atg5—Atg16 complex. This is functionally essential for autophagy (Kuma et al. 2002). The Atg16 complex has recently been shown to specify the site of LC3 lipidation for membrane biogenesis in mammalian autophagy (Fujita et al. 2008).

A second ubiquitin-like protein, Atg8, is conjugated to a membrane lipid, phosphatidylethanolamine (PE) (Ichimura et al. 2000). The C-terminal arginine 117 residue of newly synthesized Atg8 is initially proteolytically cleaved by a cysteine protease, Atg4, exposing a glycine (Kirisako et al. 2000). The glycine is then bound to the active cysteine 507 of Atg7, the same E1-like enzyme used in the Atg12—Atg5 conjugation system. The activated Atg8 is then transferred to another E2-like enzyme (Atg3) at the active cysteine 234 residue via a thioester bond. The region around cysteine 234 of Atg3 shows partial homology to the corresponding region surrounding cysteine 133 of Atg10. Eventually, Atg8 is conjugated to PE through an amide bond between the C-terminal glycine and the amino group of PE. Atg8—PE is tightly associated with membranes, being an integral membrane protein. An in vitro reconstitution of the Atg8—PE conjugation process, using purified Atg7, Atg3, and Atg8ΔR (Atg8 lacking the last arginine residue), demonstrates that

Fig. 6 (continued) glycine of Atg12 and an internal lysine residue of Atg5. Atg12—Atg5 is finally assembled with Atg16. Atg16 forms a tetramer to allow the formation of an Atg12—Atg5—Atg16 multimeric structure. The conjugation of Atg8—PE starts with the cleavage of the C-terminal arginine of Atg8 by the protease Atg4. The exposed glycine of Atg8 is then bound to the active site cysteine of the same E1-like enzyme, Atg7. The activated Atg8 is then transferred to another E2-like enzyme, Atg3. Eventually, Atg3 catalyzes the conjugation of Atg8 to form Atg8—PE. The Atg12—Atg5 conjugate might function as an E3, ubiquitin ligase-like enzyme, to promote Atg8—PE conjugation. Both the Atg12—Atg5—Atg16 complex and Atg8—PE localize to the PAS to facilitate vesicle formation. The Atg8—PE that resides on the outer face of the sequestering vesicle is released from the membrane by a second Atg4-dependent cleavage. This figure is modified from Fig. 4 of Yorimitsu and Klionsky (2005b)

Atg7 and Atg3 are minimal catalysts (Ichimura et al. 2004). Unlike the Atg12—Atg5 conjugate, Atg8—PE conjugation is a reversible process in which Atg4 liberates Atg8 from its target lipid. The liberated Atg8 is recycled and used in another conjugation reaction to allow efficient progression of autophagy and the Cvt pathway (Kirisako et al. 2000).

Both the Atg12 and the Atg8 conjugation systems are evolutionarily conserved. The mammalian homologs for each component of the yeast Atg12—Atg5 conjugation systems (Atg5, Atg7, Atg10, and Atg12) have been characterized, and they function in a similar manner to their counterparts in yeast (Mizushima et al. 1998b; Mizushima et al. 2002; Tanida et al. 1999). There is also a mammalian Atg16-like protein, Atg16L, which forms an approximately 800 kDa protein complex with the Atg12—Atg5 conjugate, again mediated by the homo-oligomerization of Atg16L (Mizushima et al. 2003). There are at least four mammalian Atg8 homologs, MAP1LC3, GATE16, GABARAP and Atg8L. All proteins possess a conserved glycine residue near their C terminus and are conjugated to PE in the same manner as occurs in yeast via the catalysts Atg4, Atg7, and Atg3 (Hemelaar et al. 2003; Kabeya et al. 2000, 2004; Tanida et al. 2002, 2003, 2006). Among them, LC3 is most abundant in autophagosomal membranes and is well established as a marker to monitor the autophagosome and autophagic activity.

During autophagosome formation, both the Atg12—Atg5—Atg16 complex and the Atg8—PE conjugate localize at the PAS (Kim et al. 2002; Suzuki et al. 2001). Electron microscopy analysis clearly shows that these two conjugates decorate the expanding phagophore (Kabeya et al. 2000; Kirisako et al. 1999; Mizushima et al. 2001, 2003). The Atg12—Atg5—Atg16 complex is mainly localized on the outer side of the phagophore and released into the cytosol before or after autophagosome completion. These observations suggest that the Atg12—Atg5—Atg16 complex might serve as a coat component to drive the expansion and/or curvature of the membrane leaflet during autophagosome formation. Recent data, however, indicate that the Atg12—Atg5 conjugate might function as an E3, ubiquitin ligase, for Atg8—PE conjugation (Fig. 6), although it is not essential for the latter process to occur (Hanada et al. 2007). In contrast, Atg8—PE displays an apparently symmetrical distribution on both sides of the phagophore. The Atg8—PE that resides on the surface that becomes the outer face of the sequestering vesicle is released from the membrane by a second Atg4-dependent cleavage, whereas the inner population remains inside the vesicle and is delivered into the vacuole/lysosome, where it is degraded (Huang et al. 2000; Kabeya et al. 2000; Kirisako et al. 1999). Accordingly, Atg8—PE is another scaffold candidate to drive membrane expansion and vesicle completion. Upon autophagy induction, Atg8 protein levels increase, and this is needed to accommodate the larger-sized autophagosome relative to the Cvt vesicle. A quantitative correlation between the amount of Atg8 and the size of the sequestering vesicle has recently been determined (Xie et al. 2008). Atg8 is also suggested to act during the expansion of the autophagosomal membrane by mediating membrane tethering and hemifusion (Nakatogawa et al. 2007), although the physiological significance of this activity is not yet known.

2.5 Atg9 and Its Cycling Systems

One of the intriguing questions concerning autophagy is the source of the lipid that is used for autophagosome formation and the mechanism used for lipid movement to the site of autophagosome assembly. Atg9 is an integral membrane protein and is thought to be a "membrane carrier" during the assembly process (He et al. 2006; Noda et al. 2000). Unlike most other Atg proteins, which primarily display single punctate localization at the PAS, Atg9 localizes to multiple punctate structures, including the PAS (Reggiori et al. 2005b; Reggiori et al. 2004a). The cycling of Atg9 between the PAS and the non-PAS punctate structures is essential for autophagosome formation. Potentially, membrane could be delivered to the PAS through this shuttling process. In yeast, several Atg9 non-PAS puncta are found to localize adjacent to or at the surface of mitochondria (Reggiori et al. 2005b). It is still unclear, however, whether Atg9 is an integral component of the mitochondrial outer membrane or the membrane component of an organelle or other structure associated with mitochondria. Moreover, a population of the Atg9 peripheral pool does not colocalize with either the PAS or mitochondria, but rather is dispersed throughout the cytosol. This portion of Atg9 is thought to be associated with membranes in the process of trafficking between the PAS and the membrane donor sites.

The anterograde movement of Atg9 to the PAS involves several Atg proteins (Fig. 7). In the absence of Atg11, the transport of Atg9 to the PAS is blocked (He et al. 2006; Shintani and Klionsky 2004b). The efficient anterograde movement of Atg9 to the PAS also involves Atg23 and Atg27, which form a cycling unit with Atg9 (Legakis et al. 2007; Yen et al. 2007). Atg23 is a peripheral membrane protein, whereas Atg27 is a type I transmembrane protein. Both of these proteins are required for the Cvt pathway and efficient autophagy. Similar to Atg9, they localize to the PAS and several other punctate structures. Current evidence suggests that Atg9, Atg23 and Atg27 are in a heterotrimeric complex, and travel together to the PAS. Based on fluorescence microscopy, the anterograde transport of these three proteins is largely interdependent. In the absence of Atg23 or Atg27, Atg9 is at multiple punctate sites other than the PAS, whereas Atg23 is dispersed throughout the cytosol without either Atg9 or Atg27.

The actin cytoskeleton also participates in Atg9 anterograde movement (He et al. 2006). Disruption of the actin cytoskeleton prevents correct targeting of Atg9 to the PAS. Moreover, an actin-related protein, Arp2, interacts with Atg9 and directly regulates the dynamics of Atg9 PAS targeting (Monastyrska et al. 2008). Arp2 is one subunit of the Arp2/3 complex, the nucleation factor of branched actin filaments. Thus, one model is that Atg11 acts as an adaptor between the cargo (Atg9 and actin), while the Arp2/3 complex provides the force to push the cargo (Atg9 and its associated membrane) away from the membrane donor and toward the forming autophagosome (He et al. 2006; Monastyrska et al. 2008; Monastyrska et al. 2006).

The retrieval of Atg9 from the PAS back to the peripheral, non-PAS sites depends on the Atg1–Atg13 kinase complex, Atg2, Atg18, and the PtdIns3K complex I (Fig. 7); the absence of any of these proteins results in the accumulation of Atg9 at the PAS

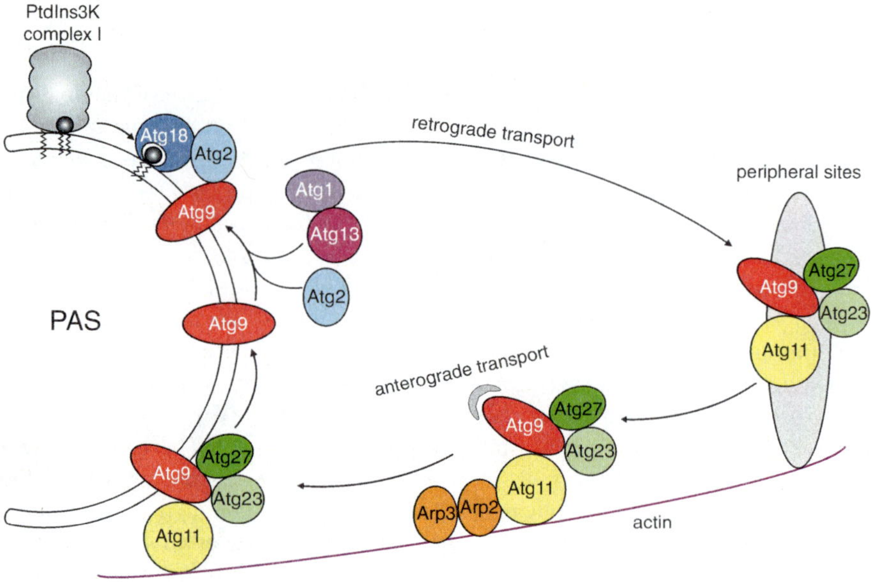

Fig. 7 Cycling of Atg9. In yeast, Atg9 cycles between the PAS and non-PAS punctate structures (peripheral sites), some of which are found to localize adjacent to or at the surface of mitochondria. The efficient anterograde movement of Atg9 to the PAS requires Atg11, Atg23, Atg27, and the actin cytoskeletion. Atg9, Atg23, and Atg27 are in a heterotrimeric complex, and their movement to the PAS is interdependent. Atg11 acts as a potential adaptor between Atg9 and actin, and between Atg9 and the Arp2/3 complex, while the latter may provide the force to push the cargo (Atg9 and its associated membrane) away from the peripheral sites and toward the PAS. The retrograde transport of Atg9 from the PAS back to the peripheral sites depends on the Atg1–Atg13 kinase complex, Atg2, Atg18, and the PtdIns3K complex I. The Atg1–Atg13 complex promotes the association of Atg2 and the PtdIns(3)P binding protein Atg18 with Atg9, and the formation of this ternary complex initiates Atg9 retrieval for another round of membrane delivery

(Reggiori et al. 2004a). Similarly, the retrieval of Atg23 and Atg27 requires the Atg1–Atg13 complex; however, only Atg23 retrieval needs a high level of Atg1 kinase activity (Legakis et al. 2007; Yen et al. 2007). Atg2 and Atg18 are two interacting peripheral membrane proteins (Suzuki et al. 2007). They can both interact with Atg9, and the interaction of Atg18 with Atg9 requires Atg2 and Atg1 (Reggiori et al. 2004a; Wang et al. 2001a). The PAS localization of Atg2 and Atg18 depends on each other, Atg1, Atg13, Atg9, and the PtdIns3K complex I. Atg18 can bind two phosphoinositides, PtdIns(3)P, and PtdIns(3,5)P_2, but only the former is essential for autophagy (Stromhaug et al. 2004). One model is that once the Atg1–Atg13 complex and Atg9 are recruited to the PAS separately, Atg1–Atg13 promotes Atg9 interaction with Atg2 and Atg18, and the formation of this ternary complex allows Atg9 to be released for another round of membrane delivery (Reggiori et al. 2004a).

Recent studies on mammalian Atg9 (mAtg9) have revealed that mAtg9 resides in a juxtanuclear region corresponding to the *trans*-Golgi network (TGN) and late endosomes (Young et al. 2006). Starvation triggers the distribution of mAtg9 from the TGN to a dispersed peripheral endosomal pool, and knockdown of Ulk1, the mammalian ortholog of Atg1, restricts mAtg9 to the TGN. These observations lead to the idea that mAtg9 traffics between the TGN and late endosomes, and that, potentially, membranes are delivered from the TGN to the forming autophagosomes.

2.6 *De Novo Vesicle Formation*

Unlike most other intracellular trafficking processes, autophagy undergoes de novo formation of double-membrane vesicles. This is a de novo process in that the sequestering vesicles do not bud from a pre-existing organelle. Instead, these vesicles are thought to form through the expansion of a membrane core of unknown origin, termed the phagophore (Mizushima et al. 2001; Noda et al. 2002; Seglen et al. 1990). Figure 8 shows a hypothetical model for de novo vesicle formation. The proposed site for vesicle formation is the phagophore assembly site (Kim et al. 2002; Suzuki et al. 2001). In yeast, the PAS is a perivacuolar site and is defined in part as the site where almost all of the Atg proteins reside, at least transiently (Suzuki et al. 2001). The PAS can be also defined as a hybrid of the phagophore and its associated Atg proteins (Xie and Klionsky 2007). In mammalian cells, colocalization of the Atg proteins has also been observed, although a comprehensive study has been lacking (Yamada et al. 2005; Young et al. 2006). In these observations, cells lack a single specialized site for autophagosome formation that is similar to the yeast PAS, and instead display multiple sites of Atg protein colocalization, possibly corresponding to multiple PAS.

Understanding the nature of the PAS is a key to studying this novel type of membrane-forming process. However, the PAS is poorly characterized. Although the role of the PAS is not fully understood, one model suggests that the PAS serves to facilitate the nucleation and/or expansion of the phagophore, the precursor of the autophagosome, through recruitment of Atg proteins (Mizushima et al. 2001; Suzuki et al. 2007). In addition, membrane has to be delivered to the phagophore; although the origin of this membrane is also not clear, it appears to include the early secretory pathway and, in yeast, the mitochondria (Reggiori et al. 2004b, 2005b).

A recent systematic analysis demonstrated that the Atg proteins depend on each other for PAS recruitment (Cheong et al. 2008; Kawamata et al. 2008; Suzuki et al. 2007). In particular, Atg11 and Atg17 act as scaffold proteins for PAS assembly, meaning that they may be the initial factors responsible for subsequent recruitment of the remaining Atg proteins. Atg11 is essential for PAS organization under vegetative conditions, whereas Atg17 (and associated proteins) plays a critical role during starvation. In cells lacking both Atg11 and Atg17, there is a complete absence of PAS localization of other Atg proteins. Starvation-induced PAS assembly, however, requires more than Atg17. In addition to the Atg1–Atg13–Atg17 ternary complex, Atg17 also interacts with two autophagy-specific proteins, Atg29 and Atg31 (Kawamata et al. 2008). Cells lacking Atg11 and any of the components in

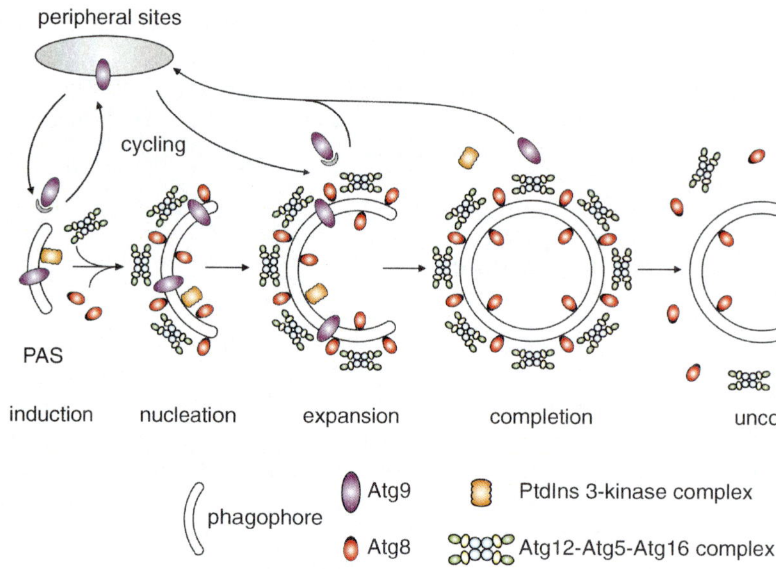

Fig. 8 Schematic depiction of double-membrane vesicle formation. The PAS serves to facilitate the nucleation and/or expansion of the phagophore, the precursor of the autophagosome, through recruitment of Atg proteins. Atg9 and the PtdIns3K complex I are recruited relatively early to the PAS, and act in membrane nucleation. Atg9 cycles between peripheral sites and the PAS; potentially, Atg9 cycling delivers lipids to the expanding membrane. The Atg12—Atg5—Atg16 complex and the Atg8—PE conjugate, components of the vesicle-forming machinery, are subsequently recruited to the PAS, and mediate the expansion of the sequestering vesicle. The Atg12—Atg5—Atg16 complex may in part behave like a coat or may function as an E3 ubiquitin ligase-like enzyme, whereas Atg8—PE acts in the elongation of the vesicle as a structural component. Before or immediately after the autophagosome is completed, most of the Atg components, including the putative coat proteins, dissociate from the vesicle; the portion of Atg8—PE on the outer surface of the vesicle is normally cleaved off by the Atg4 protease. Finally, the sequestering vesicle can fuse with the vacuole

the Atg17 complex display essentially the same phenotype as the *atg11Δ atg17Δ* double mutant, suggesting that these two complexes function as a PAS-organization center to induce the ordered recruitment of Atg proteins (Cheong et al. 2008; Kawamata et al. 2008). Although Atg1 kinase activity is not essential for PAS recruitment of other Atg proteins, it might play a role in disassembly of the PAS or the dissociation of Atg proteins from the PAS. A dynamic process of Atg protein cycling is thought to be critical to proper autophagosome expansion (Cheong et al. 2008). This concept fits with the idea that Atg1 kinase activity is related to the size of the sequestering vesicle (Noda et al. 2002).

Among the remaining Atg proteins, Atg9 is recruited relatively early to the PAS, and this also requires the function of the PtdIns3K complex I. Atg9 and the PtdIns3K complex I may play some role in membrane nucleation and facilitating the subsequent recruitment of certain Atg components, including the Atg12—Atg5—Atg16 complex and the Atg8—PE conjugate. In their absence, these two conjugates can be formed, but they become completely diffuse in the cytosol without any punctate localization. The PAS localization of Atg8—PE depends on the Atg12—Atg5–Atg16 complex (Suzuki et al. 2001).

As mentioned above, before or immediately after the autophagosome is completed, most of the Atg components dissociate from the vesicle. The sequestering vesicle must be completed before fusion with the vacuole. The Atg12—Atg5–Atg16 complex may in part behave like a coat to prevent premature fusion; the portion of Atg8—PE on the outer surface of the vesicle might also play such a role. Atg8—PE is normally cleaved off by the Atg4 protease prior to vesicle fusion. Furthermore, there may be certain unknown factors that can sense the completion of the double-membrane vesicle and trigger the disassembly of the vesicle-forming machinery. Atg1 is one possible candidate because it functions at later stages in the vesicle-forming process, such as Atg9 retrieval and Atg11 release from the PAS, and its kinase activity has been also suggested to play a role in the disassembly of the PAS or the dissociation of Atg proteins from the PAS (Cheong et al. 2008; Reggiori et al. 2004a; Yorimitsu and Klionsky 2005a).

2.7 Vesicle Docking and Fusion with the Vacuole

Once the double-membrane vesicle is formed, it is targeted to the vacuole for the fusion process. Molecular genetic studies have indicated that the machinery involved in homotypic vacuole fusion is also essential for the fusion of autophagosomes and Cvt vesicles with the vacuole (Fig. 9). This machinery includes the SNARE proteins Vam3, Vam7, Vti1 and Ykt6, the NSF Sec18, the α-SNAP Sec17, the Rab GTPase Ypt7 and the class C Vps/HOPS complex; the two recently characterized proteins Mon1 and Ccz1 are also part of the fusion machinery (Klionsky 2005; Wang and Klionsky 2003). Mon1 and Ccz1 form a complex and are critical for the Ypt7-dependent tethering/docking stage leading to the subsequent formation of the SNARE complex (Wang et al. 2003). The class C Vps/HOPS complex functions in concert with Ypt7 during the tethering/docking stage (Wang and Klionsky 2003). After fusion, the autophagosome inner single-membrane vesicle is released inside the vacuole lumen, which is termed the autophagic body.

In mammalian cells, maturation of autophagosomes includes several fusion events with vesicles originating from early and late endosomes, as well as lysosomes. Fusion with endosomes to become amphisomes allows convergence of the endocytic and autophagic pathways; subsequent fusion of autophagosomes or amphisomes with lysosomes generates autolysosomes (Berg et al. 1998; Tooze et al. 1990). In some cases where it is not possible to distinguish the precise nature of the compartment, the term "autophagic vacuoles" is used to cover all three autophagic structures: autophagosomes, amphisomes, and autolysosomes. Mammalian Vtilb is involved in the fusion of autophagsomes with multivesicular endosomes (Atlashkin et al. 2003), and the Rab GTPase Rab7 plays a role in the fusion with lysosomes (Jager et al. 2004).

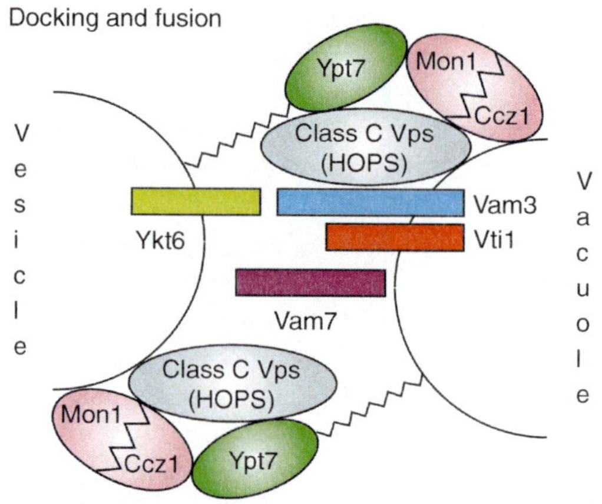

Fig. 9 Vesicle docking and fusion with the vacuole. The SNARE proteins Vam3, Vam7, Vti1, and Ykt6 function in various membrane fusion events, including the process of autophagosome fusion with the vacuole. Also shown are the Mon1–Ccz1 complex and the class C Vps/HOPS complex, which function in concert at the Ypt7-dependent tethering/docking stage. This figure is modified from Fig. 7 of Klionsky (2005)

2.8 Vesicle Breakdown and Recycling of the Resulting Macromolecules

Upon release into the vacuole, the single-membrane subvacuolar vesicle, the autophagic or Cvt body, is broken down inside the vacuolar lumen. This process depends on proper vacuole function (including vacuolar acidification) and the activity of vacuolar resident hydrolases (including Pep4 and Prb1). In addition to these factors, Atg15, a putative lipase, is also implicated at this step and seems likely to function directly in the intravacuolar lysis of the autophagic/Cvt body (Epple et al. 2001; Teter et al. 2001). Atg15 contains a lipase active-site motif, and mutations in the corresponding active site eliminate its function. Atg15 is targeted to the vacuolar lumen via the multivesicular body (MVB) pathway (Epple et al. 2003). Atg15 seems to function as a general lipase because it is also involved in the disintegration of intravacuolar MVB vesicles.

The main purpose of autophagy is to degrade cytoplasm and recycle the resulting macromolecules for the synthesis of essential components to overcome various stress conditions. Accordingly, the resulting macromolecules must be released back to the cytosol for reuse; however, little is known about this process. Atg22, a putative amino acid effluxer on the vacuolar membrane, has been found to play such role in mediating the efflux of leucine and other amino acids resulting from autophagic degradation (Yang et al. 2006). In addition, Avt3 and Avt4 seem to be part of the same family of permeases (Russnak et al. 2001). Upon elimination of all three partially redundant vacuolar effluxers, cells rapidly lose viability under starvation conditions, whereas supplementation with leucine partially restores viability.

Although a mammalian homolog of Atg22 has not been identified, homologs of Avt3 and Avt4 have been characterized as SLC36A1/LYAAT-1 (lysosomal amino acid transporter 1) (Sagne et al. 2001), and SLC36A4/LYAAT-2, respectively. How autophagy contributes to the recycling of other macromolecules, such as carbohydrate or lipids, remains unknown.

3 Conclusion

As a conserved cellular degradative pathway in eukaryotes, autophagy protects cells during various types of stress. Defects in autophagy have been linked to human diseases, indicating its crucial physiological significance. Autophagy involves dynamic membrane rearrangement for sequestration of cytoplasm and its delivery into the vacuole/lysosome. Significant breakthroughs in understanding the molecular mechanism of autophagy have been achieved from studies in yeast and other model systems. Currently, analyses of autophagy, pexophagy and the Cvt pathway in fungi have identified 31 *ATG* genes, corresponding to the unique molecular machinery that drives these pathways. However, the fundamental biochemical questions that concern the functions of Atg proteins still need to be resolved, especially those related to sequestering vesicle formation, such as the ordered vesicle assembly process and the origin of the lipid membrane. In addition, as more examples of selective types of autophagy emerge, continued studies on the specific nature of autophagy are becoming increasingly important. Yeast remains a powerful system to address these questions. Further studies of autophagy-related pathways will facilitate our understanding of the molecular mechanism and regulation of these pathways, and may allow the practical use of autophagy for therapeutic purposes.

Acknowledgments This work was supported by National Institutes of Health Public Health Service grant GM53396 to D.J.K.

References

Abeliovich H, Zhang C. Dunn WA Jr, Shokat KM, Klinosy DJ (2003) Chemical genetic analysis of Apgl reveals a non-kinase role in the induction of autophagy. Mol Biol Cell 14:477–490.
Arico S, Petiot A, Bauvy C, Dubbelhuis PF, Meijer AJ, Codogno P, Ogier-Denis E (2001) The tumor suppressor PTEN positively regulates macroautophagy by inhibiting the phosphatidylinositol 3-kinase/protein kinase B pathway. J Biol Chem 276:35243–35246
Atlashkin V, Kreykenbohm V, Eskelinen E-L, Wenzel D, Fayyazi A, Fischer von Mollard G (2003) Deletion of the SNARE vti1b in mice results in the loss of a single SNARE partner, syntaxin 8. Mol Cell Biol 23:5198–5207
Baba M, Osumi M, Scott SV, Klionsky DJ, Ohsumi Y (1997) Two distinct pathways for targeting proteins from the cytoplasm to the vacuole/lysosome. J Cell Biol 139:1687–1695
Bellu AR, Komori M, van der Klei IJ, Kiel JAKW, Veenhuis M (2001) Peroxisome biogenesis and selective degradation converge at Pex14p. J Biol Chem 276:44570–44574
Berg TO, Fengsrud M, Stromhaug PE, Berg T, Seglen PO (1998) Isolation and characterization of rat liver amphisomes. Evidence for fusion of autophagosomes with both early and late endosomes. J Biol Chem 273:21883–21892

Bjørkøy G, Lamark T, Brech A, Outzen H, Perander M, Øvervatn A, Stenmark H, Johansen T (2005) p62/SQSTM1 forms protein aggregates degraded by autophagy and has a protective effect on huntingtin-induced cell death. J Cell Biol 171:603–614

Budovskaya YV, Stephan JS, Reggiori F, Klionsky DJ, Herman PK (2004) The Ras/cAMP-dependent protein kinase signaling pathway regulates an early step of the autophagy process in *Saccharomyces cerevisiae*. J Biol Chem 279:20663–20671

Budovskaya YV, Stephan JS, Deminoff SJ, Herman PK (2005) An evolutionary proteomics approach identifies substrates of the cAMP-dependent protein kinase. Proc Natl Acad Sci USA 102:13933–13938

Carrera AC (2004) TOR signaling in mammals. J Cell Sci 117:4615–4616

Carroll AS, O'Shea EK (2002) Pho85 and signaling environmental conditions. Trends Biochem Sci 27:87–93

Chan EYW, Kir S, Tooze SA (2007) siRNA screening of the kinome identifies ULK1 as a multidomain modulator of autophagy. J Biol Chem 282:25464–25474

Chang CY, Huang W-P (2007) Atg19 mediates a dual interaction cargo sorting mechanism in selective autophagy. Mol Biol Cell 18:919–929

Cheong H, Nair U, Geng J, Klionsky DJ (2008) The Atg1 kinase complex is involved in the regulation of protein recruitment to initiate sequestering vesicle formation for nonspecific autophagy in *Saccharomyces cerevisiae*. Mol Biol Cell 19:668–681

De Virgilio C, Loewith R (2006) Cell growth control: little eukaryotes make big contributions. Oncogene 25:6392–6415

Ellson CD, Andrews S, Stephens LR, Hawkins PT (2002) The PX domain: a new phosphoinositide-binding module. J Cell Sci 115:1099–1105

Epple UD, Suriapranata I, Eskelinen E-L, Thumm M (2001) Aut5/Cvt17p, a putative lipase essential for disintegration of autophagic bodies inside the vacuole. J Bacteriol 183:5942–5955.

Epple UD, Eskelinen E-L, Thumm M (2003) Intravacuolar membrane lysis in *Saccharomyces cerevisiae*. Does vacuolar targeting of Cvt17/Aut5p affect its function? J Biol Chem 278:7810–7821

Farre JC, Manjithaya R, Mathewson RD, Subramani S (2008) PpAtg30 tags peroxisomes for turnover by selective autophagy. Dev Cell 14:365–376

Fujita N, Itoh T, Omori H, Fukuda M, Noda T, Yoshimori T (2008) The Atg16L complex specifies the site of LC3 lipidation for membrance biogenesis in autophagy. Mol Biol Cell 19:2092–2100.

Furuta S, Hidaka E, Ogata A, Yokota S, Kamata T (2004) Ras is involved in the negative control of autophagy through the class I PI3-kinase. Oncogene 23:3898–3904

Guan J, Stromhaug PE, George MD, Habibzadegah-Tari P, Bevan A, Dunn WA Jr, Klionsky DJ (2001) Cvt18/Gsa12 is required for cytoplasm-to-vacuole transport, pexophagy, and autophagy in *Saccharomyces cerevisiae* and *Pichia pastoris*. Mol Biol Cell 12:3821–3838

Gunkel K, van der Klei IJ, Barth G, Veenhuis M (1999) Selective peroxisome degradation in *Yarrowia lipolytica* after a shift of cells from acetate/oleate/ethylamine into glucose/ammonium sulfate-containing media. FEBS Lett 451:1–4

Gutierrez MG, Master SS, Singh SB, Taylor GA, Colombo MI, Deretic V (2004) Autophagy is a defense mechanism inhibiting BCG and *Mycobacterium tuberculosis* survival in infected macrophages. Cell 119:753–766

Hanada T, Noda NN, Satomi Y, Ichimura Y, Fujioka Y, Takao T, Inagaki F, Ohsumi Y (2007) The Atg12–Atg5 conjugate has a novel E3-like activity for protein lipidation in autophagy. J Biol Chem 282:37298–37302

Hanaoka H, Noda T, Shirano Y, Kato T, Hayashi H, Shibata D, Tabata S, Ohsumi Y (2002) Leaf senescence and starvation-induced chlorosis are accelerated by the disruption of an Arabidopsis autophagy gene. Plant Physiol 129:1181–1193

Hara T, Takamura A, Kishi C, Iemura S, Natsume T, Guan JL, Mizushima N (2008) FIP200, a ULK-interacting protein, is required for autophagosome formation in mammalian cells. J Cell Biol 181:497–510

He C, Song H, Yorimitsu T, Monastyrska I, Yen W-L, Legakis JE, Klionsky DJ (2006) Recruitment of Atg9 to the preautophagosomal structure by Atg11 is essential for selective autophagy in budding yeast. J Cell Biol 175:925–935

Hemelaar J, Lelyveld VS, Kessler BM, Ploegh HL (2003) A single protease, Apg4B, is specific for the autophagy-related ubiquitin-like proteins GATE-16, MAP1-LC3, GABARAP, and Apg8L. J Biol Chem 278:51841–51850
Hinnebusch AG (2005) Translational regulation of *GCN4* and the general amino acid control of yeast. Annu Rev Microbiol 59:407–450
Høyer-Hansen M, Jäättelä M (2007) AMP-activated protein kinase: a universal regulator of autophagy? Autophagy 3:381–383
Huang J, Klionsky DJ (2007) Autophagy and human disease. Cell Cycle 6:1837–1849
Huang W-P, Scott SV, Kim J, Klionsky DJ (2000) The itinerary of a vesicle component, Aut7p/Cvt5p, terminates in the yeast vacuole via the autophagy/Cvt pathways. J Biol Chem 275:5845–5851
Hutchins MU, Klionsky DJ (2001) Vacuolar localization of oligomeric α-mannosidase requires the cytoplasm to vacuole targeting and autophagy pathway components in *Saccharomyces cerevisiae*. J Biol Chem 276:20491–20498
Hutchins MU, Veenhuis M, Klionsky DJ (1999) Peroxisome degradation in *Saccharomyces cerevisiae* is dependent on machinery of macroautophagy and the Cvt pathway. J Cell Sci 112 (Pt 22):4079–4087
Ichimura Y, Kirisako T, Takao T, Satomi Y, Shimonishi Y, Ishihara N, Mizushima N, Tanida I, Kominami E, Ohsumi M, Noda T, Ohsumi Y (2000) A ubiquitin-like system mediates protein lipidation. Nature. 408:488–492
Ichimura Y, Imamura Y, Emoto K, Umeda M, Noda T, Ohsumi Y (2004) In vivo and in vitro reconstitution of Atg8 conjugation essential for autophagy. J Biol Chem 279:40584–40592
Iwata J, Ezaki J, Komatsu M, Yokota S, Ueno T, Tanida I, Chiba T, Tanaka K, Kominami E (2006) Excess peroxisomes are degraded by autophagic machinery in mammals. J Biol Chem 281:4035–4041
Jacinto E, Hall MN (2003) Tor signalling in bugs, brain and brawn. Nat Rev Mol Cell Biol 4:117–126
Jager S, Bucci C, Tanida I, Ueno T, Kominami E, Saftig P, Eskelinen E-L (2004) Role for Rab7 in maturation of late autophagic vacuoles. J Cell Sci 117:4837–4848
Jeffries TR, Dove SK, Michell RH, Parker PJ (2004) PtdIns-specific MPR pathway association of a novel WD40 repeat protein, WIPI49. Mol Biol Cell 15:2652–2663
Kabeya Y, Kamada Y, Baba M, Takikawa H, Sasaki M, Ohsumi Y (2005) Atg17 functions in cooperation with Atg1 and Atg13 in yeast autophagy. Mol Biol Cell 16:2544–2553
Kabeya Y, Mizushima N, Ueno T, Yamamoto A, Kirisako T, Noda T, Kominami E, Ohsumi Y, Yoshimori T (2000) LC3, a mammalian homologue of yeast Apg8p, is localized in autophagosome membranes after processing. EMBO J 19:5720–5728
Kabeya Y, Mizushima N, Yamamoto A, Oshitani-Okamoto S, Ohsumi Y, Yoshimori T (2004) LC3, GABARAP and GATE16 localize to autophagosomal membrane depending on form-II formation. J Cell Sci 117:2805–2812
Kamada Y, Funakoshi T, Shintani T, Nagano K, Ohsumi M, Ohsumi Y (2000) Tor-mediated induction of autophagy via an Apg1 protein kinase complex. J Cell Biol 150:1507–1513
Kawamata T, Kamada Y, Kabeya Y, Sekito T, Ohsumi Y (2008) Organization of the pre-autophagosomal structure responsible for autophagosome formation. Mol Biol Cell 19:2039–2050
Kihara A, Noda T, Ishihara N, Ohsumi Y (2001) Two distinct Vps34 phosphatidylinositol 3-kinase complexes function in autophagy and carboxypeptidase Y sorting in *Saccharomyces cerevisiae*. J Cell Biol 152:519–530
Kim I, Rodriguez-Enriquez S, Lemasters JJ (2007) Selective degradation of mitochondria by mitophagy. Arch Biochem Biophys 462:245–253
Kim J, Scott SV, Oda MN, Klionsky DJ (1997) Transport of a large oligomeric protein by the cytoplasm to vacuole protein targeting pathway. J Cell Biol 137:609–618
Kim J, Kamada Y, Stromhaug PE, Guan J, Hefner-Gravink A, Baba M, Scott SV, Ohsumi Y, Dunn WA Jr, Klionsky DJ (2001) Cvt9/Gsa9 functions in sequestering selective cytosolic cargo destined for the vacuole. J Cell Biol 153:381–396
Kim J, Huang W-P, Stromhaug PE, Klionsky DJ (2002) Convergence of multiple autophagy and cytoplasm to vacuole targeting components to a perivacuolar membrane compartment prior to *de novo* vesicle formation. J Biol Chem 277:763–773

Kirisako T, Baba M, Ishihara N, Miyazawa K, Ohsumi M, Yoshimori T, Noda T, Ohsumi Y (1999) Formation process of autophagosome is traced with Apg8/Aut7p in yeast. J Cell Biol 147:435–446

Kirisako T, Ichimura Y, Okada H, Kabeya Y, Mizushima N, Yoshimori T, Ohsumi M, Takao T, Noda T, Ohsumi Y (2000) The reversible modification regulates the membrane-binding state of Apg8/Aut7 essential for autophagy and the cytoplasm to vacuole targeting pathway. J Cell Biol 151:263–276

Klionsky DJ (2005) The molecular machinery of autophagy: unanswered questions. J Cell Sci 118:7–18

Klionsky DJ, Cregg JM, Dunn WA Jr, Emr SD, Sakai Y, Sandoval IV, Sibirny A, Subramani S, Thumm M, Veenhuis M, Ohsumi Y (2003) A unified nomenclature for yeast autophagy-related genes. Dev Cell 5:539–545

Klionsky DJ, Cueva R, Yaver DS (1992) Aminopeptidase I of *Saccharomyces cerevisiae* is localized to the vacuole independent of the secretory pathway. J Cell Biol 119:287–299

Klionsky DJ, Meijer AJ, Codogno P (2005) Autophagy and p70S6 kinase. Autophagy 1:59–61

Komatsu M, Ueno T, Waguri S, Uchiyama Y, Kominami E, Tanaka K (2007a) Constitutive autophagy: vital role in clearance of unfavorable proteins in neurons. Cell Death Differ 14:887–894

Komatsu M, Waguri S, Koike M, Sou YS, Ueno T, Hara T, Mizushima N, Iwata J, Ezaki J, Murata S, Hamazaki J, Nishito Y, Iemura S, Natsume T, Yanagawa T, Uwayama J, Warabi E, Yoshida H, Ishii T, Kobayashi A, Yamamoto M, Yue Z, Uchiyama Y, Kominami E, Tanaka K (2007b) Homeostatic levels of p62 control cytoplasmic inclusion body formation in autophagy-deficient mice. Cell 131:1149–1163

Kraft C, Deplazes A, Sohrmann M, Peter M (2008) Mature ribosomes are selectively degraded upon starvation by an autophagy pathway requiring the Ubp3p/Bre5p ubiquitin protease. Nat Cell Biol 10:602–610

Kubota H, Obata T, Ota K, Sasaki T, Ito T (2003) Rapamycin-induced translational derepression of GCN4 mRNA involves a novel mechanism for activation of the eIF2α kinase GCN2. J Biol Chem 278:20457–20460

Kuma A, Mizushima N, Ishihara N, Ohsumi Y (2002) Formation of the approximately 350-kDa Apg12–Apg5·Apg16 multimeric complex, mediated by Apg16 oligomerization, is essential for autophagy in yeast. J Biol Chem 277:18619–18625

Kuma A, Hatano M, Matsui M, Yamamoto A, Nakaya H, Yoshimori T, Ohsumi Y, Tokuhisa T, Mizushima N (2004) The role of autophagy during the early neonatal starvation period. Nature 432:1032–1036

Kundu M, Lindsten T, Yang CY, Wu J, Zhao F, Zhang J, Selak MA, Ney PA, Thompson CB (2008) Ulk1 plays a critical role in the autophagic clearance of mitochondria and ribosomes during reticulocyte maturation. Blood 112:1493–1502

Legakis JE, Yen W-L, Klionsky DJ (2007) A cycling protein complex required for selective autophagy. Autophagy 3:422–432

Levine B, Deretic V (2007) Unveiling the roles of autophagy in innate and adaptive immunity. Nat Rev Immunol 7:767–777

Levine B, Klionsky DJ (2004) Development by self-digestion: molecular mechanisms and biological functions of autophagy. Dev Cell 6:463–477

Levine B, Yuan J (2005) Autophagy in cell death: an innocent convict? J Clin Invest 115:2679–2688

Liang J, Shao SH, Xu ZX, Hennessy B, Ding Z, Larrea M, Kondo S, Dumont DJ, Gutterman JU, Walker CL, Slingerland JM, Mills GB (2007) The energy sensing LKB1–AMPK pathway regulates p27(kip1) phosphorylation mediating the decision to enter autophagy or apoptosis. Nat Cell Biol 9:218–224

Liang XH, Jackson S, Seaman M, Brown K, Kempkes B, Hibshoosh H, Levine B (1999) Induction of autophagy and inhibition of tumorigenesis by *beclin 1*. Nature 402:672–676

Loewith R, Jacinto E, Wullschleger S, Lorberg A, Crespo JL, Bonenfant D, Oppliger W, Jenoe P, Hall MN (2002) Two TOR complexes, only one of which is rapamycin sensitive, have distinct roles in cell growth control. Mol Cell 10:457–468

Martinez E, Jimenez MA, Segui-Real B, Vandekerckhove J, Sandoval IV (1997) Folding of the presequence of yeast pAPI into an amphipathic helix determines transport of the protein from the cytosol to the vacuole. J Mol Biol 267:1124–1138

Massey A, Kiffin R, Cuervo AM (2004) Pathophysiology of chaperone-mediated autophagy. Int J Biochem Cell Biol 36:2420–2434

Matsuura A, Tsukada M, Wada Y, Ohsumi Y (1997) Apg1p, a novel protein kinase required for the autophagic process in *Saccharomyces cerevisiae*. Gene 192:245–250.

Mavrakis M, Lippincott-Schwartz J, Stratakis CA, Bossis I (2006) Depletion of type IA regulatory subunit (RIα) of protein kinase A (PKA) in mammalian cells and tissues activates mTOR and causes autophagic deficiency. Hum Mol Genet 15:2962–2971

Meijer AJ, Codogno P (2006) Signalling and autophagy regulation in health, aging and disease. Mol Aspects Med 27:411–425

Meléndez A, Tallóczy Z, Seaman M, Eskelinen E-L, Hall DH, Levine B (2003) Autophagy genes are essential for dauer development and life-span extension in *C. elegans*. Science 301:1387–1391

Meley D, Bauvy C, Houben-Weerts JH, Dubbelhuis PF, Helmond MT, Codogno P, Meijer AJ (2006) AMP-activated protein kinase and the regulation of autophagic proteolysis. J Biol Chem 281:34870–34879

Mizushima N, Noda T, Yoshimori T, Tanaka Y, Ishii T, George MD, Klionsky DJ, Ohsumi M, Ohsumi Y (1998a) A protein conjugation system essential for autophagy. Nature 395:395–398

Mizushima N, Sugita H, Yoshimori T, Ohsumi Y (1998b) A new protein conjugation system in human. The counterpart of the yeast Apg12p conjugation system essential for autophagy. J Biol Chem 273:33889–33892

Mizushima N, Yamamoto A, Hatano M, Kobayashi Y, Kabeya Y, Suzuki K, Tokuhisa T, Ohsumi Y, Yoshimori T (2001) Dissection of autophagosome formation using Apg5-deficient mouse embryonic stem cells. J Cell Biol 152:657–668

Mizushima N, Yoshimori T, Ohsumi Y (2002) Mouse Apg10 as an Apg12-conjugating enzyme: analysis by the conjugation-mediated yeast two-hybrid method. FEBS Lett 532:450–454

Mizushima N, Kuma A, Kobayashi Y, Yamamoto A, Matsubae M, Takao T, Natsume T, Ohsumi Y, Yoshimori T (2003) Mouse Apg16L, a novel WD-repeat protein, targets to the autophagic isolation membrane with the Apg12-Apg5 conjugate. J Cell Sci 116:1679–88

Mizushima N, Levine B, Cuervo AM, Klionsky DJ (2008) Autophagy fights disease through cellular self-digestion. Nature 451:1069–1075

Monastyrska I, Shintani T, Klionsky DJ, Reggiori F (2006) Atg11 directs autophagosome cargoes to the PAS along actin cables. Autophagy 2:119–121

Monastyrska I, He C, Geng J, Hoppe AD, Li Z, Klionsky DJ (2008) Arp2 links autophagic machinery with the actin cytoskeleton. Mol Biol Cell 19:1962–1975

Nair U, Klionsky DJ (2005) Molecular mechanisms and regulation of specific and nonspecific autophagy pathways in yeast. J Biol Chem 280:41785–41788

Nakagawa I, Amano A, Mizushima N, Yamamoto A, Yamaguchi H, Kamimoto T, Nara A, Funao J, Nakata M, Tsuda K, Hamada S, Yoshimori T (2004) Autophagy defends cells against invading group A *Streptococcus*. Science 306:1037–1040

Nakatogawa H, Ichimura Y, Ohsumi Y (2007) Atg8, a ubiquitin-like protein required for autophagosome formation, mediates membrane tethering and hemifusion. Cell 130:165–178

Nice DC, Sato TK, Stromhaug PE, Emr SD, Klionsky DJ (2002) Cooperative binding of the cytoplasm to vacuole targeting pathway proteins, Cvt13 and Cvt20, to phosphatidylinositol 3-phosphate at the pre-autophagosomal structure is required for selective autophagy. J Biol Chem 277:30198–30207

Nobukuni T, Joaquin M, Roccio M, Dann SG, Kim SY, Gulati P, Byfield MP, Backer JM, Natt F, Bos JL, Zwartkruis FJ, Thomas G (2005) Amino acids mediate mTOR/raptor signaling through activation of class 3 phosphatidylinositol 3OH-kinase. Proc Natl Acad Sci USA 102:14238–14243

Noda T, Ohsumi Y (1998) Tor, a phosphatidylinositol kinase homologue, controls autophagy in yeast. J Biol Chem 273:3963–3966

Noda T, Kim J, Huang W-P, Baba M, Tokunaga C, Ohsumi Y, Klionsky DJ (2000) Apg9p/Cvt7p is an integral membrane protein required for transport vesicle formation in the Cvt and autophagy pathways. J Cell Biol 148:465–480
Noda T, Suzuki K, Ohsumi Y (2002) Yeast autophagosomes: de novo formation of a membrane structure. Trends Cell Biol 12:231–235
Obara K, Sekito T, Ohsumi Y (2006) Assortment of phosphatidylinositol 3-kinase complexes: Atg14p directs association of complex I to the pre-autophagosomal structure in *Saccharomyces cerevisiae*. Mol Biol Cell 17:1527–1539
Oda MN, Scott SV, Hefner-Gravink A, Caffarelli AD, Klionsky DJ (1996) Identification of a cytoplasm to vacuole targeting determinant in aminopeptidase I. J Cell Biol 132:999–1010
Ogawa M, Yoshimori T, Suzuki T, Sagara H, Mizushima N, Sasakawa C (2005) Escape of intracellular *Shigella* from autophagy. Science 307:727–731
Ohsumi Y (2001) Molecular dissection of autophagy: two ubiquitin-like systems. Nat Rev Mol Cell Biol 2:211–216
Onodera J, Ohsumi Y (2004) Ald6p is a preferred target for autophagy in yeast, *Saccharomyces cerevisiae*. J Biol Chem 279:16071–16076
Orvedahl A, Levine B (2008) Viral evasion of autophagy. Autophagy 4:280–285
Otto GP, Wu MY, Kazgan N, Anderson OR, Kessin RH (2004) Dictyostelium macroautophagy mutants vary in the severity of their developmental defects. J Biol Chem 279:15621–15629
Panaretou C, Domin J, Cockcroft S, Waterfield MD (1997) Characterization of p150, an adaptor protein for the human phosphatidylinositol (PtdIns) 3-kinase. Substrate presentation by phosphatidylinositol transfer protein to the p150·Ptdins 3-kinase complex. J Biol Chem 272:2477–2485
Paz Y, Elazar Z, Fass D (2000) Structure of GATE-16, membrane transport modulator and mammalian ortholog of autophagocytosis factor Aut7p. J Biol Chem 275:25445–25450
Reggiori F, Klionsky DJ (2002) Autophagy in the eukaryotic cell. Eukaryot Cell 1:11–21
Reggiori F, Monastyrska I, Shintani T, Klionsky DJ (2005a) The actin cytoskeleton is required for selective types of autophagy, but not nonspecific autophagy, in the yeast *Saccharomyces cerevisiae*. Mol Biol Cell 16:5843–5856
Reggiori F, Shintani T, Nair U, Klionsky DJ (2005b) Atg9 cycles between mitochondria and the pre-autophagosomal structure in yeasts. Autophagy 1:101–109
Reggiori F, Tucker KA, Stromhaug PE, Klionsky DJ (2004a) The Atg1-Atg13 complex regulates Atg9 and Atg23 retrieval transport from the pre-autophagosomal structure. Dev Cell 6:79–90
Reggiori F, Wang C-W, Nair U, Shintani T, Abeliovich H, Klionsky DJ (2004b) Early stages of the secretory pathway, but not endosomes, are required for Cvt vesicle and autophagosome assembly in *Saccharomyces cerevisiae*. Mol Biol Cell 15:2189–2204
Russnak R, Konczal D, McIntire SL (2001) A family of yeast proteins mediating bidirectional vacuolar amino acid transport. J Biol Chem 276:23849–23857
Sagne C, Agulhon C, Ravassard P, Darmon M, Hamon M, El Mestikawy S, Gasnier B, Giros B (2001) Identification and characterization of a lysosomal transporter for small neutral amino acids. Proc Natl Acad Sci USA 98:7206–7211
Sakai Y, Oku M, van der Klei IJ, Kiel JAKW (2006) Pexophagy: autophagic degradation of peroxisomes. Biochim Biophys Acta 1763:1767–1775
Schmelzle T, Beck T, Martin DE, Hall MN (2004) Activation of the RAS/cyclic AMP pathway suppresses a TOR deficiency in yeast. Mol Cell Biol 24:338–351
Scott RC, Schuldiner O, Neufeld TP (2004) Role and regulation of starvation-induced autophagy in the *Drosophila* fat body. Dev Cell 7:167–178
Scott RC, Juhasz G, Neufeld TP (2007) Direct induction of autophagy by Atg1 inhibits cell growth and induces apoptotic cell death. Curr Biol 17:1–11
Scott SV, Baba M, Ohsumi Y, Klionsky DJ (1997) Aminopeptidase I is targeted to the vacuole by a nonclassical vesicular mechanism. J Cell Biol 138:37–44
Scott SV, Guan J, Hutchins MU, Kim J, Klionsky DJ (2001) Cvt19 is a receptor for the cytoplasm-to-vacuole targeting pathway. Mol Cell 7:1131–1141
Seglen PO, Gordon PB, Holen I (1990) Non-selective autophagy. Semin Cell Biol 1:441–448

Shintani T, Klionsky DJ (2004a) Autophagy in health and disease: a double-edged sword. Science 306:990–995

Shintani T, Klionsky DJ (2004b) Cargo proteins facilitate the formation of transport vesicles in the cytoplasm to vacuole targeting pathway. J Biol Chem 279:29889–29894

Shintani T, Mizushima N, Ogawa Y, Matsuura A, Noda T, Ohsumi Y (1999) Apg10p, a novel protein-conjugating enzyme essential for autophagy in yeast. EMBO J 18:5234–5241

Shintani T, Huang W-P, Stromhaug PE, Klionsky DJ (2002) Mechanism of cargo selection in the cytoplasm to vacuole targeting pathway. Dev Cell 3:825–837

Stack JH, DeWald DB, Takegawa K, Emr SD (1995) Vesicle-mediated protein transport: regulatory interactions between the Vps15 protein kinase and the Vps34 PtdIns 3-kinase essential for protein sorting to the vacuole in yeast. J Cell Biol 129:321–334

Stenmark H, Aasland R, Driscoll PC (2002) The phosphatidylinositol 3-phosphate-binding FYVE finger. FEBS Lett 513:77–84

Stromhaug PE, Reggiori F, Guan J, Wang C-W, Klionsky DJ (2004) Atg21 is a phosphoinositide binding protein required for efficient lipidation and localization of Atg8 during uptake of aminopeptidase I by selective autophagy. Mol Biol Cell 15:3553–3566

Suzuki K, Kirisako T, Kamada Y, Mizushima N, Noda T, Ohsumi Y (2001) The pre-autophagosomal structure organized by concerted functions of *APG* genes is essential for autophagosome formation. EMBO J 20:5971–5981

Suzuki K, Kubota Y, Sekito T, Ohsumi Y (2007) Hierarchy of Atg proteins in pre-autophagosomal structure organization. Genes Cells 12:209–218

Suzuki NN, Yoshimoto K, Fujioka Y, Ohsumi Y, Inagaki F (2005) The crystal structure of plant Atg12 and its biological implication in autophagy. Autophagy 1:119–126

Tallóczy Z, Jiang W, Virgin HW IV, Leib DA, Scheuner D, Kaufman RJ, Eskelinen E-L, Levine B (2002) Regulation of starvation- and virus-induced autophagy by the eIF2α kinase signaling pathway. Proc Natl Acad Sci USA 99:190–195

Tanida I, Mizushima N, Kiyooka M, Ohsumi M, Ueno T, Ohsumi Y, Kominami E (1999) Apg7p/Cvt2p: a novel protein-activating enzyme essential for autophagy. Mol Biol Cell 10:1367–1379

Tanida I, Tanida-Miyake E, Komatsu M, Ueno T, Kominami E (2002) Human Apg3p/Aut1p homologue is an authentic E2 enzyme for multiple substrates, GATE-16, GABARAP, and MAP-LC3, and facilitates the conjugation of hApg12p to hApg5p. J Biol Chem 277: 13739–13744

Tanida I, Komatsu M, Ueno T, Kominami E (2003) GATE-16 and GABARAP are authentic modifiers mediated by Apg7 and Apg3. Biochem Biophys Res Commun 300:637–644

Tanida I, Sou YS, Minematsu-Ikeguchi N, Ueno T, Kominami E (2006) Atg8L/Apg8L is the fourth mammalian modifier of mammalian Atg8 conjugation mediated by human Atg4B, Atg7 and Atg3. FEBS J 273:2553–2562

Teter SA, Eggerton KP, Scott SV, Kim J, Fischer AM, Klionsky DJ (2001) Degradation of lipid vesicles in the yeast vacuole requires function of Cvt17, a putative lipase. J Biol Chem 276:2083–2087

Thevelein JM, de Winde JH (1999) Novel sensing mechanisms and targets for the cAMP-protein kinase A pathway in the yeast *Saccharomyces cerevisiae*. Mol Microbiol 33:904–918

Tooze J, Hollinshead M, Ludwig T, Howell K, Hoflack B, Kern H (1990) In exocrine pancreas, the basolateral endocytic pathway converges with the autophagic pathway immediately after the early endosome. J Cell Biol 111:329–345

Tuttle DL, Lewin AS, Dunn WA Jr (1993) Selective autophagy of peroxisomes in methylotrophic yeasts. Eur J Cell Biol 60:283–290

Urban J, Soulard A, Huber A, Lippman S, Mukhopadhyay D, Deloche O, Wanke V, Anrather D, Ammerer G, Riezman H, Broach JR, De Virgilio C, Hall MN, Loewith R (2007) Sch9 is a major target of TORC1 in *Saccharomyces cerevisiae*. Mol Cell 26:663–674

Veenhuis M, Douma A, Harder W, Osumi M (1983) Degradation and turnover of peroxisomes in the yeast *Hansenula polymorpha* induced by selective inactivation of peroxisomal enzymes. Arch Microbiol 134:193–203

Wang C-W, Klionsky DJ (2003) The molecular mechanism of autophagy. Mol Med 9:65–76

Wang C-W, Kim J, Huang W-P, Abeliovich H, Stromhaug PE, Dunn WA Jr, Klionsky DJ (2001a) Apg2 is a novel protein required for the cytoplasm to vacuole targeting, autophagy, and pexophagy pathways. J Biol Chem 276:30442–30451

Wang C-W, Stromhaug PE, Kauffman EJ, Weisman LS, Klionsky DJ (2003) Yeast homotypic vacuole fusion requires the Ccz1-Mon1 complex during the tethering/docking stage. J Cell Biol 163:973–985

Wang Z, Wilson WA, Fujino MA, Roach PJ (2001b) Antagonistic controls of autophagy and glycogen accumulation by Snf1p, the yeast homolog of AMP-activated protein kinase, and the cyclin-dependent kinase Pho85p. Mol Cell Biol 21:5742–5752

Xie Z, Klionsky DJ (2007) Autophagosome formation: core machinery and adaptations. Nat Cell Biol 9:1102–1109

Xie Z, Nair U, Klionsky DJ (2008) Atg8 controls phagophore expansion during autophagosome formation. Mol Biol Cell 19:3290–3298

Yamada T, Carson AR, Caniggia I, Umebayashi K, Yoshimori T, Nakabayashi K, Scherer SW (2005) Endothelial nitric-oxide synthase antisense (NOS3AS) gene encodes an autophagy-related protein (APG9-like2) highly expressed in trophoblast. J Biol Chem 280:18283–18290

Yan J, Kuroyanagi H, Kuroiwa A, Matsuda Y, Tokumitsu H, Tomoda T, Shirasawa T, Muramatsu M (1998) Identification of mouse ULK1, a novel protein kinase structurally related to *C. elegans* UNC-51. Biochem Biophys Res Commun 246:222–227

Yan J, Kuroyanagi H, Tomemori T, Okazaki N, Asato K, Matsuda Y, Suzuki Y, Ohshima Y, Mitani S, Masuho Y, Shirasawa T, Muramatsu M (1999) Mouse ULK2, a novel member of the UNC-51-like protein kinases: unique features of functional domains. Oncogene 18:5850–5859

Yang Z, Huang J, Geng J, Nair U, Klionsky DJ (2006) Atg22 recycles amino acids to link the degradative and recycling functions of autophagy. Mol Biol Cell 17:5094–5104

Yen W-L, Legakis JE, Nair U, Klionsky DJ (2007) Atg27 is required for autophagy-dependent cycling of Atg9. Mol Biol Cell 18:581–593

Yorimitsu T, Klionsky DJ (2005a) Atg11 links cargo to the vesicle-forming machinery in the cytoplasm to vacuole targeting pathway. Mol Biol Cell 16:1593–1605

Yorimitsu T, Klionsky DJ (2005b) Autophagy: molecular machinery for self-eating. Cell Death Differ 12 Suppl 2:1542–1552

Yorimitsu T, Zaman S, Broach JR, Klionsky DJ (2007) Protein kinase A and Sch9 cooperatively regulate induction of autophagy in *Saccharomyces cerevisiae*. Mol Biol Cell 18:4180–4189

Yorimitsu T, He C, Wang K, Klionsky DJ (2009) Tap42-associated protein phosphatase type 2A negatively regulates induction of autophagy. Autophagy 5:in press

Young ARJ, Chan EYW, Hu XW, Köchl R, Crawshaw SG, High S, Hailey DW, Lippincott-Schwartz J, Tooze SA (2006) Starvation and ULK1-dependent cycling of mammalian Atg9 between the TGN and endosomes. J Cell Sci 119:3888–3900

Zurita-Martinez SA, Cardenas ME (2005) Tor and cyclic AMP-protein kinase A: two parallel pathways regulating expression of genes required for cell growth. Eukaryot Cell 4:63–71

Macroautophagy Signaling and Regulation

Audrey Esclatine, Magali Chaumorcel, and Patrice Codogno

Contents

1 Introduction 34
2 Signaling Pathways 37
 2.1 TOR-Dependent Signaling Pathways 37
 2.2 mTOR-Independent Pathways 41
 2.3 Other Pathways 42
3 Autophagosome Formation 46
 3.1 Atg1 and Its Partners 46
 3.2 Atg6/Beclin 1 47
 3.3 Post-translational Modifications of Atg Proteins 50
 3.4 Transcriptional Regulation of Atg Proteins 52
 3.5 The Cytoskeleton 52
4 Maturation Step 53
 4.1 Morphology and Definition of Late Stages of Autophagy 53
 4.2 Regulation of the Maturation Events 53
 4.3 Signaling and Maturation of Autophagosomes 57
5 How Autophagy Can Be Manipulated 58
6 Conclusions 59
References 60

Abstract Macroautophagy is a vacuolar degradation pathway that terminates in the lysosomal compartment. Macroautophagy is a multistep process involving: (1) signaling events that occur upstream of the molecular machinery of autophagy; (2) molecular machinery involved in the formation of the autophagosome, the initial multimembrane-bound compartment formed in the autophagic pathway; and (3) maturation of autophagosomes, which acquire acidic and degradative capacities. In this chapter we summarize what is known about the regulation of the different steps involved in autophagy, and we also discuss how macroautophagy can be manipulated using drugs or genetic approaches that affect macroautophagy signaling, and the subsequent formation and maturation of the autophagosomes. Modulating

A. Esclatine, M. Chaumorcel, and P. Codogno (✉)
INSERM U756, Université Paris-Sud 11, Faculté de Pharmacie,
5 rue Jean-Baptiste Clément, 92290 Châtenay-Malabry, France

B. Levine et al. (eds.), *Autophagy in Infection and Immunity,*
Current Topics in Microbiology and Immunology 335,
DOI 10.1007/ 978-3-642-00302-8_2,

autophagy offers a promising new therapeutic approach to human diseases that involve macroautophagy.

Abbreviations

3-MA	3-Methyladenine
4E-BP1	Eukaryotic translational initiation factor 4E-binding protein 1
AMPK	AMP-activated protein kinase
ATG	Autophagy-related
DAP kinase	Death-associated protein kinase
DRAM	Damage-regulated autophagy modulator
DRP-1	Death-associated related protein kinase 1
eIF2α	Eukaryotic initiation factor 2 alpha
ERK	Extracellular signal-regulated protein kinase
FDA	Food and drug administration
JNK	c-Jun N-terminal kinase
LC3	Light chain 3
MAPK	Mitogen-activated protein kinase
(m)TOR	(Mammalian) target of rapamycin
PE	Phosphatidyl ethanolamine
PERK	Protein kinase R-like endoplasmic reticulum kinase
PI3K	Phosphatidylinositol 3-phosphate kinase
PKR	Double-stranded RNA-activated protein kinase
Rheb	Ras homolog enriched in brain
ROS	Reactive oxygen species
SNARE	Soluble NSF attachment protein receptors
TSC	Tuberous sclerosis complex

1 Introduction

Macroautophagy (referred to as "autophagy" below) is a mechanism conserved among eukaryotic cells that starts with the formation of a multimembrane-bound vacuole, known as an autophagosome, which ultimately fuses with the lysosomal compartment to degrade the sequestered material (Klionsky 2007; Levine and Klionsky 2004; Mizushima et al. 2008; Ohsumi 2001). The seminal discovery of *ATG* (autophagy-related) genes, originally in yeast and then in multicellular organisms, constituted an important breakthrough in our understanding of how autophagosomes are formed, and of the importance of autophagy in cell physiology and in human diseases (see the chapters by Yang and Klionsky and by Mizushima in this volume). However, it is absolutely essential to find out how autophagy is regulated if we are to grasp how it changes in response to stress, including infection. The intricacy of the regulation of autophagy and apoptosis suggests that there is a subtle dialog between these two processes that determines the fate of the cell (Codogno and Meijer 2005; Gozuacik and Kimchi 2007; Levine and Yuan 2005; Maiuri et al. 2007c).

In this chapter, we divide the regulation of autophagy into three successive levels (Fig. 1), even though the boundaries between them are not clear-cut. The first level of regulation is defined as the signaling pathways that terminate upstream of the Atg machinery. Archetypal examples of this first level of regulation are signaling pathways that terminate at mTOR (Meijer and Codogno 2006), but other signaling pathways with targets within the molecular machinery that have not yet been identified also fall into this category. Many growth factors and cytokines modulate autophagy at this level (Deretic 2006; Lum et al. 2005). The immunosuppressor rapamycin triggers autophagy by interfering with mTOR activity (Meijer and Codogno 2006). The second level of regulation involves modulating the Atg machinery by protein-protein interactions or modulating Atg activity by signaling molecules (Maiuri et al. 2007a; Pattingre and Levine 2006; Scherz-Shouval and Elazar 2007). The Beclin 1/Bcl-2 interaction (Erlich et al. 2007; Maiuri et al. 2007b; Pattingre et al. 2005) and the modification of Atg4 by reactive oxygen species fall within this category (Scherz-Shouval et al. 2007). The third level of regulation involves the late stage of autophagy (maturation and fusion with the lysosomal compartment) (Eskelinen 2005). Blockade during the late stage of autophagy is a hallmark of Alzheimer's disease (Yu et al. 2005), and of a rare cardiomyopathy known as Danon disease (Nishino et al. 2000; Tanaka et al. 2000). A given autophagy modulator may target more than one levels of autophagy; for example, starvation triggers autophagy by targeting both levels 1 and 2 (Pattingre et al. 2008). We do not yet know whether starvation can also modulate level 3.

As will be discussed in greater detail in subsequent chapters, pathogens are also able to interfere with all levels of autophagy. Herpes simplex virus 1 (HSV-1) blocks autophagy by interfering with levels 1 and 2 (Orvedahl and Levine 2008). *Listeria monocytogenes*, *Shigella,* and some other bacteria evade sequestration by manipulating level 1 and/or level 2 (Birmingham et al. 2008; Levine and Deretic 2007; Ogawa et al. 2005). Some bacteria, such as *Legionella pneumophila* and *Coxiella burnetii,* block autophagy at level 3, allowing them to avoid being trapped in the lysosome and escape degradation (Romano et al. 2007; Swanson 2006). Poliovirus probably manipulates both levels 2 and 3 to stimulate autophagy to its own benefit (Taylor and Kirkegaard 2008).

The division of autophagy into the three levels that we propose in this chapter is also important from a functional point of view. Autophagy performs two nonexclusive tasks: it sequesters cytoplasmic material and it degrades it (Mizushima 2005). In some contexts, the sequestration function is the most important, for example when harmful compounds are segregated from the cytoplasm (Komatsu et al. 2007; Rubinsztein 2006). The second purpose of autophagy is to degrade materials sequestered in the lysosomal compartment. The importance of this second function is illustrated by starvation-induced autophagy, which provides amino acids and fatty acids to maintain metabolism and ATP levels when extracellular nutrients are in short supply (Kuma et al. 2004; Lum et al. 2005). These two functions imply

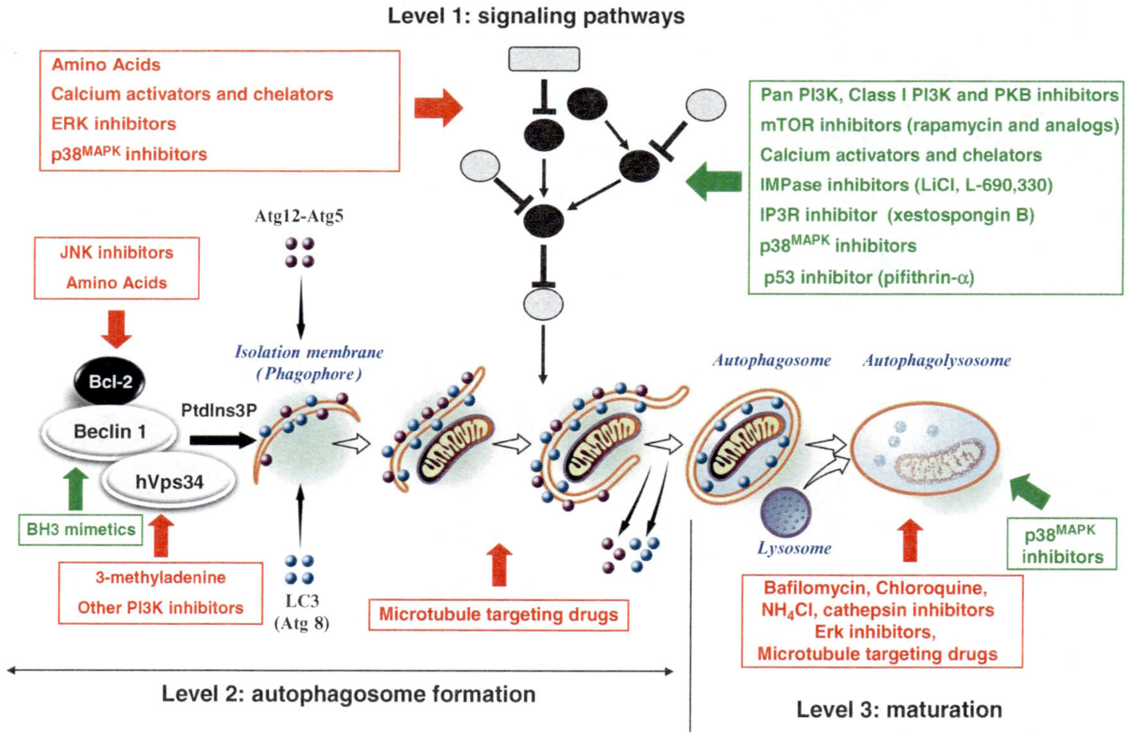

Fig. 1 Different levels of autophagy regulation. The regulation of autophagy can be divided into three levels. The first level is defined by the signaling pathways that terminate upstream of the Atg machinery involved in autophagosome formation. The second level of regulation is the Atg machinery involved in the formation of autophagosomes. The third level of regulation corresponds to the late stage of autophagy (maturation and fusion with the lysosomal compartment). Various drugs that activate (*green*) or inhibit (*red*) autophagy at the various different levels of regulation are indicated in the figure

critical regulations at levels 2 and 3, respectively. It is clear that a dysfunction at level 3 will have a greater adverse effect when ATP levels and metabolism are compromised. Level 1 is important in situations in which autophagy is stimulated, but basal autophagy is probably also finely regulated by signaling pathways, although the mechanisms involved in this are poorly defined.

In this chapter, we will describe the three levels of regulation of the autophagic pathway as shown in Fig. 1, and then discuss possible ways of manipulating autophagy at these different levels of regulation.

2 Signaling Pathways

Autophagy is known to be induced by a wide variety of stimuli (e.g., nutrient and growth factor depletion, hypoxia, drug and radiation treatment). Many signaling pathways and second messengers have been shown to regulate the activity of the Atg machinery involved in the formation of autophagosomes. This constitutes level 1 of autophagy regulation. Many of these pathways converge on the evolutionarily conserved kinase TOR (target of rapamycin). However, other autophagy signaling pathways act independently of TOR, especially in mammalian cells (e.g., the inositol phosphate pathway). Moreover, in mammalian cells, it is not clear how these signaling pathways impinge on the molecular machinery of autophagy. The aim of this section is to give an overview of the autophagy signaling that regulates the biogenesis of autophagosomes. Readers can consult recent reviews dedicated to the signaling of autophagy to obtain a more detailed description of this aspect of autophagy (Codogno and Meijer 2005; Gozuacik and Kimchi 2007).

2.1 TOR-Dependent Signaling Pathways

The kinase TOR is a major evolutionarily conserved sensor in the autophagy signaling pathway in eukaryotes, but it also regulates many other aspects of cell function, including transcription, translation, cell size and cytoskeletal organization (Schmelzle and Hall 2000). In mammals, mTOR can be included in two different complexes (Schmelzle and Hall 2000), mTORC1 and mTORC2. Although these two TOR complexes share common components, they display distinct cellular functions and phosphorylate different downstream substrates (Jacinto et al. 2004; Loewith et al. 2002). The activity of mTORC1 is regulated via the integration of many signals, including growth factors, insulin, nutrients, energy availability, and cell stressors such as hypoxia, osmotic stress, reactive oxygen species and viral infection (Corradetti and Guan 2006). mTORC1 is the only known target of the drug rapamycin, and is required for signaling to S6K and 4E-BP1. mTORC1 has recently been shown to consist of four proteins: mTOR, mLST8 (also known as GβL), proline-rich PKB/Akt substrate 40-kDa (PRAS40), and raptor (regulatory associated protein of mTOR), and it plays a major role in controlling translation and cell growth in response to nutrients. The adaptor protein mLST8 is common to both mTOR complexes. Raptor binds mTOR, S6K and 4EBP1 and facilitates mTOR phosphorylation of these molecules; but whether raptor enhances or represses mTOR kinase activity remains unclear (Hara et al. 2002; Kim et al. 2002). Unlike mTORC1, mTORC2 has some functions that cannot be inhibited by rapamycin, including the control of actin cytoskeleton dynamics (Jacinto et al. 2004; Loewith et al. 2002). The mTORC2 complex consists of mTOR, mLST8, mammalian stress-activated protein kinase-interacting protein 1 (mSin1), and rictor (for rapamycin-insensitive companion of mTOR) (Sarbassov et al. 2004).

2.1.1 mTORC1

To date, the signaling pathway including Class I PI3K/PKB and mTORC1 is the autophagy-regulating pathway that has undergone the most investigation. Class I PI3K enzymes phosphorylate PtdIns4P and PtdIns(4,5)P_2 to produce PtdIns(3,4)P_2 and PtdIns(3,4,5)P_3, which bind to Akt (also known as protein kinase B: PKB) (Brazil and Hemmings 2001). PKB is responsible for activating mTOR via tuberous sclerosis complex 1/2 (TSC1/2), GTPase-activating proteins, and Rheb, a GTPase protein (Fig. 2). When this pathway is activated by challenging receptors that recruit class I PI3K or by expressing a constitutive active form of PKB, this has an inhibitory effect on autophagy (Arico et al. 2001). The phosphatase PTEN, which hydrolyzes PtdIns(3,4,5)P_3, has a stimulatory effect on autophagy by abolishing class I PI3K/PKB inhibition (Meijer and Codogno 2004). The mTORC1 complex is also inhibited by AMP-activated kinase (AMPK), which reflects the energy status of the cell (see Sect. 2.1.3.).

Growth factors activate the class I PI3K/Akt/mTOR pathway and consequently inhibit autophagy. Historically, awareness that macroautophagic sequestration is controlled via transduction pathways emerged from pioneering studies of the effect of insulin, glucagon, and glucocorticosteroids on liver proteolysis (Deter and De Duve 1967; Hopgood et al. 1981; Mortimore and Ward 1976). In addition to the negative effect of insulin on liver protein degradation, various growth promoting factors and serum suppress autophagic proteolysis in many mammalian cell types (Blommaart et al. 1997b). In the absence of growth factors, cells are unable to take up nutrients from the extracellular medium. In this context, mTOR is inhibited and autophagy has been shown to rescue cells from death by maintaining ATP levels in starved cells (Lum et al. 2005). However, this autophagy-mediated survival mechanism is self-limiting, and persistent growth factor deprivation leads to cell death within a few weeks. This is probably due to severe degradation of essential organelles and macromolecules induced by the prolonged stimulation of autophagy.

Amino acids are the final products of the autophagic pathway, and so it is not surprising that they are able to negatively regulate autophagy and that this capacity is conserved from yeast to humans. Conversely, it has long been known that amino acid deprivation stimulates autophagy (Poso et al. 1982). However, the mechanism

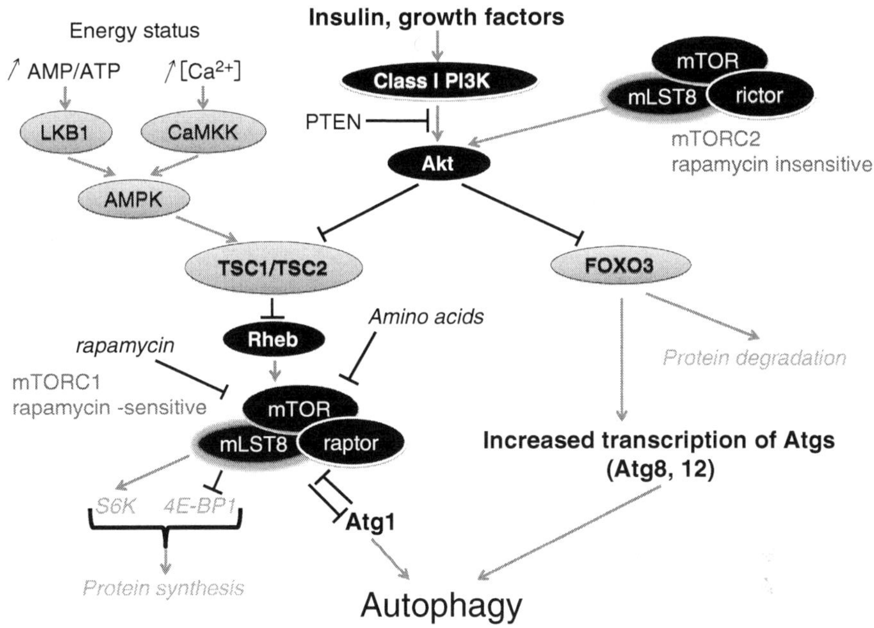

Fig. 2 Roles of the two mTOR complexes in the regulation of autophagy. mTOR exists in two different complexes, mTORC1 and mTORC2. Mammalian TORC1 (mTORC1) is composed of mLST8 (GβL) and raptor. It is sensitive to rapamycin, and immediately regulates autophagy in response to cellular stress. The mTOR signaling pathway can receive input from growth factors via class I PI3K, which activates Akt. In turn, Akt activates mTORC1 via the inhibition of TSC1/TSC2. The energy sensor AMPK (AMP-activated kinase), which is activated when the AMP/ATP ratio increases, inactivates mTORC1 by activating TSC2. Calcium signaling can also inhibit mTORC1 by activating AMPK. Atg1 Ser/Thr protein kinase functions downstream of mTORC1, but it also exercises negative feedback control over mTOR. The rapamycin-insensitive complex mTORC2 is composed of mTOR, mLST8 (GβL) and rictor; it seems to be implicated in the long-term regulation of autophagy by activating Akt, which inhibits FOXO3. FOXO3 is known to increase the transcription of several Atg genes, such as *ATG8/LC3* or *ATG12*. Molecules shown in *gray* correspond to autophagy-activating molecules, and those shown in *black* to autophagy-inhibiting molecules

by which amino acids control mTOR and inhibit autophagy is still a matter of debate (Gulati and Thomas 2007; Meijer and Codogno 2006). It has been proposed that the GTPase Rheb, which has GTPase activity controlled by the GTPase-activating protein TSC2, may integrate amino acid signaling upstream of mTOR (Sarbassov et al. 2005). Amino acids may also act at the level of the mTORC1 complex by controlling the stability of the mTOR/raptor complex (Hara et al. 2002; Kim et al. 2002). The stability of this complex is increased in cells starved of amino acids, and is correlated with the inhibition of mTOR-dependent signaling. In mammals, amino acids have recently been shown to mediate mTOR/raptor-dependent signaling by activating class III PI3K (Byfield et al. 2005; Nobukuni et al. 2005).

This signaling pathway is dependent on the increase in intracellular Ca^{2+} and the direct binding of Ca^{2+}/calmodulin to the class III PI3K/mTORC1 complex (Gulati et al. 2008). Nevertheless, in *Drosophila*, class III PI3K does not act upstream of TOR, suggesting that we must be cautious regarding the impact of amino acids in activating the different regulating complexes of autophagy (Juhasz et al. 2008).

2.1.2 mTORC2

Much less is known about the upstream regulation and function of TORC2 than for TORC1. The rictor protein directs the specificity of mTORC2 towards Akt and the FoxO3 transcription factor, and away from S6K and 4EBP1 (Guertin et al. 2006; Sarbassov et al. 2005). Akt/PKB activation blocks FoxO3 activation and autophagy, and this effect is not prevented by rapamycin (Mammucari et al. 2007). It therefore seems likely that decreased PI3K/Akt signaling activates autophagy not only through mTORC1 but also, albeit more slowly, via a transcription-dependent mechanism involving FoxO3 (Zhao et al. 2007). Indeed, in vivo in skeletal muscle, activated FoxO3 increases the transcription of two autophagy-related genes, *LC3* and *Bnip3* (a BH3-only protein), leading to the induction of autophagosome formation (Mammucari et al. 2007; Mammucari et al. 2008). Inhibition of rictor causes the translocation of FoxO3 to the nucleus and induces autophagy. In fact, FoxO3 regulates both the ubiquitin proteasome and autophagy systems during muscle atrophy.

2.1.3 AMPK

Apart from being an autophagy sensor, mTOR can also sense changes in the cellular energy via AMPK. Activation of AMPK inhibits mTOR-dependent signaling by interfering with the activity of GTPase Rheb, and with protein synthesis (Meijer and Codogno 2004). This is consistent with the switching off of ATP-dependent processes (Hardie 2004) during periods of energy crisis. In starved cells, when the AMP/ATP ratio increases, the binding of AMP to AMPK promotes its activation by the AMPK kinase LKB1 (Corradetti et al. 2004; Shaw et al. 2004). Moreover, Ca^{2+}/calmodulin-dependent kinase kinase β (CaMKK-β) has been identified as being an AMPK kinase (Hawley et al. 2005; Woods et al. 2005). The activity of AMPK is required for autophagy to be induced in response to starvation in mammalian cells (Meley et al. 2006) and in yeast (Wang et al. 2001) in a TORC1-dependent manner. Moreover, autophagy induction is also dependent on the inhibition of mTORC1 by AMPK in non starved cells in response to an increase in free cytosolic Ca^{2+} (Hoyer-Hansen et al. 2007). In this setting, the activation of AMPK and stimulation of autophagy are dependent on CaMKK-β. The induction of autophagy through AMPK activation probably also occurs in other settings, such as hypoxia (Degenhardt et al. 2006; Laderoute et al. 2006). AMPK is probably a general regulator of autophagy upstream of mTOR (Hoyer-Hansen and Jaattela 2007; Meijer and Codogno 2007). Another potential candidate of autophagy regulation downstream of AMPK is elongation factor-2 kinase (eEF-2 kinase), which controls the rate of peptide elongation (Hait et al. 2006). Activation of eEF-2 kinase increases autophagy and slows protein translation (Wu et al. 2006). The activity of eEF-2 kinase is regulated by mTOR, S6K, and AMPK (Browne et al. 2004; Browne and Proud 2002). During periods of ATP depletion, AMPK is activated and eEF-2 kinase is phosphorylated (Browne et al. 2004), leading to a balance between the inhibition of peptide elongation and the induction of autophagy. How eEF-2 kinase impinges on the molecular machinery of autophagy remains to be elucidated. Autophagy is activated by AMPK in a p53-dependent manner (Feng et al. 2005). However, the cytoplasmic form of p53 has been shown to have an inhibitory effect on autophagy (Tasdemir et al. 2008), suggesting that activation of autophagy by p53 depends on its transactivating effect on genes such as DRAM (see Sect. 4.2.5) (Crighton et al. 2006).

2.1.4 Downstream Targets of mTOR

The transcriptional factor FoxO3 acts downstream of mTORC2, as discussed above. The major downstream targets of mTORC1 are the Atg1 Ser/Thr protein kinase complex, S6 kinase (S6K) and 4E-BP1. The 4E-BP1 protein is an inhibitor of protein translation, but it is not related to autophagy. The Atg1 complex regulates various steps in autophagosome formation, but its physiological target remains to be identified (Nair and Klionsky 2005; Stephan and Herman 2006). It has been suggested that S6K, a kinase that phosphorylates ribosomal protein S6, may regulate autophagy. In *Drosophila*, S6K has been shown to contribute to stimulating autophagy, although it is not mandatory for its initiation (Scott et al. 2004). It has been suggested that in mammalian cells, S6K may contribute to the basal activity of autophagy via its feedback inhibition of the class I PI3K-dependent insulin signaling pathway (Klionsky et al. 2005). However, a recent study has shown that the rate of autophagy was not altered in the striated muscles in S6K-deficient mice (Mieulet et al. 2007), and so further studies are needed to clarify the role of S6K in autophagy.

2.2 mTOR-Independent Pathways

After LiCl treatment, autophagy is induced via the inhibition of inositol monophosphatase independently of mTOR inhibition (Sarkar et al. 2005). The depletion of free inositol and reduced levels of myo-inositol-1,4,5-phosphate (IP_3) stimulate autophagy. According to these findings, inhibition of the endoplasmic reticulum (ER) IP_3 receptor stimulates autophagy (Criollo et al. 2007). Interestingly, enhancing the level of IP_3 inhibits the autophagy induced by nutrient depletion. IP_3 is a Ca^{2+}-mobilizing second messenger that releases Ca^{2+} from the ER, and in this way may control autophagy as described above (Hoyer-Hansen et al. 2007). These findings suggest that inositol and IP_3 may regulate autophagy either via a signaling pathway parallel to mTOR, or by impinging on the molecular machinery of autophagy downstream of mTOR.

2.3 *Other Pathways*

In this section, we will summarize some of the other signaling pathways that have been shown to regulate autophagy in mammalian cells. Some of these signaling pathways engage in crosstalk with mTOR signaling, such as eIF2α kinases, sphingolipids or NF-κB, whereas the final target of other pathways remains to be identified or are independent of mTOR.

2.3.1 eIF2α Kinases

A relationship between autophagy and eIF2α phosphorylation has been shown during starvation in *Saccharomyces cerevisiae* and during starvation and viral infection in mammalian cells (Tallóczy et al. 2002). The eIF2α kinases are a family of evolutionarily conserved serine/threonine kinases that regulate stress-induced translational arrest. In yeast, GCN2, the yeast eIF2α kinase, targets the eIF2α-regulated transcriptional transactivator, GCN4, and induces autophagy in response to starvation (Tallóczy et al. 2002). In mammals, there are four distinct eIF2α kinases, GCN2, PKR, PERK, and HRI, which are activated by amino acid starvation, viral infection, ER stress, and heme depletion, respectively (Garcia et al. 2007). Thus, it is possible that various stress conditions, including ER stress and viral infection, that activate eIF2α kinases may have the ability to induce autophagy in mammalian cells.

2.3.1.1 PERK

Accumulation of misfolded proteins in the ER activates PERK (protein kinase R-like endoplasmic reticulum kinase), which phosphorylates eIF2α. During aggregate-prone protein accumulation, PERK/eIF2α phosphorylation stimulates autophagy by upregulating Atg12, and probably also by activating the Atg5–Atg12–Atg16 complex (Kouroku et al. 2007). The accumulation of misfolded proteins by autophagy default leads to neurodegenerative disorders in mouse models (Hara et al. 2006; Komatsu et al. 2006). In contrast with protein aggregate-induced autophagy, rapamycin-induced autophagy and starvation-induced autophagy are not mediated by the PERK/eIF2α pathway. This shows that eIF2α phosphorylation does not modulate all types of stress-induced autophagy (Yorimitsu and Klionsky 2007). More recently, PERK-dependent regulation has been shown to induce autophagy in human glioblastoma cells, and to lead to cell death via c-Jun N-terminal kinase (JNK) activation (Park et al. 2008). However, treatment of PERK-deficient cells with thapsigargin, an ER stressor, induces the activation of autophagy in a manner similar to that produced in wild-type cells (Ogata et al. 2006). Therefore, it is not clearly established which signaling pathway from the ER is involved in autophagy induced by ER stress.

2.3.1.2 PKR

Double-stranded RNA-dependent protein kinase (PKR), whose gene expression is upregulated by type I IFNs, is a key player in the antiviral action of interferon. When activated, PKR phosphorylates eIF2α, which blocks translation, leading to the shutoff of protein synthesis, and thereby inhibits viral replication. Not surprisingly, a wide variety of viruses have evolved strategies to counteract this (Kirkegaard et al. 2004). PKR has been shown to promote autophagy during both viral infection and starvation, although the mechanism is still unclear (Tallóczy et al. 2002; Tallóczy et al. 2006). Herpes simplex virus is unable to trigger autophagy in PKR$^{-/-}$ and Ser-51 nonphosphorylable mutant eIF2α murine embryonic fibroblasts. Furthermore, PKR and eIF2α Ser-51-dependent autophagy processes are both antagonized by the herpes simplex virus neurovirulence protein, ICP34.5. Thus, autophagy is a novel evolutionarily conserved function of the PKR pathway targeted by viral virulence gene products.

2.3.1.3 GCN2

In yeast, GCN2-dependent phosphorylation of eIF2α is necessary for the effective translation of mRNAs encoding Gcn4, a transcriptional activator of several autophagy genes, which is stimulated by nutrient starvation (Natarajan et al. 2001). Starvation-induced autophagy depends on the activity of Gcn2, which can be rescued by PKR in GCN2-disrupted yeast (Tallóczy et al. 2002). Interestingly, in GCN2-disrupted yeast, it is possible to induce autophagy by adding rapamycin, suggesting that GCN2 is not a downstream target of mTORC1.

2.3.2 MAP Kinases

The mitogen-activated (MAP) kinases are involved both in the induction of autophagy (level 1) and in the maturation of the autophagosome (level 3). In this section, we will discuss the relationship between MAP kinases and the induction of autophagy. We will define the role of MAP kinases in autophagosome maturation in Sect. 4.3.

2.3.2.1 p38^{MAPK}

The involvement of p38^{MAPK} in the control of autophagy seems to be cell-type dependent. In cultured rat hepatocytes and in flow-through perfused rat liver, amino acid-induced cell swelling caused by Na^+-dependent concentrative transport of certain amino acids inhibits autophagy independently of mTOR by activating $p38^{MAPK}$ (Haussinger et al. 1999). Similarly, blockade of $p38^{MAPK}$ signaling induces an autophagic response in colorectal cancer cells (Comes et al. 2007). In contrast, the accumulation of glial fibrillary acidic protein (GFAP) aggregates induces the activation of p38 and subsequently mTOR-dependent autophagy in Alexander disease, a rare, fatal neurological disorder (Tang et al. 2008). Finally, in murine myotubes, $p38^{MAPK}$ is not involved in the regulation of autophagy by amino acids (Tassa et al. 2003).

2.3.2.2 Extracellular Signal-Regulated Kinases

Activation of extracellular signal-regulated kinases (ERK1/2) has been shown to activate autophagy in different cell types. In response to neurotoxins, the activation of ERK stimulates autophagy in neurons (Zhu et al. 2003). In human colon cancer cells, amino acid starvation activates the ERK1/2 signaling pathway by promoting the phosphorylation of Raf-1, which reduces its kinase activity towards MEK1/2, the upstream kinase activators of ERK1/2, and thereby triggers autophagy (Pattingre et al. 2003a). Accordingly, soyasaponins induce ERK1/2-dependent autophagy in the same cell types, suggesting that the regulation of autophagy by ERK1/2 is not limited to that which occurs during nutrient starvation in intestinal cells. In MCF7 breast cancer cells, TNFα causes an increase of ERK1/2 activity, and subsequent induction of autophagy (Sivaprasad and Basu 2008).

2.3.2.3 c-Jun N-Terminal Kinases

In mouse fibroblasts, autophagy and cell death are dependent on the activation of JNK, and on the transcriptional activity of c-Jun (Yu et al. 2004). Whether the transcriptional activity of c-Jun is required to regulate autophagy remains to be carefully investigated. Indeed, the cytoplasmic target of JNK has been shown to control autophagy independently of the transcriptional activity of c-Jun (see Sect. 3).

2.3.3 Protein Kinases C

Protein kinases C (PKC) comprise a family of serine/threonine kinases that are involved in the transduction of signals for cell proliferation, differentiation, apoptosis, and angiogenesis. Two members of the PKC family have been recently involved in autophagy regulation. PKCδ constitutively suppresses autophagy through the induction of tissue transglutaminase (TG2) in pancreatic cancer cells (Akar et al. 2007). Conversely, inhibition of Ca^{2+}-dependent PKCτ prevents ER stress-induced autophagy (Sakaki et al. 2008). It seems that PKCτ activation is necessary for autophagy in response to ER stress, but not in response to amino acid starvation (Sakaki et al. 2008).

2.3.4 DAP Kinases

The death-associated protein kinase (DAPk) family contains three closely related serine/threonine kinases, known as DAPk, ZIPk, and DRP-1 (Gozuacik and Kimchi 2006). These three kinases may form multiprotein complexes that are able to induce apoptotic or autophagic cell death in response to various cellular stresses. It seems that the most studied member of the family, DAPk, is able to induce an autophagic cell death that is involved in its tumor suppressor activity. Interestingly, DRP-1 modulates both starvation-induced and IFN-induced autophagy (Inbal et al. 2002), whereas DAPk does not (Gozuacik and Kimchi 2007). Furthermore, DRP-1 is anchored in the autophagosome membrane, and may be involved in the formation of this vacuole (Inbal et al. 2002).

2.3.5 Heterotrimeric G Proteins

Previous studies have shown that a cytoplasmic heterotrimeric G_{i3} protein regulates autophagy in the human colon cancer HT-29 cell line (Ogier-Denis et al. 1995). Autophagy is stimulated when GDP is bound to the $G_{\alpha i3}$ protein (Ogier-Denis et al. 1996). It is not surprising that G_{α}-interacting protein (GAIP), a regulator of G proteins that activates the hydrolysis of GTP by the $G_{\alpha i3}$ protein, has been shown to increase the rate of autophagy (De Vries et al. 2000b; Ogier-Denis et al. 1997b). The phosphorylation of GAIP, which stimulates its GTPase activity and therefore the autophagic pathway, is dependent upon the activity of ERK1/2 (Pattingre et al. 2003a). Another G-protein regulator, AGS3 (Activator of G protein Signaling 3), which has been shown to interact with the GDP-bound form of $G_{\alpha i3}$ (De Vries et al. 2000a; Takesono et al. 1999) is involved in the control over an early step in autophagy prior to the formation of the autophagosome (Pattingre et al. 2003b).

Interestingly, $G_{\alpha i3}$ is also crucial to the antiautophagic action of insulin in mouse hepatocytes (Gohla et al. 2007). $G_{\alpha i3}$ is associated with the autophagosomal membrane in starvation-induced autophagy, and relocates to the plasma membrane in response to insulin stimulation (Gohla et al. 2007). G proteins are also implicated in autophagy and, at least in this cellular model, they have an antiautophagic role.

2.3.6 NF-κB

In cells without activated NF-κB, TNFα upregulates the expression of Beclin 1 and induces autophagy. These processes are dependent on reactive oxygen species (ROS) (Djavaheri-Mergny et al. 2006). NF-κB is an antiautophagic factor and protects cells against the action of TNFα. It has also been shown that autophagy degrades I-κB, the inhibitor of NF-κB (Xiao 2007). NF-κB is then activated and inhibits autophagy. This seems to be a regulatory feedback mechanism that prevents a burst of autophagy and autophagic cell death (Xiao 2007). We will discuss the role of ROS during autophagosome formation in Sect. 3.3.1.

2.3.7 Sphingolipids

Sphingolipids, such as ceramide and sphingosine-1-phosphate, are involved in various cellular processes, and particularly in apoptosis (Ogretmen and Hannun 2004; Spiegel and Milstien 2003). Ceramide is a pro-apoptotic signal, and sphingosine-1-phosphate modulates the apoptotic effect of ceramide. Both ceramide and sphingosine-1-phosphate are able to stimulate autophagy (Daido et al. 2004; Lavieu et al. 2007). However, their mechanisms of action are not well known. Ceramide-induced autophagy is characterized by the inhibition of Akt/PKB upstream of mTOR (Scarlatti et al. 2004). Sphingosine kinase 1 overexpression has no effect on Akt/PKB, but sphingosine-1-phosphate does seem to inhibit mTOR independently of class I PI3K (Lavieu et al. 2006).

3 Autophagosome Formation

Autophagy begins with the formation of a pre-autophagosomal sequestering cistern that subsequently gives rise to an isolation membrane or phagophore (Fengsrud et al. 2004; Suzuki and Ohsumi 2007). A class III PI3K complex containing Atg6/Beclin 1 mediates the initial nucleation of the isolation membrane. The precise origin of the membrane is still not fully understood. Atg5 has been shown to localize to the membranes of nascent phagophores, where its conjugation to Atg12 is involved in the expansion of the isolation membrane. This Atg5/Atg12 complex is required for the binding to the phagophore of Atg8/LC3 after it has been conjugated with phosphatidylethanolamine. This membrane then elongates to form the autophagosome, which is a double membrane-bound structure in the 0.5–1.5 μm range in mammalian cells. So far, 31 Atg proteins have been identified and characterized. Eighteen of these Atg proteins, Atg1–Atg10, Atg12–Atg14, Atg16–Atg18, Atg29, and Atg31, play roles in autophagosome formation (Suzuki and Ohsumi 2007). In this section, we will not attempt to provide a detailed description of the functions of all Atg proteins; instead, we will focus on the steps that can be regulated, and on the different regulatory mechanisms (protein-protein interactions, post-translational modifications, transcriptional regulation, and cytoskeleton). Readers can consult recent comprehensive reviews of the role of Atg proteins in the formation of autophagosomes (Suzuki and Ohsumi 2007; Xie and Klionsky 2007; Yoshimori and Noda 2008), as well as the chapter by Yang and Klionsky in this volume.

3.1 Atg1 and Its Partners

In yeast, autophagosomes are generated at a specific site near the vacuolar membrane, known as the preautophagosomal structure (PAS), where the Atg1 complex is recruited (Suzuki and Ohsumi 2007; Xie and Klionsky 2007). The Atg1 Ser/Thr protein kinase is contained in a dynamic protein complex with Atg17, Atg11/Cvt9, Atg13, and Vac8, the composition of which depends upon the phosphorylation status of Atg1 and Atg13 (Pattingre et al. 2008). The Atg1 complex functions downstream of TOR (see Sect. 2), and TOR signaling controls the Atg1–Atg13 interaction via the phosphorylation level of Atg13 (Klionsky 2005). Under nutrient-rich conditions, in which Atg1 and Atg13 are highly phosphorylated, Atg1 interacts with Atg17 and Atg11/Cvt9, and Atg13 is associated with Vac8. In response to starvation, dephosphorylated Atg13 interacts with Atg1 and with other partners, resulting in Atg17-dependent autophagy. Similarly, rapamycin promotes the dephosphorylation of Atg13 and the activation of Atg1. In addition, other Atgs have been found to interact with the Atg1 complex at the PAS (Suzuki and Ohsumi 2007).

Homologs of Atg1 have been found in multicellular organisms and plants, and their function in autophagy is conserved (Klionsky et al. 2003). However, the precise role of the Atg1 complex during the formation of the autophagosome in metazoans is not known, in part because several components of the Atg1 complex are not evolutionarily conserved (Atg11, Atg13, Atg17), and in part because the physiological target of the kinase activity of Atg1 remains to be identified. In mammals, two Atg1 homologs, Ulk1 and Ulk2, localize on the elongating isolation membrane under starvation conditions, and the kinase activity of Ulk1 and 2 is required to stimulate autophagy (Chan et al. 2007; Hara et al. 2008). FIP200 (a 200-kDa focal adhesion kinase family interacting protein) is a recently discovered mammalian autophagic factor which interacts with Ulk1 and 2, and is required for their phosphorylation (Hara et al. 2008). Interestingly, it has been proposed that FIP200 could be a counterpart of Atg17 in yeast (Hara et al. 2008). Ulk1 and 2 are downstream of mTORC1, and negatively feedback to it. A recent study in *Drosophila* demonstrates that dAtg1 is required to stimulate autophagy, and that it exercises negative feedback control on dTOR (Scott et al. 2007). Interestingly, cells with a high level of Atg1-dependent autophagy are eliminated by apoptosis, suggesting that the Atg1-dependent regulation of autophagy is important in order to keep autophagy in a range compatible with cell survival.

3.2 Atg6/Beclin 1

In yeast, Atg6 forms two complexes with the class III phosphatidylinositol 3-kinase (Vps34) and its regulatory factor Vps15. Complex I includes Atg14, whereas complex II contains Vps38, and both these proteins act as connectors between Atg6 and VPs34/Vps15. Complex I regulates autophagy, and complex II is required for vacuolar protein sorting of carboxypeptidase Y (Kihara et al. 2001b). Among the Atg proteins, Atg6 is relatively unique in that it is not "autophagy-specific." Beclin 1, the mammalian ortholog of Atg6, shares 24.4% identity with Atg6 in yeast. The interaction between Beclin 1 and Vps34 is conserved in mammals, and the Beclin 1/hVps34 complex is also able to bind to different partners (see below). Mammalian homologs of Atg14 and Vps38 have been recently identified (reviewed in Longetti and Tooze 2009). Interestingly, in Atg6-defective yeast, Beclin 1 is only able to restore the autophagy function of this mutant, suggesting that Beclin 1 does not regulate other lysosomal trafficking pathways (Furuya et al. 2005).

Beclin 1 was discovered in a two-hybrid screen as a Bcl-2-interacting protein (Liang et al. 1998). The Beclin 1/hVps34 complex contributes to autophagosome formation by allowing other Atg proteins to relocate to the pre-autophagosomal structure. Endogenous Beclin 1 localizes to the TGN (*trans*-Golgi network), the mitochondria, the perinuclear membrane and the endoplasmic reticulum (ER) (Kihara et al. 2001a; Pattingre et al. 2005). The Beclin 1/hVps34 interaction (and, as a result, autophagic levels) can be modulated. Beclin 1 is part of a multimolecular complex and acts as a platform, recruiting activators or repressors of Beclin 1/hVps34-dependent autophagy. Beclin 1 also has tumor-suppressive activity in breast cancer cells. Sequence and structural studies indicate that Beclin 1 has a Bcl-2-binding domain (BBD), a central coiled-coil domain (CCD), an evolutionarily conserved domain (ECD), as well as a BH3-only domain and a nuclear export signal (Furuya et al. 2005; see Fig. 3). This last domain is responsible for transporting Beclin 1 from the nucleus to the cytosol, and it is only the cytosolic form that regulates autophagy (Liang et al. 2001). The ECD is essential for Vps34 binding (Furuya et al. 2005). It has been shown recently that Beclin 1 forms a large homo-oligomer, which may contribute to its own regulation (Ku et al. 2008).

3.2.1 Negative Regulators

In addition to their key role in the regulation of apoptosis, the Bcl-2 family proteins have recently been shown to be negative regulators of autophagy (Liang et al. 1998; Pattingre et al. 2005; Shimizu et al. 2004). The Bcl-2 family proteins, Bcl-2, Bcl-xL, and Bcl-w, and to a lesser extent Mcl-1, interact with Beclin 1 and interfere with the complex formation between Vps34 and Beclin 1 (Erlich et al. 2007; Liang et al. 1998). However, Beclin 1 does not interact with the pro-apoptotic proteins of the same family, such as Bax (Liang et al. 1998). Bcl-2 proteins do not directly compete with Vps34 for binding to Beclin 1, since they bind to the Bcl-2 binding domain of Beclin 1, whereas Vps34 is thought to bind to its EC domain. JNK has recently been identified as the kinase responsible for the phosphorylation of Bcl-2 during nutrient starvation (Wei et al, 2008) or ceramide treatment (Pattingre et al., 2009). Phosphorylation of Bcl-2 occurs in the ER and leads to decreased interaction between Bcl-2 and Beclin 1, which in turn stimulates autophagy. Conversely, under nutrient-rich conditions, when autophagy is inhibited, Bcl-2 is not phosphorylated and it interacts strongly with Beclin 1. Recently, a BH3 domain that forms an amphipathic helix was identified in the Beclin 1 sequence from amino acids 108 to 127 (Oberstein et al. 2007). BH3-only proteins, including Bad and Bim, disrupt the Beclin 1-Bcl-2 (or Bcl-xL) complex and stimulate autophagy (Maiuri et al. 2007b; Oberstein et al. 2007).

Several viral proteins can also block the Beclin 1/Vps34 interaction. γ-Herpesviruses, including murine γ-herpesvirus 68, Kaposi's sarcoma-associated herpesvirus,

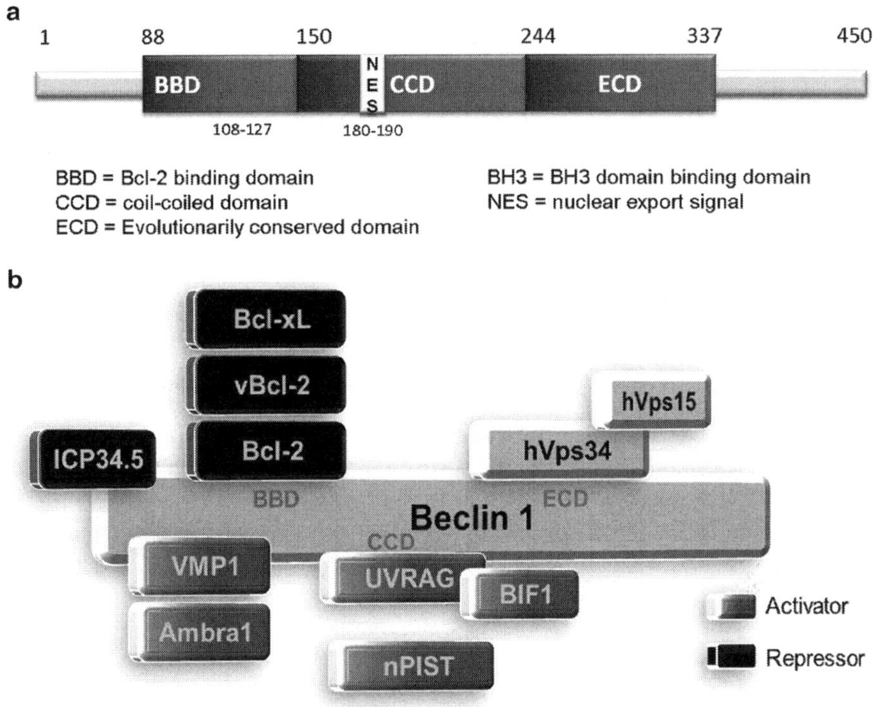

Fig. 3 Beclin 1 and its partners. **A** Schematic representation of the domains of human Beclin 1. Beclin 1 has a Bcl-2 binding domain (BBD, amino acids 88–150), a coiled coil domain (CCD, amino acids 150–244), and an evolutionarily conserved domain (ECD, amino acids 244-337). A BH3-only domain (BH3, amino acids 108–127) is present within the BBD, and a nuclear export signal (NES) is located between amino acids 180 and 190. **B** Beclin 1 functions as a platform for the formation of the complex. Bcl-2 proteins interact with the BBD of Beclin 1, UVRAG and nPIST with the CCD, and Vps34 with the ECD. The interaction between Beclin 1 and Vps34 is reinforced by UVRAG, associated with Bif-1 (activators), which upregulates autophagy. AMBRA, VMP1 and nPIST interact with Beclin 1 to stimulate autophagy (activators), whereas Beclin 1 is inhibited by Bcl-2, Bcl-xL, v-Bcl-2 and ICP34.5, which are all inhibitors of autophagy (repressors)

herpesvirus saimiri and rhesus rhadinovirus, encode viral homologs of Bcl-2 (v-Bcl-2) (Liang et al. 2008). These homologs can bind to Beclin 1 with much higher affinity than cellular Bcl-2, and the binding affinity seems to correlate directly with antiautophagic activity. It is interesting to note that these v-Bcl-2 do not have the phosphorylation sites of Bcl-2. This means that their binding to Beclin 1 cannot be modulated, and they constitutively block autophagy. ICP34.5 is a herpes simplex virus type 1 neurovirulence protein that is also able to bind to Beclin 1 and to inhibit autophagy, but it has no homology with Bcl-2 and does not bind to Beclin 1 through its Bcl-2-binding domain (Orvedahl et al. 2007). A mutant virus containing a Beclin 1-binding-deficient form of ICP34.5 fails to inhibit autophagy in neurons, and is highly neuroattenuated in mice (Orvedahl et al. 2007). Thus, Beclin 1 is targeted by several viruses to downregulate autophagy and confer pathogenicity.

3.2.2 Positive Regulators

Several proteins have been recently discovered to be active components of the pro-autophagic multimolecular complex. The UV irradiation resistance-associated gene protein (UVRAG) is a positive regulator of the Beclin 1–Vps34 complex (Liang et al. 2006, 2007). UVRAG and Beclin 1 interact directly through their coiled-coil domain by forming an α-helical bundled structure. The coiled-coil domain of Beclin 1 interacts with another protein, nPIST, which also positively regulates autophagy (Yue et al. 2002). Bax-interacting factor 1 (Bif-1) interacts with Beclin 1 through UVRAG via its SH3 domain and increases autophagosome biogenesis (Takahashi et al. 2007). During nutrient deprivation, Bif-1 accumulates in autophagosomes, where it colocalizes with LC3, Atg5, and Atg9. The activating molecule in Beclin 1-regulated autophagy (AMBRA1) positively regulates autophagy and inhibits cell proliferation (Fimia et al. 2007). Downregulation of AMBRA1 by small interfering RNA (siRNA) reduces Beclin 1-mediated autophagy levels in a manner consistent with a decrease in the association of Vps34 with Beclin 1. AMBRA1 is unique to vertebrates, and is mainly expressed in the brain, where it plays an essential role during development (Cecconi et al. 2007). Vacuole membrane protein 1 (VMP1) is a recently discovered transmembrane protein that triggers autophagosome formation in mammalian cells (Ropolo et al. 2007). VMP1 interacts with Beclin 1 through its hydrophilic C-terminal region, named the Atg domain. VMP1 also colocalizes with LC3. It has recently been shown that the small GTPase Rab5, previously known to be a regulator of early endocytosis, also interacts with and activates Vps34 in the Beclin 1–Vps34 complex, and thereby positively regulates autophagosome formation (Ravikumar et al. 2008). Rab5, Beclin 1 or Vps34 inhibition leads to decreased Atg5/Atg12 conjugation, suggesting that Rab5 acts at the autophagosome precursor stage (Ravikumar et al. 2008).

3.3 Post-translational Modifications of Atg Proteins

Post-translational protein modifications can regulate the activity of Atg proteins in the autophagic pathway as described below. The ubiquitin-like conjugations of Atg12 to Atg5 and of Atg8/LC3 to the polar head of PE are fundamental to the formation of autophagosomes (Ohsumi 2001). The oxidation of a cysteine residue near the catalytic site of Atg4 is important in regulating its effect on Atg8/LC3 (Scherz-Shouval et al. 2007). Other post-translational modifications that modulate the activity of Atg proteins, such as acetylation or ubiquitination, are now emerging (Baxter et al. 2005; Lee et al. 2008). Moreover, proteolytic cleavage may regulate the function of Atg proteins in the autophagic pathway, but also may unmask new functions of these proteins (Codogno and Meijer 2006). The discovery that the calpain-dependent cleavage of Atg5 generates an amino-terminal pro-apoptotic fragment is a promising lead in this new field (Yousefi et al. 2006).

3.3.1 Ubiquitin-Like Conjugated Systems

Two ubiquitin-like systems act sequentially in the expansion and completion of autophagosome formation. The first involves the conjugation of Atg12 to Atg5, which occurs constitutively soon after the individual proteins have been synthesized. Atg12 is activated by Atg7, a homolog of E1 enzyme, and then conjugated to Atg5 by Atg10 (E2-like enzyme). The second system involves the conjugation of Atg8/LC3 to phosphatidyl ethanolamine (PE), a component of the phospholipid bilayer, in a reaction that requires both Atg7 (E1-like) and Atg3 (E2-like). Atg5/Atg12 interact with Atg16L, which is a coiled-coil protein, to form an approximately 800 kDa complex through the homo-oligomerization of Atg16L (Mizushima et al. 2003). This complex is specifically present on isolation membranes, and is never present on mature autophagosomes. The membrane localization of Atg16L complex determines the site of LC3 lipidation. The Atg16L complex is a new type of E3-like enzyme that functions as a scaffold for LC3 lipidation on the isolation membrane (Fujita et al. 2008). The Rab small GTPases Rab33A and Rab33B, initially localized in the Golgi, specifically interact with Atg16L in a GTP-dependent manner without affecting the integrity of the Atg5–12/Atg16L complex (Itoh et al. 2008). Moreover, Rab33B also modulates autophagosome formation. One possible function of Rab33 may be to recruit the Atg5–12/Atg16L complex to the surfaces of membrane structures.

3.3.2 Atg4 and ROS

Reactive oxygen species (ROS) have been shown to regulate starvation-induced autophagy by regulating the activity of Atg4 (Scherz-Shouval et al. 2007). The protein kinase Atg4 cleaves the C terminus of Atg8/LC3 as a prerequisite for its conjugation to PE on the autophagosomal membrane. Atg4 also cleaves conjugated Atg8 and removes it from the mature autophagosome for recycling. ROS released from the mitochondria inhibit Atg4 by oxidation. Indeed, one cysteine residue located near the catalytic site of Atg4 is redox regulated. Once Atg4 has been inactivated, its substrate Atg8 can be conjugated to autophagosomes. Because ROS are short-lived molecules, it has been hypothesized that oxidation occurs only close to the mitochondria (Scherz-Shouval and Elazar 2007). Further away from the mitochondria, Atg4 will be active and therefore cleave Atg8 from the autophagosomal membrane for recycling.

3.3.3 Atg Acetylation

It has been shown recently that Sirt1, a mammalian deacetylase belonging to the sirtuin family, is necessary for autophagy (Lee et al. 2008). Sirt1 interacts directly with several Atg proteins, such as Atg5, Atg7, and Atg8, and deacetylates them. These proteins were acetylated under normal conditions, and acetylation levels were reduced by Sirt1 during starvation. A lack of Sirt1 inhibits autophagosome formation during starvation and leads to increased levels of p62. In *Drosophila melanogaster,* the activity of a second deacetylase, histone deacetylase 6 (HDAC6), a microtubule-associated deacetylase, has been linked to compensatory autophagy induced when the proteasome has been impaired (Pandey et al. 2007).

3.4 Transcriptional Regulation of Atg Proteins

Forced expression of Atg1 leads to excessive autophagy and triggers apoptosis (Scott et al. 2007). Interestingly, activation of the transcription factor E2F1, which is frequently required for apoptosis, is also involved in the expression of some genes related to the autophagic pathway (Polager et al. 2008). Activation of E2F1 by 4-hydroxytamoxifen upregulates the expression of LC3, Atg1, Atg5, and DRAM in U2OS cells containing an inducible E2F1. Moreover, E2F1 has been shown to be linked to the promoter of Beclin 1, even if the effect of E2F1 on Beclin 1 is not still clear (Weinmann et al. 2001). However, this finding supports the proposal that autophagy may be regulated at the transcriptional level. The expression of Atg5 has been shown to be increased during autophagic cell death in *bax/bak*$^{-/-}$ MEFs (Shimizu et al. 2004). Recent studies have provided evidence that the expression of some Atg proteins and of proteins related to autophagy is increased during muscle wasting (see Sect. 2.1.2) (Mammucari et al. 2007). This emerging aspect of the regulation of autophagy will probably yield important information about the impact of autophagy on cell homeostasis.

3.5 The Cytoskeleton

Cytoskeletal elements are involved at both levels 2 and 3 of autophagy regulation. Previous studies based on the inhibition of autophagy following the disruption of the network of cytokeratin intermediate filaments by okadaic acid (Blankson et al. 1995) suggested that this class of filaments could be involved in the formation of autophagosomes.

Recent studies have reported that disruption of microtubules not only slows down the maturation of autophagosomes (see Sect. 4) but also decreases the formation of autophagosomes (Fass et al. 2006; Kochl et al. 2006). These results suggest that microtubules may be implicated in the biogenesis of autophagosomes. One intriguing possibility is that the microtubule network may form a boundary preventing the undesirable fusion between phagophores and autophagosomes (Fass et al. 2006).

Cytochalasin B and D, which induce the depolymerization of microfilaments, were reported to reduce the formation of autophagosomes (Aplin et al. 1992). More recently, upregulation of F-actin polymerization has been reported during TRAIL-mediated autophagy (Han et al. 2008). Along with these observations, some elements of the autophagy machinery have been shown to interact with the actin cytoskeleton (Monastyrska et al. 2008).

4 Maturation Step

The maturation of autophagosomes is important in situations in which cells need to degrade their autophagic cargo to avoid nutrient and ATP depletion (e.g., during starvation-induced autophagy), to circumvent parasitic effects (accumulation of bacteria in the autophagosome for persistent infection), and also to prevent the accumulation of toxic compounds in autophagic compartments (generation of Aβ peptide in Alzheimer's disease). Blockade of the flux or excessive flux could both have adverse consequences for cell homeostasis. It is commonly assumed that excessive flux could be associated with self-destruction of the cell by self-digestion, but we can also envisage that excessive flux could buffer the lysosomal pH as a result of massive fusion with autophagosomes that have a neutral lumen pH. In any case, regulation of the maturation of autophagosomes (i.e., of the flux into the lysosomal compartment) is an important decision for cells in response to stress situations.

4.1 Morphology and Definition of Late Stages of Autophagy

Autophagosomes in yeast fuse with the vacuole to deliver autophagic bodies into the vacuolar lumen. Autophagic cargos become accessible to vacuolar hydrolysis after the breakdown of the membranes of the autophagic bodies by the lipase Atg15 (Epple et al. 2001; Teter et al. 2001). This scenario exists in mammalian cells, but the autophagosome can also merge with endocytic compartments (both early and late endosomes are competent at fusing with autophagosomes). Seglen has coined the evocative term "amphisome"—from the Greek *amphi* (both) plus *soma* (body)—for this unique structure (reviewed in Fengsrud et al. 2004) that has acquired acidic and degradative properties. Amphisomes correspond to intermediate autophagic vacuoles or AVi/d (AVi initial autophagic vacuoles are autophagosomes, and AVd degradative autophagic vacuoles are lysosomes; see Eskelinen 2005; Fengsrud et al. 2004 for an insightful discussion of the terminology of autophagic compartments). Fusion with lysosomes is the final fate of amphisomes.

4.2 Regulation of the Maturation Events

The maturation of autophagosomes depends on molecules that allow the autophagosomes to fuse with the vesicular compartments of the vacuolar system (endosomes, lysosomes). The late stage of autophagy is also dependent on molecules that regulate the acidification of the autophagic compartments and molecules that are involved in recycling of degraded material from the lysosomal compartment.

4.2.1 SNAREs

SNAREs (soluble NSF attachment protein receptors) are basic elements required for intracellular membrane fusion (Gurkan et al. 2007; Rothman and Wieland 1996). Depending on their role in vesicular transport, SNAREs are divided into two groups: vesicle (v-SNAREs) and target SNAREs (t-SNARES). In yeast, the vacuolar t-SNAREs Vam3 (Darsow et al. 1997) and Vti1 (Ishihara et al. 2001), are needed to complete the fusion between the autophagosome and the vacuole. The mammalian homolog of i: Vti1, Vti1b, may be involved in the late stages of autophagy, because the maturation of autophagic vacuoles is delayed in hepatocytes isolated from mice in which Vti1b has been deleted (Atlashkin et al. 2003).

4.2.2 Rab Proteins

Rab proteins are a family of monomeric GTPases necessary for vesicular transport along the exo/endocytic pathway (see for review Zerial and McBride 2001). The most compelling evidence that a Rab protein plays a role in the autophagic pathway is with Rab7. In yeast, the fusion of the autophagosome with the vacuole is dependent upon a Rab7 homolog, Ypt7 (Kirisako et al. 1999). In mammalian cells, Colombo and coworkers (Gutierrez et al. 2004), and Eskelinen and coworkers (Jager et al. 2004) have shown that Rab7 is required for the maturation of autophagosomes. However, a functional Rab7 is not mandatory for the fusion with endocytic compartments to occur. Interestingly, a functional Rab11 is required for the fusion of autophagosome and multivesicular bodies during starvation-induced autophagy in erythroleukemic cells (Fader et al. 2008). These findings suggest that fusion of specific membrane-bound compartments during the maturation of autophagosomes engage different sets of Rab proteins, and possibly associated cohort proteins. Other Rab proteins, such as Rab22 and Rab24, have a subcellular localization compatible with having a role in autophagy (Egami et al. 2005; Mesa et al. 2001; Olkkonen et al. 1993).

4.2.3 ATPases

4.2.3.1 v-ATPases

Vacuolar ATPases (v-ATPases) are ubiquitous proteins located in acidic compartments (Forgac 2007). Inhibiting the activity of v-ATPase by bafilomycin A1 or concanamycin A blocks the lysosomal pumping of H^+ and consequently inhibits lysosomal enzymes which are active at low pH. It has been proposed that bafilomycin A1 may block the late stage of autophagy by interfering with the fusion of autophagosomes with endosomes and lysosomes (Yamamoto et al. 1998). However, recent studies show that what bafilomycin A1 blocks is not fusion events in the autophagic pathway but the degradation step in lysosomes (Fass et al. 2006; Mousavi et al. 2001). Overall, the resulting effect of v-ATPase inhibition is an interruption of autophagic flux.

4.2.3.2 AAA ATPases

ATPases associated with various cellular activities proteins (AAA ATPases) are a family of proteins broadly engaged in intracellular membrane fusion (White and Lauring 2007). NSF is an AAA ATPase that binds to SNARE complexes and utilizes ATP hydrolysis to disassemble them, thus facilitating SNARE recycling. In yeast mutants lacking Sec18 (the yeast homolog of NSF), autophagosomes are formed but do not fuse with the vacuole (Ishihara et al. 2001). However it is not known whether the ATPase activity of NSF is involved in the later stages of autophagy in mammalian cells. Nevertheless, we do know that NSF activity is attenuated during starvation, which could account for the slow fusion between autophagosomes and lysosomes observed when autophagy is induced by starvation (Fass et al. 2006). SKD1 (Vps4), another AAA ATPase protein, is required for the maturation of autophagosomes (Nara et al. 2002) in mammalian cells. Vps4/Csc1, which controls the assembly of ESCRT complexes on multivesicular membrane (see below), is involved in autophagosome maturation (Rusten et al. 2007) in *Drosophila*, and in autophagosome fusion with the vacuole in yeast (Shirahama et al. 1997).

4.2.4 ESCRT and Hrs

Endosomal sorting complex required for transport (ESCRT) mediates the biogenenesis of multivesicular bodies and the sorting of proteins in the endocytic pathway (Raiborg et al. 2003). It has been recently demonstrated that the multisubunit complex ESCRT III is required for autophagosomes to fuse with multivesicular bodies to generate amphisomes, and is also involved in the fusion of autophagosomes with lysosomes (Rusten et al. 2007). ESCRT III dysfunction associated with the autophagic pathway may have important implications for understanding some neurodegenerative diseases (such as frontotemporal dementia linked to chromosome 3 and amyotropic lateral sclerosis) (Filimonenko et al. 2007; Lee et al. 2007). Hrs protein plays a major role in endosomal sorting upstream of ESCRT complexes (Raiborg et al. 2003a). Hrs contains a FYVE domain that binds specifically to PtdIns3P. It has recently been shown that Hrs facilitates the maturation of autophagosome (Tamai et al. 2007), which raises the intriguing possibility that PtdIns3P may be required for autophagosome formation via the Beclin 1 complex and its maturation via Hrs. It is interesting to note that the endosomal PtdIns(3)P 5-kinase Fab1, which uses PtdIns3P to produce PtdIns(3,5)P2, is required in *Drosophila* for amphisomes to fuse with lysosomes (Rusten et al. 2007). Since the inactivation of Fab1 in yeast causes a marked enlargement of the vacuole, which fails to acidify correctly (Yamamoto et al. 1995), the production of PtdIns(3,5)P2 may play an important part in maintaining organelle homeostasis in the autophagic pathway. The role of PIKfyve (the mammalian ortholog of Fab1) in autophagy has not been yet investigated.

4.2.5 Endo/lysosomal Proteins

4.2.5.1 LAMP-2

LAMPs (lysosomal associated membrane proteins) are a family of heavily glycosylated transmembrane endo/lysosomal proteins (Eskelinen et al. 2003). Autophagic degradation has been shown to be impaired in hepatocytes isolated from LAMP-2-deficient mice (Tanaka et al. 2000). In LAMP-2-deficient mice that reproduce a human cardiomyopathy (Danon disease) (Nishino et al. 2000), the fusion of autophagosomes with the lysosomal compartment seems to be impaired, whereas their fusion with multivesicular bodies is not. However, no defect in autophagy was observed in LAMP-2-deficient mouse fibroblasts (Eskelinen et al. 2004). Blockade in the later stage of autophagy only occurs in fibroblasts deficient in both LAMP-1 and LAMP-2. Differences in autophagic activity observed between hepatocytes and fibroblasts may be responsible for the cell-type-specific effect of LAMP-1 and -2 depletion (Eskelinen 2005).

4.2.5.2 DRAM

Damage-regulated autophagy modulator (DRAM), which encodes a 238-amino acid protein, is generally conserved through evolution but has no ortholog in yeast (Crighton et al. 2006). DRAM is a direct target of p53. The protein is a multispanning transmembrane protein present in the lysosome. DRAM may regulate the late stage of autophagy, but surprisingly it also controls autophagosome formation (Crighton et al. 2006). This suggests a possible new paradigm in which feedback signals from the lysosomes control the early stages of autophagy.

4.2.5.3 Recycling Molecules

Two categories of lysosomal recycling molecules can be distinguished. The first category consists of the lysosomal proteins that recycle entities needed for the ongoing autophagic pathway. Proteins such as DRAM and Fab1, which were discussed in preceding sections, may fall into this category, although this has not been conclusively demonstrated. The second category includes lysosomal transporters that recycle nutrients generated by the lysosomal degradation of macromolecules. Several transporters in the lysosomal membrane have been shown to recycle amino acids, monosaccharides, or lipids (reviewed in Lloyd 1996). Atg22 was recently identified as an amino acid transporter in the vacuole membrane of *S. cerevisiae* (Yang et al. 2006). Atg22, which regulates the final stage of autophagy (i.e., recycling from the lysosomal/vacuolar compartment), is crucial for maintaining cell survival during nutrient starvation (Yang et al. 2006).

4.2.6 Microtubules

As discussed in Sect. 3, microtubules are involved in the formation of autophagosomes. Originally, the involvement of microtubules was demonstrated in the later steps of the autophagic pathway. The destabilization of microtubules by either vinblastine (Hoyvik et al. 1991) or nocodazole (Aplin et al. 1992) blocks the maturation of autophagosomes, whereas their stabilization by taxol increases the fusion between autophagic vacuoles and lysosomes (Yu and Marzella 1986). Subsequent findings have confirmed the role played by microtubules in fusion with the acidic compartment (Jahreiss et al. 2008; Kochl et al. 2006; Webb et al. 2004). Autophagosomes move bidirectionally along microtubules, and their centripetal movement is dependent on the motor protein dynein (Jahreiss et al. 2008; Ravikumar et al. 2005; Webb et al. 2004). Two types of fusion have been documented (Jahreiss et al. 2008), including (1) complete fusion of the autophagosome with the lysosome; and (2) transfer of material from the autophagosome to the lysosomal compartment following a kiss-and-run fusion process in which two separate vesicles are maintained. However, it has been reported that autophagosome fusion with lysosomes is microtubule-independent during starvation-induced autophagy (Fass et al. 2006). Under these conditions, autophagosomes are formed in the vicinity of lysosomes and the fusion of vesicles may be independent of microtubules.

4.3 Signaling and Maturation of Autophagosomes

4.3.1 MAPKs

Protein kinases of the MAPK kinase family (JNK, p38, ERK1/2) have been shown to regulate autophagy at both level 1 and level 2. However, p38 and ERK1/2 are probably also involved in regulating the late stage of autophagy. A recent report shows that the activation of ERK1/2 promotes the formation of large autolysosomes (Corcelle et al. 2006). Thus, activating ERK1/2 activates both the formation and maturation of autophagosomes. A protein involved in regulating the activity of trimeric G_{i3} protein has been shown to act downstream of ERK1/2, and to regulate the early stages of autophagy (Pattingre et al. 2003a). However, the target of ERK1/2 in the regulation of the late stage of autophagy remains to be identified. As in its inhibitory effect on the early stage of autophagy in hepatocytes (Haussinger et al. 1999), p38 has also been shown to have an inhibitory effect on the maturation of autophagosomes (Corcelle et al. 2007). Here too, the targets of p38 in the early and late stages of autophagy are not yet identified.

5 How Autophagy Can Be Manipulated

Understanding how autophagy can be manipulated is important for potential therapeutic applications of autophagy. In this section we will focus on drugs that act at different stages of autophagy (Rubinsztein et al. 2007). Of course, autophagy can be manipulated by genetic approaches. Atg-knockout mice and Atg knockdown by RNA interference-based methods are of fundamental importance in identifying the function of autophagy in various physiological and pathophysiological situations (Levine and Kroemer 2008). Genetic approaches are important not only for investigating the autophagic machinery (level 2 of regulation), but also for investigating the signaling of autophagy (regulation level 1). For example, knockdown of mTOR partners (raptor and rictor) led to the discovery that mTOR complex 2, but not mTOR complex 1, is involved in the regulation of autophagy during the atrophy of skeletal muscle (Mammucari et al. 2007).

One of the drugs most often used to stimulate autophagy is the immunosuppressive agent rapamycin (Meijer and Codogno 2006). Rapamycin targets the kinase TOR by binding to the 12 kDa immunopilin FKBP12. The rapamycin-FKBP12 complex inhibits mTORC1. It should be noted that chronic treatment with rapamycin has an inhibitory effect on the supposedly "rapamycin-insensitive" mTORC2 (Rubinsztein et al. 2007). The role of autophagy in the effects observed with analogs of rapamycin (CCI-779, RAD001, and AP23573), which are currently used in clinical trials, remains to be carefully investigated (Faivre et al. 2006). Drugs that act on signaling elements upstream of mTOR are also useful for manipulating autophagy (Meijer and Codogno 2006). However, caution is called for with regard to the specificity of some of these drugs when investigating their effect on autophagy. For example, AICAR, an AMPK activator, has an inhibitory effect on autophagy that seems to be independent of AMPK (Meley et al. 2006; Samari and Seglen 1998).

Autophagy can be manipulated independently of mTOR using drugs that act on the myo- inositol phosphate metabolism (Sarkar et al. 2005). Lithium chloride, sodium valproate and carbamazepine, which lower the levels of myo-inositol-1,4,5-triphosphate (IP_3), induce autophagy. Accordingly, xestospongin B, an inhibitor of the IP_3 receptor, is a potent inducer of autophagy (Criollo et al. 2007).

Recently, screens for drugs regulating autophagy have been undertaken. The first strategy employed was to identify compounds that enhance the growth-inhibitory effects of rapamycin in yeast (these compounds were named SMERs, for small-molecule enhancers) (Sarkar et al. 2007). Three of these SMERs were shown to induce autophagy independently of rapamycin in mammalian cells. One of the major points of interest of these SMERs is that they do not have the immunosuppressive effect of rapamycin. Other strategies were based on screening a library of compounds (Williams et al. 2008; Zhang et al. 2007). One of these screens identified seven drugs that had already received FDA approval for the treatment of human diseases (schizophrenia, cardiovascular disorders) (Zhang et al. 2007). Interestingly, some of these drugs are known to act on Ca^{2+} channels and the Ca^{2+} current. The screening of FDA-approved drug libraries also identified minoxidil (a K^+ ATP channel opener) and clonidine (a G_i signaling activator) as activators of autophagy (Williams et al. 2008; Zhang et al. 2007).

One of the drugs most often used to inhibit autophagy is 3-methyladenine (3-MA) (Seglen and Gordon 1982). 3-MA inhibits the formation of autophagosomes by interfering with the activity of hVps34 in the Beclin 1 complex (Petiot et al. 2000). Thus, 3-MA inhibits autophagy by acting at level 2 of autophagy regulation. In fact, 3-MA is a PI3K inhibitor and a similar effect on autophagy has also been observed with two other PI3K inhibitors, wortmannin and LY294002 (Blommaart et al. 1997a). However, these inhibitors also interfere with the activity of class I PI3K and other kinases. For example, 3-MA inhibits the activity of p38MAPK and JNK (Tolkovsky et al. 2002). The design of specific inhibitors for the different classes of PI3K would be important to enable us to target the various different enzymes in this family more specifically. These enzymes sometimes have opposite effects on the regulation of autophagy (Petiot et al. 2000). Interestingly, the activity of the Beclin 1 complex in autophagy can also be manipulated via the Beclin 1/ Bcl-2 interaction. Pharmacological BH3-mimetic compounds, such as ABT737, stimulate autophagy by competitively disrupting the interaction between the BH3-domain of Beclin 1 and Bcl-2 (or Bcl-xL) (Maiuri et al. 2007b).

Specific modulation of level 3 of autophagy regulation requires the targeting of entities that only control this part of the autophagic pathway. This would exclude microtubules (which are also involved at level 2, unless the target is dynein, a microtubule motor that is not involved in level 2). It also rules out targeting the signaling pathways identified so far at level 3, because p38 and ERK1/2 can also be involved at level 1 in some cell types. Some endo/lysosomal proteins are not likely good candidates, because DRAM is probably also involved at level 2. So far the most specific targets identified in level 3 are inhibitors of v-ATPase (bafilomycin A1, concanamycin A) and the dominant-negative form of Rab7. Brefeldin A, which blocks several membrane fusion events in the exocytic and endocytic pathways by interfering with the exchange of GDP for GTP on Arf, does not interrupt the autophagic pathway (Ogier-Denis et al. 1997a; Purhonen et al. 1997). This suggests that guanine nucleotide exchange factors containing a Sec7 domain, a target of brefeldin A, are not involved in membrane fusion events in the autophagic pathway. Although this is not discussed in this chapter, it should be pointed out that inhibiting lysosomal activity by cathepsin inhibitors or lysosomotropic agents such as chloroquine is also a valuable tool for blocking the late stage of autophagy (Amaravadi et al. 2007; Boya et al. 2005). It has been suggested that chloroquine treatment may block cytoprotective autophagy in tumor cells that resist chemotherapy (Amaravadi and Thompson 2007). Combining an autophagy inhibitor such as chloroquine with apoptosis-inducing chemotherapies may lead to improved tumor regression and reduced tumor recurrence.

6 Conclusions

Some progress has been made in understanding how autophagy is regulated. There are many ways to regulate the formation of autophagosomes. Besides the long-known (although still not completely defined) mTOR complex 1, which occurs upstream of the Atg1/Ulk1 and Ulk2 complexes, it has now been established that the Beclin 1 complex (which can also be designated PI3K complex I by analogy with the two

PI3K complexes in yeast) and Atg4 are also possible sites for regulation by signaling molecules. We cannot exclude the possibility that there are other possible points within the molecular machinery at which autophagy can be regulated. The machinery responsible for the maturation of autophagosomes and their fusion with acidic compartments is also regulated by several entities, including ATPase and GTPase activities, SNARE, MAPK signaling, and cytoskeletal motor proteins. However, more investigation is required to understand the roles of these different regulation systems in the late stages of autophagy. The tight regulation of the different stages of autophagy is a safety procedure that allows cells to retain control over a self-eating process that ends up in a "suicide bag," to use Christian de Duve's vivid description of lysosomes (De Duve and Wattiaux 1966).

The use of drugs, RNA interference and gene invalidation provide various ways to manipulate the autophagic pathway in order to study its role in different physiological and pathophysiological situations. The recent development of screening for drugs that modulate autophagy offers new perspectives for therapeutic interventions in human disease (Sarkar et al. 2007; Zhang et al. 2007), and is revealing new regulatory circuits during level 1 of the autophagic pathway (Williams et al. 2008). Moreover, drugs that interfere with level 3 provide a new way to modulate autophagy in the context of cancer therapy (Amaravadi and Thompson 2007). A future challenge will be to design drugs that specifically target level 2 of autophagy regulation.

The dialog between pathogens and autophagy, which will be discussed elsewhere in this issue of *Current Topics in Microbiology and Immunology*, depends at least partially on the ability of these microorganisms to exploit the vast repertoire of autophagy-regulating mechanisms to introduce "flats and sharps" in order to modulate the full musical score of autophagy.

Acknowledgments Work in P. Codogno's laboratory is supported by institutional funding from The Institut National de la Santé et de la Recherche Médicale (INSERM), Université Paris-Sud 11, and grants from the Agence Nationale de la Recherche (ANR to A.E.) and the Association pour la Recherche sur le Cancer (ARC to P.C.). M.C. is the recipient of a Ph.D. fellowship from the French Ministry of Research.

References

Akar U, Ozpolat B, Mehta K, Fok J, Kondo Y, Lopez-Berestein G (2007) Tissue transglutaminase inhibits autophagy in pancreatic cancer cells. Mol Cancer Res 5:241–249

Amaravadi RK, Thompson CB (2007) The roles of therapy-induced autophagy and necrosis in cancer treatment. Clin Cancer Res 13:7271–7279

Amaravadi RK, Yu D, Lum JJ, Bui T, Christophorou MA, Evan GI, Thomas-Tikhonenko A, Thompson CB (2007) Autophagy inhibition enhances therapy-induced apoptosis in a Myc-induced model of lymphoma. J Clin Invest 117:326–336

Aplin A, Jasionowski T, Tuttle DL, Lenk SE, Dunn WA Jr (1992) Cytoskeletal elements are required for the formation and maturation of autophagic vacuoles. J Cell Physiol 152:458–466

Arico S, Petiot A, Bauvy C, Dubbelhuis PF, Meijer AJ, Codogno P, Ogier-Denis E (2001) The tumor suppressor PTEN positively regulates macroautophagy by inhibiting the phosphatidylinositol 3-kinase/protein kinase B pathway. J Biol Chem 276:35243–35246

Atlashkin V, Kreykenbohm V, Eskelinen EL, Wenzel D, Fayyazi A, Fischer von Mollard G (2003) Deletion of the SNARE vti1b in mice results in the loss of a single SNARE partner, syntaxin 8. Mol Cell Biol 23:5198–5207

Baxter BK, Abeliovich H, Zhang X, Stirling AG, Burlingame AL, Goldfarb DS (2005) Atg19p ubiquitination and the cytoplasm to vacuole trafficking pathway in yeast. J Biol Chem 280:39067–39076

Birmingham CL, Canadien V, Kaniuk NA, Steinberg BE, Higgins DE, Brumell JH (2008) Listeriolysin O allows *Listeria monocytogenes* replication in macrophage vacuoles. Nature 451:350–354

Blankson H, Holen I, Seglen PO (1995) Disruption of the cytokeratin cytoskeleton and inhibition of hepatocytic autophagy by okadaic acid. Exp Cell Res 218:522–530

Blommaart EF, Krause U, Schellens JP, Vreeling-Sindelarova H, Meijer AJ (1997a) The phosphatidylinositol 3-kinase inhibitors wortmannin and LY294002 inhibit autophagy in isolated rat hepatocytes. Eur J Biochem 243:240–246

Blommaart EF, Luiken JJ, Meijer AJ (1997b) Autophagic proteolysis: control and specificity. Histochem J 29:365–385

Boya P, Gonzalez-Polo RA, Casares N, Perfettini JL, Dessen P, Larochette N, Metivier D, Meley D, Souquere S, Yoshimori T, Pierron G, Codogno P, Kroemer G (2005) Inhibition of macroautophagy triggers apoptosis. Mol Cell Biol 25:1025–1040

Brazil DP, Hemmings BA (2001) Ten years of protein kinase B signalling: a hard Akt to follow. Trends Biochem Sci 26:657–664

Browne GJ, Proud CG (2002) Regulation of peptide-chain elongation in mammalian cells. Eur J Biochem 269:5360–5368

Browne GJ, Finn SG, Proud CG (2004) Stimulation of the AMP-activated protein kinase leads to activation of eukaryotic elongation factor 2 kinase and to its phosphorylation at a novel site, serine 398. J Biol Chem 279:12220–12231

Byfield MP, Murray JT, Backer JM (2005) hVps34 is a nutrient-regulated lipid kinase required for activation of p70 S6 kinase. J Biol Chem 280:33076–33082

Cecconi F, Di Bartolomeo S, Nardacci R, Fuoco C, Corazzari M, Giunta L, Romagnoli A, Stoykova A, Chowdhury K, Fimia GM, Piacentini M (2007) A novel role for autophagy in neurodevelopment. Autophagy 3:506–508

Chan EY, Kir S, Tooze SA (2007) siRNA screening of the kinome identifies ULK1 as a multidomain modulator of autophagy. J Biol Chem 282:25464–25474

Codogno P, Meijer AJ (2005) Autophagy and signaling: their role in cell survival and cell death. Cell Death Differ 12 Suppl 2:1509–1518

Codogno P, Meijer AJ (2006) Atg5: more than an autophagy factor. Nat Cell Biol 8:1045

Comes F, Matrone A, Lastella P, Nico B, Susca FC, Bagnulo R, Ingravallo G, Modica S, Lo Sasso G, Moschetta A, Guanti G, Simone C (2007) A novel cell type-specific role of p38alpha in the control of autophagy and cell death in colorectal cancer cells. Cell Death Differ 14:693–702

Corcelle E, Nebout M, Bekri S, Gauthier N, Hofman P, Poujeol P, Fenichel P, Mograbi B (2006) Disruption of autophagy at the maturation step by the carcinogen lindane is associated with the sustained mitogen-activated protein kinase/extracellular signal-regulated kinase activity. Cancer Res 66:6861–6870

Corcelle E, Djerbi N, Mari M, Nebout M, Fiorini C, Fenichel P, Hofman P, Poujeol P, Mograbi B (2007) Control of the autophagy maturation step by the MAPK ERK and p38: lessons from environmental carcinogens. Autophagy 3:57–59

Corradetti MN, Guan KL (2006) Upstream of the mammalian target of rapamycin: do all roads pass through mTOR? Oncogene 25:6347–6360

Corradetti MN, Inoki K, Bardeesy N, DePinho RA, Guan KL (2004) Regulation of the TSC pathway by LKB1: evidence of a molecular link between tuberous sclerosis complex and Peutz-Jeghers syndrome. Genes Dev 18:1533–1538

Crighton D, Wilkinson S, O'Prey J, Syed N, Smith P, Harrison PR, Gasco M, Garrone O, Crook T, Ryan KM (2006) DRAM, a p53-induced modulator of autophagy, is critical for apoptosis. Cell 126:121–134

Criollo A, Maiuri MC, Tasdemir E, Vitale I, Fiebig AA, Andrews D, Molgo J, Diaz J, Lavandero S, Harper F, Pierron G, di Stefano D, Rizzuto R, Szabadkai G, Kroemer G (2007) Regulation of autophagy by the inositol trisphosphate receptor. Cell Death Differ 14:1029–1039
Daido S, Kanzawa T, Yamamoto A, Takeuchi H, Kondo Y, Kondo S (2004) Pivotal role of the cell death factor BNIP3 in ceramide-induced autophagic cell death in malignant glioma cells. Cancer Res 64:4286–4293
Darsow T, Rieder SE, Emr SD (1997) A multispecificity syntaxin homologue, Vam3p, essential for autophagic and biosynthetic protein transport to the vacuole. J Cell Biol 138:517–529
De Duve C, Wattiaux R (1966) Functions of lysosomes. Annu Rev Physiol 28:435–492
De Vries L, Fischer T, Tronchere H, Brothers GM, Strockbine B, Siderovski DP, Farquhar MG (2000a) Activator of G protein signaling 3 is a guanine dissociation inhibitor for Galpha i subunits. Proc Natl Acad Sci USA 97:14364–14369
De Vries L, Zheng B, Fischer T, Elenko E, Farquhar MG (2000b) The regulator of G protein signaling family. Annu Rev Pharmacol Toxicol 40:235–271
Degenhardt K, Mathew R, Beaudoin B, Bray K, Anderson D, Chen G, Mukherjee C, Shi Y, Gelinas C, Fan Y, Nelson DA, Jin S, White E (2006) Autophagy promotes tumor cell survival and restricts necrosis, inflammation, and tumorigenesis. Cancer Cell 10:51–64
Deretic V (2006) Autophagy as an immune defense mechanism. Curr Opin Immunol 18:375–382
Deter RL, De Duve C (1967) Influence of glucagon, an inducer of cellular autophagy, on some physical properties of rat liver lysosomes. J. Cell Biol. 33:437–449
Djavaheri-Mergny M, Amelotti M, Mathieu J, Besancon F, Bauvy C, Souquere S, Pierron G, Codogno P (2006) NF-kappaB activation represses tumor necrosis factor-alpha-induced autophagy. J Biol Chem 281:30373–30382
Egami Y, Kiryu-Seo S, Yoshimori T, Kiyama H (2005) Induced expressions of Rab24 GTPase and LC3 in nerve-injured motor neurons. Biochem Biophys Res Commun 337:1206–1213
Epple UD, Suriapranata I, Eskelinen EL, Thumm M (2001) Aut5/Cvt17p, a putative lipase essential for disintegration of autophagic bodies inside the vacuole. J Bacteriol 183:5942–5955
Erlich S, Mizrachy L, Segev O, Lindenboim L, Zmira O, Adi-Harel S, Hirsch JA, Stein R, Pinkas-Kramarski R (2007) Differential interactions between Beclin 1 and Bcl-2 family members. Autophagy 3:561–568
Eskelinen EL (2005) Maturation of autophagic vacuoles in mammalian cells. Autophagy 1:1–10
Eskelinen EL, Tanaka Y, Saftig P (2003) At the acidic edge: emerging functions for lysosomal membrane proteins. Trends Cell Biol 13:137–145
Eskelinen EL, Schmidt CK, Neu S, Willenborg M, Fuertes G, Salvador N, Tanaka Y, Lullmann-Rauch R, Hartmann D, Heeren J, von Figura K, Knecht E, Saftig P (2004) Disturbed cholesterol traffic but normal proteolytic function in LAMP-1/LAMP-2 double-deficient fibroblasts. Mol Biol Cell 15:3132–3145
Fader CM, Sanchez D, Furlan M, Colombo MI (2008) Induction of autophagy promotes fusion of multivesicular bodies with autophagic vacuoles in k562 cells. Traffic 9:230–250
Faivre S, Kroemer G, Raymond E (2006) Current development of mTOR inhibitors as anticancer agents. Nat. Rev. Drugs Discov. 5:671–688
Fass E, Shvets E, Degani I, Hirschberg K, Elazar Z (2006) Microtubules support production of starvation-induced autophagosomes but not their targeting and fusion with lysosomes. J Biol Chem 281:36303–36316
Feng Z, Zhang H, Levine AJ, Jin S (2005) The coordinate regulation of the p53 and mTOR pathways in cells. Proc Natl Acad Sci USA 102:8204–8209
Fengsrud M, Lunde Sneve M, Øverbye A, Seglen PO (2004) Structural aspects of mammalian autophagy. In: Klionsky DJ (ed) Autophagy. Landes Bioscience, Georgetown, TX, pp 11–25
Filimonenko M, Stuffers S, Raiborg C, Yamamoto A, Malerod L, Fisher EM, Isaacs A, Brech A, Stenmark H, Simonsen A (2007) Functional multivesicular bodies are required for autophagic clearance of protein aggregates associated with neurodegenerative disease. J Cell Biol 179:485–500
Fimia GM, Stoykova A, Romagnoli A, Giunta L, Di Bartolomeo S, Nardacci R, Corazzari M, Fuoco C, Ucar A, Schwartz P, Gruss P, Piacentini M, Chowdhury K, Cecconi F (2007) Ambra1 regulates autophagy and development of the nervous system. Nature 447:1121–1125

Forgac M (2007) Vacuolar ATPases: rotary proton pumps in physiology and pathophysiology. Nat Rev Mol Cell Biol 8:917–929
Fujita N, Itoh T, Omori H, Fukuda M, Noda T, Yoshimori T (2008) The Atg16L complex specifies the site of LC3 lipidation for membrane biogenesis in autophagy. Mol Biol Cell 19:2092–2100
Furuya N, Yu J, Byfield M, Pattingre S, Levine B (2005) The evolutionarily conserved domain of Beclin 1 is required for Vps34 binding, autophagy, and tumor suppressor function. Autophagy 1:46–52
Garcia MA, Meurs EF, Esteban M (2007) The dsRNA protein kinase PKR: virus and cell control. Biochimie 89:799–811
Gohla A, Klement K, Piekorz RP, Pexa K, vom Dahl S, Spicher K, Dreval V, Haussinger D, Birnbaumer L, Nurnberg B (2007) An obligatory requirement for the heterotrimeric G protein Gi3 in the antiautophagic action of insulin in the liver. Proc Natl Acad Sci USA 104:3003–3008
Gozuacik D, Kimchi A (2006) DAPk protein family and cancer. Autophagy 2:74–79
Gozuacik D, Kimchi A (2007) Autophagy and cell death. Curr Top Dev Biol 78:217
Guertin DA, Stevens DM, Thoreen CC, Burds AA, Kalaany NY, Moffat J, Brown M, Fitzgerald KJ, Sabatini DM (2006) Ablation in mice of the mTORC components raptor, rictor, or mLST8 reveals that mTORC2 is required for signaling to Akt-FOXO and PKCalpha, but not S6K1. Dev Cell 11:859–871
Gulati P, Thomas G (2007) Nutrient sensing in the mTOR/S6K1 signalling pathway. Biochem Soc Trans 35:236–238
Gulati P, Gaspers LD, Dann SG, Joaquin M, Nobukuni T, Natt F, Kozma SC, Thomas AP, Thomas G (2008) Amino acids activate mTOR complex 1 via Ca(2+)/CaM signaling to hVps34. Cell Metab 7:456–465
Gurkan C, Koulov AV, Balch WE (2007) An evolutionary perspective on eukaryotic membrane trafficking. Adv Exp Med Biol 607:73–83
Gutierrez MG, Munafo DB, Beron W, Colombo MI (2004) Rab7 is required for the normal progression of the autophagic pathway in mammalian cells. J Cell Sci 117:2687–2697
Hait WN, Wu H, Jin S, Yang JM (2006) Elongation factor-2 kinase: its role in protein synthesis and autophagy. Autophagy 2:294–296
Han J, Hou W, Goldstein LA, Lu C, Stolz DB, Yin XM, Rabinowich H (2008) Involvement of protective autophagy in TRAIL resistance of apoptosis defective tumor cells. J Biol Chem 283:19665–19677
Hara K, Maruki Y, Long X, Yoshino K, Oshiro N, Hidayat S, Tokunaga C, Avruch J, Yonezawa K (2002) Raptor, a binding partner of target of rapamycin (TOR), mediates TOR action. Cell 110:177–189
Hara T, Nakamura K, Matsui M, Yamamoto A, Nakahara Y, Suzuki-Migishima R, Yokoyama M, Mishima K, Saito I, Okano H, Mizushima N (2006) Suppression of basal autophagy in neural cells causes neurodegenerative disease in mice. Nature 441:885–889
Hara T, Takamura A, Kishi C, Iemura S, Natsume T, Guan JL, Mizushima N (2008) FIP200, a ULK-interacting protein, is required for autophagosome formation in mammalian cells. J Cell Biol 181:497–510
Hardie DG (2004) The AMP-activated protein kinase pathway: new players upstream and downstream. J Cell Sci 117:5479–5487
Haussinger D, Schliess F, Dombrowski F, Vom Dahl S (1999) Involvement of p38MAPK in the regulation of proteolysis by liver cell hydration. Gastroenterology 116:921–935
Hawley SA, Pan DA, Mustard KJ, Ross L, Bain J, Edelman AM, Frenguelli BG, Hardie DG (2005) Calmodulin-dependent protein kinase kinase-b is an alternative upstream kinase for AMP-activated protein kinase. Cell Metabol 2:9–19
Hopgood MF, Clark MG, Ballard FJ (1981) Stimulation by glucocorticoids of protein degradation in hepatocyte monolayers. Biochem J 196:33–40
Hoyer-Hansen M, Bastholm L, Szyniarowski P, Campanella M, Szabadkai G, Farkas T, Bianchi K, Fehrenbacher N, Elling F, Rizzuto R, Mathiasen IS, Jaattela M (2007) Control of macroautophagy by calcium, calmodulin-dependent kinase kinase-beta, and Bcl-2. Mol Cell 25:193–205

Hoyer-Hansen M, Jaattela M (2007) AMP-activated protein kinase: a universal regulator of autophagy? Autophagy 3:381–383

Hoyvik H, Gordon PB, Berg TO, Stromhaug PE, Seglen PO (1991) Inhibition of autophagic-lysosomal delivery and autophagic lactolysis by asparagine. J Cell Biol 113:1305–1312

Inbal B, Bialik S, Sabanay I, Shani G, Kimchi A (2002) DAP kinase and DRP-1 mediate membrane blebbing and the formation of autophagic vesicles during programmed cell death. J Cell Biol 157:455–468

Ishihara N, Hamasaki M, Yokota S, Suzuki K, Kamada Y, Kihara A, Yoshimori T, Noda T, Ohsumi Y (2001) Autophagosome requires specific early Sec proteins for its formation and NSF/SNARE for vacuolar fusion. Mol Biol Cell 12:3690–3702

Itoh T, Fujita N, Kanno E, Yamamoto A, Yoshimori T, Fukuda M (2008) Golgi-resident small GTPase Rab33B interacts with Atg16L and modulates autophagosome formation. Mol Biol Cell 19:2916–2925

Jacinto E, Loewith R, Schmidt A, Lin S, Ruegg MA, Hall A, Hall MN (2004) Mammalian TOR complex 2 controls the actin cytoskeleton and is rapamycin insensitive. Nat Cell Biol 6:1122–1128

Jager S, Bucci C, Tanida I, Ueno T, Kominami E, Saftig P, Eskelinen EL (2004) Role for Rab7 in maturation of late autophagic vacuoles. J Cell Sci 117:4837–4848

Jahreiss L, Menzies FM, Rubinsztein DC (2008) The itinerary of autophagosomes: from peripheral formation to kiss-and-run fusion with lysosomes. Traffic 9:574–587

Juhasz G, Hill JH, Yan Y, Sass M, Baehrecke EH, Backer JM, Neufeld TP (2008) The class III PI(3)K Vps34 promotes autophagy and endocytosis but not TOR signaling in Drosophila. J Cell Biol 181:655–666

Kihara A, Kabeya Y, Ohsumi Y, Yoshimori T (2001a) Beclin-phosphatidylinositol 3-kinase complex functions at the trans-Golgi network. EMBO Rep 2:330–335

Kihara A, Noda T, Ishihara N, Ohsumi Y (2001b) Two distinct Vps34 phosphatidylinositol 3-kinase complexes function in autophagy and carboxypeptidase Y sorting in Saccharomyces cerevisiae. J Cell Biol 152:519–530

Kim DH, Sarbassov DD, Ali SM, King JE, Latek RR, Erdjument-Bromage H, Tempst P, Sabatini DM (2002) mTOR interacts with raptor to form a nutrient-sensitive complex that signals to the cell growth machinery. Cell 110:163–175

Kirisako T, Baba M, Ishihara N, Miyazawa K, Ohsumi M, Yoshimori T, Noda T, Ohsumi Y (1999) Formation process of autophagosome is traced with Apg8/Aut7p in yeast. J Cell Biol 147:435–446

Kirkegaard K, Taylor MP, Jackson WT (2004) Cellular autophagy: surrender, avoidance and subversion by microorganisms. Nat Rev Microbiol 2:301–314

Klionsky DJ (2005) The molecular machinery of autophagy: unanswered questions. J Cell Sci 118:7–18

Klionsky DJ (2007) Autophagy: from phenomenology to molecular understanding in less than a decade. Nat Rev Mol Cell Biol 8:931

Klionsky DJ, Cregg JM, Dunn WA Jr, Emr SD, Sakai Y, Sandoval IV, Sibirny A, Subramani S, Thumm M, Veenhuis M, Ohsumi Y (2003) A unified nomenclature for yeast autophagy-related genes. Dev Cell 5:539–545

Klionsky DJ, Meijer AJ, Codogno P (2005) Autophagy and p70S6 kinase. Autophagy 1:59–60; discussion 60–51

Kochl R, Hu XW, Chan EY, Tooze SA (2006) Microtubules facilitate autophagosome formation and fusion of autophagosomes with endosomes. Traffic 7:129–145

Komatsu M, Waguri S, Chiba T, Murata S, Iwata J, Tanida I, Ueno T, Koike M, Uchiyama Y, Kominami E, Tanaka K (2006) Loss of autophagy in the central nervous system causes neurodegeneration in mice. Nature 441:880–884

Komatsu M, Ueno T, Waguri S, Uchiyama Y, Kominami E, Tanaka K (2007) Constitutive autophagy: vital role in clearance of unfavorable proteins in neurons. Cell Death Differ 14:887–894

Kouroku Y, Fujita E, Tanida I, Ueno T, Isoai A, Kumagai H, Ogawa S, Kaufman RJ, Kominami E, Momoi T (2007) ER stress (PERK/eIF2alpha phosphorylation) mediates the polyglutamine-induced LC3 conversion, an essential step for autophagy formation. Cell Death Differ 14:230–239

Ku B, Woo J-S, Liang C, Lee KH, Jung JU, Oh BH (2008) An insight into the mechanistic role of Beclin 1 and its inhibition by prosurvival Bcl-2 family proteins. Autophagy 4:519–520

Kuma A, Hatano M, Matsui M, Yamamoto A, Nakaya H, Yoshimori T, Ohsumi Y, Tokuhisa T, Mizushima N (2004) The role of autophagy during the early neonatal starvation period. Nature 432:1032–1036

Laderoute KR, Amin K, Calaoagan JM, Knapp M, Le T, Orduna J, Foretz M, Viollet B (2006) 5′-AMP-activated protein kinase (AMPK) is induced by low-oxygen and glucose deprivation conditions found in solid-tumor microenvironments. Mol Cell Biol 26:5336–5347

Lavieu G, Scarlatti F, Sala G, Carpentier S, Levade T, Ghidoni R, Botti J, Codogno P (2006) Regulation of autophagy by sphingosine kinase 1 and its role in cell survival during nutrient starvation. J Biol Chem 281:8518–8527

Lavieu G, Scarlatti F, Sala G, Levade T, Ghidoni R, Botti J, Codogno P (2007) Is autophagy the key mechanism by which the sphingolipid rheostat controls the cell fate decision? Autophagy 3:45–47

Lee IH, Cao L, Mostoslavsky R, Lombard DB, Liu J, Bruns NE, Tsokos M, Alt FW, Finkel T (2008) A role for the NAD-dependent deacetylase Sirtl in the regulation of autophagy. Proc Natl Acad Sci USA 105:3374–3379

Lee JA, Beigneux A, Ahmad ST, Young SG, Gao FB (2007) ESCRT-III dysfunction causes autophagosome accumulation and neurodegeneration. Curr Biol 17:1561–1567

Levine B, Deretic V (2007) Unveiling the roles of autophagy in innate and adaptive immunity. Nat Rev Immunol 7:767–777

Levine B, Klionsky DJ (2004) Development by self-digestion: molecular mechanisms and biological functions of autophagy. Dev Cell 6:463–477

Levine B, Kroemer G (2008) Autophagy in the pathogenesis of disease. Cell 132:27–42

Levine B, Yuan J (2005) Autophagy in cell death: an innocent convict? J Clin Invest 115:2679–2688

Liang C, Feng P, Ku B, Dotan I, Canaani D, Oh BH, Jung JU (2006) Autophagic and tumour suppressor activity of a novel Beclin 1-binding protein UVRAG. Nat Cell Biol 8:688–699

Liang C, Feng P, Ku B, Oh BH, Jung JU (2007) UVRAG: a new player in autophagy and tumor cell growth. Autophagy 3:69–71

Liang C, Xiaofei E, Jung JU (2008) Downregulation of autophagy by herpesvirus Bcl-2 homologs. Autophagy 4:268–272

Liang XH, Kleeman LK, Jiang HH, Gordon G, Goldman JE, Berry G, Herman B, Levine B (1998) Protection against fatal Sindbis virus encephalitis by beclin, a novel Bcl-2-interacting protein. J Virol 72:8586–8596

Liang XH, Yu J, Brown K, Levine B (2001) Beclin 1 contains a leucine-rich nuclear export signal that is required for its autophagy and tumor suppressor function. Cancer Res 61:3443–3449

Lloyd JB (1996) Metabolite efflux and influx across the lysosome membrane. Subcell Biochem 27:361–386

Loewith R, Jacinto E, Wullschleger S, Lorberg A, Crespo JL, Bonenfant D, Oppliger W, Jenoe P, Hall MN (2002) Two TOR complexes, only one of which is rapamycin sensitive, have distinct roles in cell growth control. Mol Cell 10:457–468

Longetti A, Tooze SA (2009) Vesicular trafficking and autophagosome formation. Cell Death Differ Online April 17.

Lum JJ, DeBerardinis RJ, Thompson CB (2005) Autophagy in metazoans: cell survival in the land of plenty. Nat Rev Mol Cell Biol 6:439–448

Maiuri MC, Criollo A, Tasdemir E, Vicencio JM, Tajeddine N, Hickman JA, Geneste O, Kroemer G (2007a) BH3-only proteins and BH3 mimetics induce autophagy by competitively disrupting the interaction between Beclin 1 and Bcl-2/Bcl-X(L). Autophagy 3:374–376

Maiuri MC, Le Toumelin G, Criollo A, Rain JC, Gautier F, Juin P, Tasdemir E, Pierron G, Troulinaki K, Tavernarakis N, Hickman JA, Geneste O, Kroemer G (2007b) Functional and physical interaction between Bcl-X(L) and a BH3-like domain in Beclin-1. Embo J 26:2527–2539

Maiuri MC, Zalckvar E, Kimchi A, Kroemer G (2007c) Self-eating and self-killing: crosstalk between autophagy and apoptosis. Nat Rev Mol Cell Biol 8:741–752

Mammucari C, Milan G, Romanello V, Masiero E, Rudolf R, Del Piccolo P, Burden SJ, Di Lisi R, Sandri C, Zhao J, Goldberg AL, Schiaffino S, Sandri M (2007) FoxO3 controls autophagy in skeletal muscle in vivo. Cell Metab 6:458–471

Mammucari C, Schiaffino S, Sandri M (2008) Downstream of Akt: FoxO3 and mTOR in the regulation of autophagy in skeletal muscle. Autophagy 4:524–526

Meijer AJ, Codogno P (2004) Regulation and role of autophagy in mammalian cells. Int J Biochem Cell Biol 36:2445–2462

Meijer AJ, Codogno P (2006) Signalling and autophagy regulation in health, aging and disease. Mol Aspects Med 27:411–425

Meijer AJ, Codogno P (2007) AMP-activated protein kinase and autophagy. Autophagy 3:238

Meley D, Bauvy C, Houben-Weerts JH, Dubbelhuis PF, Helmond MT, Codogno P, Meijer AJ (2006) AMP-activated protein kinase and the regulation of autophagic proteolysis. J Biol Chem 281:34870–34879

Mesa R, Salomon C, Roggero M, Stahl PD, Mayorga LS (2001) Rab22a affects the morphology and function of the endocytic pathway. J Cell Sci 114:4041–4049

Mieulet V, Roceri M, Espeillac C, Sotiropoulos A, Ohanna M, Oorschot V, Klumperman J, Sandri M, Pende M (2007) S6 kinase inactivation impairs growth and translational target phosphorylation in muscle cells maintaining proper regulation of protein turnover. Am J Physiol Cell Physiol 293:C712–C722

Mizushima N (2005) The pleiotropic role of autophagy: from protein metabolism to bactericide. Cell Death Differ 12 Suppl 2:1535–1541

Mizushima N, Kuma A, Kobayashi Y, Yamamoto A, Matsubae M, Takao T, Natsume T, Ohsumi Y, Yoshimori T (2003) Mouse Apg16L, a novel WD-repeat protein, targets to the autophagic isolation membrane with the Apg12-Apg5 conjugate. J Cell Sci 116:1679–1688

Mizushima N, Levine B, Cuervo AM, Klionsky DJ (2008) Autophagy fights disease through cellular self-digestion. Nature 451:1069–1075

Monastyrska I, He C, Geng J, Hoppe AD, Li Z, Klionsky DJ (2008) Arp2 links autophagic machinery with the actin cytoskeleton. Mol Biol Cell 19:1962–1975

Mortimore GE, Ward WF (1976) Behavior of the lysosomal system during organ perfusion. An inquiry into the mechanism of hepatic proteolysis. Front Biol 45:157–184

Mousavi SA, Kjeken R, Berg TO, Seglen PO, Berg T, Brech A (2001) Effects of inhibitors of the vacuolar proton pump on hepatic heterophagy and autophagy. Biochim Biophys Acta 1510:243–257

Nair U, Klionsky DJ (2005) Molecular mechanisms and regulation of specific and nonspecific autophagy pathways in yeast. J Biol Chem 280:41785–41788

Nara A, Mizushima N, Yamamoto A, Kabeya Y, Ohsumi Y, Yoshimori T (2002) SKD1 AAA ATPase-dependent endosomal transport is involved in autolysosome formation. Cell Struct Funct 27:29–37

Natarajan K, Meyer MR, Jackson BM, Slade D, Roberts C, Hinnebusch AG, Marton MJ (2001) Transcriptional profiling shows that Gcn4p is a master regulator of gene expression during amino acid starvation in yeast. Mol Cell Biol 21:4347–4368

Nishino I, Fu J, Tanji K, Yamada T, Shimojo S, Koori T, Mora M, Riggs JE, Oh SJ, Koga Y, Sue CM, Yamamoto A, Murakami N, Shanske S, Byrne E, Bonilla E, Nonaka I, DiMauro S, Hirano M (2000) Primary LAMP-2 deficiency causes X-linked vacuolar cardiomyopathy and myopathy (Danon disease). Nature 406:906–910

Nobukuni T, Joaquin M, Roccio M, Dann SG, Kim SY, Gulati P, Byfield MP, Backer JM, Natt F, Bos JL, Zwartkruis FJ, Thomas G (2005) Amino acids mediate mTOR/raptor signaling through activation of class 3 phosphatidylinositol 3OH-kinase. Proc Natl Acad Sci USA 102:14238–14243

Oberstein A, Jeffrey PD, Shi Y (2007) Crystal structure of the Bcl-XL-Beclin 1 peptide complex: Beclin 1 is a novel BH3-only protein. J Biol Chem 282:13123–13132

Ogata M, Hino S, Saito A, Morikawa K, Kondo S, Kanemoto S, Murakami T, Taniguchi M, Tanii I, Yoshinaga K, Shiosaka S, Hammarback JA, Urano F, Imaizumi K (2006) Autophagy is activated for cell survival after endoplasmic reticulum stress. Mol Cell Biol 26:9220–9231

Ogawa M, Yoshimori T, Suzuki T, Sagara H, Mizushima N, Sasakawa C (2005) Escape of intracellular Shigella from autophagy. Science 307:727–731

Ogier-Denis E, Couvineau A, Maoret JJ, Houri JJ, Bauvy C, De Stefanis D, Isidoro C, Laburthe M, Codogno P (1995) A heterotrimeric Gi3-protein controls autophagic sequestration in the human colon cancer cell line HT-29. J Biol Chem 270:13–16

Ogier-Denis E, Houri JJ, Bauvy C, Codogno P (1996) Guanine nucleotide exchange on heterotrimeric Gi3 protein controls autophagic sequestration in HT-29 cells. J Biol Chem 271: 28593–28600

Ogier-Denis E, Bauvy C, Houri JJ, Codogno P (1997a) Evidence for a dual control of macroautophagic sequestration and intracellular trafficking of N-linked glycoproteins by the trimeric G(i3) protein in HT-29 cells. Biochem Biophys Res Commun 235:166–170

Ogier-Denis E, Petiot A, Bauvy C, Codogno P (1997b) Control of the expression and activity of the Galpha-interacting protein (GAIP) in human intestinal cells. J Biol Chem 272:24599–24603

Ogretmen B, Hannun YA (2004) Biologically active sphingolipids in cancer pathogenesis and treatment. Nat Rev Cancer 4:604–616

Ohsumi Y (2001) Molecular dissection of autophagy: two ubiquitin-like systems. Nat Rev Mol Cell Biol 2:211–216

Olkkonen VM, Dupree P, Killisch I, Lutcke A, Zerial M, Simons K (1993) Molecular cloning and subcellular localization of three GTP-binding proteins of the rab subfamily. J Cell Sci 106 (Pt 4):1249–1261

Orvedahl A, Alexander D, Talloczy Z, Sun Q, Wei Y, Zhang W, Burns D, Leib DA, Levine B (2007) HSV-1 ICP34.5 confers neurovirulence by targeting the Beclin 1 autophagy protein. Cell Host Microbe 1:23–35

Orvedahl A, Levine B (2008) Viral evasion of autophagy. Autophagy 4:280–285

Pandey UB, Nie Z, Batlevi Y, McCray BA, Ritson GP, Nedelsky NB, Schwartz SL, DiProspero NA, Knight MA, Schuldiner O, Padmanabhan R, Hild M, Berry DL, Garza D, Hubbert CC, Yao TP, Baehrecke EH, Taylor JP (2007) HDAC6 rescues neurodegeneration and provides an essential link between autophagy and the UPS. Nature 447:859–863

Park MA, Yacoub A, Sarkar D, Emdad L, Rahmani M, Spiegel S, Koumenis C, Graf M, Curiel DT, Grant S, Fisher PB, Dent P (2008) PERK-dependent regulation of MDA-7/IL-24-induced autophagy in primary human glioma cells. Autophagy 4:513–515

Pattingre S, Bauvu C, Carpentier S, Levade T, Levine B, Codogno P (2009) Role of JNK1-dependent phosphorylation in ceramide-induced macroautophagy. J Biol Chem 284:2719–2728

Pattingre S, Levine B (2006) Bcl-2 inhibition of autophagy: a new route to cancer? Cancer Res 66:2885–2888

Pattingre S, Bauvy C, Codogno P (2003a) Amino acids interfere with the ERK1/2-dependent control of macroautophagy by controlling the activation of Raf-1 in human colon cancer HT-29 cells. J Biol Chem 278:16667–16674

Pattingre S, De Vries L, Bauvy C, Chantret I, Cluzeau F, Ogier-Denis E, Vandewalle A, Codogno P (2003b) The G-protein regulator AGS3 controls an early event during macroautophagy in human intestinal HT-29 cells. J Biol Chem 278: 20995–21002

Pattingre S, Tassa A, Qu X, Garuti R, Liang XH, Mizushima N, Packer M, Schneider MD, Levine B (2005) Bcl-2 antiapoptotic proteins inhibit Beclin 1-dependent autophagy. Cell 122:927–939

Pattingre S, Espert L, Biard-Piechaczyk M, Codogno P (2008) Regulation of macroautophagy by mTOR and Beclin 1 complexes. Biochimie 90:313–323

Petiot A, Ogier-Denis E, Blommaart EF, Meijer AJ, Codogno P (2000) Distinct classes of phosphatidylinositol 3′-kinases are involved in signaling pathways that control macroautophagy in HT-29 cells. J Biol Chem 275:992–998

Polager S, Ofir M, Ginsberg D (2008) E2F1 regulates autophagy and the transcription of autophagy genes. Oncogene 27:4860–4864

Poso AR, Wert JJ Jr, Mortimore GE (1982) Multifunctional control of amino acids of deprivation-induced proteolysis in liver. Role of leucine. J Biol Chem 257:12114–12120

Purhonen P, Pursiainen K, Reunanen H (1997) Effects of brefeldin A on autophagy in cultured rat fibroblasts. Eur J Cell Biol 74:63–67

Raiborg C, Rusten TE, Stenmark H (2003) Protein sorting into multivesicular endosomes. Curr Opin Cell Biol 15:446–455

Ravikumar B, Acevedo-Arozena A, Imarisio S, Berger Z, Vacher C, O'Kane CJ, Brown SD, Rubinsztein DC (2005) Dynein mutations impair autophagic clearance of aggregate-prone proteins. Nat Genet 37:771–776

Ravikumar B, Imarisio S, Sarkar S, O'Kane CJ, Rubinsztein DC (2008) Rab5 modulates aggregation and toxicity of mutant huntingtin through macroautophagy in cell and fly models of Huntington disease. J Cell Sci 121:1649–1660

Romano PS, Gutierrez MG, Beron W, Rabinovitch M, Colombo MI (2007) The autophagic pathway is actively modulated by phase II Coxiella burnetii to efficiently replicate in the host cell. Cell Microbiol 9:891–909

Ropolo A, Grasso D, Pardo R, Sacchetti ML, Archange C, Lo Re A, Seux M, Nowak J, Gonzalez CD, Iovanna JL, Vaccaro MI (2007) The pancreatitis-induced vacuole membrane protein 1 triggers autophagy in mammalian cells. J Biol Chem 282:37124–37133

Rothman JE, Wieland FT (1996) Protein sorting by transport vesicles. Science 272:227–234

Rubinsztein DC (2006) The roles of intracellular protein-degradation pathways in neurodegeneration. Nature 443:780–786

Rubinsztein DC, Gestwicki JE, Murphy LO, Klionsky DJ (2007) Potential therapeutic applications of autophagy. Nat Rev Drug Discov 6:304–312

Rusten TE, Vaccari T, Lindmo K, Rodahl LM, Nezis IP, Sem-Jacobsen C, Wendler F, Vincent JP, Brech A, Bilder D, Stenmark H (2007) ESCRTs and Fab1 regulate distinct steps of autophagy. Curr Biol 17:1817–1825

Sakaki K, Wu J, Kaufman RJ (2008) Protein kinase C-theta is required for autophagy in response to stress in the endoplasmic reticulum. J Biol Chem 283:15370–15380

Samari HR, Seglen PO (1998) Inhibition of hepatocytic autophagy by adenosine, aminoimidazole-4-carboxamide riboside, and N6-mercaptopurine riboside. Evidence for involvement of amp-activated protein kinase. J Biol Chem 273:23758–23763

Sarbassov DD, Ali SM, Kim DH, Guertin DA, Latek RR, Erdjument-Bromage H, Tempst P, Sabatini DM (2004) Rictor, a novel binding partner of mTOR, defines a rapamycin-insensitive and raptor-independent pathway that regulates the cytoskeleton. Curr Biol 14:1296–1302

Sarbassov DD, Ali SM, Sabatini DM (2005) Growing roles for the mTOR pathway. Curr Opin Cell Biol 17:596–603

Sarkar S, Floto RA, Berger Z, Imarisio S, Cordenier A, Pasco M, Cook LJ, Rubinsztein DC (2005) Lithium induces autophagy by inhibiting inositol monophosphatase. J Cell Biol 170:1101–1111

Sarkar S, Perlstein EO, Imarisio S, Pineau S, Cordenier A, Maglathlin RL, Webster JA, Lewis TA, O'Kane CJ, Schreiber SL, Rubinsztein DC (2007) Small molecules enhance autophagy and reduce toxicity in Huntington's disease models. Nat Chem Biol 3:331–338

Scarlatti F, Bauvy C, Ventruti A, Sala G, Cluzeaud F, Vandewalle A, Ghidoni R, Codogno P (2004) Ceramide-mediated macroautophagy involves inhibition of protein kinase B and up-regulation of beclin 1. J Biol Chem 279:18384–18391

Scherz-Shouval R, Elazar Z (2007) ROS, mitochondria and the regulation of autophagy. Trends Cell Biol 17:422–427

Scherz-Shouval R, Shvets E, Fass E, Shorer H, Gil L, Elazar Z (2007) Reactive oxygen species are essential for autophagy and specifically regulate the activity of Atg4. Embo J 26:1749–1760

Schmelzle T, Hall MN (2000) TOR, a central controller of cell growth. Cell 103:253–262

Scott RC, Schuldiner O, Neufeld TP (2004) Role and regulation of starvation-induced autophagy in the *Drosophila* fat body. Dev Cell 7:167–178

Scott RC, Juhasz G, Neufeld TP (2007) Direct induction of autophagy by Atg1 inhibits cell growth and induces apoptotic cell death. Curr Biol 17:1–11

Seglen PO, Gordon PB (1982) 3-Methyladenine: specific inhibitor of autophagic/lysosomal protein degradation in isolated rat hepatocytes. Proc Natl Acad Sci USA 79:1889–1892

Shaw RJ, Bardeesy N, Manning BD, Lopez L, Kosmatka M, DePinho RA, Cantley LC (2004) The LKB1 tumor suppressor negatively regulates mTOR signaling. Cancer Cell 6:91–99

Shimizu S, Kanaseki T, Mizushima N, Mizuta T, Arakawa-Kobayashi S, Thompson CB, Tsujimoto Y (2004) Role of Bcl-2 family proteins in a non-apoptotic programmed cell death dependent on autophagy genes. Nat Cell Biol 6:1221–1228

Shirahama K, Noda T, Ohsumi Y (1997) Mutational analysis of Csc1/Vps4p: involvement of endosome in regulation of autophagy in yeast. Cell Struct Funct 22:501–509

Sivaprasad U, Basu A (2008) Inhibition of ERK attenuates autophagy and potentiates tumor necrosis factor-alpha-induced cell death in MCF-7 cells. J Cell Mol Med 12:1265–1271

Spiegel S, Milstien S (2003) Sphingosine-1-phosphate: an enigmatic signalling lipid. Nat Rev Mol Cell Biol 4:397–407

Stephan JS, Herman PK (2006) The regulation of autophagy in eukaryotic cells: do all roads pass through Atg1? Autophagy 2:146–148

Suzuki K, Ohsumi Y (2007) Molecular machinery of autophagosome formation in yeast, Saccharomyces cerevisiae. FEBS Lett 581:2156–2161

Swanson MS (2006) Autophagy: eating for good health. J Immunol 177:4945–4951

Takahashi Y, Coppola D, Matsushita N, Cualing HD, Sun M, Sato Y, Liang C, Jung JU, Cheng JQ, Mul JJ, Pledger WJ, Wang HG (2007) Bif-1 interacts with Beclin 1 through UVRAG and regulates autophagy and tumorigenesis. Nat Cell Biol 9:1142–1151

Takesono A, Cismowski MJ, Ribas C, Bernard M, Chung P, Hazard S, 3rd, Duzic E, Lanier SM (1999) Receptor-independent activators of heterotrimeric G-protein signaling pathways. J Biol Chem 274:33202–33205

Tallóczy Z, Jiang W, Virgin HW, Leib DA, Scheuner D, Kaufman RJ, Eskelinen EL, Levine B (2002) Regulation of starvation- and virus-induced autophagy by the eIF2alpha kinase signaling pathway. Proc Natl Acad Sci USA 99:190–195

Tallóczy Z, Virgin HW, Levine B (2006) PKR-dependent autophagic degradation of herpes simplex virus type 1. Autophagy 2:24–29

Tamai K, Tanaka N, Nara A, Yamamoto A, Nakagawa I, Yoshimori T, Ueno Y, Shimosegawa T, Sugamura K (2007) Role of Hrs in maturation of autophagosomes in mammalian cells. Biochem Biophys Res Commun 360:721–727

Tanaka Y, Guhde G, Suter A, Eskelinen EL, Hartmann D, Lullmann-Rauch R, Janssen PM, Blanz J, von Figura K, Saftig P (2000) Accumulation of autophagic vacuoles and cardiomyopathy in LAMP-2-deficient mice. Nature 406:902–906

Tang G, Yue Z, Talloczy Z, Hagemann T, Cho W, Messing A, Sulzer DL, Goldman JE (2008) Autophagy induced by Alexander Disease-mutant GFAP accumulation is regulated by p38/MAPK and mTOR signaling pathways. Hum Mol Genet 17:1540–1555

Tasdemir E, Maiuri MC, Galluzzi L, Vitale I, Djavaheri-Mergny M, D'Amelio M, Criollo A, Morselli E, Zhu C, Harper F, Nannmark U, Samara C, Pinton P, Vicencio JM, Carnuccio R, Moll UM, Madeo F, Paterlini-Brechot P, Rizzuto R, Szabadkai G, Pierron G, Blomgren K, Tavernarakis N, Codogno P, Cecconi F, Kroemer G (2008) Regulation of autophagy by cytoplasmic p53. Nat Cell Biol 10:676–687

Tassa A, Roux MP, Attaix D, Bechet DM (2003) Class III phosphoinositide 3-kinase-Beclin1 complex mediates the amino acid-dependent regulation of autophagy in C2C12 myotubes. Biochem J 376:577–586

Taylor MP, Kirkegaard K (2008) Potential subversion of autophagosomal pathway by picornaviruses. Autophagy 4:286–289

Teter SA, Eggerton KP, Scott SV, Kim J, Fischer AM, Klionsky DJ (2001) Degradation of lipid vesicles in the yeast vacuole requires function of Cvt17, a putative lipase. J Biol Chem 276:2083–2087

Tolkovsky AM, Xue L, Fletcher GC, Borutaite V (2002) Mitochondrial disappearance from cells: a clue to the role of autophagy in programmed cell death and disease? Biochimie 84:233–240

Wang Z, Wilson WA, Fujino MA, Roach PJ (2001) Antagonistic controls of autophagy and glycogen accumulation by Snf1p, the yeast homolog of AMP-activated protein kinase, and the cyclin-dependent kinase Pho85p. Mol Cell Biol 21:5742–5752

Webb JL, Ravikumar B, Rubinsztein DC (2004) Microtubule disruption inhibits autophagosome-lysosome fusion: implications for studying the roles of aggresomes in polyglutamine diseases. Int J Biochem Cell Biol 36:2541–2550

Wei Y, Pattingre S, Sinha S, Bassik M, Levine B (2008) JNK1-mediated phosphorylation of Bcl-2 regulates starvation-induced autophagy. Mol Cell 30:678–688

Weinmann AS, Bartley SM, Zhang T, Zhang MQ, Farnham PJ (2001) Use of chromatin immunoprecipitation to clone novel E2F target promoters. Mol Cell Biol 21:6820–6832
White SR, Lauring B (2007) AAA + ATPases: achieving diversity of of function with conserved machinery. Traffic 8:1657–1667
Williams A, Sarkar S, Cuddon P, Ttofi EK, Saiki S, Siddiqi FH, Jahreiss L, Fleming A, Pask D, Goldsmith P, O'Kane CJ, Floto RA, Rubinsztein DC (2008) Novel targets for Huntington's disease in an mTOR-independent autophagy pathway. Nat Chem Biol 4:295–305
Woods A, Dickerson K, Heath R, Hong SP, Momcilovic M, Johnstone SR, Carlson M, Carling D (2005) Ca2+/calmodulin-dependent protein kinase kinase-b acts upstream of AMP-activated protein kinase. Cell Metabol 2:21–33
Wu H, Yang JM, Jin S, Zhang H, Hait WN (2006) Elongation factor-2 kinase regulates autophagy in human glioblastoma cells. Cancer Res 66:3015–3023
Xiao G (2007) Autophagy and NF-kappaB: fight for fate. Cytokine Growth Factor Rev 18:233–243
Xie Z, Klionsky DJ (2007) Autophagosome formation: core machinery and adaptations. Nat Cell Biol 9:1102–1109
Yamamoto A, DeWald DB, Boronenkov IV, Anderson RA, Emr SD, Koshland D (1995) Novel PI(4)P 5-kinase homologue, Fab1p, essential for normal vacuole function and morphology in yeast. Mol Biol Cell 6:525–539
Yamamoto A, Tagawa Y, Yoshimori T, Moriyama Y, Masaki R, Tashiro Y (1998) Bafilomycin A1 prevents maturation of autophagic vacuoles by inhibiting fusion between autophagosomes and lysosomes in rat hepatoma cell line, H-4-II-E cells. Cell Struct Funct 23:33–42
Yang Z, Huang J, Geng J, Nair U, Klionsky DJ (2006) Atg22 recycles amino acids to link the degradative and recycling functions of autophagy. Mol Biol Cell 17:5094–5104
Yorimitsu T, Klionsky DJ (2007) Eating the endoplasmic reticulum: quality control by autophagy. Trends Cell Biol 17:279–285
Yoshimori T, Noda T (2008) Toward unraveling membrane biogenesis in mammalian autophagy. Curr Opin Cell Biol 20:401–407
Yousefi S, Perozzo R, Schmid I, Ziemiecki A, Schaffner T, Scapozza L, Brunner T, Simon HU (2006) Calpain-mediated cleavage of Atg5 switches autophagy to apoptosis. Nat Cell Biol 8:1124–1132
Yu L, Alva A, Su H, Dutt P, Freundt E, Welsh S, Baehrecke EH, Lenardo MJ (2004) Regulation of an ATG7-beclin 1 program of autophagic cell death by caspase-8. Science 304:1500–1502
Yu QC, Marzella L (1986) Modification of lysosomal proteolysis in mouse liver with taxol. Am J Pathol 122:553–561
Yu WH, Cuervo AM, Kumar A, Peterhoff CM, Schmidt SD, Lee JH, Mohan PS, Mercken M, Farmery MR, Tjernberg LO, Jiang Y, Duff K, Uchiyama Y, Naslund J, Mathews PM, Cataldo AM, Nixon RA (2005) Macroautophagy: a novel beta-amyloid peptide-generating pathway activated in Alzheimer's disease. J Cell Biol 171:87–98
Yue Z, Horton A, Bravin M, DeJager PL, Selimi F, Heintz N (2002) A novel protein complex linking the delta 2 glutamate receptor and autophagy: implications for neurodegeneration in lurcher mice. Neuron 35:921–933
Zerial M, McBride H (2001) Rab proteins as membrane organizers. Nat Rev Mol Cell Biol 2:107–117
Zhang L, Yu J, Pan H, Hu P, Hao Y, Cai W, Zhu H, Yu AD, Xie X, Ma D, Yuan J (2007) Small molecule regulators of autophagy identified by an image-based high-throughput screen. Proc Natl Acad Sci USA 104:19023–19028
Zhao J, Brault JJ, Schild A, Cao P, Sandri M, Schiaffino S, Lecker SH, Goldberg AL (2007) FoxO3 coordinately activates protein degradation by the autophagic/lysosomal and proteasomal pathways in atrophying muscle cells. Cell Metab 6:472–483
Zhu JH, Guo F, Shelburne J, Watkins S, Chu CT (2003) Localization of phosphorylated ERK/MAP kinases to mitochondria and autophagosomes in Lewy body diseases. Brain Pathol 13:473–481

Physiological Functions of Autophagy

Noboru Mizushima

Contents

1 Introduction 71
2 Physiological Functions of Autophagy 73
2.1 Maintenance of the Amino Acid Pool 73
2.2 Intracellular Quality Control 74
2.3 Selective Degradation by Autophagy 75
2.4 Development and Cell Death 76
2.5 Tumor Suppression 77
2.6 Anti-Aging 78
3 Conclusion 79
References 79

Abstract The field of autophagy research has advanced rapidly in recent years, with important discoveries made in relation to both molecular mechanisms and physiological functions. Initially, autophagy was thought to be primarily a response to starvation. Although this might be true in lower eukaryotes, this catabolic process exerts various physiological functions in higher eukaryotes. This review summarizes the physiological roles of autophagy in amino acid pool maintenance, intracellular quality control, development, cell death, tumor suppression and anti-aging.

1 Introduction

Intracellular protein degradation systems can be roughly classified into two groups: selective and nonselective. Selective degradation is primarily carried out by the ubiquitin-proteasome system, whereas most nonselective degradation occurs

N. Mizushima
Department of Physiology and Cell Biology, Tokyo Medical and Dental University, Tokyo, 113-8519, Japan
e-mail: nmizu.phy2@tmd.ac.jp

B. Levine et al. (eds.), *Autophagy in Infection and Immunity,*
Current Topics in Microbiology and Immunology 335,
DOI 10.1007/ 978-3-642-00302-8_3, © Springer-Verlag Berlin Heidelberg 2009

in the lysosome. The degradation of cytoplasmic components in the lysosomes is generically referred to as autophagy (Cuervo 2004; Klionsky 2007; Mizushima 2007; Levine and Kroemer 2008; Mizushima et al. 2008). The three types of autophagy are macroautophagy, microautophagy, and chaperone-mediated autophagy (CMA). In CMA, substrate proteins are specifically recognized by chaperones and directly transported across the lysosomal membrane to the lumen (Cuervo 2004). Microautophagy has been proposed to occur by invagination of the lysosomal membrane into the lumen; however, its molecular mechanisms remain unknown. This chapter focuses on macroautophagy, having been extensively studied and closely related to immunology and microbiology.

Macroautophagy (referred to as autophagy hereafter) is mediated by a unique organelle called the autophagosome (Fig. 1). Upon autophagy induction, a portion of the cytoplasm is enclosed by an autophagosome. Not only cytosolic proteins but also organelles such as mitochondria and endoplasmic reticulum are often sequestered into autophagosomes. The outer membrane of the autophagosome then fuses with the lysosome, allowing lysosomal enzymes to degrade the sequestered cytoplasmic materials. Since bulk cytoplasm is sequestered by autophagosomes, autophagy is usually considered a nonselective random degradation system. However, recent studies have indicated that some proteins and organelles are selectively degraded by autophagy, an issue discussed later in this chapter.

Recently, the molecular mechanisms of autophagy have become increasingly clear. Breakthroughs came from genetic analyses performed in yeast, and at least 18 autophagy-related genes have been identified in this organism so far. Since most of these genes are conserved in higher eukaryotes, studies using reverse genetic approaches have been carried out in various organisms. These studies have demonstrated that, although autophagy is a simple membrane-mediated process, it has a wide variety of physiological and pathophysiological roles.

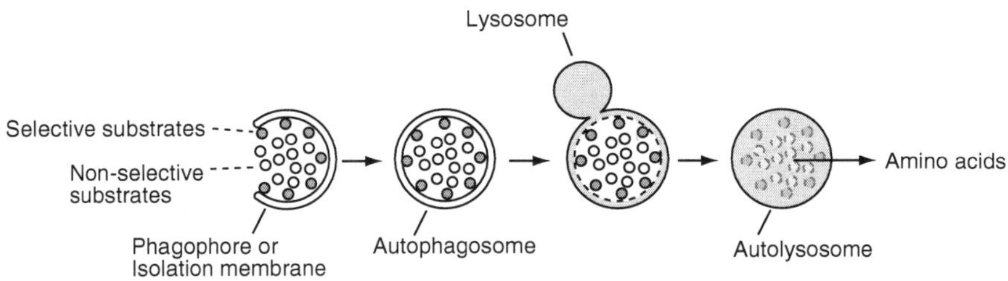

Fig. 1 Schematic model of macroautophagy. A portion of cytoplasm is enclosed by a phagophore or isolation membrane to form an autophagosome. While most substrates are enclosed nonselectively, some proteins such as p62 are selectively recognized by the autophagosome membrane (via LC3 in the case of p62). The outer membrane of the autophagosome subsequently fuses with the lysosome, and the internal materials are degraded. The resulting amino acids are delivered back into the cytosol for reuse or further metabolism

2 Physiological Functions of Autophagy

2.1 Maintenance of the Amino Acid Pool

An evolutionarily conserved role of autophagy is adaptation to starvation through the generation of amino acids inside cells. Under normal conditions, the intracellular amino acid pool can be maintained by the proteasome, which continuously degrades cytoplasmic proteins (Vabulas and Hartl 2005). In contrast, this pool is largely maintained by autophagy during starvation. In yeast, autophagy is suppressed to undetectable levels under growing conditions, but is rapidly upregulated during nitrogen starvation (Takeshige et al. 1992). The levels of amino acids in autophagy-deficient yeast cells are lower than those in wild-type cells during starvation. Likewise, autophagy is immediately activated in cultured mammalian cells following amino acid withdrawal. Accordingly, autophagy has been shown to be crucial for surviving starvation in *Sacharomyces cerevisiae* (Tsukada and Ohsumi 1993), *Dictyostelium discoideum* (Otto et al. 2003), *Drosophila melanogaster* (Scott et al. 2004) and *Caenorhabditis elegans* (Kang et al. 2007). Moreover, autophagy is upregulated in most tissues except the nervous tissues in starved mice (Mizushima et al. 2004). It is also upregulated shortly after birth, when nutrient supply from the placenta is abruptly terminated (Kuma et al. 2004). Mice lacking Atg5, an essential factor in autophagosome formation, die about 12 h after birth. The amino acid levels of these mice are normal at birth but immediately decrease thereafter. These experiments illustrate the important role of autophagy in response to starvation.

The maintenance of the amino acid pool during starvation is important for the production of subsets of proteins needed for adaptation to starvation conditions. Starved yeast cells upregulate the synthesis of chaperones and enzymes for amino acid synthesis, which is severely affected in autophagy-defective mutants (Onodera and Ohsumi 2005). This defect might account for the low survival rate of these mutants under starvation conditions (Tsukada and Ohsumi 1993).

Amino acids produced by autophagy can also be utilized for energy production. Although amino acids are generally considered poor fuel sources, they can be metabolized through the tricarboxylic acid (TCA) cycle to produce energy. This is particularly apparent in muscle (Shimomura et al. 2004), but may also be the case in cultured cells (Lum et al. 2005). Indeed, the autophagy-deficient phenotype of IL-3-dependent cells can be restored by methylpyruvate, a cell-permeable substrate for the TCA, supporting the idea that autophagy produces energy. In addition to direct energy production, amino acids generated by autophagy can be used for gluconeogenesis, which is an important physiological response to starvation. One of the well-known pathways is the glucose-alanine cycle. Under starvation conditions, peripheral tissues such as muscle secrete alanine, which is converted to glucose in the liver. However, the extent that autophagy contributes to this pathway remains unknown.

2.2 Intracellular Quality Control

Although massive induction following nutrient withdrawal is the most prominent feature of autophagy, this process also occurs constitutively at low levels under normal growth conditions. The primary role of this basal autophagy is seemingly not the maintenance of the amino acid pool, because the intracellular amino acid pool is not affected in autophagy-deficient cells and animals as long as sufficient nutrients are available (Kuma et al. 2004; Komatsu et al. 2005; Onodera and Ohsumi 2005). However, recent mouse genetic studies have revealed that basal autophagy is quite important for intracellular quality control through the constitutive turnover of cytoplasmic components. Abnormal ubiquitinated proteins and organelles accumulate in the cytoplasm immediately after the deletion of autophagy genes such as *Atg5* and *Atg7* in the liver (Komatsu et al. 2005; Hara et al. 2006) and nervous system (Hara et al. 2006; Komatsu et al. 2006). These proteins are present in both the diffuse cytosolic and intracellular protein-aggregated forms. Interestingly, ubiquitinated proteins accumulate extensively in the liver, neurons and some endocrine glands, and much more slightly in other tissues such as the skeletal muscle, heart, and kidney (Hara et al. 2006). Since protein turnover in the liver is very high, it is not surprising that basal autophagy is very important in this organ. However, the reason underlying the critical importance of autophagy in the brain remains unclear. Autophagic activity in the brain is very low and is not induced during starvation (Mizushima et al. 2004; Nixon et al. 2005). Quality control may be more important in nondividing, quiescent cells than in rapidly dividing cells. In agreement with this concept, primary mouse embryonic fibroblasts prepared from $Atg5^{-/-}$ mice do not show protein aggregates at early phases, but do so in later senescent phases. Intracellular accumulation of abnormal proteins is also observed in *Atg7*-deficient *Drosophila* (Juhasz et al. 2007).

Basal autophagy is apparently crucial for cellular homeostasis. Liver-specific $Atg7^{-/-}$ mice develop hepatomegaly and hepatic failure (Komatsu et al. 2005), and neural cell-specific *Atg5* and *Atg7* knockout mice show neurodegeneration accompanied by progressive motor deficits (Hara et al. 2006; Komatsu et al. 2006). These phenotypes are cell autonomous because Purkinje cell-specific *Atg5* and *Atg7* knockout mice show Purkinje cell degeneration (Komatsu et al. 2007a; Nishiyama et al. 2007). Abnormal proteins and organelles are also detected in heart-specific *Atg5*-deficient mice (Nakai et al. 2007). Therefore, autophagy serves as a housekeeper under normal conditions in order to prevent cell degeneration, particularly in the nervous tissue, even if animals do not express disease-associated mutant (aggregate-prone) proteins.

Although the accumulation of protein aggregates and autophagic vacuoles is a hallmark of many neurodegenerative diseases such as Alzheimer's disease (Okamoto et al. 1991; Cataldo et al. 1996), polyglutamine (CAG) repeat diseases (Petersen et al. 2001; Ravikumar et al. 2002), and Parkinson's disease (Anglade et al. 1997), it remains unknown whether autophagy is indeed involved in the pathogenesis of these diseases. In some familial neurodegenerative diseases such as amyotrophic lateral sclerosis-like motor disease and frontotemporal dementia, the causative mutations in dynein and *CHMP2B* do indeed affect autophagosome–lysosome fusions, which should impair autophagic clearance of abnormal proteins (Ravikumar et al. 2005; Filimonenko et al. 2007; Lee et al. 2007). Irrespective of whether the autophagy defect is the direct cause or not, autophagy could be a good therapeutic target in these neurodegenerative diseases (Rubinsztein 2006). Inhibitors of Tor, a potent endogenous suppressor of autophagy, have been shown to be effective in attenuating symptoms in fly and mouse Huntington disease models (Ravikumar et al. 2004). In addition, other molecules such as lithium and trehalose were shown to modulate autophagy (Sarkar et al. 2005; Sarkar et al. 2007a; Zhang et al. 2007a). Furthermore, small-molecule enhancers of the cytostatic effects of rapamycin (SMERs) enhance autophagy in an mTOR-independent manner, and accelerate the clearance of mutant huntingtin and α-synuclein in a fly Huntington disease model (Sarkar et al. 2007b). Finally, lithium, which induces autophagy by inhibiting inositol monophosphatase independently of mTOR, delays progression of amyotrophic lateral sclerosis in humans (Fornai et al. 2008). To achieve the maximum effect, combination therapy using lithium and mTOR inhibitors has been proposed (Sarkar et al. 2008). Since most neurodegenerative diseases progress slowly, slight modulation of autophagy could produce dramatic effects on disease prognosis.

2.3 Selective Degradation by Autophagy

Whether abnormal proteins and inclusion bodies are selectively degraded by autophagy has been a continuing issue for debate. In the temporary controlled liver-specific *Atg5* knockout model, a loss of autophagy first leads to the accumulation of diffuse ubiquitinated proteins in the cytosol followed by the generation of inclusion bodies (Hara et al. 2006). This suggests that the accumulation of inclusion bodies in autophagy-deficient models is a secondary phenomenon, and large inclusions are not primary substrates. If protein turnover is generally impaired by random autophagy, proteins would have more opportunities to be damaged, misfolded, ubiquitinated and finally aggregated.

However, these studies do not rule out the possibility that oligomerized misfolded proteins or ubiquitinated proteins might be selectively incorporated into autophagosomes. Recently, it has been proposed that p62/SQSTM1 may serve as an adaptor protein for mediating the binding of ubiquitinated proteins by autophagosomes (Bjørkøy et al. 2005; Pankiv et al. 2007). Apart from the known functions of p62 in various signaling pathways (Wooten et al. 2006; Moscat et al. 2007), it can also bind both ubiquitin and LC3. Therefore, the LC3-p62 complex on the inner membrane of an autophagosome may recruit ubiquitinated proteins into autophagosomes (Fig. 1). However, the extent to which this pathway contributes to the degradation of ubiquitinated proteins under normal conditions remains unclear. A recent study showed that K63-ubiquitinated proteins accumulate in $p62^{-/-}$ mouse brain, though this may be due to the reduced activity of a K63-deubiquitinating enzyme called cylindromatosis tumor suppressor (CYLD), which seems to be independent of autophagy (Wooten et al. 2008).

p62 is mainly degraded by autophagy together with LC3, but accumulates excessively in autophagy-deficient cells (Wang et al. 2006; Komatsu et al. 2007b; Nakai et al. 2007). Interestingly, the maintenance of p62 expression at certain levels by autophagy is critically important for cellular homeostasis. The liver enlargement and dysfunction found in liver-specific *Atg7* knockout mice are significantly rescued by simultaneous ablation of p62 (Komatsu et al. 2007b). Ubiquitin-positive inclusion bodies are not generated in $Atg7^{-/-}p62^{-/-}$ mouse liver, suggesting that the excess amount of p62 accounts for the generation of the inclusions and hepatocyte damage. However, deletion of p62 does not alter the clinical course of neural cell-specific *Atg7* knockout mice (Komatsu et al. 2007b). Therefore, basal autophagy is important for the degradation of not only p62 and its interacting proteins but also other proteins.

Selective degradation by autophagy has been also demonstrated for yeast Ald6 (Onodera and Ohsumi 2004), peroxisomes (Luiken et al. 1992; Iwata et al. 2006), mitochondria (Kim et al. 2007), ribosomes (Kraft et al. 2008) and invading bacteria (Levine and Deretic 2007; Schmid and Münz 2007).

2.4 Development and Cell Death

Autophagy has been reported to be important for the development of various organisms. For example, yeast autophagy mutants are defective in spore formation during starvation (Tsukada and Ohsumi 1993), and autophagy mutants of *D. discoideum* are defective in multicellular development (Otto et al. 2003). Dauer formation is also affected in *Caenorhabditis elegans* autophagy mutants (Melendez et al. 2003). These findings might suggest that autophagy is important for nutrient mobilization during these remodeling processes because these developmental events occur under starvation conditions. However, this idea may be reconsidered based on a recent *Drosophila* study (Juhasz et al. 2007). Previous studies of *Drosophila* revealed that several autophagy mutants show premature death from the third larval to the pupal stages (Juhasz et al. 2003; Scott et al. 2004). It could thus be interpreted that larval tissues are degraded to produce nutrients for generating adult tissues in a pupa. Indeed, in dying larval tissues such as the salivary glands, massive autophagy is observed (Baehrecke 2003). However, recently generated *Atg7*-deficient *Drosophila* have been shown to be viable, although autophagy is virtually suppressed and adult flies are sensitive to nutrient and oxidative stresses (Juhasz et al. 2007). The pupal period is extended but the larval-adult midgut transition proceeds normally. The previously reported *Drosophila* mutants such as the *Atg1* mutant may have defects beyond autophagy. Therefore, some other pathways may compensate for the defective protein breakdown in the autophagy-defective mutant during metamorphosis.

Autophagy has also been thought to be a type of cell death-inducing process, especially during development. It is sometimes referred to as "type 2 cell death" or "autophagic cell death." During development, autophagy occurs in dying cells in various embryonic tissues (Baehrecke 2005; Debnath et al. 2005; Levine and Yuan 2005). However, the role of autophagy in cell death execution has been an issue of great controversy, while that of autophagy in cell survival has been well documented. In apoptosis-deficient cells, autophagy contributes to cell death induced by apoptogenic stimuli such as genotoxic stress and staurosporine, and a caspase inhibitor (z-VAD) (Shimizu et al. 2004; Yu et al. 2004). There have been no lines of evidence that autophagy induces cell death during physiological development in mammals because *Atg5*$^{-/-}$ and *Atg7*$^{-/-}$ mice are born grossly normal at birth. However, a recent *Drosophila* study revealed that autophagy is indeed required for the complete degradation of a dying salivary gland in a pupa (Berry and Baehrecke 2007). Cells in the salivary glands rapidly die and whole salivary glands are degraded after pupa formation. This tissue destruction is at least partially mediated by autophagy, because the suppression of several *ATG* genes leads to incomplete degradation (Berry and Baehrecke 2007). Since caspase inhibition also partially suppresses cell death, both apoptosis and autophagic degradation may function in the rapid destruction of the salivary gland. Thus, the physiological role of autophagy in cell death is rather complicated and depends on the presenting situation. The term "autophagic cell death" may not be appropriate in certain cases, even if autophagy is detected in dying cells.

The role of autophagy in mammalian development has not been well understood because mice deficient for *Atg5* or *Atg7* can survive embryogenesis (Kuma et al. 2004; Komatsu et al. 2005). However, these studies overlook the requirement of autophagy during early developmental stages, when maternally inherited proteins remain in the cytoplasm of knockout oocytes. Indeed, autophagy is activated shortly after fertilization, which is essential for preimplantation development (Tsukamoto et al. 2008).

Autophagy is also involved in another step of cell death. Cells undergoing apoptosis expose phosphatidylserine (PS) at the cell surface, which is recognized by phagocytes. However, autophagy-defective cells cannot expose PS efficiently due to low levels of cellular ATP, resulting in the failure of dead cell clearance (Qu et al. 2007).

2.5 Tumor Suppression

The role of autophagy in tumorigenesis and cancer progression has been discussed for a long time. It may be considered that autophagy is also important for the survival of tumor cells, just like normal cells. However, many studies have suggested that autophagy instead acts as a tumor suppressor (Hippert et al. 2006; Jin and White 2007; Levine 2007; Mathew et al. 2007a).

The first genetic linkage between autophagy and cancer was indicated by a study on Beclin 1. Beclin 1 is a mammalian counterpart of yeast Atg6/Vps30, which is a component of the autophagy-related PI3–kinase complex (Liang et al. 1999; Kihara et al. 2001b; Kihara et al. 2001a). Beclin 1 was originally identified as an interacting partner of an antiapoptotic protein, Bcl-2 (Liang et al. 1998). Accordingly, Bcl-2 negatively regulates autophagy by binding with Beclin 1 (Pattingre et al. 2005). Importantly, Beclin 1 is monoallelically deleted in 40–75% of sporadic human breast and ovarian cancers (Liang et al. 1999). In addition, *beclin 1*$^{+/-}$ mice develop spontaneous tumors such as lung cancer, hepatocellular carcinoma and lymphoma (Qu et al. 2003; Yue et al. 2003). These studies revealed that Beclin 1 is a novel tumor suppressor. Recently, UVRAG (a Beclin 1-interacting protein; Ionov et al. 2004; Liang et al. 2006) and Atg4C were also shown to have tumor-suppressive roles (Mariño et al. 2007). Therefore, autophagy likely has a protective role against tumorigenesis.

There are two (not mutually exclusive) hypotheses as to why defective autophagy causes tumors. The first is that tumorigenesis is induced by an inflammatory response. As discussed above, autophagy is basically a protective mechanism and its deficiency causes necrotic cell death if the apoptotic pathway is also compromised. This phenomenon is particularly important under metabolic stress (ischemia and nutrient starvation) conditions, which is often observed at the center of solid tumors (Degenhardt et al. 2006). The resulting necrotic cell death induces an inflammatory response, which in turn promotes secondary tumorigenesis (Degenhardt et al. 2006). The second hypothesis is that autophagy can protect cells from genomic damage in a cell-autonomous manner (Karantza-Wadsworth et al. 2007; Mathew et al. 2007b). A high rate of genomic damage and instability were observed in *beclin 1*$^{+/-}$ and *Atg5*$^{-/-}$ kidney (Mathew et al. 2007b) and mammary epithelial cells during metabolic stress (Karantza-Wadsworth et al. 2007). Why autophagy is beneficial for genome protection is not fully understood. One possible explanation is that deficiency in autophagy, which is important for mitochondrial quality control, causes the accumulation of damaged mitochondria (Komatsu et al. 2006; Kim et al. 2007; Zhang et al. 2007b; Twig et al. 2008). In yeast, reactive oxygen species (ROS) tend to accumulate in autophagy-defective cells (Xiong et al. 2007). Also in mouse liver, an oxidative transcription factor, Nrf2, is activated if autophagy is impaired (Komatsu et al. 2007b). Such oxidative stress may promote DNA damage and ultimately tumorigenesis.

2.6 Anti-Aging

As discussed above, autophagy is involved in both nutrient regulation and intracellular quality control. Therefore, it can be assumed that an excess of nutrients suppresses intracellular clearance, while mild starvation promotes such clearance. Indeed, an inverse relationship between autophagy and aging has frequently been postulated (Bergamini et al. 2004; Dröge 2004; Levine and Klionsky 2004). Caloric restriction is the most effective method for extending the life spans of various species from yeast to mammals. The precise mechanisms underlying such extension are not completely understood, but autophagy could be one of the effectors. Autophagy or general protein degradation is upregulated during calorie restriction in rodents and worms (Ward 1988; Donati et al. 2001; Bergamini et al. 2004; Mörck and Pilon 2006). Furthermore, genetic studies have also suggested that autophagy is indeed important for extending life span. A *C. elegans* mutant lacking the insulin signaling gene *daf-2* shows a life-extension phenotype, which is suppressed if autophagy-related genes are simultaneously silenced (Melendez et al. 2003; Hars et al. 2007; Hansen et al. 2008). Similarly, the life-extending effect of calorie restriction is also impaired in *atg* gene-knockdown worms (Jia and Levine 2007; Hansen et al. 2008). Although both dietary restriction and *daf-2* mutation upregulate autophagy, their underlying mechanism appears to be different. Dietary restriction-induced autophagy requires the transcription factor Pha-4/FoxA, whereas *daf-2* mutation-induced autophagy requires neither Pha-4 nor Daf-16/FoxO (Hansen et al. 2008). Thus, autophagy induction may not be sufficient to extend life span, and a parallel pathway through Daf-16/FoxO is also required. The NAD-dependent deacetylase Sirt1, which is an evolutionally conserved regulator of life span, was also shown to play an important role in autophagy induction both in vitro and in vivo through the deacetylation of several Atg proteins by Sirt1 (Lee et al. 2008). Therefore, there appear to be complicated interrelations between autophagy and life span determination.

In aged cells, various damaged proteins are accumulated, which could be (at least partially) due to the seemingly reduced activity of both macroautophagy and CMA with age (Bergamini et al. 2004; Terman, 2006; Martinez-Vicente and Cuervo 2007). Therefore, autophagic activity may be a critical determinant of life span, and its modulation will be a potential strategy against aging.

3 Conclusion

Autophagy has a wide variety of physiological roles due to its several different modes (induced vs. constitutive) (Mizushima 2005) and steps (sequestration vs. degradation) (Mizushima 2007). In addition to the topics described in this review, recent studies have also demonstrated that autophagy is important for protection against heart failure and liver and muscle diseases. Furthermore, as this book highlights, autophagy is important for the degradation of not only self-proteins but also various microbes. Paradoxically, "degradation" is not the main function of autophagy in some circumstances. The autophagosome can also be used as a special site for the survival and replication of subsets of bacteria and viruses. Thus, as described in this book, autophagy has various roles in infection and immunology.

Acknowledgments This work was supported in part by Grants-in-Aid for Scientific Research from the Ministry of Education, Culture, Sports, Science and Technology of Japan. The author also thanks the Kato Memorial Bioscience Foundation and Toray Science Foundation for financial support.

References

Anglade P, Vyas S, Javoy-Agid F, Herrero MT, Michel PP, Marquez J, Mouatt-Prigent A, Ruberg M, Hirsch EC, Agid Y (1997) Apoptosis and autophagy in nigral neurons of patients with Parkinson's disease. Histol Histopathol 12:25–31

Baehrecke EH (2003) Autophagic programmed cell death in *Drosophila*. Cell Death Differ 10:940–945

Baehrecke EH (2005) Opinion: Autophagy: dual roles in life and death? Nat Rev Mol Cell Biol 6:505–510

Bergamini E, Cavallini G, Donati A, Gori Z (2004) The role of macroautophagy in the ageing process, anti-ageing intervention and age-associated diseases. Int J Biochem Cell Biol 36:2392–2404

Berry DL, Baehrecke EH (2007) Growth arrest and autophagy are required for salivary gland cell degradation in *Drosophila*. Cell 131:1137–1148

Bjørkøy G, Lamark T, Brech A, Outzen H, Perander M, Øvervatn A, Stenmark H, Johansen T (2005) p62/SQSTM1 forms protein aggregates degraded by autophagy and has a protective effect on huntingtin-induced cell death. J Cell Biol 171:603–614

Cataldo AM, Hamilton DJ, Barnett JL, Paskevich PA, Nixon RA (1996) Properties of the endosomal-lysosomal system in the human central nervous system: disturbances mark most neurons in populations at risk to degenerate in Alzheimer's disease. J Neurosci 16:186–199

Cuervo AM (2004) Autophagy: in sickness and in health. Trends Cell Biol 14:70–77

Debnath J, Baehrecke EH, Kroemer G (2005) Does autophagy contribute to cell death? Autophagy 1:66–74

Degenhardt K, Mathew R, Beaudoin B, Bray K, Anderson D, Chen G, Mukherjee C, Shi Y, Gelinas C, Fan Y, Nelson DA, Jin S, White E (2006) Autophagy promotes tumor cell survival and restricts necrosis, inflammation, and tumorigenesis. Cancer Cell 10:51–64

Donati A, Cavallini G, Paradiso C, Vittorini S, Pollera M, Gori Z, Bergamini E (2001) Age-related changes in the regulation of autophagic proteolysis in rat isolated hepatocytes. J Gerontol A Biol Sci Med Sci 56:B288–B293

Dröge W (2004) Autophagy and aging: importance of amino acid levels. Mech Ageing Dev 125:161–168

Filimonenko M, Stuffers S, Raiborg C, Yamamoto A, Malerod L, Fisher EM, Isaacs A, Brech A, Stenmark H, Simonsen A (2007) Functional multivesicular bodies are required for autophagic clearance of protein aggregates associated with neurodegenerative disease. J Cell Biol 179:485–500

Fornai F, Longone P, Cafaro L, Kastsiuchenka O, Ferrucci M, Manca ML, Lazzeri G, Spalloni A, Bellio N, Lenzi P, Modugno N, Siciliano G, Isidoro C, Murri L, Ruggieri S, Paparelli A (2008) Lithium delays progression of amyotrophic lateral sclerosis. Proc Natl Acad Sci USA 105:2052–2057

Hansen M, Chandra A, Mitic LL, Onken B, Driscoll M, Kenyon C (2008) A role for autophagy in the extension of lifespan by dietary restriction in *C. elegans*. PLoS Genet 4:e24

Hara T, Nakamura K, Matsui M, Yamamoto A, Nakahara Y, Suzuki-Migishima R, Yokoyama M, Mishima K, Saito I, Okano H, Mizushima N (2006) Suppression of basal autophagy in neural cells causes neurodegenerative disease in mice. Nature 441:885–889

Hars ES, Qi H, Ryazanov AG, Jin S, Cai L, Hu C, Liu LF (2007) Autophagy regulates ageing in *C. elegans*. Autophagy 3:93–95

Hippert MM, O'Toole PS, Thorburn A (2006) Autophagy in cancer: good, bad, or both? Cancer Res 66:9349–9351

Ionov Y, Nowak N, Perucho M, Markowitz S, Cowell JK (2004) Manipulation of nonsense mediated decay identifies gene mutations in colon cancer cells with microsatellite instability. Oncogene 23:639–645

Iwata J, Ezaki J, Komatsu M, Yokota S, Ueno T, Tanida I, Chiba T, Tanaka K, Kominami E (2006) Excess peroxisomes are degraded by autophagic machinery in mammals. J Biol Chem 281:4035–4041

Jia K, Levine B (2007) Autophagy is required for dietary restriction-mediated life span extension in C. elegans. Autophagy 3:597–599

Jin S, White E (2007) Role of autophagy in cancer: management of metabolic stress. Autophagy 3:28–31

Juhasz G, Erdi B, Sass M, Neufeld TP (2007) Atg7-dependent autophagy promotes neuronal health, stress tolerance, and longevity but is dispensable for metamorphosis in *Drosophila*. Genes Dev 21:3061–3066

Juhasz G, Csikos G, Sinka R, Erdelyi M, Sass M (2003) The *Drosophila* homolog of Aut1 is essential for autophagy and development. FEBS Lett 543:154–158

Kang C, You YJ, Avery L (2007) Dual roles of autophagy in the survival of *Caenorhabditis elegans* during starvation. Genes Dev 21:2161–2171

Karantza-Wadsworth V, Patel S, Kravchuk O, Chen G, Mathew R, Jin S, White E (2007) Autophagy mitigates metabolic stress and genome damage in mammary tumorigenesis. Genes Dev 21:1621–1635

Kihara A, Noda T, Ishihara N, Ohsumi Y (2001a) Two distinct Vps34 phosphatidylinositol 3-kinase complexes function in autophagy and carboxypeptidase Y sorting in *Saccharomyces cerevisiae*. J Cell Biol 152:519–530

Kihara A, Kabeya Y, Ohsumi Y, Yoshimori T (2001b) Beclin-phosphatidylinositol 3-kinase complex functions at the *trans*-Golgi network. EMBO Reports 2:330–335

Kim I, Rodriguez-Enriquez S, Lemasters JJ (2007) Selective degradation of mitochondria by mitophagy. Arch Biochem Biophys 462:245–253

Klionsky DJ (2007) Autophagy: from phenomenology to molecular understanding in less than a decade. Nat Rev Mol Cell Biol 8:931–937

Komatsu M, Wang QJ, Holstein GR, Friedrich VL Jr, Iwata J, Kominami E, Chait BT, Tanaka K, Yue Z (2007a) Essential role for autophagy protein Atg7 in the maintenance of axonal homeostasis and the prevention of axonal degeneration. Proc Natl Acad Sci USA 104:14489–14494

Komatsu M, Waguri S, Chiba T, Murata S, Iwata JI, Tanida I, Ueno T, Koike M, Uchiyama Y, Kominami E, Tanaka K (2006) Loss of autophagy in the central nervous system causes neurodegeneration in mice. Nature 441:880–884

Komatsu M, Waguri S, Ueno T, Iwata J, Murata S, Tanida I, Ezaki J, Mizushima N, Ohsumi Y, Uchiyama Y, Kominami E, Tanaka K, Chiba T (2005) Impairment of starvation-induced and constitutive autophagy in Atg7-deficient mice. J Cell Biol 169:425–434

Komatsu M, Waguri S, Koike M, Sou YS, Ueno T, Hara T, Mizushima N, Iwata JI, Ezaki J, Murata S, Hamazaki J, Nishito Y, Iemura SI, Natsume T, Yanagawa T, Uwayama J, Warabi E, Yoshida H, Ishii T, Kobayashi A, Yamamoto M, Yue Z, Uchiyama Y, Kominami E, Tanaka K (2007b) Homeostatic levels of p62 control cytoplasmic inclusion body formation in autophagy-deficient mice. Cell 131:1149–1163

Kraft C, Deplazes A, Sohrmann M, Peter M (2008) Mature ribosomes are selectively degraded upon starvation by an autophagy pathway requiring the Ubp3p/Bre5p ubiquitin protease. Nat Cell Biol 10:602–610

Kuma A, Hatano M, Matsui M, Yamamoto A, Nakaya H, Yoshimori T, Ohsumi Y, Tokuhisa T, Mizushima N (2004) The role of autophagy during the early neonatal starvation period. Nature 432:1032–1036

Lee IH, Cao L, Mostoslavsky R, Lombard DB, Liu J, Bruns NE, Tsokos M, Alt FW, Finkel T (2008) A role for the NAD-dependent deacetylase Sirt1 in the regulation of autophagy. Proc Natl Acad Sci USA 105:3374–3379

Lee JA, Beigneux A, Ahmad ST, Young SG, Gao FB (2007) ESCRT-III dysfunction causes autophagosome accumulation and neurodegeneration. Curr Biol 17:1561–1567

Levine B (2007) Cell biology: autophagy and cancer. Nature 446:745–747

Levine B, Klionsky DJ (2004) Development by self-digestion: molecular mechanisms and biological functions of autophagy. Dev Cell 6:463–477

Levine B, Yuan J (2005) Autophagy in cell death: an innocent convict? J Clin Invest 115:2679–2688

Levine B, Deretic V (2007) Unveiling the roles of autophagy in innate and adaptive immunity. Nat Rev Immunol 7:767–777

Levine B, Kroemer G (2008) Autophagy in the pathogenesis of disease. Cell 132:27–42

Liang C, Feng P, Ku B, Dotan I, Canaani D, Oh BH, Jung JU (2006) Autophagic and tumour suppressor activity of a novel Beclin1-binding protein UVRAG. Nat Cell Biol 8:688–699

Liang XH, Jackson S, Seaman M, Brown K, Kempkes B, Hibshoosh H, Levine B (1999) Induction of autophagy and inhibition of tumorigenesis by *beclin 1*. Nature 402:672–676

Liang XH, Kleeman LK, Jiang HH, Gordon G, Goldman JE, Berry G, Herman B, Levine B (1998) Protection against fatal Sindbis virus encephalitis by Beclin, a novel Bcl-2-interacting protein. J Virol 72:8586–8596

Luiken JJ, van den Berg M, Heikoop JC, Meijer AJ (1992) Autophagic degradation of peroxisomes in isolated rat hepatocytes. FEBS Lett 304:93–97

Lum JJ, Bauer DE, Kong M, Harris MH, Li C, Lindsten T, Thompson CB (2005) Growth factor regulation of autophagy and cell survival in the absence of apoptosis. Cell 120:237–248

Mörck C, Pilon M (2006) *C. elegans* feeding defective mutants have shorter body lengths and increased autophagy. BMC Dev Biol 6:39

Mariño G, Salvador-Montoliu N, Fueyo A, Knecht E, Mizushima N, Lopez-Otin C (2007) Tissue-specific autophagy alterations and increased tumorigenesis in mice deficient in ATG4C/autophagin-3. J Biol Chem 282:18573–18583

Martinez-Vicente M, Cuervo AM (2007) Autophagy and neurodegeneration: when the cleaning crew goes on strike. Lancet Neurol 6:352–361

Mathew R, Karantza-Wadsworth V, White E (2007a) Role of autophagy in cancer. Nat Rev Cancer 7:961–967

Mathew R, Kongara S, Beaudoin B, Karp CM, Bray K, Degenhardt K, Chen G, Jin S, White E (2007b) Autophagy suppresses tumor progression by limiting chromosomal instability. Genes Dev 21:1367–1381

Melendez A, Tallóczy Z, Seaman M, Eskelinen E-L, Hall DH, Levine B (2003) Autophagy genes are essential for Dauer development and life-span extension in C. elegans. Science 301:1387–1391

Mizushima N (2005) The pleiotropic role of autophagy: from protein metabolism to bactericide. Cell Death Differ 12 Suppl 2:1535–1541

Mizushima N (2007) Autophagy: process and function. Genes Dev 21:2861–2873

Mizushima N, Levine B, Cuervo AM, Klionsky DJ (2008) Autophagy fights disease through cellular self-digestion. Nature 451:1069–1075

Mizushima N, Yamamoto A, Matsui M, Yoshimori T, Ohsumi Y (2004) In vivo analysis of autophagy in response to nutrient starvation using transgenic mice expressing a fluorescent autophagosome marker. Mol Biol Cell 15:1101–1111

Moscat J, Diaz-Meco MT, Wooten MW (2007) Signal integration and diversification through the p62 scaffold protein. Trends Biochem Sci 32:95–100

Nakai A, Yamaguchi O, Takeda T, Higuchi Y, Hikoso S, Taniike M, Omiya S, Mizote I, Matsumura Y, Asahi M, Nishida K, Hori M, Mizushima N, Otsu K (2007) The role of autophagy in cardiomyocytes in the basal state and in response to hemodynamic stress. Nat Med 13:619–624

Nishiyama J, Miura E, Mizushima N, Watanabe M, Yuzaki M (2007) Aberrant membranes and double-membrane structures accumulate in the axons of Atg5-null Purkinje cells before neuronal death. Autophagy 3:591–596

Nixon RA, Wegiel J, Kumar A, Yu WH, Peterhoff C, Cataldo A, Cuervo AM (2005) Extensive involvement of autophagy in Alzheimer disease: an immuno-electron microscopy study. J Neuropathol Exp Neurol 64:113–122

Okamoto K, Hirai S, Iizuka T, Yanagisawa T, Watanabe M (1991) Reexamination of granulovacuolar degeneration. Acta Neuropathol (Berl) 82:340–345

Onodera J, Ohsumi Y (2004) Ald6p is a preferred target for autophagy in yeast, *Saccharomyces cerevisiae*. J Biol Chem 279:16071–16076

Onodera J, Ohsumi Y (2005) Autophagy is required for maintenance of amino acid levels and protein synthesis under nitrogen starvation. J Biol Chem 280:31582–31586

Otto GP, Wu MY, Kazgan N, Anderson OR, Kessin RH (2003) Macroautophagy is required for multicellular development of the social amoeba *Dictyostelium discoideum*. J Biol Chem 278:17636–17645

Pankiv S, Hoyvarde Clausen T, Lamark T, Brech A, Bruun JA, Outzen H, Øvervatn A, Bjørkøy G, Johansen T (2007) p62/SQSTM1 binds directly to Atg8/LC3 to facilitate degradation of ubiquitinated protein aggregates by autophagy. J Biol Chem 282:24131–24145

Pattingre S, Tassa A, Qu X, Garuti R, Liang XH, Mizushima N, Packer M, Schneider MD, Levine B (2005) Bcl-2 antiapoptotic proteins inhibit Beclin 1-dependent autophagy. Cell 122:927–939

Petersen A, Larsen KE, Behr GG, Romero N, Przedborski S, Brundin P, Sulzer D (2001) Expanded CAG repeats in exon 1 of the Huntington's disease gene stimulate dopamine-mediated striatal neuron autophagy and degeneration. Hum Mol Genet 10:1243–1254

Qu X, Zou Z, Sun Q, Luby-Phelps K, Cheng P, Hogan RN, Gilpin C, Levine B (2007) Autophagy gene-dependent clearance of apoptotic cells during embryonic development. Cell 128:931–946

Qu X, Yu J, Bhagat G, Furuya N, Hibshoosh H, Troxel A, Rosen J, Eskelinen E-L, Mizushima N, Ohsumi Y, Cattoretti G, Levine B (2003) Promotion of tumorigenesis by heterozygous disruption of the beclin 1 autophagy gene. J Clin Invest 112:1809–1820

Ravikumar B, Duden R, Rubinsztein DC (2002) Aggregate-prone proteins with polyglutamine and polyalanine expansions are degraded by autophagy. Hum Mol Genet 11:1107–1117

Ravikumar B, Acevedo-Arozena A, Imarisio S, Berger Z, Vacher C, O'Kane CJ, Brown SD, Rubinsztein DC (2005) Dynein mutations impair autophagic clearance of aggregate-prone proteins. Nat Genet 37:771–776

Ravikumar B, Vacher C, Berger Z, Davies JE, Luo S, Oroz LG, Scaravilli F, Easton DF, Duden R, O'Kane CJ, Rubinsztein DC (2004) Inhibition of mTOR induces autophagy and reduces toxicity of polyglutamine expansions in fly and mouse models of Huntington disease. Nat Genet 36:585–595

Rubinsztein DC (2006) The roles of intracellular protein-degradation pathways in neurodegeneration. Nature 443:780–786

Sarkar S, Davies JE, Huang Z, Tunnacliffe A, Rubinsztein DC (2007a) Trehalose, a novel mTOR-independent autophagy enhancer, accelerates the clearance of mutant huntingtin and alpha-synuclein. J Biol Chem 282:5641–5652

Sarkar S, Krishna G, Imarisio S, Saiki S, O'Kane CJ, Rubinsztein DC (2008) A rational mechanism for combination treatment of Huntington's disease using lithium and rapamycin. Hum Mol Genet 17:170–178

Sarkar S, Floto RA, Berger Z, Imarisio S, Cordenier A, Pasco M, Cook LJ, Rubinsztein DC (2005) Lithium induces autophagy by inhibiting inositol monophosphatase. J Cell Biol 170:1101–1111

Sarkar S, Perlstein EO, Imarisio S, Pineau S, Cordenier A, Maglathlin RL, Webster JA, Lewis TA, O'Kane CJ, Schreiber SL, Rubinsztein DC (2007b) Small molecules enhance autophagy and reduce toxicity in Huntington's disease models. Nat Chem Biol 3:331–338

Schmid D, Münz C (2007) Innate and adaptive immunity through autophagy. Immunity 27:11–21

Scott RC, Schuldiner O, Neufeld TP (2004) Role and regulation of starvation-induced autophagy in the *Drosophila* fat body. Dev Cell 7:167–178

Shimizu S, Kanaseki T, Mizushima N, Mizuta T, Arakawa-Kobayashi S, Thompson CB, Tsujimoto Y (2004) Role of Bcl-2 family proteins in a non-apoptotic programmed cell death dependent on autophagy genes. Nat Cell Biol 6:1221–1228

Shimomura Y, Murakami T, Nakai N, Nagasaki M, Harris RA (2004) Exercise promotes BCAA catabolism: effects of BCAA supplementation on skeletal muscle during exercise. J Nutr 134:1583S–1587S

Takeshige K, Baba M, Tsuboi S, Noda T, Ohsumi Y (1992) Autophagy in yeast demonstrated with proteinase-deficient mutants and conditions for its induction. J Cell Biol 119:301–311

Terman A (2006) Catabolic insufficiency and aging. Ann N Y Acad Sci 1067:27–36

Tsukada M, Ohsumi Y (1993) Isolation and characterization of autophagy-defective mutants of *Saccharomyces cerevisiae*. FEBS Lett 333:169–174

Tsukamoto S, Kuma A, Murakami M, Kishi C, Yamamoto A, Mizushima N (2008) Autophagy is essential for preimplantation development of mouse embryos. Science 321:117–120

Twig G, Elorza A, Molina AJ, Mohamed H, Wikstrom JD, Walzer G, Stiles L, Haigh SE, Katz S, Las G, Alroy J, Wu M, Py BF, Yuan J, Deeney JT, Corkey BE, Shirihai OS (2008) Fission and selective fusion govern mitochondrial segregation and elimination by autophagy. EMBO J 27:433–446

Vabulas RM, Hartl FU (2005) Protein synthesis upon acute nutrient restriction relies on proteasome function. Science 310:1960–1963

Wang QJ, Ding Y, Kohtz S, Mizushima N, Cristea IM, Rout MP, Chait BT, Zhong Y, Heintz N, Yue Z (2006) Induction of autophagy in axonal dystrophy and degeneration. J Neurosci 26:8057–8068

Ward WF (1988) Food restriction enhances the proteolytic capacity of the aging rat liver. J Gerontol 43:B121–124

Wooten MW, Hu X, Babu JR, Seibenhener ML, Geetha T, Paine MG, Wooten MC (2006) Signaling, polyubiquitination, trafficking, and inclusions: sequestosome 1/p62's role in neurodegenerative disease. J Biomed Biotechnol 2006:62079

Wooten MW, Geetha T, Babu JR, Seibenhener ML, Peng J, Cox N, Diaz-Meco MT, Moscat J (2008) Essential role of sequestosome 1/p62 in regulating accumulation of Lys^{63}-ubiquitinated proteins. J Biol Chem 283:6783–6789

Xiong Y, Contento AL, Bassham DC (2007) Disruption of autophagy results in constitutive oxidative stress in *Arabidopsis*. Autophagy 3:257–258

Yu L, Alva A, Su H, Dutt P, Freundt E, Welsh S, Baehrecke EH, Lenardo MJ (2004) Regulation of an ATG7-beclin 1 program of autophagic cell death by caspase-8. Science 304:1500–1502

Yue Z, Jin S, Yang C, Levine AJ, Heintz N (2003) Beclin 1, an autophagy gene essential for early embryonic development, is a haploinsufficient tumor suppressor. Proc Natl Acad Sci USA 100:15077–15082

Zhang L, Yu J, Pan H, Hu P, Hao Y, Cai W, Zhu H, Yu AD, Xie X, Ma D, Yuan J (2007a) Small molecule regulators of autophagy identified by an image-based high-throughput screen. Proc Natl Acad Sci USA 104:19023–19028

Zhang Y, Qi H, Taylor R, Xu W, Liu LF, Jin S (2007b) The role of autophagy in mitochondria maintenance: characterization of mitochondrial functions in autophagy-deficient *S. cerevisiae* strains. Autophagy 3:337–346

Autophagy and Lymphocyte Homeostasis

Heather H. Pua and You-Wen He

Contents

1 Introduction 86
2 Regulation of T Cell Homeostasis by Cytokines and MHC/Peptide Ligands 86
  2.1 Cytokines in T Cell Homeostasis 87
  2.2 TCR-MHC Interactions in T Cell Homeostasis 87
3 Regulation of T Cell Homeostasis by the Intrinsic and Extrinsic Apoptotic Pathways 88
  3.1 Extrinsic Pathway Apoptosis in T Cell Homeostasis 88
  3.2 Intrinsic Apoptosis Pathway in T Cell Homeostasis 89
4 Autophagy Induction in T Lymphocytes 90
5 Autophagy: Dual Roles in T Cell Survival and Death 92
  5.1 Autophagy Contributes to Homeostatic T Cell Survival In Vivo 92
  5.2 Autophagy Contributes to T Cell Death 94
  5.3 Autophagy as a Paradoxical Mediator of Both Survival and Death in T Cells? 95
6 Autophagy in T Cell Proliferation 95
  6.1 Autophagosome Formation in T Cell Proliferation 96
  6.2 A Role for Autophagy Genes in T Cell Proliferation Outside of Autophagosome Formation? 97
7 Autophagy in B Lymphocytes 97
8 Conclusion 98
References 99

Abstract Lymphocyte homeostasis is tightly regulated in vivo by various factors including cytokines, antigens, and costimulatory signals. Central to this regulation is the intricate balance between survival and apoptosis determined by pro- and antiapoptotic factors, including Bcl-2/Bcl-xL of the Bcl-2 family in the intrinsic death pathway and Fas/FADD of the TNF death receptor superfamily in the extrinsic death pathway. Recent studies have identified a critical role for autophagy, a well-conserved catabolic process in eukaryotic cells, in T and B lymphocyte homeostasis. Autophagy is essential for mature T lymphocyte survival and proliferation. In addition, autophagy can promote T cell death in defined physiologic or pathologic

H.H. Pua and Y.-W. He (✉)
Department of Immunology, Duke University Medical Center, Box 3010, Durham, NC, 27710, USA
e-mail: he000004@mc.duke.edu

B. Levine et al. (eds.), *Autophagy in Infection and Immunity,*
Current Topics in Microbiology and Immunology 335,
DOI 10.1007/ 978-3-642-00302-8_4,

conditions. Autophagy also contributes to the survival of subsets of B lymphocytes, including developing pre-B cells as well as B1 B cells in vivo. Thus, autophagy represents a novel pathway regulating both developing and mature lymphocytes. Future studies are required to investigate the role of autophagy in regulating T and B cell homeostasis during immune responses to pathogens, as well as to define the mechanisms by which autophagy regulates lymphocyte death and survival.

1 Introduction

Lymphocytes are essential regulators and effectors of the adaptive immune system. In collaboration with innate and antigen-presenting immune cells, T cells and B cells provide protective immune responses to a diversity of pathogens and form long-lived immunologic memory. In order to maintain a competent and robust immune system, the peripheral pool of mature lymphocytes is tightly regulated by a careful balance of cell production, survival, death and proliferation. The physiologically relevant balance of these processes is defined as lymphocyte homeostasis (Crowley et al. 2008; Jameson 2002; Marrack et al. 2000; Plas et al. 2002). In healthy adult animals, homeostasis stabilizes the population of mature T and B cells. During periods of induced or naturally occurring lymphopenia, lymphocyte homeostasis can also restore T and B cell numbers (Agenes and Freitas 1999; Beutner and MacDonald 1998; Cabatingan et al. 2002; Ernst et al. 1999; Goldrath and Bevan 1999a; Miller and Stutman 1984; Viret et al. 1999).

The regulation of lymphocyte expansion and death during immune responses also contributes critically to the maintenance of homeostasis in the adaptive immune system. Although effector T cells can expand up to 1,000-fold during the primary immune response (Blattman et al. 2002; Murali-Krishna et al. 1998), the programmed death of a majority of these cells limits the accumulation of total T cell numbers (Badovinac and Harty 2006). Regulated contraction of effector cells also ensures that the peripheral T cell repertoire remains diverse in preparation for future antigen encounters. These homeostatic mechanisms provide a large and diverse lymphocyte population in secondary lymphoid organs while limiting pathology due to the dysregulation of T or B cell numbers. Extensive studies have identified many factors that regulate peripheral lymphocyte homeostasis (Crowley et al. 2008; Jameson 2002; Marrack et al. 2000; Plas et al. 2002). Recent studies suggest that autophagy, a fundamental intracellular process, provides a novel mechanism regulating T lymphocyte homeostasis (Pua and He 2007).

2 Regulation of T Cell Homeostasis by Cytokines and MHC/Peptide Ligands

T cell homeostasis is regulated throughout the lifespan of T lymphocytes. Robust T cell development in the thymus in young animals is required for the production of mature T cells to populate secondary lymphoid organs. The homeostatic survival and proliferation of mature T lymphocytes is profoundly regulated in peripheral lymphoid organs. This regulation depends on critical environmental signals that help to define the mature T cell niche (Jameson 2002; Marrack et al. 2000; Plas et al. 2002). Although multiple essential extrinsic factors may contribute to T cell homeostasis, attention has been largely focused on the role of cytokines and the T cell receptor (TCR) in the regulation of T cell development as well as the maintenance of naïve and memory T cell pools.

2.1 Cytokines in T Cell Homeostasis

Cytokines are small secreted proteins that act on cellular targets to promote diverse biological responses, including survival, differentiation, proliferation, and migration. A group of cytokines sharing the common γc chain receptor subunit including interleukin-2 (IL-2), IL-7, and IL-15 support T cell survival and play important roles in regulating T cell homeostasis (Boise et al. 1995; He and Malek 1998; Rathmell et al. 2001; Vella et al. 1997). Many T cell subsets, including double negative (DN) thymocytes, single positive (SP) thymocytes, naïve T cells and memory T cells, express IL-7Rα on their surface (Goldrath et al. 2002; Sudo et al. 1993; Tan et al. 2001). IL-15Rβ is also expressed at a high level on $CD8^+$ memory T cells (Zhang et al. 1998b). Consistent with this expression pattern, thymocyte development as well as the pool of mature T cells is severely compromised in IL-$7^{-/-}$ mice, IL-$7R^{-/-}$ mice, or mice treated with an antagonistic antibody for the IL-7R (Schluns et al. 2000; Sudo et al. 1993; Tan et al. 2001). In naïve and memory T cells, this reflects a role for IL-7 both in promoting proliferation as well as supporting the survival of cells in mice with filled peripheral T cell compartments (Goldrath et al. 2002; Kondrack et al. 2003; Lenz et al. 2004; Schluns et al. 2000; Seddon et al. 2003; Seddon and Zamoyska 2002; Tan et al. 2001; Tan et al. 2002). In parallel, IL-$15^{-/-}$ or IL-$15R^{-/-}$ animals have significant defects in $CD8^+$ memory T cells with prominent alterations in the proliferation and long-term maintenance of these memory cells (Becker et al. 2002; Goldrath et al. 2002; Judge et al. 2002; Kennedy et al. 2000; Lodolce et al. 1998; Tan et al. 2002).

Importantly, IL-7 and IL-15 not only promote T cell survival and proliferation but they also limit the size of the peripheral T cell pool, helping to define a homeostatic niche for T cells in the spleen and lymph nodes. Administration of recombinant IL-7 and transgenic overexpression of IL-7 both increase T cell numbers in vivo, suggesting that competition for endogenous cytokine helps to limit the population of mature T cells (El Kassar et al. 2004; Kieper et al. 2002; Morrissey et al. 1991). Transgenic overexpression of IL-15 can similarly expand target lymphocyte populations, suggesting that its expression is also limiting in vivo, though the activity of IL-15 is more selective within the T cell compartment, predominantly affecting memory $CD8^+$ T cells (Fehniger et al. 2001; Marks-Konczalik et al. 2000).

2.2 TCR–MHC Interactions in T Cell Homeostasis

TCR–major histocompatability complex (MHC) interactions play a critical role in the positive and negative selection of thymocytes during development as well as the response of mature T cells to foreign antigens in the context of infection or immunization (Goldrath and Bevan 1999b). However, the interaction of the TCR with MHC molecules also plays a role in naïve T cell homeostasis. Both the grafting of wild type thymi as well as the adoptive transfer of T lymphocytes into $MHCI^{-/-}$ and $MHCII^{-/-}$ hosts reveal a critical role for these molecules in the proliferation and expansion of T cells in response to lymphopenia (Bender et al. 1999; Brocker 1997; Goldrath and Bevan 1999a; Kirberg et al. 1997; Nesic and Vukmanovic 1998). Consistent with productive homeostatic TCR–MHC interactions, a low level of MHC-dependent constitutive phosphorylation of the TCR CD3ζ chain occurs in mature T cells, suggesting tonic signaling in the periphery (Dorfman and Germain 2002; van Oers et al. 1994; Witherden et al. 2000). In addition, the inducible deletion of TCRα chain expression results in a profound reduction in the half-life of $CD8^+$ $TCR^{-/-}$ T cells when compared to undeleted controls (Polic et al. 2001). Interestingly, memory $CD4^+$ and $CD8^+$ T cell homeostatic proliferation as well as long-term survival appears to be independent of MHC-peptide complexes, though tonic stimulation may be required to maintain the functional capacity of memory cells (Kassiotis et al. 2002; Lau et al. 1994; Murali-Krishna et al. 1999; Swain et al. 1999).

3 Regulation of T Cell Homeostasis by the Intrinsic and Extrinsic Apoptotic Pathways

Apoptosis is defined as a programmed form of cellular death that results in the orderly destruction of cell viability with characteristic morphologic features including nuclear condensation, DNA cleavage, membrane blebbing, cell shrinkage, and mitochondrial depolarization (Danial and Korsmeyer 2004). Although there are many initiators and executioners of programmed cell death, central to most forms of apoptosis is the activation of a well-conserved family of cysteine proteases termed caspases. Caspases can be activated by either the intrinsic or extrinsic cell death pathway, and both forms of apoptosis have a role in T cell biology (Zhang et al. 2005a).

3.1 Extrinsic Pathway Apoptosis in T Cell Homeostasis

Classically, the initiation of the extrinsic apoptotic pathway depends on the ligation of tumor necrosis factor (TNF) receptor superfamily death receptors, and results in caspase 8 activation through adaptor molecules including Fas-associated death domain (FADD) and tumor necrosis factor receptor-associated death domain (TRADD). In thymocytes, death receptor activity may contribute to the elimination of cells that fail to rearrange the TCRβ chain early in development as well as auto-reactive SP thymocytes undergoing negative selection, although the relative contribution of individual death receptors is limited and remains somewhat controversial

(Corazza et al. 2004; Cretney et al. 2003; Diehl et al. 2004; Erickson et al. 1994; Kishimoto et al. 1998; Lamhamedi-Cherradi et al. 2003; Newton et al. 2000; Page 1999; Page et al. 1998; Pfeffer et al. 1993; Wang et al. 2001). While in vitro models of activation-induced cell death generated much interest in a possible role for the death receptor Fas in effector T cell death, Fas and TNF signaling are not required for the normal contraction of an acute immune response in vivo (Nguyen et al. 2000; Reich et al. 2000). However, the death receptor tumor necrosis factor-related apoptosis-inducing ligand (TRAIL) contributes to cell death during the secondary expansion of $CD8^+$ T cells that are primed in the absence of $CD4^+$ help, as may occur in certain pathologic conditions such as HIV infection (Janssen et al. 2005). In addition, recent data has demonstrated an important role for death receptors and extrinsic apoptosis in maintaining T cell homeostasis during chronic infections (Hughes et al. 2008; Weant et al. 2008).

3.2 *Intrinsic Apoptosis Pathway in T Cell Homeostasis*

Extensive studies have demonstrated that T lymphocyte homeostasis is critically regulated by the intrinsic apoptotic pathway in vivo. In the intrinsic apoptosis pathway, cellular stresses, which classically include genotoxic damage and growth factor withdrawal, lead to the activation of B cell leukemia/lymphoma 2 (Bcl-2) family members, permeabilization of the mitochondria, and downstream caspase activity. The important role of the intrinsic pathway in T cell biology has largely been demonstrated through examining the role of pro- and antiapoptotic Bcl-2 family proteins.

Three antiapoptotic proteins of the Bcl-2 family, including Bcl-2, Bcl-xL, and Mcl-1, have an important role in T cell homeostasis in vivo. The loss of either Bcl-2 or Mcl-1 expression in developing T cells results in an approximately fivefold reduction in the number of thymocytes and peripheral T cells in mice (Dzhagalov et al. 2008; Matsuzaki et al. 1997; Nakayama et al. 1993; Opferman et al. 2003; Veis et al. 1993). *Bcl-2*$^{-/-}$ and *Mcl-1*$^{-/-}$ thymocytes and mature T cells demonstrate enhanced rates of cellular death under conditions that initiate the intrinsic apoptosis pathway, such as cytokine withdrawal, dexamethasone exposure, or irradiation. The regulated expression of both Bcl-2 and Mcl-1 provides a link between pro-survival cytokine signaling and the intrinsic apoptosis pathway in T cell homeostasis. The expressions of both Bcl-2 and Mcl-1 can be upregulated in T cells by IL-7 (Opferman et al. 2003; Vivien et al. 2001; von Freeden-Jeffry et al. 1997), and the transgenic overexpression of Bcl-2 rescues the development of T cells in the thymus as well as the homeostatic survival of mature T cells in the spleen and lymph nodes of IL-7R$\alpha^{-/-}$ mice (Maraskovsky et al. 1997).

Consistent with its expression in double positive (DP) thymocytes but not naïve resting T cells, the antiapoptotic protein Bcl-xL has a role in promoting DP cell survival (Broome et al. 1995; Motoyama et al. 1995; Zhang and He 2005). Interestingly, although Bcl-xL is highly upregulated during T cell activation, its expression is not required for T cell responses or memory cell formation in vivo

(Zhang and He 2005). The role of Mcl-1 and Bcl-2 during immune responses and in memory formation remains largely unexamined due to the developmental defects in genetic models examined to date.

Pro-apoptotic Bcl-2 family members also have an important role in regulating T cell homeostasis. Bim, Bax, and Bak all contribute to thymocyte death in models examining negative selection in vitro and in vivo (Bouillet et al. 2002; Rathmell et al. 2002; Villunger et al. 2004). In the periphery, although the multidomain pro-apoptotic family members have largely redundant roles in T cell homeostasis (Knudson and Korsmeyer 1997; Knudson et al. 1995; Rathmell et al. 2002), *Bax*$^{-/-}$*, *Bak*$^{-/-}$* mice and bone marrow chimeras demonstrate an expanded T cell compartment with defects in negative selection as well as impaired sensitivity to intrinsic pathway apoptosis in vitro (Lindsten et al. 2000; Rathmell et al. 2002). In these mice, the number of effector/memory T cells is increased and they accumulate due to a defect in the death of effector cells after immune responses. Bim also plays a critical role in the death of activated cells, as expanded antigen-specific *Bim*$^{-/-}$ T cells persist in vivo after an immune response due to an impairment of the contraction phase (Hildeman et al. 2002; Pellegrini et al. 2003; Wojciechowski et al. 2006).

4 Autophagy Induction in T Lymphocytes

Autophagy is a well-conserved catabolic process in eukaryotic cells with diverse functional roles both outside and within the immune system (Klionsky and Emr 2000; Levine and Kroemer 2008; Mizushima et al. 2002; Schmid and Munz 2007). Defined by the de novo formation of specialized double-membrane vesicles within the cytoplasm of cells, an explosion of research examining the molecular and cellular regulation of autophagy has followed the recent discovery of a network of genes required for autophagosome formation (Thumm et al. 1994; Tsukada and Ohsumi 1993). Although T lymphocytes are small cells with limited cytoplasmic volume, T lineage cells do express essential autophagy genes and form autophagosomes. Transmission electron micrograph studies have identified characteristic double-membrane and cytoplasm-containing autophagic vacuoles in primary human and mouse T cells (Fig. 1) (Espert et al. 2006; Gerland et al. 2004; Jia et al. 1997; Li et al. 2006; Pua et al. 2007).

Autophagic activity may also be regulated at different developmental stages in T lymphocytes. Within the thymus, transcription of the autophagy genes *Atg5*, *beclin 1,* and *LC3* is most robustly detected early in thymocyte development within the DN faction of cells (Pua et al. 2007). Studies in a *beclin 1-GFP* bacterial artificial chromosome transgenic line also report varying protein levels of this autophagy gene, with detectible expression in DN1, DN2, DN3, and SP cells but not DN4 and DP thymocytes (Arsov et al. 2008). In addition, although there is a low level of constitutive autophagy in mature resting $CD4^+$ and $CD8^+$ T cells (Pua et al. 2007), autophagic activity increases in activated T cells. Stimulation of T cells with anti-CD3 in vitro for 24-48 h induces autophagy as measured by electron microscopy, LC3 processing, as well as GFP-LC3 puncta formation (Li et al. 2006; Pua et al. 2007). Chronic restimulation of human CD8⁺ T cells with mitogen in vitro also results in a fourfold increase in the number of cells with autophagic vacuoles by electron microscopy after 14 weeks of culture (Gerland et al. 2004).

The signaling pathways regulating the induction of autophagy in T lymphocytes are not well characterized. Nevertheless, some interesting clues have emerged (Fig. 1). As mentioned above, the activation of T cells by either mitogens or TCR stimulation results in the induction of autophagy (Gerland et al. 2004; Li et al. 2006; Pua et al. 2007). In differentiated effector CD4⁺ T cell cultures, increases in the number of cells containing multiple GFP-LC3⁺ autophagic puncta after anti-CD3 stimulation can be inhibited by JNK inhibitors or 3-methyladenine (3-MA) and enhanced by rapamycin and zVAD (Li et al. 2006). These results suggest that target of rapamycin (TOR), a central regulator of cell growth and metabolism, as well as Class III phosphatidylinositol 3-kinase (PI3K) and MAP kinase pathways may all provide important signals in T cell autophagy. Continued investigations will be required to further define the signaling pathways downstream of the antigen receptor that are necessary for autophagy induction. However, the TCR is not the only receptor associated with autophagy in T lymphocytes. The ability of HIV envelope glycoprotein to induce autophagosome formation through the receptor CXCR4 in CD4⁺ T cells suggests that G-protein-coupled receptors in general or chemokine receptors in specific may regulate autophagy in this lymphocyte population (Espert et al. 2006). Understanding the regulation of autophagosome formation in T cells is particularly important to determining the biologic function of autophagy in lymphocytes.

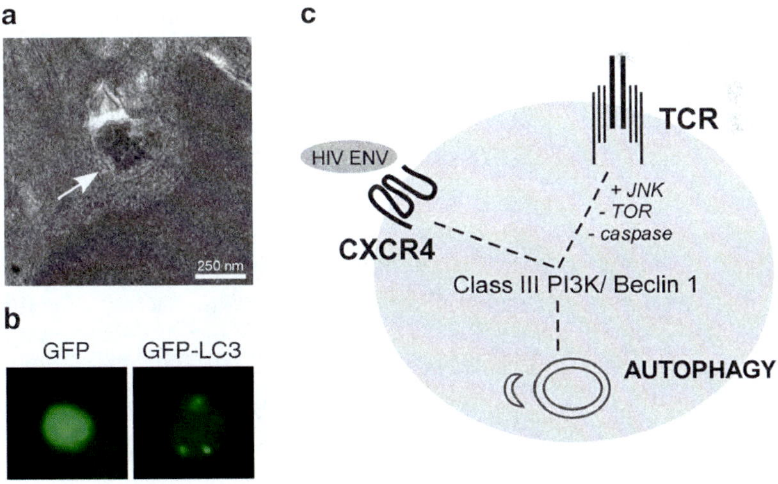

Fig. 1a–c Autophagosomes form in primary T lymphocytes and can be identified by transmission electron microscopy (**a**) and LC3 localization (**b**). The induction of autophagy can be regulated in T lymphocytes by signaling pathways downstream of the T cell receptor and chemokine receptor (**c**). The image in **a** was originally published in the *Journal of Experimental Medicine* (Pua et al. 2007) published by Rockefeller University Press, and the image in **b** is unpublished data from Pua and He

5 Autophagy: Dual Roles in T Cell Survival and Death

Autophagy was first identified as a pathway contributing to cell survival in yeast (Thumm et al. 1994; Tsukada and Ohsumi 1993). However, the relationship between autophagy, cell survival, and cell death is complex in higher eukaryotes. Deletion of essential autophagy genes in the neonate, the central nervous system, and the liver impairs cellular survival, demonstrating a clear cytoprotective role for autophagy (Hara et al. 2006; Komatsu et al. 2006; Komatsu et al. 2005; Kuma et al. 2004). Yet, in other model systems, autophagy may contribute actively to cell death. In particular, autophagy has been demonstrated to promote cell death during salivary gland generation in *Drosophila* as well as in cells where classical apoptotic pathways are disabled (Baehrecke 2005; Berry and Baehrecke 2007; Codogno and Meijer 2005). Given the dynamic and essential regulation of T cell survival during development, peripheral homeostasis and immune responses in vivo, it is particularly important to determine the roles of autophagy in primary T lymphocytes (Fig. 2).

5.1 Autophagy Contributes to Homeostatic T Cell Survival In Vivo

To investigate the role of autophagy in T cells in vivo, two mouse genetic model systems have been employed. Since mice lacking the essential autophagy genes *Atg5* or *Atg7* die within the first 24–48 h of birth (Komatsu et al. 2005; Kuma et al. 2004), the role of autophagy in T cells has been studied using $Atg5^{-/-}$ fetal liver chimeric mice as well as $Atg7^{f/f}$ Lck-Cre mice with a conditional deletion early in thymocyte development (Pua et al. 2007; Pua et al. 2009). In both mice, there are subtle but potentially significant changes within the thymus. In $Atg5^{-/-}$ fetal liver chimeras, although there are no obvious perturbations in the relative percentage of developing thymocyte subsets, there is ~50% reduction in the total number of thymocytes in autophagy-deficient chimeric mice (Pua et al. 2007). In $Atg7^{f/f}$ Lck-Cre mice, there is also a modest reduction in thymocyte numbers with a specific decrease in SP cells (Pua et al. 2009). Interestingly, thymocytes from $Atg5^{-/-}$ and $Atg7^{-/-}$ T cells display no increase in ex vivo apoptosis (Pua et al. 2007; Pua et al. 2009). Therefore, more detailed studies will be required to determine the function of autophagy in thymocytes, though it seems possible that $Atg5^{-/-}$ hematopoietic chimeras have a defect in progenitor cells and/or autophagy-deficient thymocytes may have a proliferation defect which limits cell numbers in vivo (see below).

The effects of deleting essential autophagy genes in T lineage cells are more dramatic in the periphery. There is a substantial reduction in peripheral T cell numbers in the spleen and lymph nodes of $Atg5^{-/-}$ and $Atg7^{f/f}$ Lck-Cre mice (Pua et al. 2007; Pua et al. 2009). Consistent with a role for autophagy in the homeostatic survival of mature T cells, $Atg5^{-/-}$ and $Atg7^{-/-}$ T cells have enhanced rates of apoptosis (Pua et al. 2007, 2009). In contrast to $Bax^{-/-}Bak^{-/-}$ hematopoietic progenitor cell lines that require autophagy to survive after growth factor withdrawal (Lum et al. 2005),

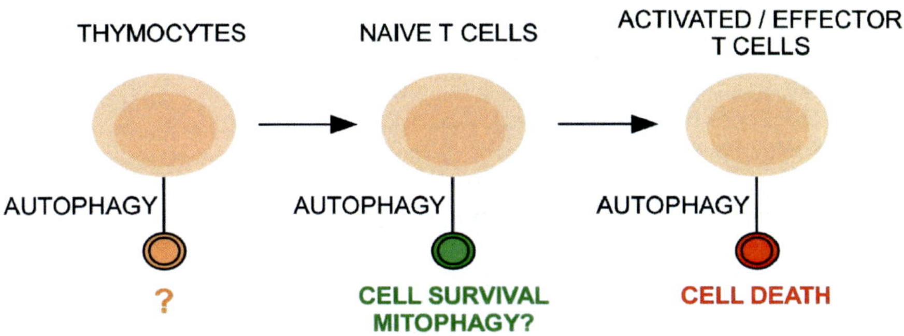

Fig. 2 The function of autophagosome formation varies depending on the developmental and activation status of T lymphocytes. Autophagy appears to largely promote survival in naïve T lymphocytes while potentially contributing to cell death in activated or effector T cells

autophagy-deficient primary T cells demonstrate an in vitro survival defect in the presence of abundant pro-survival IL-7 cytokine (Pua and He, unpublished observation). Therefore although peripheral T cell homeostasis requires IL-7 (El Kassar et al. 2004; Goldrath et al. 2002; Kieper et al. 2002; Morrissey et al. 1991; Schluns et al. 2000; Seddon and Zamoyska 2002; Sudo et al. 1993; Tan et al. 2001), the role of autophagy in T lymphocyte survival goes beyond providing an alternative survival pathway in a limited cytokine environment.

How then might autophagy contribute functionally to the survival of mature T cells? An alternative pro-survival function for autophagy depends upon its ability to degrade long-lived proteins and whole organelles that become toxic or damaging to cells. Both *Atg5*$^{-/-}$ and *Atg7*$^{-/-}$ mature T cells have enhanced mitochondrial content when compared to control cells (Pua et al. 2009). This defect appears to result from an inability to clear mitochondria by autophagy. Interestingly, the mitochondrial content in T lymphocytes is developmentally regulated. The exit of T lymphocytes from the thymus marks a transition from high mitochondrial content in thymocytes to low mitochondrial content in mature T cells in normal mice, and this change depends on autophagy (Pua et al. 2009). The autophagic sequestration of mitochondria, termed "mitophagy," which has been identified in both yeast (Baba et al. 1994; Kissova et al. 2007) and mammalian cells (Reipert et al. 1995; Takano-Ohmuro et al. 2000), results in functionally significant changes in mitochondrial content within cells. In vitro studies have demonstrated that induction of autophagy results in mitochondrial clearance while the inhibition of autophagy leads to an increase in mitochondrial numbers (Colell et al. 2007; Ravikumar et al. 2006; Terman et al. 2003). Much interest has focused on the role of autophagy in the clearance of organelles in epithelial cells during lens development and in erythrocytes during hematopoiesis, though the role of autophagy in these processes remains unclear given the normal clearance of lens and erythrocyte organelles in *Atg5*$^{-/-}$ mice (Bassnett 2002; Kent et al. 1966; Matsui et al. 2006; Takano-Ohmuro et al. 2000). However, the Bcl-2 family protein Nix has recently been demonstrated to play a critical role in mitochondrial clearance by autophagy in developing erythrocytes (Sandoval et al. 2008; Schweers et al. 2007). Along with results in T lymphocytes (Pua et al. 2009), these findings should continue to spark interest in the role of autophagic organelle clearance in the biology of cells of the hematopoietic system.

Consistent with a protective role for mitophagy in cells, mitochondrial dysfunction including osmotic swelling (Nowikovsky et al. 2007) and loss of mitochondrial potential (Elmore et al. 2001; Priault et al. 2005; Rodriguez-Enriquez et al. 2006; Sandoval et al. 2008) induce mitophagy. In addition, the ability to clear mitochondria by autophagy induction in cell culture and after pro-apoptotic stimuli prevents cell death, suggesting that mitochondrial clearance is an important pro-survival mechanism downstream of autophagosome formation (Colell et al. 2007; Ravikumar et al. 2006; Terman et al. 2003). Further supporting this notion, hepatocytes not only have increased mitochondrial content after the induced deletion of *Atg7* but also deformed mitochondria, suggesting organelle dysfunction (Komatsu et al. 2005). In autophagy-deficient T lymphocytes, enhanced mitochondrial content is associated with increased reactive oxygen species (ROS) production, apoptosis and imbalanced expression of pro- and antiapoptotic Bcl-2 family members (Pua et al. 2009). These results point toward an important pro-survival role for autophagy in clearing mitochondria, organelles that regulate T cell death during homeostasis through the intrinsic apoptosis pathway in vivo. Continued work is needed to determine the signals that regulate mitophagy in T cells. It will also be interesting to determine whether the clearance of mitochondria is the primary function of autophagy in naïve T cells, or whether there are other important autophagic functions in T lymphocytes. Finally, the specific signals or cellular stresses that contribute to death in autophagy-deficient T cells need to be determined to better understand the role of autophagy in the larger program of T cell homeostasis.

5.2 *Autophagy Contributes to T Cell Death*

In contrast to its pro-survival role in naive T cells during homeostasis, there is also evidence that autophagy contributes to cell death in T lymphocytes under specific physiologic and pathologic conditions. In particular, autophagy may contribute to cell death in activated T cells, reflecting a dynamic role for autophagy in T cell biology. Consistent with the induction of autophagy in T cells after mitogen or TCR stimulation (Gerland et al. 2004; Li et al. 2006; Pua et al. 2007), there appears to be an important functional role for autophagy in activated T cells. In a model of T cell senescence after chronic restimulation, the percentage of autophagosome-positive cells decreases within 48 h after restimulation (Gerland et al. 2004). The authors of this study suggest that the accumulation of autophagic vesicles in T cells may increase the susceptibility of these cells to cell death after stimulation. This raises the interesting possibility that autophagy could contribute to death in aged memory or chronically restimulated effector T cells. Genetic models capable of addressing this question will lead to important advances in investigating the role of autophagy in T cells under various stimulation conditions.

Two additional reports have also suggested that the induction of autophagy may lead to cell death. In an immortalized T helper cell line, silencing the essential autophagy genes *Atg7* or *beclin 1* leads to enhanced cell survival after IL-2 growth factor withdrawal in vitro (Li et al. 2006). It will be interesting to determine how autophagy may contribute to activated or effector T cell survival decisions, and whether this result coincides with an important function for autophagy in the regulation of effector T cell survival in vivo. Finally, pathologic cell death in T lymphocytes may also be regulated by autophagy. HIV envelope-mediated bystander killing of $CD4^+$ human T cells depends on autophagic activity in these cells, as the siRNA-mediated knockdown of essential autophagy genes inhibits the cellular toxicity (Espert et al. 2006) (see chapter xx). Whether this effect results from the co-opting of a normal regulatory pathway in T cells or the pathologic activation of the T cell autophagic systems remains to be determined. In addition, the effects of HIV on T cells may be complex, like many host-pathogen interactions, given a recent report demonstrating an inhibition of autophagy in T cells with active virus infection (Zhou and Spector 2008).

5.3 Autophagy as a Paradoxical Mediator of Both Survival and Death in T Cells?

The dual role of autophagy in T cell survival and death likely results from the complex regulation of both T cell survival decisions and autophagy. In particular, the levels or targets of autophagic activity in T cells may vary depending on cell type and stimulation, altering the consequence of autophagic induction. In addition, the effects of autophagy may vary substantially among T cell subsets and alter the functional importance of this process in T cells. It is also possible that regulation of molecular interactions between autophagic and apoptotic machinery dictate the cellular outcome of autophagy induction in T cells. Atg5 and FADD (Pyo et al. 2005) as well as Beclin 1 and Bcl-2/Bcl-xL (Liang et al. 1998; Pattingre et al. 2005) interact, and both of these molecules play an important role in T cell biology. Given the extensive interest in apoptosis in T lymphocytes, these cells may provide an excellent model to study physiologically relevant interactions between the autophagy and apoptosis pathways.

6 Autophagy in T Cell Proliferation

Consistent with the induction of autophagy in activated T cells (Gerland et al. 2004; Li et al. 2006; Pua et al. 2007), autophagy-deficient T cells have defects in proliferation. $Atg5^{-/-}$ and $Atg7^{-/-}$ T cells fail to proliferate optimally after TCR crosslinking despite the normal upregulation of early activation markers (Pua et al. 2007, 2009), suggesting that the autophagy machinery may actively participate in TCR-induced proliferation (Fig. 3). Given that the development of many other cell types appears to be normal in autophagy-deficient mice, autophagy is not universally required for cell proliferation.

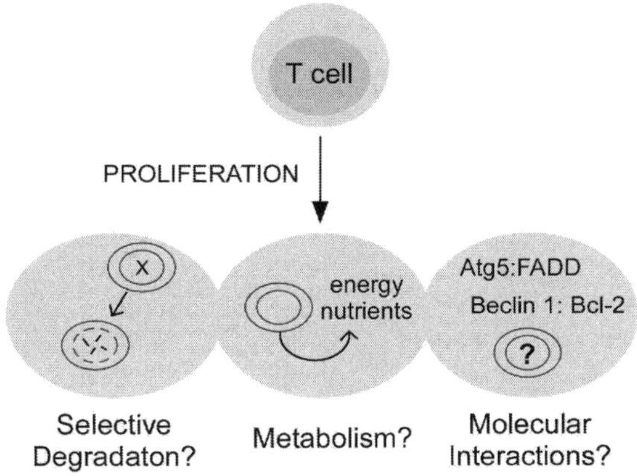

Fig. 3 Autophagy genes contribute to TCR-mediated T cell proliferation in vitro. Although the function of autophagy in T cell proliferation is not known, it may depend on the selective degradation of cellular components, autophagy-dependent metabolic function and/or the molecular interaction of essential autophagy genes with other signaling pathways

6.1 *Autophagosome Formation in T Cell Proliferation*

How might the formation of autophagosomes function in cell cycle progression in activated T lymphocytes? First, autophagosomes may degrade key regulatory factors that normally inhibit T cell proliferation. Although the sequestration of cytosol during starvation-induced autophagy is though to be nonspecific, there is some evidence that the autophagy-dependent degradation of specific proteins has biological consequences. The selective degradation of catalases by autophagy has been shown to contribute to death after caspase inhibition (Yu et al. 2004). In addition, heat shock protein 90 regulates the autophagy-dependent degradation of the NF-κB pathway kinase IKK in human T cell leukemia virus-transformed T cells (Qing et al. 2006, 2007; Yan et al. 2007). This result suggests that degradation of critical signaling molecules by autophagy could play a role in TCR-mediated signaling pathways and will provide an interesting area for future investigation.

Second, autophagy may provide a key intracellular nutrient supply during T cell proliferation. In proliferating T cells, the TCR-mediated calcium-dependent activation of adenosine monophosphate-activated protein kinase (AMPK) suggests that TCR-mediated signals anticipate enhanced energy demand (Tamas et al. 2006). AMPK activation in cardiac myocytes during ischemia/reperfusion (Matsui et al. 2007), in human cancer cell lines through calcium signals (Hoyer-Hansen et al. 2007), and in human cancer cell lines after nutrient deprivation (Liang et al. 2007) result in the induction of autophagy. Although the effectors downstream of AMPK which activate autophagy in the previous experimental systems have not been determined, AMPK classically inhibits TOR (Wullschleger et al. 2005), a known negative regulator of autophagosome formation. Pharmacologic inhibition of the TOR pathway with rapamycin in activated T cells leads to the induction of autophagy (Li et al. 2006), consistent with a functional role for autophagosome formation in a larger program of cellular growth and metabolism.

6.2 A Role for Autophagy Genes in T Cell Proliferation Outside of Autophagosome Formation?

Although the defective proliferative responses in both *Atg5*$^{-/-}$ and *Atg7*$^{-/-}$ T cells suggest that autophagic function is required for optimal TCR-induced proliferation, it is possible that some essential autophagy genes have a role in T cell proliferation outside of autophagosome formation. Atg5 interacts with the death domain of the death receptor adaptor protein FADD, and this interaction has functional consequence in HeLa cells during interferon-γ-induced cell death (Pyo et al. 2005). Like *Atg5*$^{-/-}$ T cells (Pua et al. 2007), *FADD*$^{-/-}$ or FADD dominant-negative expressing T cells have reduced TCR-induced proliferation with normal upregulation of early activation markers (Beisner et al. 2003; Newton et al. 1998, 2001; Walsh et al. 1998; Zhang et al. 1998a, 2001, 2005b). Although the precise mechanism by which FADD regulates the cell cycle is not known, it will be interesting to determine whether Atg5 and FADD interact in primary T cells to affect as yet unidentified downstream signaling pathways independent of autophagy. Alternatively, the cell cycle arrest observed in *FADD*$^{-/-}$ T cells may be due to alterations in autophagic activity. Overexpression of the death domain of FADD or stimulation through the death receptor TRAIL in epithelial cells results in the induction of autophagy (Thorburn et al. 2005). Therefore, it is possible that TCR-associated autophagic induction depends on the activity of FADD or death receptors in T cells.

A second autophagy-associated apoptotic molecule, Bcl-2, regulates T cell proliferation. Bcl-2 deficiency inhibits early cell cycle progression in multiple cell types (Zinkel et al. 2006), and Bcl-2 dose inversely correlates with proliferation in T cells (Linette et al. 1996). The autophagy protein Beclin 1 was first identified as a Bcl-2 interacting protein in mammals (Liang et al. 1999), and the interaction between Beclin 1 and Bcl-2 has important physiologic consequences both for the induction of autophagy as well as the execution of other cellular functions, including cell death (Pattingre et al. 2005; Shimizu et al. 2004). Therefore, it will be interesting to determine whether Beclin 1 and Bcl-2 cooperate to regulate autophagy-dependent or autophagy-independent pathways for T cell proliferation.

7 Autophagy in B Lymphocytes

Although fewer studies have investigated autophagy in B cells, the available data demonstrates an important role for autophagy in this second lymphocyte lineage. As in T cells, deletion of the essential autophagy gene *Atg5* leads to selective survival defects in B cells in mouse models. However, the stages impacted in T and B cells differ. While thymocyte development appears normal in *Atg5*$^{-/-}$ fetal liver hematopoietic chimeric mice, the development of B cells is altered in the bone marrow of these mice (Pua et al. 2007). The number of pre- and immature B cells is reduced in *Atg5*$^{-/-}$ chimeras, and this reduction correlates with an increase in the rate of apoptosis in autophagy-deficient developing B cells (Miller et al. 2008). Why the loss of the autophagy gene *Atg5* affects the pre-B but not the pro-B cell compartment will be an interesting area of future investigation, likely providing physiologically relevant information about both B cell development as well as the functions of autophagy.

In *Atg5*$^{-/-}$ chimeric mice, there is also a statistically significant reduction in mature B cells in the spleen and lymph nodes (Miller et al. 2008; Pua et al. 2007). However when *Atg5* is conditionally deleted in B cells using a CD19-Cre transgene, no reduction in developing B cells or mature B-2 cells is observed (Miller et al. 2008). In this model system, only the population of innate cells like self-renewing B-1a B cells are reduced in number. Therefore the essential autophagy gene *Atg5* is not required acutely for conventional B cell survival when cells maintain normal autophagic function early in development.

This work on the role of B cell autophagy raises many important questions for future investigation. When and how is autophagic activity regulated in B lymphocytes? Analysis of the expression of a *beclin 1-GFP* transgene in B cell subsets suggests that at least this autophagy-associated protein is differentially expressed in B cell subsets (Arsov et al. 2008). Recent work in B cell lines and mouse splenic B cells demonstrates an increase in autophagosome formation 36 h after IgM crosslinking (Watanabe et al. 2008). Therefore, signaling through antigen receptors in both B and T lymphocytes may result in the induction of autophagy.

Future investigations are needed to address the function of autophagy in B lymphocytes as well as whether autophagy may contribute to programmed death within these cells. Finally, there is good preliminary evidence to suggest that autophagy may play an important role not only in normal B cell homeostasis but also in pathologic B cell processes. In a c-Myc-induced model of B cell lymphoma, inhibition of autophagosome maturation results in the enhanced clearance of tumor cells during chemotherapy (Amaravadi et al. 2007). Therefore, understanding the function of autophagy in B lymphocytes may ultimately have therapeutic consequences for human disease.

8 Conclusion

Although the importance of autophagy in T and B lymphocytes is somewhat unexpected, the critical role of autophagic activity in many organ systems highlights the breadth and complexity of this pathway both within and outside of the immune system. Not only is autophagy activity in lymphocytes regulated, but it contributes to diverse cellular processes including survival, death, and proliferation. This reflects complex activity depending on B or T cell stage as well as activation status, and likely will lead to important roles for autophagy in both normal physiologic activity as well as the pathology of lymphocytes in the adaptive immune system. Hopefully, continued work in this exciting area will result in both fundamental advances in understanding the function of autophagy in cells of the immune system as well as an application of that knowledge to diseases including cancer, autoimmunity, and host-pathogen interactions.

References

Agenes F, Freitas AA (1999) Transfer of small resting B cells into immunodeficient hosts results in the selection of a self-renewing activated B cell population. J Exp Med 189:319–330

Amaravadi RK, Yu D, Lum JJ, Bui T, Christophorou MA, Evan GI, Thomas-Tikhonenko A, Thompson CB (2007) Autophagy inhibition enhances therapy-induced apoptosis in a Myc-induced model of lymphoma. J Clin Invest 117:326–336

Arsov I, Li X, Matthews G, Coradin J, Hartmann B, Simon AK, Sealfon SC, Yue Z (2008) BAC-mediated transgenic expression of fluorescent autophagic protein Beclin 1 reveals a role for Beclin 1 in lymphocyte development. Cell Death Differ 15:1385–1395

Baba M, Takeshige K, Baba N, Ohsumi Y (1994) Ultrastructural analysis of the autophagic process in yeast: detection of autophagosomes and their characterization. J Cell Biol 124:903–913

Badovinac VP, Harty JT (2006) Programming, demarcating, and manipulating CD8+ T-cell memory. Immunol Rev 211:67–80

Baehrecke EH (2005) Autophagy: dual roles in life and death? Nat Rev Mol Cell Biol 6:505–510

Bassnett S (2002) Lens organelle degradation. Exp Eye Res 74:1–6

Becker TC, Wherry EJ, Boone D, Murali-Krishna K, Antia R, Ma A, Ahmed R (2002) Interleukin 15 is required for proliferative renewal of virus-specific memory CD8 T cells. J Exp Med 195:1541–1548

Beisner DR, Chu IH, Arechiga AF, Hedrick SM, Walsh CM (2003) The requirements for Fas-associated death domain signaling in mature T cell activation and survival. J Immunol 171:247–256

Bender J, Mitchell T, Kappler J, Marrack P (1999) CD4+ T cell division in irradiated mice requires peptides distinct from those responsible for thymic selection. J Exp Med 190:367–374

Berry DL, Baehrecke EH (2007) Growth arrest and autophagy are required for salivary gland cell degradation in Drosophila. Cell 131:1137–1148

Beutner U, MacDonald HR (1998) TCR-MHC class II interaction is required for peripheral expansion of CD4 cells in a T cell-deficient host. Int Immunol 10:305–310

Blattman JN, Antia R, Sourdive DJ, Wang X, Kaech SM, Murali-Krishna K, Altman JD, Ahmed R (2002) Estimating the precursor frequency of naive antigen-specific CD8 T cells. J Exp Med 195:657–664

Boise LH, Minn AJ, June CH, Lindsten T, Thompson CB (1995) Growth factors can enhance lymphocyte survival without committing the cell to undergo cell division. Proc Natl Acad Sci USA 92:5491–5495

Bouillet P, Purton JF, Godfrey DI, Zhang LC, Coultas L, Puthalakath H, Pellegrini M, Cory S, Adams JM, Strasser A (2002) BH3-only Bcl-2 family member Bim is required for apoptosis of autoreactive thymocytes. Nature 415:922–926

Brocker T (1997) Survival of mature CD4 T lymphocytes is dependent on major histocompatibility complex class II-expressing dendritic cells. J Exp Med 186:1223–1232

Broome HE, Dargan CM, Krajewski S, Reed JC (1995) Expression of Bcl-2, Bcl-x, and Bax after T cell activation and IL-2 withdrawal. J Immunol 155:2311–2317

Cabatingan MS, Schmidt MR, Sen R, Woodland RT (2002) Naive B lymphocytes undergo homeostatic proliferation in response to B cell deficit. J Immunol 169:6795–6805

Codogno P, Meijer AJ (2005) Autophagy and signaling: their role in cell survival and cell death. Cell Death Differ 12(Suppl 2):1509–1518

Colell A, Ricci JE, Tait S, Milasta S, Maurer U, Bouchier-Hayes L, Fitzgerald P, Guio-Carrion A, Waterhouse NJ, Li CW et al. (2007) GAPDH and autophagy preserve survival after apoptotic cytochrome c release in the absence of caspase activation. Cell 129:983–997

Corazza N, Brumatti G, Jakob S, Villunger A, Brunner T (2004) TRAIL and thymocyte apoptosis: not so deadly? Cell Death Differ 11(Suppl 2):S213–S215

Cretney E, Uldrich AP, Berzins SP, Strasser A, Godfrey DI, Smyth MJ (2003) Normal thymocyte negative selection in TRAIL-deficient mice. J Exp Med 198:491–496

Crowley JE, Scholz JL, Quinn Iii WJ, Stadanlick JE, Treml JF, Treml LS, Hao Y, Goenka R, O'Neill PJ, Matthews AH et al (2008) Homeostatic control of B lymphocyte subsets. Immunol Res 42:75–83

Danial NN, Korsmeyer SJ (2004) Cell death: critical control points. Cell 116:205–219

Diehl GE, Yue HH, Hsieh K, Kuang AA, Ho M, Morici LA, Lenz LL, Cado D, Riley LW, Winoto A (2004) TRAIL-R as a negative regulator of innate immune cell responses. Immunity 21:877–889

Dorfman JR, Germain RN (2002) MHC-dependent survival of naive T cells? A complicated answer to a simple question. Microbes Infect 4:547–554

Dzhagalov I, Dunkle A, He YW (2008) The anti-apoptotic Bcl-2 family member Mcl-1 promotes T lymphocyte survival at multiple stages. J Immunol 181:521–528

El Kassar N, Lucas PJ, Klug DB, Zamisch M, Merchant M, Bare CV, Choudhury B, Sharrow SO, Richie E, Mackall CL et al. (2004) A dose effect of IL-7 on thymocyte development. Blood 104:1419–1427

Elmore SP, Qian T, Grissom SF, Lemasters JJ (2001) The mitochondrial permeability transition initiates autophagy in rat hepatocytes. Faseb J 15:2286–2287

Erickson SL, de Sauvage FJ, Kikly K, Carver-Moore K, Pitts-Meek S, Gillett N, Sheehan KC, Schreiber RD, Goeddel DV, Moore MW (1994) Decreased sensitivity to tumour-necrosis factor but normal T-cell development in TNF receptor-2-deficient mice. Nature 372:560–563

Ernst B, Lee DS, Chang JM, Sprent J, Surh CD (1999) The peptide ligands mediating positive selection in the thymus control T cell survival and homeostatic proliferation in the periphery. Immunity 11:173–181

Espert L, Denizot M, Grimaldi M, Robert-Hebmann V, Gay B, Varbanov M, Codogno P, Biard-Piechaczyk M (2006) Autophagy is involved in T cell death after binding of HIV-1 envelope proteins to CXCR4. J Clin Invest 116:2161–2172

Fehniger TA, Suzuki K, Ponnappan A, VanDeusen JB, Cooper MA, Florea SM, Freud AG, Robinson ML, Durbin J, Caligiuri MA (2001) Fatal leukemia in interleukin 15 transgenic mice follows early expansions in natural killer and memory phenotype $CD8^+$ T cells. J Exp Med 193:219–231

Gerland LM, Genestier L, Peyrol S, Michallet MC, Hayette S, Urbanowicz I, Ffrench P, Magaud JP, Ffrench M (2004) Autolysosomes accumulate during in vitro $CD8^+$ T-lymphocyte aging and may participate in induced death sensitization of senescent cells. Exp Gerontol 39:789–800

Goldrath AW, Bevan MJ (1999a) Low-affinity ligands for the TCR drive proliferation of mature $CD8^+$ T cells in lymphopenic hosts. Immunity 11:183–190

Goldrath AW, Bevan MJ (1999b) Selecting and maintaining a diverse T-cell repertoire. Nature 402:255–262

Goldrath AW, Sivakumar PV, Glaccum M, Kennedy MK, Bevan MJ, Benoist C, Mathis D, Butz EA (2002) Cytokine requirements for acute and basal homeostatic proliferation of naive and memory CD8+ T cells. J Exp Med 195:1515–1522

Hara T, Nakamura K, Matsui M, Yamamoto A, Nakahara Y, Suzuki-Migishima R, Yokoyama M, Mishima K, Saito I, Okano Het al. (2006) Suppression of basal autophagy in neural cells causes neurodegenerative disease in mice. Nature 441:885–889

He YW, Malek TR (1998) The structure and function of gamma c-dependent cytokines and receptors: regulation of T lymphocyte development and homeostasis. Crit Rev Immunol 18:503–524

Hildeman DA, Zhu Y, Mitchell TC, Bouillet P, Strasser A, Kappler J, Marrack P (2002) Activated T cell death in vivo mediated by proapoptotic bcl-2 family member bim. Immunity 16:759–767

Hoyer-Hansen M, Bastholm L, Szyniarowski P, Campanella M, Szabadkai G, Farkas T, Bianchi K, Fehrenbacher N, Elling F, Rizzuto Ret al. (2007) Control of macroautophagy by calcium, calmodulin-dependent kinase kinase-beta, and Bcl-2. Mol Cell 25:193–205

Hughes PD, Belz GT, Fortner KA, Budd RC, Strasser A, Bouillet P (2008) Apoptosis regulators Fas and Bim cooperate in shutdown of chronic immune responses and prevention of autoimmunity. Immunity 28:197–205

Jameson SC (2002) Maintaining the norm: T-cell homeostasis. Nat Rev Immunol 2:547–556

Janssen EM, Droin NM, Lemmens EE, Pinkoski MJ, Bensinger SJ, Ehst BD, Griffith TS, Green DR, Schoenberger SP (2005) $CD4^+$ T-cell help controls $CD8^+$ T-cell memory via TRAIL-mediated activation-induced cell death. Nature 434:88–93

Jia L, Dourmashkin RR, Allen PD, Gray AB, Newland AC, Kelsey SM (1997) Inhibition of autophagy abrogates tumour necrosis factor alpha induced apoptosis in human T-lymphoblastic leukaemic cells. Br J Haematol 98:673–685

Judge AD, Zhang X, Fujii H, Surh CD, Sprent J (2002) Interleukin 15 controls both proliferation and survival of a subset of memory-phenotype CD8(+) T cells. J Exp Med 196:935–946

Kassiotis G, Garcia S, Simpson E, Stockinger B (2002) Impairment of immunological memory in the absence of MHC despite survival of memory T cells. Nat Immunol 3:244–250

Kennedy MK, Glaccum M, Brown SN, Butz EA, Viney JL, Embers M, Matsuki N, Charrier K, Sedger L, Willis CRet al. (2000) Reversible defects in natural killer and memory CD8 T cell lineages in interleukin 15-deficient mice. J Exp Med 191:771–780

Kent G, Minick OT, Volini FI, Orfei E (1966) Autophagic vacuoles in human red cells. Am J Pathol 48:831–857

Kieper WC, Tan JT, Bondi-Boyd B, Gapin L, Sprent J, Ceredig R, Surh CD (2002) Overexpression of interleukin (IL)-7 leads to IL-15-independent generation of memory phenotype $CD8^+$ T cells. J Exp Med 195:1533–1539

Kirberg J, Berns A, von Boehmer H (1997) Peripheral T cell survival requires continual ligation of the T cell receptor to major histocompatibility complex-encoded molecules. J Exp Med 186:1269–1275

Kishimoto H, Surh CD, Sprent J (1998) A role for Fas in negative selection of thymocytes in vivo. J Exp Med 187:1427–1438

Kissova I, Salin B, Schaeffer J, Bhatia S, Manon S, Camougrand N (2007) Selective and non-selective autophagic degradation of mitochondria in yeast. Autophagy 3:329–336

Klionsky DJ, Emr SD (2000) Autophagy as a regulated pathway of cellular degradation. Science 290:1717–1721

Knudson CM, Korsmeyer SJ (1997) Bcl-2 and Bax function independently to regulate cell death. Nat Genet 16:358–363

Knudson CM, Tung KS, Tourtellotte WG, Brown GA, Korsmeyer SJ (1995) Bax-deficient mice with lymphoid hyperplasia and male germ cell death. Science 270:96–99

Komatsu M, Waguri S, Ueno T, Iwata J, Murata S, Tanida I, Ezaki J, Mizushima N, Ohsumi Y, Uchiyama Y et al. (2005) Impairment of starvation-induced and constitutive autophagy in Atg7-deficient mice. J Cell Biol 169:425–434

Komatsu M, Waguri S, Chiba T, Murata S, Iwata J, Tanida I, Ueno T, Koike M, Uchiyama Y, Kominami E et al (2006) Loss of autophagy in the central nervous system causes neurodegeneration in mice. Nature 441:880–884

Kondrack RM, Harbertson J, Tan JT, McBreen ME, Surh CD, Bradley LM (2003) Interleukin 7 regulates the survival and generation of memory CD4 cells. J Exp Med 198:1797–1806

Kuma A, Hatano M, Matsui M, Yamamoto A, Nakaya H, Yoshimori T, Ohsumi Y, Tokuhisa T, Mizushima N (2004) The role of autophagy during the early neonatal starvation period. Nature 432:1032–1036

Lamhamedi-Cherradi SE, Zheng SJ, Maguschak KA, Peschon J, Chen YH (2003) Defective thymocyte apoptosis and accelerated autoimmune diseases in $TRAIL^{-/-}$ mice. Nat Immunol 4:255–260

Lau LL, Jamieson BD, Somasundaram T, Ahmed R (1994) Cytotoxic T-cell memory without antigen. Nature 369:648–652

Lenz DC, Kurz SK, Lemmens E, Schoenberger SP, Sprent J, Oldstone MB, Homann D (2004) IL-7 regulates basal homeostatic proliferation of antiviral $CD4^+$ T cell memory. Proc Natl Acad Sci USA 101:9357–9362

Levine B, Kroemer G (2008) Autophagy in the pathogenesis of disease. Cell 132:27–42

Li C, Capan E, Zhao Y, Zhao J, Stolz D, Watkins SC, Jin S, Lu B (2006) Autophagy is induced in CD4+ T cells and important for the growth factor-withdrawal cell death. J Immunol 177:5163–5168
Liang J, Shao SH, Xu ZX, Hennessy B, Ding Z, Larrea M, Kondo S, Dumont DJ, Gutterman JU, Walker CLet al (2007) The energy sensing LKB1-AMPK pathway regulates p27(kip1) phosphorylation mediating the decision to enter autophagy or apoptosis. Nat Cell Biol 9:218–224
Liang XH, Kleeman LK, Jiang HH, Gordon G, Goldman JE, Berry G, Herman B, Levine B (1998) Protection against fatal Sindbis virus encephalitis by Beclin, a novel Bcl-2-interacting protein. J Virol 72:8586–8596
Liang XH, Jackson S, Seaman M, Brown K, Kempkes B, Hibshoosh H, Levine B (1999) Induction of autophagy and inhibition of tumorigenesis by *beclin* 1. Nature 402:672–676
Lindsten T, Ross AJ, King A, Zong WX, Rathmell JC, Shiels HA, Ulrich E, Waymire KG, Mahar P, Frauwirth Ket al (2000) The combined functions of proapoptotic Bcl-2 family members bak and bax are essential for normal development of multiple tissues. Mol Cell 6:1389–1399
Linette GP, Li Y, Roth K, Korsmeyer SJ (1996) Cross talk between cell death and cell cycle progression: BCL-2 regulates NFAT-mediated activation. Proc Natl Acad Sci USA 93:9545–9552
Lodolce JP, Boone DL, Chai S, Swain RE, Dassopoulos T, Trettin S, Ma A (1998) IL-15 receptor maintains lymphoid homeostasis by supporting lymphocyte homing and proliferation. Immunity 9:669–676
Lum JJ, Bauer DE, Kong M, Harris MH, Li C, Lindsten T, Thompson CB (2005) Growth factor regulation of autophagy and cell survival in the absence of apoptosis. Cell 120:237–248
Maraskovsky E, O'Reilly LA, Teepe M, Corcoran LM, Peschon JJ, Strasser A (1997) Bcl-2 can rescue T lymphocyte development in interleukin-7 receptor-deficient mice but not in mutant rag-$1^{-/-}$ mice. Cell 89:1011–1019
Marks-Konczalik J, Dubois S, Losi JM, Sabzevari H, Yamada N, Feigenbaum L, Waldmann TA, Tagaya Y (2000) IL-2-induced activation-induced cell death is inhibited in IL-15 transgenic mice. Proc Natl Acad Sci USA 97:11445–11450
Marrack P, Bender J, Hildeman D, Jordan M, Mitchell T, Murakami M, Sakamoto A, Schaefer BC, Swanson B, Kappler J (2000) Homeostasis of alpha beta TCR^+ T cells. Nat Immunol 1:107–111
Matsui M, Yamamoto A, Kuma A, Ohsumi Y, Mizushima N (2006) Organelle degradation during the lens and erythroid differentiation is independent of autophagy. Biochem Biophys Res Commun 339:485–489
Matsui Y, Takagi H, Qu X, Abdellatif M, Sakoda H, Asano T, Levine B, Sadoshima J (2007) Distinct roles of autophagy in the heart during ischemia and reperfusion: roles of AMP-activated protein kinase and Beclin 1 in mediating autophagy. Circ Res 100:914–922
Matsuzaki Y, Nakayama K, Tomita T, Isoda M, Loh DY, Nakauchi H (1997) Role of bcl-2 in the development of lymphoid cells from the hematopoietic stem cell. Blood 89:853–862
Miller BC, Zhao Z, Stephenson LM, Cadwell K, Pua HH, Lee HK, Mizushima NN, Iwasaki A, He YW, Swat Wet al. (2008) The autophagy gene *ATG5* plays an essential role in B lymphocyte development. Autophagy 4:309–314
Miller RA, Stutman O (1984) T cell repopulation from functionally restricted splenic progenitors: 10,000-fold expansion documented by using limiting dilution analyses. J Immunol 133:2925–2932
Mizushima N, Ohsumi Y, Yoshimori T (2002) Autophagosome formation in mammalian cells. Cell Struct Funct 27:421–429
Morrissey PJ, Conlon P, Charrier K, Braddy S, Alpert A, Williams D, Namen AE, Mochizuki D (1991) Administration of IL-7 to normal mice stimulates B-lymphopoiesis and peripheral lymphadenopathy. J Immunol 147:561–568
Motoyama N, Wang F, Roth KA, Sawa H, Nakayama K, Negishi I, Senju S, Zhang Q, Fujii S et al. (1995) Massive cell death of immature hematopoietic cells and neurons in Bcl-x-deficient mice. Science 267:1506–1510
Murali-Krishna K, Altman JD, Suresh M, Sourdive DJ, Zajac AJ, Miller JD, Slansky J, Ahmed R (1998) Counting antigen-specific CD8 T cells: a reevaluation of bystander activation during viral infection. Immunity 8:177–187
Murali-Krishna K, Lau LL, Sambhara S, Lemonnier F, Altman J, Ahmed R (1999) Persistence of memory CD8 T cells in MHC class I-deficient mice. Science 286:1377–1381

Nakayama K, Nakayama K, Negishi I, Kuida K, Shinkai Y, Louie MC, Fields LE, Lucas PJ, Stewart V, Alt FWet al. (1993) Disappearance of the lymphoid system in Bcl-2 homozygous mutant chimeric mice. Science 261:1584–1588

Nesic D, Vukmanovic S (1998) MHC class I is required for peripheral accumulation of $CD8^+$ thymic emigrants. J Immunol 160:3705–3712

Newton K, Harris AW, Bath ML, Smith KG, Strasser A (1998) A dominant interfering mutant of FADD/MORT1 enhances deletion of autoreactive thymocytes and inhibits proliferation of mature T lymphocytes. EMBO J 17:706–718

Newton K, Harris AW, Strasser A (2000) FADD/MORT1 regulates the pre-TCR checkpoint and can function as a tumour suppressor. EMBO J 19:931–941

Newton K, Kurts C, Harris AW, Strasser A (2001) Effects of a dominant interfering mutant of FADD on signal transduction in activated T cells. Curr Biol 11:273–276

Nguyen LT, McKall-Faienza K, Zakarian A, Speiser DE, Mak TW, Ohashi PS (2000) TNF receptor 1 (TNFR1) and CD95 are not required for T cell deletion after virus infection but contribute to peptide-induced deletion under limited conditions. Eur J Immunol 30:683–688

Nowikovsky K, Reipert S, Devenish RJ, Schweyen RJ (2007) Mdm38 protein depletion causes loss of mitochondrial K^+/H^+ exchange activity, osmotic swelling and mitophagy. Cell Death Differ 14:1647–1656

Opferman JT, Letai A, Beard C, Sorcinelli MD, Ong CC, Korsmeyer SJ (2003) Development and maintenance of B and T lymphocytes requires antiapoptotic MCL-1. Nature 426:671–676

Page DM (1999) Cutting edge: thymic selection and autoreactivity are regulated by multiple coreceptors involved in T cell activation. J Immunol 163:3577–3581

Page DM, Roberts EM, Peschon JJ, Hedrick SM (1998) TNF receptor-deficient mice reveal striking differences between several models of thymocyte negative selection. J Immunol 160:120–133

Pattingre S, Tassa A, Qu X, Garuti R, Liang XH, Mizushima N, Packer M, Schneider MD, Levine B (2005) Bcl-2 antiapoptotic proteins inhibit Beclin 1-dependent autophagy. Cell 122:927–939

Pellegrini M, Belz G, Bouillet P, Strasser A (2003) Shutdown of an acute T cell immune response to viral infection is mediated by the proapoptotic Bcl-2 homology 3-only protein Bim. Proc Natl Acad Sci USA 100:14175–14180

Pfeffer K, Matsuyama T, Kundig TM, Wakeham A, Kishihara K, Shahinian A, Wiegmann K, Ohashi PS, Kronke M, Mak TW (1993) Mice deficient for the 55 kd tumor necrosis factor receptor are resistant to endotoxic shock, yet succumb to L monocytogenes infection. Cell 73:457–467

Plas DR, Rathmell JC, Thompson CB (2002) Homeostatic control of lymphocyte survival: potential origins and implications. Nat Immunol 3:515–521

Polic B, Kunkel D, Scheffold A, Rajewsky K (2001) How alpha beta T cells deal with induced TCR alpha ablation. Proc Natl Acad Sci USA 98:8744–8749

Priault M, Salin B, Schaeffer J, Vallette FM, di Rago JP, Martinou JC (2005) Impairing the bioenergetic status and the biogenesis of mitochondria triggers mitophagy in yeast. Cell Death Differ 12:1613–1621

Pua HH, He YW (2007) Maintaining T lymphocyte homeostasis: another duty of autophagy. Autophagy 3:266–267

Pua HH, Dzhagalov I, Chuck M, Mizushima N, He YW (2007) A critical role for the autophagy gene *Atg5* in T cell survival and proliferation. J Exp Med 204:25–31

Pua HH, Komatsu M, He YW (2009) Autophagy is essential for mitochondrial clearance in mature T lymphocytes. J Immunol 182:4046–4055

Pyo JO, Jang MH, Kwon YK, Lee HJ, Jun JI, Woo HN, Cho DH, Choi B, Lee H, Kim JHet al. (2005) Essential roles of Atg5 and FADD in autophagic cell death: dissection of autophagic cell death into vacuole formation and cell death. J Biol Chem 280:20722–20729

Qing G, Yan P, Xiao G (2006) Hsp90 inhibition results in autophagy-mediated proteasome-independent degradation of IkappaB kinase (IKK). Cell Res 16:895–901

Qing G, Yan P, Qu Z, Liu H, Xiao G (2007) Hsp90 regulates processing of NF-kappaB2 p100 involving protection of NF-kappaB-inducing kinase (NIK) from autophagy-mediated degradation. Cell Res 17:520–530

Rathmell JC, Farkash EA, Gao W, Thompson CB (2001) IL-7 enhances the survival and maintains the size of naive T cells. J Immunol 167:6869–6876

Rathmell JC, Lindsten T, Zong WX, Cinalli RM, Thompson CB (2002) Deficiency in Bak and Bax perturbs thymic selection and lymphoid homeostasis. Nat Immunol 3:932–939

Ravikumar B, Berger Z, Vacher C, O'Kane CJ, Rubinsztein DC (2006) Rapamycin pre-treatment protects against apoptosis. Hum Mol Genet 15:1209–1216

Reich A, Korner H, Sedgwick JD, Pircher H (2000) Immune down-regulation and peripheral deletion of CD8 T cells does not require TNF receptor–ligand interactions nor CD95 (Fas, APO-1). Eur J Immunol 30:678–682

Reipert S, Berry J, Hughes MF, Hickman JA, Allen TD (1995) Changes of mitochondrial mass in the hemopoietic stem cell line FDCP-mix after treatment with etoposide: a correlative study by multiparameter flow cytometry and confocal and electron microscopy. Exp Cell Res 221:281–288

Rodriguez-Enriquez S, Kim I, Currin RT, Lemasters JJ (2006) Tracker dyes to probe mitochondrial autophagy (mitophagy) in rat hepatocytes. Autophagy 2:39–46

Sandoval H, Thiagarajan P, Dasgupta SK, Schumacher A, Prchal JT, Chen M, Wang J (2008) Essential role for Nix in autophagic maturation of erythroid cells. Nature 454:232–235

Schluns KS, Kieper WC, Jameson SC, Lefrancois L (2000) Interleukin-7 mediates the homeostasis of naive and memory CD8 T cells in vivo. Nat Immunol 1:426–432

Schmid D, Munz C (2007) Innate and adaptive immunity through autophagy. Immunity 27:11–21

Schweers RL, Zhang J, Randall MS, Loyd MR, Li W, Dorsey FC, Kundu M, Opferman JT, Cleveland JL, Miller JL et al. (2007) NIX is required for programmed mitochondrial clearance during reticulocyte maturation. Proc Natl Acad Sci USA 104:19500–19505

Seddon B, Zamoyska R (2002) TCR and IL-7 receptor signals can operate independently or synergize to promote lymphopenia-induced expansion of naive T cells. J Immunol 169:3752–3759

Seddon B, Tomlinson P, Zamoyska R (2003) Interleukin 7 and T cell receptor signals regulate homeostasis of CD4 memory cells. Nat Immunol 4:680–686

Shimizu S, Kanaseki T, Mizushima N, Mizuta T, Arakawa-Kobayashi S, Thompson CB, Tsujimoto Y (2004) Role of Bcl-2 family proteins in a non-apoptotic programmed cell death dependent on autophagy genes. Nat Cell Biol 6:1221–1228

Sudo T, Nishikawa S, Ohno N, Akiyama N, Tamakoshi M, Yoshida H, Nishikawa S (1993) Expression and function of the interleukin 7 receptor in murine lymphocytes. Proc Natl Acad Sci USA 90:9125–9129

Swain SL, Hu H, Huston G (1999) Class II-independent generation of CD4 memory T cells from effectors. Science 286:1381–1383

Takano-Ohmuro H, Mukaida M, Kominami E, Morioka K (2000) Autophagy in embryonic erythroid cells: its role in maturation. Eur J Cell Biol 79:759–764

Tamas P, Hawley SA, Clarke RG, Mustard KJ, Green K, Hardie DG, Cantrell DA (2006) Regulation of the energy sensor AMP-activated protein kinase by antigen receptor and Ca^{2+} in T lymphocytes. J Exp Med 203:1665–1670

Tan JT, Dudl E, LeRoy E, Murray R, Sprent J, Weinberg KI, Surh CD (2001) IL-7 is critical for homeostatic proliferation and survival of naive T cells. Proc Natl Acad Sci USA 98:8732–8737

Tan JT, Ernst B, Kieper WC, LeRoy E, Sprent J, Surh CD (2002) Interleukin (IL)-15 and IL-7 jointly regulate homeostatic proliferation of memory phenotype $CD8^+$ cells but are not required for memory phenotype CD4+ cells. J Exp Med 195:1523–1532

Terman A, Dalen H, Eaton JW, Neuzil J, Brunk UT (2003) Mitochondrial recycling and aging of cardiac myocytes: the role of autophagocytosis. Exp Gerontol 38:863–876

Thorburn J, Moore F, Rao A, Barclay WW, Thomas LR, Grant KW, Cramer SD, Thorburn A (2005) Selective inactivation of a Fas-associated death domain protein (FADD)-dependent apoptosis and autophagy pathway in immortal epithelial cells. Mol Biol Cell 16:1189–1199

Thumm M, Egner R, Koch B, Schlumpberger M, Straub M, Veenhuis M, Wolf DH (1994) Isolation of autophagocytosis mutants of *Saccharomyces cerevisiae*. FEBS Lett 349:275–280

Tsukada M, Ohsumi Y (1993) Isolation and characterization of autophagy-defective mutants of *Saccharomyces cerevisiae*. FEBS Lett 333:169–174

van Oers NS, Killeen N, Weiss A (1994) ZAP-70 is constitutively associated with tyrosine-phosphorylated TCR zeta in murine thymocytes and lymph node T cells. Immunity 1:675–685

Veis DJ, Sorenson CM, Shutter JR, Korsmeyer SJ (1993) Bcl-2-deficient mice demonstrate fulminant lymphoid apoptosis, polycystic kidneys, and hypopigmented hair. Cell 75:229–240
Vella A, Teague TK, Ihle J, Kappler J, Marrack P (1997) Interleukin 4 (IL-4) or IL-7 prevents the death of resting T cells: stat6 is probably not required for the effect of IL-4. J Exp Med 186:325–330
Villunger A, Marsden VS, Zhan Y, Erlacher M, Lew AM, Bouillet P, Berzins S, Godfrey DI, Heath WR, Strasser A (2004) Negative selection of semimature CD4(+)8(–)HSA^{+} thymocytes requires the BH3-only protein Bim but is independent of death receptor signaling. Proc Natl Acad Sci USA 101:7052–7057
Viret C, Wong FS, Janeway CA Jr (1999) Designing and maintaining the mature TCR repertoire: the continuum of self-peptide:self-MHC complex recognition. Immunity 10:559–568
Vivien L, Benoist C, Mathis D (2001) T lymphocytes need IL-7 but not IL-4 or IL-6 to survive in vivo. Int Immunol 13:763–768
von Freeden-Jeffry U, Solvason N, Howard M, Murray R (1997) The earliest T lineage-committed cells depend on IL-7 for Bcl-2 expression and normal cell cycle progression. Immunity 7:147–154
Walsh CM, Wen BG, Chinnaiyan AM, O'Rourke K, Dixit VM, Hedrick SM (1998) A role for FADD in T cell activation and development. Immunity 8:439–449
Wang EC, Thern A, Denzel A, Kitson J, Farrow SN, Owen MJ (2001) DR3 regulates negative selection during thymocyte development. Mol Cell Biol 21:3451–3461
Watanabe K, Ichinose S, Hayashizaki K, Tsubata T (2008) Induction of autophagy by B cell antigen receptor stimulation and its inhibition by costimulation. Biochem Biophys Res Commun 374:274–281
Weant AE, Michalek RD, Khan IU, Holbrook BC, Willingham MC, Grayson JM (2008) Apoptosis regulators Bim and Fas function concurrently to control autoimmunity and CD8^{+} T cell contraction. Immunity 28:218–230
Witherden D, van Oers N, Waltzinger C, Weiss A, Benoist C, Mathis D (2000) Tetracycline-controllable selection of CD4(+) T cells: half-life and survival signals in the absence of major histocompatibility complex class II molecules. J Exp Med 191:355–364
Wojciechowski S, Jordan MB, Zhu Y, White J, Zajac AJ, Hildeman DA (2006) Bim mediates apoptosis of CD127(lo) effector T cells and limits T cell memory. Eur J Immunol 36:1694–1706
Wullschleger S, Loewith R, Oppliger W, Hall MN (2005) Molecular organization of target of rapamycin complex 2. J Biol Chem 280:30697–30704
Yan P, Qing G, Qu Z, Wu CC, Rabson A, Xiao G (2007) Targeting autophagic regulation of NFkappaB in HTLV-I transformed cells by geldanamycin: implications for therapeutic interventions. Autophagy 3:600–603
Yu L, Alva A, Su H, Dutt P, Freundt E, Welsh S, Baehrecke EH, Lenardo MJ (2004) Regulation of an ATG7-beclin 1 program of autophagic cell death by caspase-8. Science 304:1500–1502
Zhang J, Cado D, Chen A, Kabra NH, Winoto A (1998a) Fas-mediated apoptosis and activation-induced T-cell proliferation are defective in mice lacking FADD/Mort1. Nature 392:296–300
Zhang J, Kabra NH, Cado D, Kang C, Winoto A (2001) FADD-deficient T cells exhibit a disaccord in regulation of the cell cycle machinery. J Biol Chem 276:29815–29818
Zhang N, He YW (2005) The antiapoptotic protein Bcl-xL is dispensable for the development of effector and memory T lymphocytes. J Immunol 174:6967–6973
Zhang N, Hartig H, Dzhagalov I, Draper D, He YW (2005a) The role of apoptosis in the development and function of T lymphocytes. Cell Res 15:749–769
Zhang X, Sun S, Hwang I, Tough DF, Sprent J (1998b) Potent and selective stimulation of memory-phenotype CD8^{+} T cells in vivo by IL-15. Immunity 8:591–599
Zhang Y, Rosenberg S, Wang H, Imtiyaz HZ, Hou YJ, Zhang J (2005b) Conditional Fas-associated death domain protein (FADD): GFP knockout mice reveal FADD is dispensable in thymic development but essential in peripheral T cell homeostasis. J Immunol 175:3033–3044
Zhou D, Spector SA (2008) Human immunodeficiency virus type-1 infection inhibits autophagy. AIDS 22:695–699
Zinkel S, Gross A, Yang E (2006) BCL2 family in DNA damage and cell cycle control. Cell Death Differ 13:1351–1359

Autophagy and Innate Recognition Systems

Michal Caspi Tal and Akiko Iwasaki

Contents

1 Introduction 108
2 Autophagy and Viral Sensing: Patrolling the Fort 109
2.1 Innate Viral Recognition 109
2.2 Autophagy in Innate Viral Recognition Through TLRs 110
2.3 The Role of Atg5/Autophagy in TLR9 Signaling in pDCs 111
2.4 Negative Regulation of RLRs via the Atg5–Atg12 Conjugate 112
3 Interplay Among Innate Recognition, Induction of Autophagy and Phagocytosis: Securing the Fort 113
3.1 Phagocytosis 113
3.2 Convergence of Autophagy and Phagocytosis 113
4 Cytokine Feedback: Beating the Battle Drum 117
5 Concluding Remarks 117
References 118

Abstract Autophagy is an ancient, highly conserved pathway responsible for the lysosomal degradation of cytosolic constituents and organelles that is critical in maintaining cellular homeostasis. Recent studies have illustrated an important interplay between autophagy and the innate immune system. Signaling through innate pattern recognition receptors leads to the induction of autophagy. Autophagy is utilized by the innate immune cells to survey for virus infection through delivery of cytosolic viral replication complexes to the endosomal viral sensors. In another case, key molecules in the autophagy pathway were found to negatively regulate cytosolic sensors of RNA viruses. Moreover, it has recently become apparent that the autophagic machinery is utilized by phagocytic cells for efficient phagocytosis and clearance of extracellular pathogens. These studies shed light on the possibility that molecules classically thought to be dedicated to the process of autophagy may

M.C. Tal and A. Iwasaki (✉)
Department of Immunobiology, Yale University School of Medicine, 300 Cedar Street, New Haven, CT, 06520, USA
e-mail: akiko.iwasaki@yale.edu

B. Levine et al. (eds.), *Autophagy in Infection and Immunity,*
Current Topics in Microbiology and Immunology 335,
DOI 10.1007/ 978-3-642-00302-8_5,

function in important physiological processes independent of autophagy, whereby the double-membrane structures form within the cytosol to enclose organelles and long-lived proteins. In this chapter, we will highlight key findings relevant to the role of the autophagic machinery in the innate immune system.

1 Introduction

Elegant studies performed in yeast have laid the foundation for our current understanding of the molecular components involved in autophagy. These studies identified more than 30 autophagy-related (*ATG*) gene products that are involved in multiple aspects of the process of autophagy (Klionsky et al. 2003). The molecular details of the process of autophagy have been covered by others extensively in elegant reviews (Klionsky and Emr 2000; Ohsumi 2001). Here, we briefly describe elements of autophagy relevant to the discussions pertaining to this chapter. Two ubiquitin-like conjugation systems are necessary for autophagosome formation (Mizushima et al. 1998). The first involves the Atg5-Atg12 conjugation complex, which localizes to the isolation membrane. Atg7, an E1-like enzyme, activates Atg12 and then Atg12 is transferred to the Atg10, E2-like enzyme. The C-terminal glycine residue of Atg12 is finally linked through covalent binding to the lysine residue of Atg5. The Atg12–Atg5 complex noncovalently binds to Atg16, allowing association with the isolation outer membrane. This complex dissociates from the membrane before autophagosome completion. The second conjugation system involves coupling of Atg8 (or microtubule-associated protein 1 light chain 3, LC3, in mammals) to phosphatidylethanolamine (PE), which allows its incorporation into the isolation membrane. LC3 is a soluble cytoplasmic protein that is proteolytically cleaved by Atg4, activated by Atg7, and then transferred to Atg3. Finally, LC3 is covalently linked to PE. These LC3-PE proteins remain coupled to both sides of the autophagosomal membrane (Ichimura et al. 2000; Kabeya et al. 2000). The autophagosome ultimately fuses with lysosomes and the luminal contents are degraded. Alternatively, autophagosomes can also fuse with endosomes to form amphisomes (Berg et al. 1998). After degradation, the proteins are recycled through lysosomal transporters.

All cells are capable of undergoing autophagy; however, the extent of constitutive autophagy is dependent on both cell type and organ type (Mizushima et al. 2004). Autophagy is also inducible under certain conditions, the best characterized of which is starvation. Under starvation conditions, autophagy provides an alternate source of nutrients and amino acids from the degradation of existing organelles and proteins (Levine and Klionsky 2004), and therefore serves as an important survival pathway during physiological stress. Manipulations of the *ATG* genes and their upstream regulatory molecules in the past decade have unveiled critical roles for autophagy not only in cellular homeostasis but also in host defense (Levine and Deretic 2007). The clearance of intracellular pathogens via autophagy followed by degradation in the lysosome is referred to as "xenophagy" and presents the most direct example of autophagy contributing to host defense. Autophagy is used by the immune cells to deliver cytosolic viral genetic materials to the endosomal receptors for the purpose of innate recognition (Lee et al. 2007). In addition, autophagy is utilized by antigen-presenting cells to deliver cytosolic viral antigens to the MHC class II loading compartment, which serves to stimulate adaptive immunity (Schmid and Munz 2007). In this chapter, we will provide an overview of the recent literature examining the relationship between autophagy and innate recognition of pathogens, and highlight key emerging concepts that provide fundamental links between these two highly evolutionarily conserved processes.

2 Autophagy and Viral Sensing: Patrolling the Fort

2.1 Innate Viral Recognition

Viruses represent a major class of pathogens that cause significant diseases in bacteria, plants, and animals. Unlike other pathogens, viruses rely entirely on the host cell machinery for their own synthesis and replication. The innate immune system utilizes pattern recognition receptors (PRRs) to recognize molecular signatures that are unique to pathogens, pathogen-associated molecular patterns (PAMPs) (Janeway 1989). Owing to the host origin of virion components, the mechanism by which viruses are recognized by the immune system has remained unclear until recently. Studies in the past decade have revealed that the unique features associated with the nature—e.g., double-stranded (ds)RNA, single-stranded (ss)RNA or (ds) DNA—and/or the locations (e.g., endosomal) of viral genomes are detected as a molecular signature by the PRRs and comprise the cell's viral reconnaissance unit (see below for more details). Once alerted, viral PRRs stimulate the expression of type I interferons (IFNs), which are the most potent known antiviral factors, capable of limiting the replication and spread of most viruses (Taniguchi and Takaoka 2002). In addition to IFN production, viral recognition also evokes a proinflammatory cytokine and chemokine response, which is required for the initiation and subsequent coordination of innate and adaptive antiviral immunity (Takeuchi and Akira 2007). In general terms, these antiviral PRRs can be divided into two groups that function in parallel. The Toll-like receptors (TLRs), including TLR3, TLR7, TLR8 and TLR9, survey the endosomal environment for viral nucleic acids. These TLRs are located in the endosome and recognize dsRNA (TLR3) (Alexopoulou et al. 2001), single-stranded (ss)RNA (TLR7 (Diebold et al. 2004; Lund et al. 2004) and TLR8 (Heil et al. 2004)) and dsDNA (TLR9) (Krug et al. 2004; Lund et al. 2003) associated with virus genomes or replication intermediates. The second group of PRRs are located in the cytosol and patrol for the presence of viral nucleic acids. These belong to two families: the retinoic-acid-inducible gene (RIG-I)-like receptor (RLR) family, consisting of RIG-I and melanoma differentiation-associated gene 5 (MDA-5), and the sensor(s) of DNA (Ishii et al. 2006; Stetson and Medzhitov 2006), including DNA-dependent activator of IRFs (DAI) (Takaoka 2007). RIG-I recognizes 5′-triphosphate RNA (Hornung et al. 2006; Pichlmair et al. 2006), which is only present in the viral genomes of certain ssRNA viruses, not in the mammalian genome. MDA-5 recognizes the viral RNA of picornaviruses and synthetic dsRNA such as polyinosine-polycytidic acid (poly I:C) through a ligand that is yet to be determined (Gitlin 2006; Kato 2006). These endosomal and cytosolic pattern recognition systems ensure broad surveillance of viruses that have RNA or DNA genomes and/or replication intermediates, and subsequently elicit the production of the antiviral type I IFNs. The critical difference between these pathways lies in their cell-type specific expression and compartmentalization. Most cells rely solely on cytosolic recognition of viruses, which occurs after a cell has been infected. Cytosolic sensors, through the local secretion of type I IFNs, induce an antiviral state in the infected and the neighboring cells. On the other hand, a subset of dendritic cells (DCs) known as plasmacytoid dendritic cells (pDCs) relies exclusively on TLR7 and TLR9 to detect the genomes and related structures of ssRNA and dsDNA viruses before becoming fully infected. pDCs serve as professional viral sensors that are specialized in viral detection and vigorous secretion of type I IFNs, allowing them to induce a local and systemic antiviral state in the infected host (Liu 2005). Autophagy is emerging as an essential component of antiviral defense that is intimately linked to both families of innate viral sensors.

2.2 *Autophagy in Innate Viral Recognition Through TLRs*

The recognition of viral RNA or DNA by the endosomal TLR7 or TLR9, respectively, induces type I IFN and cytokine production by pDCs. This was thought to occur exclusively upon endocytosis of virions and uncoating of capsid within the lysosomes, whereby the viral genomes become accessible to the TLRs (Fig. 1a). We recently demonstrated that this pathway is not the only mechanism by which pDCs detect the presence of viruses. We found that IFN-α and cytokine production following recognition of vesicular stomatitis virus (VSV) and Sendai virus by TLR7 requires replicating virus, suggesting that pDCs recognize viral replicative intermediates of cytosolic origin. We hypothesized that autophagy could deliver cytosolic viral replication intermediates to the endosomes where TLR7 resides (Fig. 1b). Because *Atg5*$^{-/-}$ mice are neonatally lethal (Kuma et al. 2004), we used *Atg5*$^{-/-}$ neonatal liver chimeras. Indeed, pDCs lacking *Atg5* were defective in sensing VSV through TLR7, and mice lacking *Atg5* in hematopoietic compartments failed to respond to VSV infection in vivo (Lee et al. 2007). Thus, our study demonstrates that autophagy plays a critical role in the innate recognition of ssRNA viruses via delivery of viral replication intermediates from the cytosol to the endosome within which TLR7 resides. Autophagy is utilized by pDCs to bridge the compartmental separation of ligand (in cytosol) and sensor (in endosomes) by ferrying cytosolic viral replication intermediates into endosomal compartments containing TLR7, thus facilitating the detection of viral nucleic acids. An intriguing question that emerges from these findings relates to how TLR7 distinguishes self from non-self

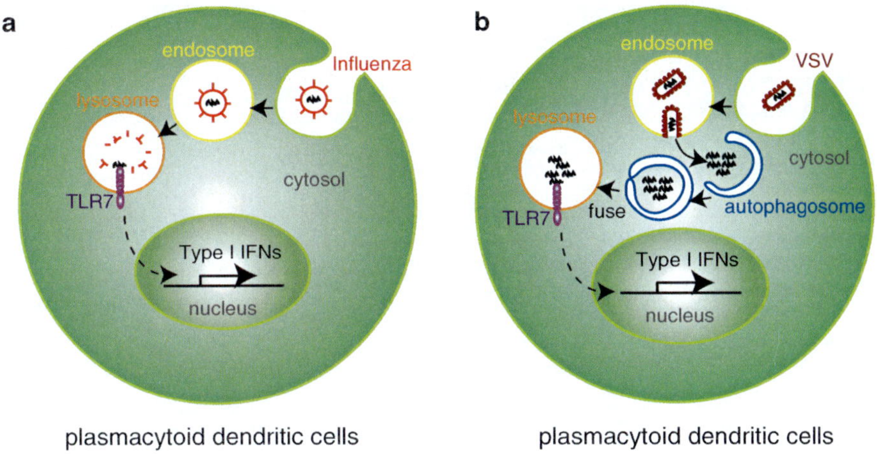

Fig. 1a–b Recognition of virus-associated molecular patterns by plasmacytoid dendritic cells. **a** Viruses such as influenza are recognized in the lysosomes of pDCs via TLR7 following endocytosis of virions. The molecular signature of influenza virus is provided by the endocytic location of the viral genomic ssRNA. **b** In contrast, VSV recognition via TLR7 requires cytosolic virus replication intermediates, which are transported to the lysosome by means of autophagy. This figure is used with permission from Iwasaki A. "Role of autophagy in innate viral recognition." Autophagy 3:354–6, 2007

RNA, if not through compartmental separation as had been previously thought. TLR7 does not seem to distinguish viral or mammalian RNA based on studies using synthetic and purified nucleic acids (Diebold et al. 2006). Our study implies that there may be other ligands of TLR7 that are naturally produced during ssRNA virus infection, such as the replication intermediates, that serve as potent agonists for TLR7. In contrast to recognition of VSV or SeV, TLR7-dependent recognition of influenza virus and TLR9-dependent recognition of a dsDNA virus, herpes simplex virus (HSV), have been shown to be unaffected by UV irradiation, heat inactivation and formaldehyde fixation (Asselin-Paturel et al. 2001; Diebold et al. 2004; Eloranta and Alm 1999; Krug et al. 2004; Lund et al. 2003). Thus, it appears that viral replication intermediates are required for the complete stimulation of TLR7 upon infection with ssRNA viruses that replicate within the cytosol; however, further study will be required to determine whether such products are required for robust innate immune stimulation following infection with viruses harboring distinct genome content and/or replication strategies.

2.3 *The Role of Atg5/Autophagy in TLR9 Signaling in pDCs*

While both TLR9 and TLR7 signals depend on the adaptor protein myeloid differentiation primary response gene (88) (MyD88), recent evidence suggests the existence of two bifurcating pathways emanating from these TLRs. The first pathway leads to the activation of nuclear factor κB (NF-κB), IFN regulatory factor 5 (IRF5) and mitogen-activated protein kinase (MAPK) activation (Takaoka et al. 2005). The second pathway leads to the activation of type I IFN genes through phosphorylation of IRF7 (Honda et al. 2005). Honda and Taniguchi termed these two distinct signaling complexes the cytoplasmic transductional-transcriptional processors for proinflammatory cytokines and IFNs (Honda and Taniguchi 2006). While both pathways depend on MyD88 (Lund et al. 2003), interleukin-1 receptor-associated kinase 4 (IRAK4) (Honda et al. 2004), UNC93B (Tabeta et al. 2006) and TNF receptor-associated factor 6 (TRAF6), the latter pathway leading to IFN activation also requires additional molecules including IRAK1 (Uematsu et al. 2005), TRAF3 (Hacker et al. 2006; Oganesyan et al. 2006), IκB kinase α (IKKα) (Hoshino et al. 2006), osteopontin (Shinohara et al. 2006) and pDC triggering receptors expressed on myeloid cells (TREM) (Watarai et al. 2008). The mechanism by which these two types of signals are mediated through TLR7/9 is unknown. Our recent work also demonstrated that Atg5 is required for secretion of type I IFNs but not proinflammatory cytokines in pDCs stimulated with CpG or HSV-1 (Lee et al. 2007). These data indicated that while Atg5 is not required for the recognition of CpG or HSV-1, it is required for TLR9 signaling leading to type I IFNs but not proinflammatory cytokines. The defect in *Atg5*$^{-/-}$ pDCs is at the level of transcription of IFN genes (Lee and Iwasaki, unpublished observations). These data indicated that in addition to the previously described molecules, Atg5 plays a key role in TLR9 signaling leading to the transcription of type I IFN genes in pDCs. Whether a similar requirement for Atg5 in IFN production applies also to other cell types and other TLRs remains an important question, as does the molecular mechanism by which Atg5 and autophagy regulate TLR signaling.

2.4 Negative Regulation of RLRs via the Atg5–Atg12 Conjugate

TLR-mediated pathogen recognition is of vital importance in specialized viral sensors, pDCs. However, most other cell types utilize cytosolic RLRs such as RIG-I and MDA-5 to detect the presence of viral invaders (Foy et al. 2005; Yoneyama et al. 2005; Yoneyama et al. 2004). In light of this fundamental difference in viral sensing, it is interesting to note that the role of autophagy in viral recognition is strikingly different in pDCs versus non-pDCs. A recent report indicated that innate recognition in mouse embryonic fibroblasts (MEFs) of VSV via RIG-I and poly I:C via MDA-5 pathways is regulated by the Atg5-Atg12 conjugate (Jounai et al. 2007). The authors showed that *Atg5*$^{-/-}$ and *Atg7*$^{-/-}$ MEFs (both lacking the Atg5-Atg12 conjugate) produced enhanced type I IFNs in response to VSV and poly I:C. Thus, in contrast to pDCs, fibroblasts, which rely on cytosolic sensors of viral replication, appear to utilize molecules involved in autophagy to repress type I IFN responses. The Atg5-Atg12 conjugate blocked RLR activity by binding to RIG-I, MDA-5, and the adaptor protein IFN-β promoter stimulator 1 (IPS-1) [also known as mitochondrial antiviral signaling (MAVS) (Seth et al. 2005), Cardif (Meylan et al. 2005), or virus-induced signaling adaptor (VISA) (Xu et al. 2005)] through the caspase activation and recruitment domain (CARD). Thus, Atg5-Atg12 may exert a noncanonical function in the constitutive regulation of RLRs (Takeshita et al. 2008). However, since the experimental system used in this study led not only to the deficiency in Atg5-Atg12 conjugate but also to the complete inhibition of the autophagic process, it remains a possibility that the observed phenotype may also reflect the lack of autophagy in addition to Atg5-Atg12 conjugate-mediated inhibition of RLRs.

3 Interplay Among Innate Recognition, Induction of Autophagy and Phagocytosis: Securing the Fort

Considering the emerging evidence suggesting shared elements and cooperation between phagocytosis and autophagy, it is important to explore a broader view of the intersection between these two pathways. It is known that both autophagy and phagocytosis play important roles both in scavenging intracellular versus extracellular elements for degradation and in host defense. However, investigations into the potential overlap between these two pathways of lysosomal degradation have only just begun.

3.1 Phagocytosis

While all cells need to ingest nutrients from extracellular sources, several specialized cell types have evolved the ability to specifically engulf microbes or apoptotic cells through phagocytosis, thus serving an essential role in host defense and in maintaining tissue homeostasis. These cells, known as phagocytes, include macrophages, neutrophils, and DCs. Phagocytosis is an actin-dependent process of engulfing solid particles by the cell membrane to form an internal vesicle, or phagosome. Phagosomes undergo a sequence of fission and fusion events resulting in phagosomal maturation. Ultimately, phagosomes fuse with lysosomes and the ingested materials are degraded by hydrolases that act within the acidified environment of the lysosome. Phagocytosis can be further characterized as constitutive or inducible. For example, phagocytes will take up many types of extracellular particles varying in size and charge. However, if a particular cargo stimulates PRR activation through TLRs, phagosomal maturation is accelerated and (Blander and Medzhitov 2004) antigen processing for MHC class II is enhanced (Blander and Medzhitov 2006). Phagocytes express a variety of cell surface PRRs, allowing them to survey the extracellular space around them for both inert (through constitutive phagocytosis) and pathogenic (through induced phagocytosis) invaders.

3.2 Convergence of Autophagy and Phagocytosis

Striking similarities exist between phagosome and autophagosome formation and maturation. After all, phagocytosis and autophagy are both processes of ingestion, where the cargo or pathogen resides in extracellular or cytosolic spaces, respectively. It is tempting to speculate on the nature of the relationship between autophagy and phagocytosis and the degree to which these two processes are intertwined. Below, we will describe studies that demonstrated that TLR signals induce (1) autophagy to elicit autophagosome-mediated clearance of cytosolic pathogens or (2) recruitment of the machinery of autophagy to enhance phagosome fusion to the lysosome for clearance of extracellular pathogens.

3.2.1 TLR Signals Induce Autophagy to Clear Intracellular Pathogens

Here, we consider the evidence for the intersections between autophagy, innate recognition, and phagocytosis. The advantages of crosstalk between these pathways are readily apparent as the induction of autophagy in response to pathogen detection would mediate both xenophagy and increased transfer of cytosolic antigen into endosomal recognition compartments. Macrophages are considered "professional phagocytes" responsible for clearance of both apoptotic host cells and microbial invaders. Recently it was shown that LPS stimulation of RAW 264.7 macrophage-like cells induced LC3 punctate structure formation and increased levels of LC3II (lipidated form of LC3 capable of incorporating into the phagophore), both of which are indicative of autophagy (Xu et al. 2007). TLR4 signaling is mediated via two different sets of adaptors commonly referred to as the MyD88-dependent pathway (which utilizes the MyD88 and Toll-interleukin 1 receptor (TIR) domain-containing adaptor protein (TIRAP) adaptors) versus the MyD88-independent pathway (which utilizes Toll/IL-1 receptor domain-containing adaptor inducing IFN-β (TRIF) and TRIF-related adaptor molecule (TRAM) adaptors) (Takeda and Akira 2005). TLR4 and MyD88 dominant-negative mutant expression and TLR4, MyD88 and TRIF siRNA were used to demonstrate that the induction of autophagy following 16 h of LPS stimulation was TRIF dependent and MyD88 independent (Xu et al. 2007). The prolonged timescale described for this induction of autophagy, in comparison to the standard 1-2 h starvation or pharmacological induction of autophagy by rapamycin, suggests that LPS-mediated autophagy may involve an element of transcriptional regulation.

In order to examine the effect of LPS-induced autophagy on antimicrobial defense, the authors examined mycobacterial infection. *Mycobacterium tuberculosis* wields a double-pronged sword against PI(3)P to halt phagosome maturation in its attempt to evade immune destruction. The first assault is by glycosylated phosphatidylinositol lipoarabinomannan secretion, which leads to reduced recruitment of Vps34 and subsequent generation of PI(3)P (Vergne et al. 2003). Renegade PI(3)P that manages to escape the first assault is then targeted by SapM, a PI(3)P phosphatase, thereby preventing maturation of the phagosome (Vergne et al. 2005). Thus, mycobacteria-containing phagosomes fail to mature for lack of PI(3)P. Previous studies have demonstrated that induction of autophagy by starvation, rapamycin or by expression of a p47 resistance GTPase (LRG47) results in colocalization of the mycobacteria phagosomes with autophagosomes, leading to the elimination of mycobacteria (Gutierrez et al. 2004; Singh et al. 2006). Xu et al. found that LPS-induced autophagy also led to the colocalization of mycobacterium-containing phagosomes and autophagosomes (Xu et al. 2007). However, this occurred in a MyD88-independent but TRIF-dependent manner. An interesting parallel between TLR4-induced autophagosome fusion to the lysosome and TLR-induced phagosome fusion to the lysosome is that they both require p38 MAPK activity. In the case of TLR-induced phagosome maturation, cargo containing TLR agonists triggers signals in a phagosome-autonomous manner, specifically marking that particular phagosome for an inducible mode of maturation, which is blocked by p38 MAPK inhibitors (Blander and Medzhitov 2004). It is intriguing to speculate that phagosome-to-lysosome and autophagosome-to-lysosome fusion share common machinery that require p38 MAPK for activation.

In another study, Deretic and colleagues surveyed various TLR agonists for autophagy induction using RAW 264.7 macrophages and found that TLR3, TLR4, and TLR7 induce an approximately twofold increase in GFP-LC3 puncta per cell (Delgado et al. 2008). This study focused on TLR7-induced autophagy, as the effect was more significant than TLR3 or TLR4 activation. LC3 puncta formation occurred in a TLR7- and MyD88-dependent manner 4 h after stimulation with ssRNA or imiquimod (Delgado et al. 2008). Induction of autophagy via TLR7 signaling also resulted in a decrease in mycobacterial survival, which was abrogated by siRNA knockdown of *beclin 1* (Delgado et al. 2008). These data suggested that TLR7-mediated clearance of mycobacteria was solely mediated by autophagy. Since IFN-γ-mediated clearance of mycobacteria is also dependent on autophagy, mediated though LRG47 (Singh et al. 2006), it would be interesting to determine the relationship between TLR7- and IFN-γ-induced autophagy in mycobacterial clearance in macrophages.

A major question that remains unanswered is how TLR signals lead to the induction of autophagy. It appears that both TRIF- (Xu et al. 2007) and MyD88- (Delgado et al. 2008) dependent signals can induce autophagy in macrophages, depending on the TLR ligand used. In addition, do the autophagosomes induced by TLRs form randomly in the cytosol or is there a targeting signal? If the purpose of the formation of autophagosomes through TLRs is to clear intracellular pathogens, selective enclosure of pathogen or pathogen-containing vesicles by the autophagosome would be more beneficial to the host. Dissecting these molecular mechanisms will provide the next important advance in moving this field forward.

3.2.2 TLR-Induced Phagocytosis Utilize Machinery of Autophagy to Clear Extracellular Pathogens

The studies on TLR4- and TLR7-mediated induction of autophagy identify a means by which TLR-activation triggers autophagy, which in turn enhances autophagosome fusion with the arrested phagosomes containing mycobacteria. An elegant and provocative report by Sanjuan et al. (2007) demonstrated that molecules involved in autophagy are utilized by macrophages to facilitate fusion of phagosomes with lysosomes. This study showed that LPS, imiquimod, and CpG all induced macrophages to form LC3 punctate structures throughout the cytosol as early as 1 h after stimulation. However, upon phagocytosis of bacteria and yeast, only the cargo-containing phagosomes rapidly recruited Beclin 1 and LC3, which was followed by fusion with the lysosome and acidification. This was not observed in phagosomes containing latex beads that were not associated with TLR agonists. The recruitment of LC3 to phagosomes containing zymosan (recognized by TLR2 and Dectin-1) or *Escherichia coli* (recognized by TLR2 and TLR4) was found to be independent of MyD88 and p38 MAPK, indicating that signals mediated by Dectin-1 (zymosan) and TRIF (*E. coli*) can also result in the recruitment of LC3 to phagosomes. Live *Saccharomyces cerevisiae* engulfed by *Atg7* knockout macrophages showed a marked survival advantage over yeast taken up by wild-type macrophages. Importantly, no double-membrane structure typical of autophagosomes was ever detected around the phagosomes. Collectively, this study revealed that in contrast to the other reports, TLR signaling during conventional phagocytosis of an extracellular organism utilizes the autophagy machinery to associate LC3 rapidly with phagosomes, without the formation of conventional autophagosomes.

This concept prompts us to re-examine certain definitions and lines of reasoning that have shaped this field. First, there is the startling demonstration of LC3 recruitment to phagosomes that remain with a single lipid bilayer membrane as opposed to the double lipid bilayer that sets autophagosomes apart (Sanjuan et al. 2007). While this is considered autophagy-independent recruitment of LC3, it is still dependent on Beclin 1 and Atg7. How then should we define autophagy experimentally? Are LC3 punctate structures or LC3 lipidation patterns and the dependency on essential proteins in the autophagic pathway such as Atg5 and Atg7 sufficient to evaluate autophagy without showing degradation of long-lived cytosolic proteins? If electron microscopy does not show a double membrane surrounding the phagosome, but Beclin 1, Atg5, Atg7, and LC3 are shown to be necessary, is this a utilization of the autophagic machinery by phagocytosis, or is this perhaps evidence that phagocytosis and autophagy are convergent pathways?

It may be necessary to revisit the definition of autophagy as specifically involving the envelopment by a double membrane. In 1963, a century after the discovery of phagocytosis, Christian de Duve coined the term autophagy to describe a different type of cellular "eating," in this case "self-eating." This name was chosen to portray the observations of a vesicle that contained sequestered cytoplasmic elements including organelles in various stages of degradation, and to distinguish this phenomenon from heterophagosomes (phagosomes) that engulfed extracellular matter. In his 1966 review on the functions of lysosomes, de Duve speculated on a cellular digestive system that could originate with either autophagy or phagocytosis and then feed into the lysosomes for degradation. While a multitude of reports and images of autophagic vesicles had already accumulated by that time, only a few of them depicted a double membrane. This led to the idea that an early double membrane of preformed cytomembrane enveloped around cytoplasmic constituents and rapid degradation of the inner membrane left only a single membrane on the vesicle (De Duve and Wattiaux 1966). Later, evidence for the existence of an amphisome, a fusion product between autophagosomes and endosomes, was provided by demonstrating the cleavage of lactose sequestered in autophagosomes by endocytosed β-galactosidase during lysosomal inhibition (Gordon and Seglen 1988).

It is plausible that what has been hitherto considered two parallel processes are actually two facets of a convergent process. Future research in this area is expected to further reveal fundamental relationships between autophagy and phagocytosis.

4 Cytokine Feedback: Beating the Battle Drum

Cytokine-mediated signaling during the initiation of an immune response allows the coordination of an appropriate defense pathway suitable for protecting against a plurality of potential invaders. It is worth noting that Th1- and Th2-polarizing cytokines differentially affect autophagy. It has been reported that IFN-γ (Th1) can induce autophagy (Gutierrez et al. 2004; Inbal et al. 2002), while IL-4 and IL-13 (Th2) have an inhibitory effect on autophagy (Harris et al. 2007). Autophagy bypasses mycobacterium-mediated arrest of phagosomal maturation targeting Vps34 upon induction by IFN-γ and LRG-47 GTPase, providing a mechanism by which Th1 cells trigger phagosomal maturation in macrophages (Gutierrez et al. 2004; Singh et al. 2006). In contrast, IL-4 and IL-13 inhibit both IFN-γ-induced and starvation-induced autophagy. These cytokines stimulate the Akt pathway in macrophages, resulting in the activation of mTOR and the inhibition of starvation-induced autophagy (Harris et al. 2007). It will be important to determine the regulatory effects exerted by other important groups of cytokines, including IL-17 (produced by Th17 cells), transforming growth factor (TGF)-β, IL-10 (suppressive cytokines), type I IFNs (antiviral cytokines), IL-1, and IL-18 (released upon inflammasome activation), on autophagy and on autophagy-mediated immune effector functions described throughout this chapter. For instance, if xenophagy plays an important role in viral clearance, it would not be surprising if autophagy is induced by type I IFNs. What is unclear currently is the precise mechanism of regulation of autophagy by cytokines. Such complexities are exemplified by the fact that NF-κB, which activates the transcription of numerous genes including proinflammatory cytokines and IFNs, has been shown to downregulate autophagy (Djavaheri-Mergny et al. 2006). In addition, selective autophagy has been shown to mediate degradation of IKK and NF-κB-inducing kinase (NIK), which are essential activators of NF-κB through the canonical and noncanonical pathways, respectively (Qing et al. 2007; Qing et al. 2006). Future studies must delineate the molecular pathways involved in the cytokine regulation of autophagy and its relevance in antiviral and other antimicrobial defense in vivo.

5 Concluding Remarks

Multiple layers of host defense are triggered in response to viral infection; however, viruses multiply rapidly, and can quickly evolve to evade immune recognition and elimination. Gaining an understanding, at the molecular level, of the innate immune pathways that are stimulated by viral infection and the mechanisms by which such factors catalyze protective immunity will shed new light on critical immune pathways and provide an opportunity for therapeutic intervention. For example, dsRNA-dependent protein kinase (PKR) is an elongation and initiation factor 2α (eIF2α) kinase that is activated by dsRNA. Intriguingly, in addition to inhibiting translation of mRNAs, phosphorylated eIF2α also induces autophagy. It was demonstrated that HSV-1 infection triggers PKR-mediated induction of autophagy (Talloczy et al. 2002). However, autophagy is blocked by the HSV-1 neurovirulence factor ICP34.5 through the inhibition of both PKR signaling and Beclin 1 function (Orvedahl et al. 2007). Although virus recognition might lead to the induction of autophagy, the role of autophagy in the outcome of infection can be multifold. Some viruses may manage to inhibit autophagy (Talloczy et al. 2002), or might even harness it as a replication platform (Jackson et al. 2005), while others may be successfully eliminated by xenophagy (Liang et al. 1998). There are enumerable factors to consider before declaring a winner or a loser in the battle over autophagy between pathogen and host, and undoubtedly many of these examples and mechanisms still await discovery.

Acknowledgments We wish to thank Ruslan Medzhitov, Yosuke Kumamoto, Joseph M. Thompson, and Brian Yordy for critical reading of the manuscript. We would like to acknowledge all of our colleagues for their contributions and apologize to those whose work we could not cite. This work was supported by Public Health Service grants from the NIH/NIAID; R01 AI 054359, R01 AI 062428, R01 AI 064705 (to A.I.), and Virology Training grant T32 (to M.T.). A.I. is an Investigator of the Pathogenesis of Infectious Diseases through the Burroughs Wellcome Fund.

References

Alexopoulou L, Holt AC, Medzhitov R, Flavell RA (2001) Recognition of double-stranded RNA and activation of NF-kappaB by Toll-like receptor 3. Nature 413:732–738

Asselin-Paturel C, Boonstra A, Dalod M, Durand I, Yessaad N, Dezutter-Dambuyant C, Vicari A, O'Garra A, Biron C, Briere F, Trinchieri G (2001) Mouse type I IFN-producing cells are immature APCs with plasmacytoid morphology. Nat Immunol 2:1144–1150

Berg TO, Fengsrud M, Stromhaug PE, Berg T, Seglen PO (1998) Isolation and characterization of rat liver amphisomes. Evidence for fusion of autophagosomes with both early and late endosomes. J Biol Chem 273:21883–21892

Blander JM, Medzhitov R (2004) Regulation of phagosome maturation by signals from Toll-like receptors. Science 304:1014–1018

Blander JM, Medzhitov R (2006) Toll-dependent selection of microbial antigens for presentation by dendritic cells. Nature 440:808–812

De Duve C, Wattiaux R (1966) Functions of lysosomes. Annu Rev Physiol 28:435–492

Delgado MA, Elmaoued RA, Davis AS, Kyei G, Deretic V (2008) Toll-like receptors control autophagy. EMBO J 27:1110–1121

Diebold SS, Kaisho T, Hemmi H, Akira S, Reis e Sousa C (2004) Innate antiviral responses by means of TLR7-mediated recognition of single-stranded RNA. Science 303:1529–1531

Diebold SS, Massacrier C, Akira S, Paturel C, Morel Y, Reis e Sousa C (2006) Nucleic acid agonists for Toll-like receptor 7 are defined by the presence of uridine ribonucleotides. Eur J Immunol 36:3256–3267

Djavaheri-Mergny M, Amelotti M, Mathieu J, Besancon F, Bauvy C, Souquere S, Pierron G, Codogno P (2006) NF-kappaB activation represses tumor necrosis factor-alpha-induced autophagy. J Biol Chem 281:30373–30382

Eloranta ML, Alm GV (1999) Splenic marginal metallophilic macrophages and marginal zone macrophages are the major interferon-alpha/beta producers in mice upon intravenous challenge with herpes simplex virus. Scand J Immunol 49:391–394

Foy E, Li K, Sumpter R Jr, Loo YM, Johnson CL, Wang C, Fish PM, Yoneyama M, Fujita T, Lemon SM, Gale M Jr (2005) Control of antiviral defenses through hepatitis C virus disruption of retinoic acid-inducible gene-I signaling. Proc Natl Acad Sci USA 102:2986–2991
Gitlin L (2006) Essential role of mda-5 in type I IFN responses to polyriboinosinic: polyribocytidylic acid and encephalomyocarditis picornavirus. Proc Natl Acad Sci USA 103:8459
Gordon PB, Seglen PO (1988) Prelysosomal convergence of autophagic and endocytic pathways. Biochem Biophys Res Commun 151:40
Gutierrez MG, Master SS, Singh SB, Taylor GA, Colombo MI, Deretic V (2004) Autophagy is a defense mechanism inhibiting BCG and Mycobacterium tuberculosis survival in infected macrophages. Cell 119:753–766
Hacker H, Redecke V, Blagoev B, Kratchmarova I, Hsu LC, Wang GG, Kamps MP, Raz E, Wagner H, Hacker G, Mann M, Karin M (2006) Specificity in Toll-like receptor signalling through distinct effector functions of TRAF3 and TRAF6. Nature 439:204–207
Harris J, De Haro SA, Master SS, Keane J, Roberts EA, Delgado M, Deretic V (2007) T helper 2 cytokines inhibit autophagic control of intracellular Mycobacterium tuberculosis. Immunity 27:505–517
Heil F, Hemmi H, Hochrein H, Ampenberger F, Kirschning C, Akira S, Lipford G, Wagner H, Bauer S (2004) Species-specific recognition of single-stranded RNA via toll-like receptor 7 and 8. Science 303:1526–1529
Honda K, Taniguchi T (2006) IRFs: master regulators of signalling by Toll-like receptors and cytosolic pattern-recognition receptors. Nat Rev Immunol 6:644–658
Honda K, Yanai H, Mizutani T, Negishi H, Shimada N, Suzuki N, Ohba Y, Takaoka A, Yeh WC, Taniguchi T (2004) Role of a transductional-transcriptional processor complex involving MyD88 and IRF-7 in Toll-like receptor signaling. Proc Natl Acad Sci USA 101:15416–15421
Honda K, Yanai H, Negishi H, Asagiri M, Sato M, Mizutani T, Shimada N, Ohba Y, Takaoka A, Yoshida N, Taniguchi T (2005) IRF-7 is the master regulator of type-I interferon-dependent immune responses. Nature 434:772–777
Hornung V, Ellegast J, Kim S, Brzózka K, Jung A, Kato H, Poeck H, Akira S, Conzelmann KK, Schlee M, Endres S, Hartmann G (2006) 5′-Triphosphate RNA is the ligand for RIG-I. Science 314:994–997
Hoshino K, Sugiyama T, Matsumoto M, Tanaka T, Saito M, Hemmi H, Ohara O, Akira S, Kaisho T (2006) IkappaB kinase-alpha is critical for interferon-alpha production induced by Toll-like receptors 7 and 9. Nature 440:949–953
Ichimura Y, Kirisako T, Takao T, Satomi Y, Shimonishi Y, Ishihara N, Mizushima N, Tanida I, Kominami E, Ohsumi M, Noda T, Ohsumi Y (2000) A ubiquitin-like system mediates protein lipidation. Nature 408:488–492
Inbal B, Bialik S, Sabanay I, Shani G, Kimchi A (2002) DAP kinase and DRP-1 mediate membrane blebbing and the formation of autophagic vesicles during programmed cell death. J Cell Biol 157:455
Ishii KJ, Coban C, Kato H, Takahashi K, Torii Y, Takeshita F, Ludwig H, Sutter G, Suzuki K, Hemmi H, Sato S, Yamamoto M, Uematsu S, Kawai T, Takeuchi O, Akira S (2006) A Toll-like receptor-independent antiviral response induced by double-stranded B-form DNA. Nat Immunol 7:40–48
Jackson WT, Giddings TH Jr, Taylor MP, Mulinyawe S, Rabinovitch M, Kopito RR, Kirkegaard K (2005) Subversion of cellular autophagosomal machinery by RNA viruses. PLoS Biol 3:156
Janeway CA Jr (1989) Approaching the asymptote? Evolution and revolution in immunology. Cold Spring Harb Symp Quant Biol 54 Pt 1:1–13
Jounai N, Takeshita F, Kobiyama K, Sawano A, Miyawaki A, Xin KQ, Ishii KJ, Kawai T, Akira S, Suzuki K, Okuda K (2007) The Atg5-Atg12 conjugate associates with innate antiviral immune responses. Proc Natl Acad Sci USA 104:14050–14055
Kabeya Y, Mizushima N, Ueno T, Yamamoto A, Kirisako T, Noda T, Kominami E, Ohsumi Y, Yoshimori T (2000) LC3, a mammalian homologue of yeast Apg8p, is localized in autophagosome membranes after processing. Embo J 19:5720–5728
Kato H (2006) Differential roles of MDA5 and RIG-I helicases in the recognition of RNA viruses. Nature 441:101

Klionsky DJ, Emr SD (2000) Autophagy as a regulated pathway of cellular degradation. Science 290:1717–1721

Klionsky DJ, Cregg JM, Dunn WA Jr, Emr SD, Sakai Y, Sandoval IV, Sibirny A, Subramani S, Thumm M, Veenhuis M, Ohsumi Y (2003) A unified nomenclature for yeast autophagy-related genes. Dev Cell 5:539–545

Krug A, Luker GD, Barchet W, Leib DA, Akira S, Colonna M (2004) Herpes simplex virus type 1 activates murine natural interferon-producing cells through toll-like receptor 9. Blood 103: 1433–1437

Kuma A, Hatano M, Matsui M, Yamamoto A, Nakaya H, Yoshimori T, Ohsumi Y, Tokuhisa T, Mizushima N (2004) The role of autophagy during the early neonatal starvation period. Nature 432:1032–1036

Lee HK, Lund JM, Ramanathan B, Mizushima N, Iwasaki A (2007) Autophagy-dependent viral recognition by plasmacytoid dendritic cells. Science 315:1398–1401

Levine B, Deretic V (2007) Unveiling the roles of autophagy in innate and adaptive immunity. Nat Rev Immunol 7:767–777

Levine B, Klionsky DJ (2004) Development by self-digestion: molecular mechanisms and biological functions of autophagy. Dev Cell 6:463

Liang XH, Kleeman LK, Jiang HH, Gordon G, Goldman JE, Berry G, Herman B, Levine B (1998) Protection against fatal Sindbis virus encephalitis by Beclin, a novel Bcl-2-interacting protein. J Virol 72:8586–8596

Liu YJ (2005) IPC: professional type 1 interferon-producing cells and plasmacytoid dendritic cell precursors. Annu Rev Immunol 23:275–306

Lund J, Sato A, Akira S, Medzhitov R, Iwasaki A (2003) Toll-like receptor 9-mediated recognition of herpes simplex virus-2 by plasmacytoid dendritic cells. J Exp Med 198:513–520

Lund JM, Alexopoulou L, Sato A, Karow M, Adams NC, Gale NW, Iwasaki A, Flavell RA (2004) Recognition of single-stranded RNA viruses by Toll-like receptor 7. Proc Natl Acad Sci USA 101:5598–5603

Meylan E, Curran J, Hofmann K, Moradpour D, Binder M, Bartenschlager R, Tschopp J (2005) Cardif is an adaptor protein in the RIG-I antiviral pathway and is targeted by hepatitis C virus. Nature 437:1167–1172

Mizushima N, Noda T, Yoshimori T, Tanaka Y, Ishii T, George MD, Klionsky DJ, Ohsumi M, Ohsumi Y (1998) A protein conjugation system essential for autophagy. Nature 395:395–398

Mizushima N, Yamamoto A, Matsui M, Yoshimori T, Ohsumi Y (2004) In vivo analysis of autophagy in response to nutrient starvation using transgenic mice expressing a fluorescent autophagosome marker. Mol Biol Cell 15:1101–1111

Oganesyan G, Saha SK, Guo B, He JQ, Shahangian A, Zarnegar B, Perry A, Cheng G (2006) Critical role of TRAF3 in the Toll-like receptor-dependent and -independent antiviral response. Nature 439:208–211

Ohsumi Y (2001) Molecular dissection of autophagy: two ubiquitin-like systems. Nat Rev Mol Cell Biol 2:211

Orvedahl A, Alexander D, Talloczy Z, Sun Q, Wei Y, Zhang W, Burns D, Leib DA, Levine B (2007) HSV-1 ICP34.5 confers neurovirulence by targeting the Beclin 1 autophagy protein. Cell Host Microbe 1:23

Pichlmair A, Schulz O, Tan CP, Naslund TI, Liljestrom P, Weber F, Reis e Sousa C (2006) RIG-I-mediated antiviral responses to single-stranded RNA bearing 5′-phosphates. Science 314:997–1001

Qing G, Yan P, Xiao G (2006) Hsp90 inhibition results in autophagy-mediated proteasome-independent degradation of IκB kinase (IKK). Cell Res 16:895

Qing G, Yan P, Qu Z, Liu H, Xiao G (2007) Hsp90 regulates processing of NF-κB2 p100 involving protection of NF-κB-inducing kinase (NIK) from autophagy-mediated degradation. Cell Res 17:520

Sanjuan MA, Dillon CP, Tait SW, Moshiach S, Dorsey F, Connell S, Komatsu M, Tanaka K, Cleveland JL, Withoff S, Green DR (2007) Toll-like receptor signalling in macrophages links the autophagy pathway to phagocytosis. Nature 450:1253–1257

Schmid D, Munz C (2007) Innate and adaptive immunity through autophagy. Immunity 27:11–21

Seth RB, Sun L, Ea CK, Chen ZJ (2005) Identification and characterization of MAVS, a mitochondrial antiviral signaling protein that activates NF-kappaB and IRF-3. Cell 122: 669–682

Shinohara ML, Lu L, Bu J, Werneck MB, Kobayashi KS, Glimcher LH, Cantor H (2006) Osteopontin expression is essential for interferon-alpha production by plasmacytoid dendritic cells. Nat Immunol 7:498–506

Singh SB, Davis AS, Taylor GA, Deretic V (2006) Human IRGM induces autophagy to eliminate intracellular mycobacteria. Science 313:1438–1441

Stetson DB, Medzhitov R (2006) Recognition of cytosolic DNA activates an IRF3-dependent innate immune response. Immunity 24:93–103

Tabeta K, Hoebe K, Janssen EM, Du X, Georgel P, Crozat K, Mudd S, Mann N, Sovath S, Goode J, Shamel L, Herskovits AA, Portnoy DA, Cooke M, Tarantino LM, Wiltshire T, Steinberg BE, Grinstein S, Beutler B (2006) The Unc93b1 mutation 3d disrupts exogenous antigen presentation and signaling via Toll-like receptors 3, 7 and 9. Nat Immunol 7:156–164

Takaoka A (2007) DAI (DLM-1/ZBP1) is a cytosolic DNA sensor and an activator of innate immune response. Nature 448:501

Takaoka A, Yanai H, Kondo S, Duncan G, Negishi H, Mizutani T, Kano S, Honda K, Ohba Y, Mak TW, Taniguchi T (2005) Integral role of IRF-5 in the gene induction programme activated by Toll-like receptors. Nature 434:243–249

Takeda K, Akira S (2005) Toll-like receptors in innate immunity. Int Immunol 17:1

Takeshita F, Kobiyama K, Miyawaki A, Jounai N, Okuda K (2008) The non-canonical role of Atg family members as suppressors of innate antiviral immune signaling. Autophagy 4:67–69

Takeuchi O, Akira S (2007) Recognition of viruses by innate immunity. Immunol Rev 220:214

Talloczy Z, Jiang W, Virgin HW, Leib DA, Scheuner D, Kaufman RJ, Eskelinen EL, Levine B (2002) Regulation of starvation- and virus-induced autophagy by the eIF2alpha kinase signaling pathway. Proc Natl Acad Sci USA 99:190–195

Taniguchi T, Takaoka A (2002) The interferon-alpha/beta system in antiviral responses: a multimodal machinery of gene regulation by the IRF family of transcription factors. Curr Opin Immunol 14:111–116.

Uematsu S, Sato S, Yamamoto M, Hirotani T, Kato H, Takeshita F, Matsuda M, Coban C, Ishii KJ, Kawai T, Takeuchi O, Akira S (2005) Interleukin-1 receptor-associated kinase-1 plays an essential role for Toll-like receptor (TLR)7- and TLR9-mediated interferon-{alpha} induction. J Exp Med 201:915–923

Vergne I, Chua J, Deretic V (2003) Tuberculosis toxin blocking phagosome maturation inhibits a novel $^{Ca2+}$/calmodulin-PI3K hVPS34 cascade. J Exp Med 198:653

Vergne I, Chua J, Lee HH, Lucas M, Belisle J, Deretic V (2005) Mechanism of phagolysosome biogenesis block by viable *Mycobacterium tuberculosis*. Proc Natl Acad Sci USA 102:4033

Watarai H, Sekine E, Inoue S, Nakagawa R, Kaisho T, Taniguchi M (2008) PDC-TREM, a plasmacytoid dendritic cell-specific receptor, is responsible for augmented production of type I interferon. Proc Natl Acad Sci USA 105:2993–2998

Xu LG, Wang YY, Han KJ, Li LY, Zhai Z, Shu HB (2005) VISA is an adapter protein required for virus-triggered IFN-beta signaling. Mol Cell 19:727–740

Xu Y, Jagannath C, Liu XD, Sharafkhaneh A, Kolodziejska KE, Eissa NT (2007) Toll-like receptor 4 is a sensor for autophagy associated with innate immunity. Immunity 27:135–144

Yoneyama M, Kikuchi M, Natsukawa T, Shinobu N, Imaizumi T, Miyagishi M, Taira K, Akira S, Fujita T (2004) The RNA helicase RIG-I has an essential function in double-stranded RNA-induced innate antiviral responses. Nat Immunol 5:730–737

Yoneyama M, Kikuchi M, Matsumoto K, Imaizumi T, Miyagishi M, Taira K, Foy E, Loo YM, Gale M Jr, Akira S, Yonehara S, Kato A, Fujita T (2005) Shared and unique functions of the DExD/H-box helicases RIG-I, MDA5, and LGP2 in antiviral innate immunity. J Immunol 175:2851–2858

Autophagy in MHC Class II Presentation of Endogenous Antigens

Monique Gannagé and Christian Münz

Contents

1 Introduction 124
2 Source Proteins of Peptide Ligands For MHC Class II Molecules 125
3 Intracellular Antigen Processing onto MHC Class II Proteins 127
4 Macroautophagy as a Source of MHC Class II Ligands 128
4.1 Model Antigens 129
4.2 Tumor Antigens 130
4.3 Intracellular Pathogens as a Source of Macroautophagy-Dependent MHC Class II Antigens 131
4.4 Self-Antigens as a Source of MHC Class II Epitopes After Macroautophagy 132
4.5 Macroautophagy in Professional Antigen-Presenting Cells 133
4.6 Macroautophagy in Epithelial Cells 133
5 Conclusions 136
References 136

Abstract Macroautophagy is a catabolic process for the lysosomal turnover of cell organelles and protein aggregates. Lysosomal degradation products are displayed by major histocompatibility class II molecules to $CD4^+$ T cells in the steady state for tolerance induction and during infections to mount adaptive immune responses. It has recently been shown that macroautophagy substrates can also give rise to MHC class II ligands. We review here the breadth of antigens that may utilize this pathway and the possible implications of this alternate route to MHC class II antigen presentation for immunity and tolerance. Based on this discussion, it is apparent that the regulation of macroautophagy may be beneficial in various disease settings in order to enhance adaptive immune responses or to reduce autoimmunity.

M. Gannagé and C. Münz (✉)
Viral Immunobiology, Institute of Experimental Immunology, University Hospital of Zürich, Winterthurerstrasse 190, CH-8057 Zürich, Switzerland
e-mail: christian.muenz@usz.ch

B. Levine et al. (eds.), *Autophagy in Infection and Immunity,*
Current Topics in Microbiology and Immunology 335,
DOI 10.1007/ 978-3-642-00302-8_6,

1 Introduction

The adaptive immune system monitors, via its T cells, the protein degradation products of cells in order to check for deviations from the steady-state peptide repertoire under display. $CD8^+$ T cells monitor peptide presentation on major histocompatibility complex (MHC) class I molecules and $CD4^+$ T cells monitor peptide presentation by MHC class II molecules. MHC class I and II molecules are thought to present peptide products of proteasomal and lysosomal proteolysis, respectively (Trombetta and Mellman 2005). MHC class I and II ligands are loaded in different intracellular compartments. Proteasomal products are transported into the endoplasmic reticulum (ER) via the transporter associated with antigen processing (TAP), and get primarily loaded in the ER onto newly synthesized MHC class I molecules, which then travel to the cell surface for $CD8^+$ T cell immune surveillance. In contrast, MHC class II molecules are loaded in late endosomal compartments with intravesicular membranes, termed MHC class II-containing compartments (MIIC). MHC class II molecules are directed there by the invariant chain (Ii), a chaperone that associates with newly synthesized MHC class II molecules in the ER and prevents premature peptide binding, and also facilitates the vesicular transport to the MIIC. In the MIIC, antigen and Ii are degraded by lysosomal hydrolases, and then the last remnant of Ii, the class II-associated Ii chain peptide (CLIP), is exchanged for a high-affinity peptide by the chaperone HLA-DM. The stable MHC class II/peptide complex is successively released to the cell surface for $CD4^+$ T cell stimulation. Since the proteasome is located in the cytosol and nucleus, and the MIIC is part of the endosomal system, MHC class I displays mainly fragments of cytosolic and nuclear antigens, while MHC class II presents mainly peptides from secreted and membrane proteins. However, with the discovery of multiple pathways leading to the access of endocytosed proteins to the cytosol for cross-presentation on MHC class I molecules (Groothuis and Neefjes 2005) and of autophagic delivery of cytoplasmic constituents to late endosomes, followed by lysosomal proteolysis for MHC class II presentation (Schmid and Münz 2007), the cellular localization has become less important, and the degradation behavior of the individual antigen is emerging as a more important determinant of MHC class I and II presentation to $CD8^+$ and $CD4^+$ T cells. We will discuss in the following chapter how research on autophagy, which has blossomed in the recent past, enriches our understanding of what can be seen by the adaptive immune system.

When we speak of autophagy, we must consider several pathways that lead to the lysosomal degradation of cytoplasmic constituents. These include primarily macroautophagy, microautophagy and chaperone-mediated autophagy (Mizushima and Klionsky 2007). In this chapter, we will primarily discuss macroautophagy and chaperone-mediated autophagy, since the evidence for microautophagy in higher eukaryotes is sparse, and an involvement of this pathway in MHC presentation has not yet been described. During macroautophagy, around 30 essential autophagy-related gene (*Atg*) products participate in the formation of a double-membrane-surrounded vesicle, the so-called autophagosome, and its fusion with late endosomes and lysosomes. While other chapters in this book will discuss the mechanistic basis of this process in more detail (see the first three chapters of this volume), we will briefly outline the two ubiquitin-like systems involved in autophagosome formation (Ohsumi 2001), since these are primarily utilized by immunologists to monitor macroautophagy and to selectively interfere with it.

One of these systems uses Atg8, which in mammals is a family of proteins with four members: LC3, GABARAP, GATE-16 and ApgL (Hemelaar et al. 2003). Atg8/LC3 is processed by the Atg4 protease to expose a C-terminal glycine residue, and is then coupled in a ubiquitin-like reaction (assisted by the E1-activating enzyme Atg7 and the E2-conjugating enzyme Atg3) to the lipid phosphatidylethanolamine (PE) on the inside and outside of the forming autophagosome membrane, called the isolation membrane. The second ubiquitin-like system couples Atg12 to Atg5, catalyzed by Atg7 and the E2-conjugating enzyme Atg10. The Atg12/Atg5 heterodimer then associates with Atg16L and decorates the outer surface of the isolation membrane. Once the cup-shaped isolation membrane closes around its cargo to form the completed autophagosome, the Atg12/Atg5/Atg16L complex dissociates from it and a substantial portion of the Atg8/LC3 is also cleaved by Atg4 from the outer membrane, but Atg8/LC3 on the inner membrane stays associated with the autophagosome and is degraded with the inner membrane in autolysosomes.

In contrast to macroautophagy, which requires the formation of a separate vesicle, chaperone-mediated autophagy imports its substrates directly into the lysosome (Agarraberes and Dice 2001). For this purpose, a signal peptide in the protein substrate is recognized by a cytosolic HSC70 chaperone and then imported via LAMP-2A into the lysosome with the help of a lysosomal HSC70 chaperone. While the pentamer KFERQ from RNAse A was originally described as the prototypic signal peptide for directing substrates to chaperone-mediated autophagy, a more degenerate sequence of five amino acids including a glutamine preceded or followed by one or two basic and one or two hydrophobic residues has been found to efficiently mediate lysosomal import (Chiang and Dice 1988). Not too surprisingly, these essential catabolic processes grouped under the term "autophagy" are involved in many biological functions (Levine and Kroemer 2008; Mizushima et al. 2008), but we will focus on their role in alarming and tolerizing the mammalian immune system, primarily T cells.

2 Source Proteins of Peptide Ligands For MHC Class II Molecules

One way to investigate if MHC molecules present fragments of autophagic substrates to T cells is to analyze their natural peptide ligands after affinity purification of MHC molecules and acid extraction of their small molecular weight cargo (Rammensee et al. 1999). Interestingly, up to 30% of MHC class I ligands originate from cytosolic and nuclear proteins (Dengjel et al. 2005). Of these, some proteins preferentially give rise to MHC class I ligands and some primarily to MHC class II ligands. For example, cyclins with protein half-lives of around 15 min (Barette et al. 2001; Germain et al. 2000; Singer et al. 1999) have been found to be source proteins for at least ten natural peptide ligands of mouse and human MHC class I molecules, but have only been found to be a source of MHC class II ligands on one occasion (Rammensee et al. 1999). Similarly, ornithine decarboxylase with a half-life of 12 min (Dice and Goldberg 1975) contains two natural ligands, which have been eluted from human MHC class I molecules (Rammensee et al. 1999). On the other hand, MHC class II ligands seem to be preferentially generated from long-lived proteins, and this protein pool has been described as being preferentially turned over by autophagy (Henell et al. 1987). HSC70- and HSP70-derived peptides are two- to threefold more frequently found as natural ligands of MHC class II molecules than MHC class I molecules, and these chaperones have half-lives of between 4 and 20 h (Jiang et al. 2001; Landry et al. 1991; Li and Duncan 1995). In addition, glyceraldehyde-3-phosphate dehydrogenase (GAPDH), a protein with an extraordinarily long half-life of 130 h (Dice and Goldberg 1975), is a prominent source of MHC class II ligands, and GAPDH peptides have been eluted from five human MHC class II molecules, but not yet from MHC class I molecules (Dengjel et al. 2005; Rammensee et al. 1999). In addition, GAPDH has been described as a substrate of chaperone-mediated autophagy (Aniento et al. 1993) and has been isolated from autophagosomes (Fengsrud et al. 2000). Therefore, long-lived proteins, which are preferential substrates of autophagy, more frequently give rise to natural MHC class II ligands, and peptides from GAPDH, a known autophagy substrate, have been eluted from several MHC class II molecules.

In addition to potential and proven autophagy substrates, MHC class II ligands from proteins of the molecular machinery of macroautophagy have also been eluted from MHC class II molecules. Two peptides of Atg8/LC3 ($MAP1LC3B_{93-109}$ and $MAP1LC3B_{93-110}$) have been found on HLA-DR4 and/or HLA-DR1 of an Epstein-Barr virus (EBV)-transformed B cell line (Dengjel et al. 2005). In addition, $H2\text{-}A^{g7}$, a diabetogenic mouse MHC class II molecule, was found to present a peptide from another mammalian Atg8 member, $GABARAP_{29-45}$, on the surface of a mouse B cell lymphoma cell line (Suri et al. 2008). Interestingly, a potentially autoreactive $CD4^+$ T cell clone against the latter peptide was isolated from pancreas-draining lymph nodes, identifying GABARAP as a potential autoantigen in type I diabetes in mice (Suri et al. 2008). These results suggest that as well as being presented on MHC class II molecules, mammalian Atg8 proteins may also become autoantigens in susceptible genetic backgrounds.

Finally, proteins may also make themselves destined for autophagy by inhibiting their degradation through the other main proteolytic machinery in cells, the proteasome (Ciechanover 2005). Along these lines, the nuclear antigen 1 of EBV (EBNA1) encodes a glycine alanine (GA) repeat domain that impairs proteasomal degradation in *cis*, and at the same time prolongs EBNA1's half-life (Hoyt et al. 2006; Levitskaya et al. 1995; Levitskaya et al. 1997; Tellam et al. 2001). Deletion of the GA repeat domain enhances proteasomal degradation and MHC class I presentation of EBNA1 (Lee et al. 2004; Levitskaya et al. 1997). In contrast, long-lived, full-length EBNA1 is intracellularly processed for MHC class II presentation via macroautophagy (Münz et al. 2000; Paludan et al. 2005). Similarly, long-lived influenza matrix protein 1 (MP1) is presented on MHC class II after intracellular processing (Jaraquemada et al. 1990). However, when it is destabilized for more efficient proteasomal degradation via N-end rule modification, MHC class II presentation is severely compromised (Gueguen and Long 1996). Finally, the nuclear antigen RAD23, which is involved in nucleotide excision repair, contains a UBA2 domain that blocks proteasomal processing and extends the protein half-life of RAD23 (Heessen et al. 2005). Interestingly, MHC class II molecules of cells that were starved to upregulate autophagy presented elevated levels of a RAD23-derived peptide ligand along with other peptides of nuclear and cytosolic sources, while the presentation of membrane and secretory proteins was not increased (Dengjel et al. 2005). Altogether, these data suggest that proteins that are protected from proteasomal degradation and therefore have long half-lives or proteins that are targeted to autophagosomes as part of the molecular machinery involved in the generation of these vesicles are substrates of macroautophagy and are preferentially presented onto MHC class II molecules to $CD4^+$ T cells.

3 Intracellular Antigen Processing onto MHC Class II Proteins

In addition to the MHC class II ligand isolation studies that identified the presentation of cytosolic and nuclear antigens, some $CD4^+$ T cell specificities recognize their antigens after intracellular processing. This fact was primarily documented with in vitro MHC class II presentation assays to $CD4^+$ T cell clones or lines, whereby the antigen is expressed either in target cells together with the cognate MHC class II allele or in bystander cells that are mismatched in their MHC both to the T cell specificity and to target cells with the correct MHC class II molecules, which are mixed with the antigen-expressing cells in the culture. If only the coexpression of antigen and restricting MHC class II molecule is recognized, an endogenous route rather than release and re-uptake is suggested for antigen processing onto MHC class II. However, autophagy has only been investigated and documented as the pathway by which the respective antigen gains access to MHC class II loading compartments for a few of these antigens.

Four separate intracellular pathways for MHC class II presentation have now been postulated. The first pathway involves MHC class II loading with peptide epitopes of secreted or membrane proteins that are co-translationally inserted into the ER, travel with MHC class II molecules to MIICs for loading, and are degraded along this route (either in the ER itself or by lysosomal hydrolysis). Several viral and targeted model antigens have been shown to follow this proteasome- and TAP-independent intracellular pathway of membrane, secreted and ER antigens. Prominent examples of this pathway are influenza hemagglutinin (Aichinger et al. 1997; Kittlesen et al. 1993), the glycoprotein poison tail of vesicular stomatitis virus (Bartido et al. 1995), ER-retained immunoglobulin λ-light chain (Weiss and Bogen 1989; Weiss and Bogen 1991), and secreted as well as ER-retained hen egg white lysozyme (Bonifaz et al. 1999; Brooks et al. 1991; Brooks and McCluskey 1993; Dissanayake et al. 2005).

A second pathway provides access for cytosolic antigens to MHC class II loading via TAP, and most likely requires proteasomal processing of the antigen in the cytosol. Influenza hemagglutinin peptides expressed in the cytosol after transfection (Malnati et al. 1992), and hemagglutinin and neuraminidase that escape from endosomes in cross-presenting dendritic cells (Tewari et al. 2005) use this route for MHC class II presentation to $CD4^+$ T cells.

The other two pathways do not utilize ER import mechanisms but directly target late endosomal and lysosomal compartments for MHC class II loading. The third pathway of intracellular antigen processing for MHC class II presentation is proteasome dependent but TAP independent. Ovalbumin, introduced into the cytosol by hyperosmotic pinosome lysis, (Mukherjee et al. 2001), cytosolic and membrane-bound forms of the MHC class II α-chain H2-Eα (Dani et al. 2004; Mukherjee et al. 2001), and the autoantigens glutamate decarboxylase 65 (GAD65) as well as the mutant immunoglobulin κ-light chain SMA (Lich et al. 2000; Zhou et al. 2005) have been shown to use this pathway. MHC class II presentation of GAD65 and SMA was increased after overexpression of the transporter for chaperone-mediated autophagy LAMP-2A (Zhou et al. 2005). Therefore, chaperone-mediated autophagy might import proteasomal products into lysosomal compartments for MHC class II presentation.

The fourth and last pathway for intracellular antigen processing onto MHC class II is proteasome and TAP independent. In addition to the cytosolic, secretory and membrane proteins processed for MHC class II presentation by the pathways outlined above, nuclear and mitochondrial antigens have also been found to be delivered for MHC class II loading by an intracellular route (Le Roy et al. 2002; Paludan et al. 2005; Qi et al. 2000). Furthermore, some cytosolic proteins are also processed for MHC class II presentation by this pathway that relies on lysosomal hydrolysis and targets substrates outside the secretory pathway (Chen et al. 1998; Jacobson et al. 1989; Jaraquemada et al. 1990). A long protein half-life seems beneficial in order to enter this pathway (Gueguen and Long 1996; Paludan et al. 2005). Even so, it is currently not known how measles virus matrix and nucleocapsid (Jacobson et al. 1989), hepatitis C virus core protein (Chen et al. 1998), and influenza matrix protein 1 (Gueguen and Long 1996; Jaraquemada et al. 1990) access lysosomal proteolysis for MHC class II presentation. EBNA1 reaches MIICs via macroautophagy for $CD4^+$ T cell stimulation (Paludan et al. 2005). In addition, other cytosolic antigens, including neomycin phosphotransferase II (Nimmerjahn et al. 2003) and complement C5 (Brazil et al. 1997), are also presented on MHC class II molecules after autophagy. Therefore, multiple pathways seem to give intracellular antigens access to MHC class II presentation, and probably intracellular location as well as susceptibility to proteasomal versus lysosomal degradation largely determine which pathway will be used by a given antigen for loading onto MHC class II molecules.

4 Macroautophagy as a Source of MHC Class II Ligands

We have previously demonstrated that macroautophagy is a constitutive pathway in professional antigen-presenting cells (dendritic cells and B cells) as well as in epithelial cells. In these cells, autophagosomes constitutively fuse with MHC class II compartments, delivering their contents to the MHC class II loading machinery. Hence, autophagosome cargo is a new source of endogenous MHC class II epitopes in the steady state as well as in pathogenic situations. Several studies have now implicated this pathway in the endogenous processing of MHC class II epitopes for model antigens and (more relevant physiologically) for some tumor or viral antigens (Fig. 1).

4.1 Model Antigens

We used two model antigens, a viral antigen—the matrix protein 1 (MP1) of influenza virus, and a tumor antigen—the cancer testis antigen NY-ESO-1, to generate two fusion proteins with the autophagosome membrane protein Atg8/LC3 (MP1-LC3

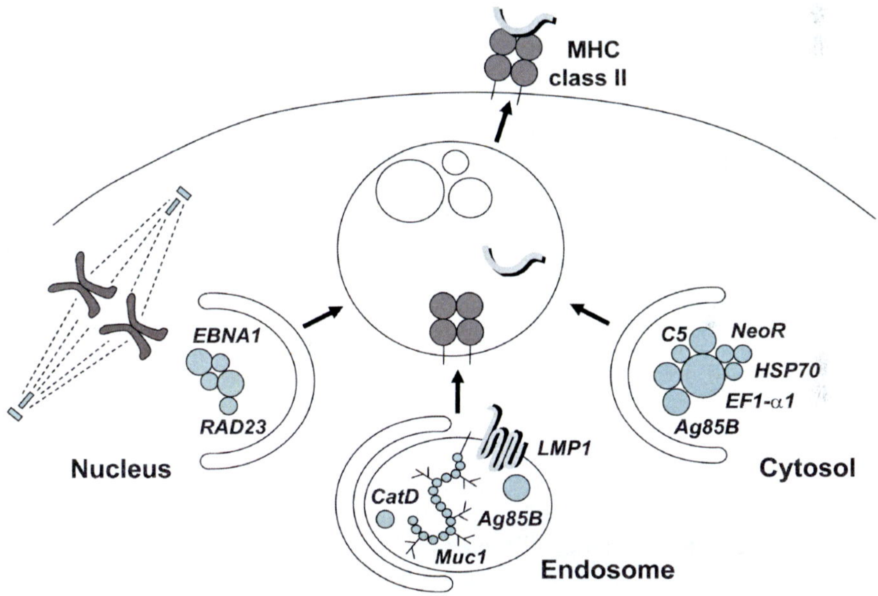

Fig. 1 Examples of proteins that may gain access to MHC class II presentation via macroautophagy. The viral nuclear antigen EBNA1 (Paludan et al. 2005), the endosomal tumor antigen Muc1 (Dorfel et al. 2005) and the cytosolic model proteins NeoR (Nimmerjahn et al. 2003) and C5 (Brazil et al. 1997) have been shown to be recognized by CD4+ T cells after macroautophagy. The *M. tb* antigen Ag85B is also processed onto MHC class II after macroautophagy, but it is unclear if this happens from its endosomal or cytosolic pool (Chinnaswamy Jagannath, personal communications). The self-proteins RAD23 (nucleus), cathepsin D (endosome), HSP70 and EF1-a1 (cytosol) were significantly better presented on MHC class II molecules after starvation-induced macroautophagy (Dengjel et al. 2005). Finally, the viral membrane protein LMP1 has been found to be a substrate of macroautophagy (Lee and Sugden 2008), and its presentation on MHC class II is frequently recognized by CD4+ T cells (Marshall et al. 2003; Münz et al. 2000)

and NY-ESO-1-LC3, respectively; Schmid et al. 2007, and unpublished data). Since Atg8/LC3 is degraded with the inner membrane in autolysosomes, our idea was to target these antigens to the autophagy compartment and therefore enhance their MHC class II presentation. Indeed, by expressing NY-ESO-1-LC3 and MP1-LC3 in epithelial cells, we observed a significant increase (five- to tenfold) in IFN-γ secretion by antigen-specific $CD4^+$ T cell clones compared to the unmodified proteins NY-ESO-1 and MP1, respectively. Therefore, by targeting these two antigens to the macroautophagy compartment, we were able to show a significant enhancement in MHC class II presentation by accessing macroautophagy.

It has been suggested that another cytosolic model antigen, neomycin phosphotransferase II (NeoR), is delivered to MHC class II-containing compartments by macroautophagy (Nimmerjahn et al. 2003). In this study, the recognition of a $CD4^+$ T cell epitope derived from NeoR was significantly reduced by pharmacological inhibitors of macroautophagy with 3-methyladenine or low concentrations of wortmannin. As a control, the exogenous presentation of NeoR was not affected by these pharmacological inhibitors, and neither was the $CD8^+$ T cell recognition of control antigens. Similarly, overexpression of complement C5 in mouse macrophages has been shown to elicit $CD4^+$ T cell recognition (Brazil et al. 1997). This recognition was sensitive to 3-methyladenine inhibition, while external addition of C5 could be efficiently processed for MHC class II presentation even in the presence of 3-methyladenine. These examples document the efficient processing of cytosolic antigens for $CD4^+$ T cell stimulation after macroautophagy.

In addition to these model antigens, evidence is mounting that a number of tumor- and pathogen-derived antigens are delivered to the MIIC after macroautophagy, and this will be reviewed in the next two sections.

4.2 Tumor Antigens

The role that macroautophagy plays during tumorigenesis has been a matter of debate. It remains unclear if macroautophagy acts as a tumor suppressor and/or a tumor pro-survival pathway. Nevertheless, tumor antigens have been shown to enter this pathway for the generation of tumor-specific $CD4^+$ T cell epitopes. One of the first tumor antigens to be described as being processed by macroautophagy was mucin gene 1 (*MUC1*). Dorfel and coworkers demonstrated that in *MUC1*-transfected dendritic cells, the processing of a $CD4^+$ T cell epitope of MUC1 was dependent on the macroautophagy pathway (Dorfel et al. 2005). Indeed, by inhibiting the pathway using class III PI3K inhibitors (3-methyladenine or wortmannin), they could significantly reduce the proliferation of MUC1-specific $CD4^+$ T cell lines, while the $CD8^+$ T cell response to the same antigen was unchanged.

It is tempting to speculate that other tumor antigens can also be processed by macroautophagy. Indeed, there are an increasing number of MHC class II epitopes derived from melanoma antigens that are described as being processed via an endogenous route onto MHC class II molecules (Godefroy et al. 2006; Zarour et al.

2000). Cytosolic melanoma antigens such as Melan-A/MART-1 or MAGE3 that lack a sorting signal to melanosomes and are not expressed on the cell surface could represent a potential pool of macroautophagy substrates. Clearly more studies are needed to assess which tumor antigens can gain access to MHC class II presentation and CD4$^+$ T cell stimulation. A better understanding of tumor immune surveillance by CD4$^+$ T cells after macroautophagy could provide insights into how adaptive tumor immune responses are influenced by macroautophagy regulation in tumors.

4.3 *Intracellular Pathogens as a Source of Macroautophagy-Dependent MHC Class II Antigens*

Several intracellular pathogens interfere with the autophagic machinery. While some pathogens can be degraded by macroautophagy, like *Rickettsia conorii* (Walker et al. 1997), others like *Shigella flexneri* (Ogawa et al. 2005) have evolved to evade this catabolic pathway, or make possibly use of autophagosomal membranes for their own replication (Jackson et al. 2005). Despite this growing evidence for the involvement of macroautophagy in innate immune response to intracellular pathogens, there are only a few studies suggesting a direct link between the adaptive immune response and macroautophagic clearance of bacteria, parasites and viruses.

4.3.1 *Mycobacterium tuberculosis*

A recent study demonstrated that induction of macroautophagy by rapamycin or starvation significantly enhanced MHC class II presentation of the immunodominant Ag85B antigen on *M. tuberculosis*-infected macrophages and dendritic cells (Chinnaswamy Jagannath, personal communications). This mechanism is an active process, occurring only during live *M. tuberculosis* infection, and is specifically inhibited by silencing of the essential autophagy gene *beclin 1* (the mammalian ortholog of yeast *ATG6*). Moreover, confocal immunofluorescence analysis of infected macrophages demonstrated that mycobacteria localize to autophagosomes upon rapamycin treatment. Furthermore, rapamycin pretreatment of *M. tuberculosis*-infected dendritic cells significantly enhanced the efficiency of vaccination by adoptive DC transfer, and protected from *M. tuberculosis* challenge in mice. This interesting work describes for the first time how the manipulation of macroautophagy can improve vaccine strategies.

4.3.2 Epstein–Barr Virus (EBV)

Two latent antigens of EBV, the nuclear antigen EBNA1 and the membrane protein LMP1, have been described as being at least partially degraded by macroautophagy. LMP1 (latent membrane protein 1), the main oncogene of the virus, has recently been shown to induce macroautophagy in a dose-dependent manner (Lee and Sugden 2008). In this recent work, Sugden and colleagues demonstrated that LMP1 accumulated upon the inhibition of two essential autophagy genes, *beclin 1* and *ATG7*. Therefore, LMP1 regulates its own clearance via macroautophagy, but further studies are needed to evaluate if CD4$^+$ T cell epitopes of LMP1, which are frequently recognized in healthy virus carriers by IFN-γ- and/or IL-10-producing CD4$^+$ T cells (Münz et al. 2000); (Marshall et al. 2003), can be processed via this pathway.

Another latent protein, EBV nuclear antigen 1 (EBNA1), the most consistently recognized CD4$^+$ T cell antigen of latent Epstein-Barr virus infection, gains access to MHC class II loading compartments via macroautophagy. Indeed, our group has demonstrated that in EBV-transformed B cells and EBNA1-transfected cell lines, EBNA1 is degraded via lysosomal degradation after macroautophagy (Paludan et al. 2005). This pathway is also at least partially responsible for delivering this antigen for MHC class II presentation. Upon the inhibition of macroautophagy by siRNA-mediated silencing of *ATG12* or the pharmacological inhibition of macroautophagy with 3-methyladenine, antigen presentation of distinct epitopes to EBNA1-specific CD4$^+$ T cell clones was significantly reduced, while CD8$^+$ T cell recognition of another nuclear EBV antigen was not affected. This study was the first demonstration of MHC class II presentation of physiological levels of a pathogen-derived antigen via the autophagic route.

4.4 Self-Antigens as a Source of MHC Class II Epitopes After Macroautophagy

The influence of macroautophagy on the MHC class II self-ligandome has been investigated in an elegant study (Dengjel et al. 2005). Indeed, upon starvation-induced macroautophagy, mass spectrometric analysis of natural HLA-DR ligands from a human B lymphoblastoid cell line showed a significant change in the amount of presented peptides derived from cytosolic and nuclear proteins, while MHC class II presentation of membrane and secretory proteins was unchanged. Interestingly, the presentations of four proteins were particularly elevated (HSP70, RAD23, elongation factor-1α and cathepsin D). As we discussed above, two of these, HSP70 and RAD23, are proteins with long half-lives and are therefore good candidates for macroautophagy substrates. Further studies are needed to investigate the possible implications of self-protein presentation as the basis for autoantigen recognition during autoimmune diseases or for allogeneic transplant rejections where upregulated macroautophagy could enhance the MHC class II presentation of self-epitopes on tissue epithelial cells.

While MHC class II presentation of antigens via macroautophagy has been described for a number of in vitro systems, little information is available on the role of this pathway in vivo. In the remaining parts of this chapter, we will speculate on the role(s) that macroautophagic delivery of proteins to MHC class II presentation may play during immune responses and steady-state tolerance in professional antigen-presenting cells and in epithelial cells.

4.5 *Macroautophagy in Professional Antigen-Presenting Cells*

4.5.1 Tolerance Induction by DCs

Immature DCs were found to contain significant numbers of autophagosomes. Since the interaction of naïve T cells or T cell precursors with immature DCs has been described as being a mechanism of peripheral and central tolerance induction in vivo (Probst et al. 2005); (Steinman et al. 2003), macroautophagy could contribute to these tolerance-inducing mechanisms by continuously delivering self-antigen to MHC class II loading compartments. DCs may enhance their efficiency of self-antigen presentation on MHC class II molecules through macroautophagic mechanisms for tolerance induction.

4.5.2 Immunological Synapse

Given the growing evidence for the role of endogenous (self-) peptide in triggering the activation of TCR for the recognition of an agonist peptide (Krogsgaard et al. 2005), it is tempting to speculate that autophagy may participate in the delivery of endogenous self-epitopes to the cell surface of dendritic cells and enhance $CD4^+$ T cell priming to agonist peptides. Indeed, in the so-called "pseudodimer model" of T cell activation (Hailman and Allen 2004; Krogsgaard and Davis 2005), dimerization of two TCRs occurs through a high-affinity interaction with an agonist MHC/peptide complex together with a low-affinity interaction with a self-peptide MHC/peptide complex, which both contribute to stabilizing the T cell/target cell interaction. Interestingly, among MHC class II self-epitopes that accumulate at the immunological synapse (Krogsgaard et al. 2005), some of them (like HSC70) are long-lived proteins that are suggested to be macroautophagy substrates (Dengjel et al. 2005). In addition, infected DCs may use macroautophagy to display pathogen-derived peptides via MHC class II on their surface. Further studies are needed to determine the consequences of macroautophagy silencing in dendritic cells, both in the steady state as well as during intracellular pathogen invasion and oncogenic transformation. It will be interesting to see if endogenous MHC class II pathways can also contribute to the adaptive immune response in these cells that have a high phagocytic capacity.

4.6 *Macroautophagy in Epithelial Cells*

In epithelial cells that have diminished phagocytic capacity, macroautophagy may be even more important to the loading of MHC class II molecules due to its delivery of proteins to MHC class II loading compartments, and may play major roles in both immune surveillance of inflamed tissues and central and peripheral tolerance in the thymus and secondary lymphoid organs.

4.6.1 Possible Roles of Autophagy in Central Tolerance

The thymus is involved in the education of the T cell compartment. In this process, medullary thymic epithelial cells autoreactive T cell specificities (Klein and Kyewski 2000). They do so for both CD4+ and CD8+ T cells. In order to perform this task comprehensively, they should express and present both epithelial self-antigens and also a broad spectrum of proteins that T cells may encounter in the periphery. The importance of thymic peripheral antigen expression has been clearly demonstrated by mutations in the transcription factor Aire. In human diseases and in mouse models in which the *Aire* gene is mutated, a dramatic reduction in the efficiency of negative selection is responsible for the susceptibility to autoimmune disease (Mathis and Benoist 2007). Medullary thymic epithelial cells (mTECs) express a wide range of tissue-specific antigens that are dependent on Aire for expression at this site. Although the classical view of negative selection of CD4+ T cells implicates thymic dendritic cells, it is tempting to speculate that, just as for MHC class I-restricted thymocytes, endogenous processing of MHC class II-presented self-epitopes in mTECs may participate in the selection of MHC class II-restricted thymocytes. This hypothesis is reinforced by the fact that Aire expression is very weak in dendritic cells.

Interestingly, in GFP-LC3 transgenic mice, which allow for the investigation of fluorescently labeled autophagosomes in vivo, the thymic epithelia was described as a tissue of high macroautophagic activity in the steady state, independent of starvation (Mizushima et al. 2004). This study initially raised the possibility that macroautophagy in thymic epithelial cells contributes to T cell education and central tolerance. Indeed, a recent study demonstrated that selective disruption of the essential mouse macroautophagy gene *ATG5* in thymic epithelial cells led to the altered selection of certain MHC class II-restricted CD4+ T cell specificities and to colitis and multiorgan inflammation (Nedjic et al. 2008). Interestingly, alterations in the MHC class II-presented peptide repertoire seemed to be the basis for these differences in positive and negative T cell selection. Thus, the macroautophagy pathway in thymic epithelial cells may contribute to self-antigen loading onto MHC class II molecules for T cell selection and may be essential for the generation of a self-tolerant T cell repertoire in vivo.

4.6.2 Peripheral Tolerance

4.6.2.1 Extrathymic Expression of Aire

A recent report has described the presence of extrathymic expression of Aire and Aire-dependent peripheral tissue antigens in peripheral lymph nodes. The authors demonstrated that peripheral tolerance can be induced by this stromal cell population of the lymph node cortex, and that this cell population presented intestinal self-antigens to naïve CD8 T cells (Lee et al. 2007). Given the epithelial origin of this particular stromal cell population, it will be interesting to analyze the level of autophagy in these cells and further evaluate a possible link between macroautophagy and the induction of peripheral tolerance to self-reactive CD4$^+$ T cell clones that have escaped thymic deletion.

4.6.2.2 Association Between Macroautophagy Regulation and Crohn's Disease

A possible role for macroautophagy in the maintenance of immune homeostasis in the gut has been highlighted by recent genome-wide association studies on Crohn's disease. The results of these studies have implicated mutations in an essential autophagy gene, *ATG16L1* (Hampe et al. 2007; Rioux et al. 2007). A significant association with Crohn's disease risk was found for patients bearing a variant of *ATG16L1* expressing the coding SNP (T300A), and two recent studies discussed in detail elsewhere in this volume describe intestinal pathology in mice harboring mutations in *Atg16L* (see the chapter by Cadwell et al. in this volume). The importance of intestinal epithelial cells (IECs) in maintaining gut homeostasis is notable; indeed, IECs must deal with a massive antigenic challenge from commensal bacteria and try to keep gut-associated lymphocytes in a state of immunological hyporesponsiveness. Since CD4$^+$ T cells play a key role in the pathogenesis of Crohn's disease and represent the vast majority of activated mononucleated cells invading the gut, compromised macroautophagy in IECs could participate in immune deregulation, leading to uncontrolled activation of intraepithelial CD4$^+$ T lymphocytes. Further studies are required to determine if deregulation of macroautophagy in Crohn's disease causes a defect in the adaptive response or in innate immunity against gut commensals.

4.6.2.3 Autophagy and the Allogeneic Response

Recent advances in the field of allogeneic recognition consider TCR interactions with self or with allogeneic MHC molecules to be comparable. Indeed, it now appears that in the majority of cases, such as for "conventional recognition," alloreactive T cells recognize both the peptide and the backbone of the foreign MHC molecule (Felix and Allen 2007). Therefore, macroautophagy could play a role in the allogeneic response by presenting self-peptides at the cell surfaces of dendritic cells and epithelial cells during either hematopoietic stem cell or solid organ transplantation.

In graft versus host disease (GVHD), epithelial cells of the gut, the skin and the liver are the main targets of allogeneic T cells. During the GVHD-induced cytokine storm, these cells upregulate MHC class II molecules and therefore present self-epitopes, also of endogenous origin, on their MHC class II molecules. This renders them susceptible to recognition by alloreactive CD4$^+$ T cell cells. Interestingly, it has been suggested that TNF-α and IFN-γ, two major stimuli present during the acute phase of GVHD, increase macroautophagy (Djavaheri-Mergny et al. 2006; Gutierrez et al. 2004).

In organ transplantation, rapamycin treatment, a known stimulator of macroautophagy via repression of the mTor pathway, is used therapeutically as an immunosuppressive regimen. The exact mechanism of action of rapamycin is not well understood, but some mouse studies have demonstrated that rapamycin-conditioned dendritic cells can expand Foxp3$^+$ T regulatory cells and promote organ transplant tolerance (Turnquist et al. 2007). It is tempting to speculate that rapamycin-enhanced macroautophagy may lead to increased self-peptide presentation on MHC class II molecules, thereby facilitating regulatory T cell expansion.

These examples demonstrate that self-antigen and commensal antigen presentation on MHC class II molecules via macroautophagy may contribute to peripheral tolerance, and unfortunately also to autoimmunity in various settings. A better understanding of the beneficial and detrimental contributions of this pathway in vivo, as well as its regulatory mechanisms, should allow us to develop therapeutic interventions.

5 Conclusions

The cargo of autophagosomes has now been shown to reach MHC class II loading compartments quite efficiently, and therefore must be considered a new potential source of endogenous MHC class II epitopes for CD4$^+$ T cells during adaptive immune responses as well as tolerance induction. The targeting of this pathway could be beneficial in order to enhance immune responses against intracellular pathogens and tumors, as well as to downmodulate autoimmunity. However, a better understanding of the degree to which antigen processing for MHC class II presentation via macroautophagy contributes to tolerance and immunity is required in order to select disease states in which enhanced macroautophagy might be beneficial without breaking tolerance.

Acknowledgments Our work is supported by the Arnold and Mabel Beckman Foundation, the Alexandrine and Alexander Sinsheimer Foundation, the Burroughs Wellcome Fund, the Dana Foundation's Neuroimmunology Program, the National Cancer Institute (R01CA108609 and R01CA101741), the National Institute of Allergy and Infectious Diseases (RFP-NIH-NIAID-DAIDS-BAA-06–19), the Foundation for the National Institutes of Health (Grand Challenges in Global Health), the Starr Foundation (to C.M.), and an Institutional Clinical and Translational Science Award (to the Rockefeller University Hospital).

References

Agarraberes FA, Dice JF (2001) Protein translocation across membranes. Biochim Biophys Acta 1513:1–24

Aichinger G, Karlsson L, Jackson MR, Vestberg M, Vaughan JH, Teyton L, Lechler RI, Peterson PA (1997) Major histocompatibility complex class II-dependent unfolding, transport, and degradation of endogenous proteins. J Biol Chem 272:29127–29136

Aniento F, Roche E, Cuervo AM, Knecht E (1993) Uptake and degradation of glyceraldehyde-3-phosphate dehydrogenase by rat liver lysosomes. J Biol Chem 268:10463–10470

Barette C, Jariel-Encontre I, Piechaczyk M, Piette J (2001) Human cyclin C protein is stabilized by its associated kinase cdk8, independently of its catalytic activity. Oncogene 20:551–562

Bartido SM, Diment S, Reiss CS (1995) Processing of a viral glycoprotein in the endoplasmic reticulum for class II presentation. Eur J Immunol 25:2211–2219

Bonifaz LC, Arzate S, Moreno J (1999) Endogenous and exogenous forms of the same antigen are processed from different pools to bind MHC class II molecules in endocytic compartments. Eur J Immunol 29:119–131

Brazil MI, Weiss S, Stockinger B (1997) Excessive degradation of intracellular protein in macrophages prevents presentation in the context of major histocompatibility complex class II molecules. Eur J Immunol 27:1506–1514

Brooks A, Hartley S, Kjer-Nielsen L, Perera J, Goodnow CC, Basten A, McCluskey J (1991) Class II-restricted presentation of an endogenously derived immunodominant T-cell determinant of hen egg lysozyme. Proc Natl Acad Sci USA 88:3290–3294

Brooks AG, McCluskey J (1993) Class II-restricted presentation of a hen egg lysozyme determinant derived from endogenous antigen sequestered in the cytoplasm or endoplasmic reticulum of the antigen presenting cells. J Immunol 150:3690–3697

Chen M, Shirai M, Liu Z, Arichi T, Takahashi H, Nishioka M (1998) Efficient class II major histocompatibility complex presentation of endogenously synthesized hepatitis C virus core protein by Epstein–Barr virus-transformed B-lymphoblastoid cell lines to CD4+ T cells. J Virol 72:8301–8308

Chiang HL, Dice JF (1988) Peptide sequences that target proteins for enhanced degradation during serum withdrawal. J Biol Chem 263:6797–6805

Ciechanover A (2005) Intracellular protein degradation: from a vague idea through the lysosome and the ubiquitin–proteasome system and onto human diseases and drug targeting. Cell Death Differ 12:1178–1190

Dani A, Chaudhry A, Mukherjee P, Rajagopal D, Bhatia S, George A, Bal V, Rath S, Mayor S (2004) The pathway for MHCII-mediated presentation of endogenous proteins involves peptide transport to the endo-lysosomal compartment. J Cell Sci 117:4219–4230

Dengjel J, Schoor O, Fischer R, Reich M, Kraus M, Muller M, Kreymborg K, Altenberend F, Brandenburg J, Kalbacher H Brock R, Driessen C, Rammensee HG, Stevanovic S (2005) Autophagy promotes MHC class II presentation of peptides from intracellular source proteins. Proc Natl Acad Sci USA 102:7922–7927

Dice JF, Goldberg AL (1975) A statistical analysis of the relationship between degradative rates and molecular weights of proteins. Arch Biochem Biophys 170:213–219

Dissanayake SK, Tuera N, Ostrand-Rosenberg S (2005) Presentation of endogenously synthesized MHC class II-restricted epitopes by MHC class II cancer vaccines is independent of transporter associated with Ag processing and the proteasome. J Immunol 174:1811–1819

Djavaheri-Mergny M, Amelotti M, Mathieu J, Besancon F, Bauvy C, Souquere S, Pierron G, Codogno P (2006) NF-kappaB activation represses tumor necrosis factor-alpha-induced autophagy. J Biol Chem 281:30373–30382

Dorfel D, Appel S, Grunebach F, Weck MM, Muller MR, Heine A, Brossart P (2005) Processing and presentation of HLA class I and II epitopes by dendritic cells after transfection with in vitro transcribed MUC1 RNA. Blood 105:3199–3205

Felix NJ, Allen PM (2007) Specificity of T-cell alloreactivity. Nat Rev Immunol 7:942–953

Fengsrud M, Raiborg C, Berg TO, Stromhaug PE, Ueno T, Erichsen ES, Seglen PO (2000) Autophagosome-associated variant isoforms of cytosolic enzymes. Biochem J 352:773–781

Germain D, Russell A, Thompson A, Hendley J (2000) Ubiquitination of free cyclin D1 is independent of phosphorylation on threonine 286. J Biol Chem 275:12074–12079

Godefroy E, Scotto L, Souleimanian NE, Ritter G, Old LJ, Jotereau F, Valmori D, Ayyoub M (2006) Identification of two Melan-A CD4+ T cell epitopes presented by frequently expressed MHC class II alleles. Clin Immunol 121:54–62

Groothuis TA, Neefjes J (2005) The many roads to cross-presentation. J Exp Med 202:1313–1318

Gueguen M, Long EO (1996) Presentation of a cytosolic antigen by major histocompatibility complex class II molecules requires a long-lived form of the antigen. Proc Natl Acad Sci USA 93:14692–14697

Gutierrez MG, Master SS, Singh SB, Taylor GA, Colombo MI, Deretic V (2004) Autophagy is a defense mechanism inhibiting BCG and *Mycobacterium tuberculosis* survival in infected macrophages. Cell 119:753–766

Hailman E, Allen PM (2004) Self help for T cells. Nat Immunol 5:780–781.

Hampe J, Franke A, Rosenstiel P, Till A, Teuber M, Huse K, Albrecht M, Mayr G, De La Vega FM, Briggs J Gunther S, Prescott NJ, Onnie CM, Hasler R, Sipos B, Folsch UR, Lengauer T, Platzer M, Mathew CG, Krawczak M, Schreiber S (2007) A genome-wide association scan of nonsynonymous SNPs identifies a susceptibility variant for Crohn disease in Atg16l1. Nat Genet 39:207–211

Heessen S, Masucci MG, Dantuma NP (2005) The UBA2 domain functions as an intrinsic stabilization signal that protects Rad23 from proteasomal degradation. Mol Cell 18:225–235

Hemelaar J, Lelyveld VS, Kessler BM, Ploegh HL (2003) A single protease, Apg4B, is specific for the autophagy-related ubiquitin-like proteins GATE-16, MAP1-LC3, GABARAP, and Apg8L. J Biol Chem 278:51841–51850

Henell F, Berkenstam A, Ahlberg J, Glaumann H (1987) Degradation of short- and long-lived proteins in perfused liver and in isolated autophagic vacuoles-lysosomes. Exp Mol Pathol 46:1–14

Hoyt MA, Zich J, Takeuchi J, Zhang M, Govaerts C, Coffino P (2006) Glycine-alanine repeats impair proper substrate unfolding by the proteasome. Embo J 25:1720–1729

Jackson WT, Giddings TH Jr, Taylor MP, Mulinyawe S, Rabinovitch M, Kopito RR, Kirkegaard K (2005) Subversion of cellular autophagosomal machinery by RNA viruses. PLoS Biol 3:e156

Jacobson S, Sekaly RP, Jacobson CL, McFarland HF, Long EO (1989) HLA class II-restricted presentation of cytoplasmic measles virus antigens to cytotoxic T cells. J Virol 63:1756–1762

Jaraquemada D, Marti M, Long EO (1990) An endogenous processing pathway in vaccinia virus-infected cells for presentation of cytoplasmic antigens to class II-restricted T cells. J Exp Med 172:947–954

Jiang J, Ballinger CA, Wu Y, Dai Q, Cyr DM, Hohfeld J, Patterson C (2001) CHIP is a U-box-dependent E3 ubiquitin ligase: identification of Hsc70 as a target for ubiquitylation. J Biol Chem 276:42938–42944

Kittlesen DJ, Brown LR, Braciale VL, Sambrook JP, Gething MJ, Braciale TJ (1993) Presentation of newly synthesized glycoproteins to CD4+ T lymphocytes. An analysis using influenza hemagglutinin transport mutants. J Exp Med 177:1021–1030

Klein L, Kyewski B (2000) Self-antigen presentation by thymic stromal cells: a subtle division of labor. Curr Opin Immunol 12:179–186

Krogsgaard M, Davis MM (2005) How T cells "see" antigen. Nat Immunol 6:239–245

Krogsgaard M, Li QJ, Sumen C, Huppa JB, Huse M, Davis MM (2005) Agonist/endogenous peptide-MHC heterodimers drive T cell activation and sensitivity. Nature 434:238–243

Landry J, Chretien P, Laszlo A, Lambert H (1991) Phosphorylation of HSP27 during development and decay of thermotolerance in Chinese hamster cells. J Cell Physiol 147:93–101

Le Roy E, Baron M, Faigle W, Clement D, Lewinsohn DM, Streblow DN, Nelson JA, Amigorena S, Davignon JL (2002) Infection of APC by human cytomegalovirus controlled through recognition of endogenous nuclear immediate early protein 1 by specific CD4+ T lymphocytes. J Immunol 169:1293–1301

Lee DY, Sugden B (2008) The latent membrane protein 1 oncogene modifies B-cell physiology by regulating autophagy. Oncogene 27:2833–2842

Lee JW, Epardaud M, Sun J, Becker JE, Cheng AC, Yonekura AR, Heath JK, Turley SJ (2007) Peripheral antigen display by lymph node stroma promotes T cell tolerance to intestinal self. Nat Immunol 8:181–190

Lee SP, Brooks JM, Al-Jarrah H, Thomas WA, Haigh TA, Taylor GS, Humme S, Schepers A, Hammerschmidt W, Yates JL Rickinson AB, Blake NW (2004) CD8 T cell recognition of endogenously expressed Epstein-Barr virus nuclear antigen 1. J Exp Med 199:1409–1420
Levine B, Kroemer G (2008) Autophagy in the pathogenesis of disease. Cell 132:27–42
Levitskaya J, Coram M, Levitsky V, Imreh S, Steigerwald-Mullen PM, Klein G, Kurilla MG, Masucci MG (1995) Inhibition of antigen processing by the internal repeat region of the Epstein-Barr virus nuclear antigen-1. Nature 375:685–688
Levitskaya J, Sharipo A, Leonchiks A, Ciechanover A, Masucci MG (1997) Inhibition of ubiquitin/proteasome-dependent protein degradation by the Gly-Ala repeat domain of the Epstein-Barr virus nuclear antigen 1. Proc Natl Acad Sci USA 94:12616–12621
Li D, Duncan RF (1995) Transient acquired thermotolerance in Drosophila, correlated with rapid degradation of Hsp70 during recovery. Eur J Biochem 231:454–465
Lich JD, Elliott JF, Blum JS (2000) Cytoplasmic processing is a prerequisite for presentation of an endogenous antigen by major histocompatibility complex class II proteins. J Exp Med 191:1513–1524
Malnati MS, Marti M, LaVaute T, Jaraquemada D, Biddison W, DeMars R, Long EO (1992) Processing pathways for presentation of cytosolic antigen to MHC class II-restricted T cells. Nature 357:702–704
Marshall NA, Vickers MA, Barker RN (2003) Regulatory T cells secreting IL-10 dominate the immune response to EBV latent membrane protein 1. J Immunol 170:6183–6189
Mathis D, Benoist C (2007) A decade of AIRE. Nat Rev Immunol 7:645–650
Mizushima N, Klionsky DJ (2007) Protein turnover via autophagy: implications for metabolism. Annu Rev Nutr 27:19–40
Mizushima N, Levine B, Cuervo AM, Klionsky DJ (2008) Autophagy fights disease through cellular self-digestion. Nature 451:1069–1075
Mizushima N, Yamamoto A, Matsui M, Yoshimori T, Ohsumi Y (2004) In vivo analysis of autophagy in response to nutrient starvation using transgenic mice expressing a fluorescent autophagosome marker. Mol Biol Cell 15:1101–1111
Mukherjee P, Dani A, Bhatia S, Singh N, Rudensky AY, George A, Bal V, Mayor S, Rath S (2001) Efficient presentation of both cytosolic and endogenous transmembrane protein antigens on MHC class II is dependent on cytoplasmic proteolysis. J Immunol 167:2632–2641
Münz C, Bickham KL, Subklewe M, Tsang ML, Chahroudi A, Kurilla MG, Zhang D, O'Donnell M, Steinman RM (2000) Human $CD4^+$ T lymphocytes consistently respond to the latent Epstein–Barr virus nuclear antigen EBNA1. J Exp Med 191:1649–1660
Nedjic J, Alchinger M, Emmerich J, Mizushima N, Klein L (2008) Autophagy in thymic epithelium shapes the T-cell repertoire and is essential for tolerance. Nature 455:396–400
Nimmerjahn F, Milosevic S, Behrends U, Jaffee EM, Pardoll DM, Bornkamm GW, Mautner J (2003) Major histocompatibility complex class II-restricted presentation of a cytosolic antigen by autophagy. Eur J Immunol 33:1250–1259
Ogawa M, Yoshimori T, Suzuki T, Sagara H, Mizushima N, Sasakawa C (2005) Escape of intracellular *Shigella* from autophagy. Science 307:727–731
Ohsumi Y (2001) Molecular dissection of autophagy: two ubiquitin-like systems. Nat Rev Mol Cell Biol 2:211–216
Paludan C, Schmid D, Landthaler M, Vockerodt M, Kube D, Tuschl T, Münz C (2005) Endogenous MHC class II processing of a viral nuclear antigen after autophagy. Science 307:593–596
Probst HC, McCoy K, Okazaki T, Honjo T, van den Broek M (2005) Resting dendritic cells induce peripheral $CD8^+$ T cell tolerance through PD-1 and CTLA-4. Nat Immunol 6:280–286
Qi L, Rojas JM, Ostrand-Rosenberg S (2000) Tumor cells present MHC class II-restricted nuclear and mitochondrial antigens and are the predominant antigen presenting cells in vivo. J Immunol 165:5451–5461
Rammensee HG, Bachmann J, Emmerich NP, Bachor OA, Stevanovic S (1999) SYFPEITHI: database for MHC ligands and peptide motifs. Immunogenetics 50:213–219
Rioux JD, Xavier RJ, Taylor KD, Silverberg MS, Goyette P, Huett A, Green T, Kuballa P, Barmada MM, Datta LW, Shugart YY, Griffiths AM, Targan SR, Ippoliti AF, Bernard EJ, Mei

L, Nicolae DL, Regueiro M, Schumm LP, Steinhart AH, Rotter JI, Duerr RH, Cho JH, Daly MJ, Brant SR (2007) Genome-wide association study identifies new susceptibility loci for Crohn disease and implicates autophagy in disease pathogenesis. Nat Genet 39:596–604

Schmid D, Münz C (2007) Innate and adaptive immunity through autophagy. Immunity 26:11–21

Schmid D, Pypaert M, Münz C (2007) MHC class II antigen loading compartments continuously receive input from autophagosomes. Immunity 26:79–92

Singer JD, Gurian-West M, Clurman B, Roberts JM (1999) Cullin-3 targets cyclin E for ubiquitination and controls S phase in mammalian cells. Genes Dev 13:2375–2387

Steinman RM, Hawiger D, Nussenzweig MC (2003) Tolerogenic dendritic cells. Annu Rev Immunol 21:685–711

Suri A, Walters JJ, Rohrs HW, Gross ML, Unanue ER (2008) First signature of islet β-cell-derived naturally processed peptides selected by diabetogenic class II MHC molecules. J Immunol 180:3849–3856

Tellam J, Sherritt M, Thomson S, Tellam R, Moss DJ, Burrows SR, Wiertz E, Khanna R (2001) Targeting of EBNA1 for rapid intracellular degradation overrides the inhibitory effects of the Gly-Ala repeat domain and restores $CD8^+$ T cell recognition. J Biol Chem 276:33353–33360

Tewari MK, Sinnathamby G, Rajagopal D, Eisenlohr LC (2005) A cytosolic pathway for MHC class II-restricted antigen processing that is proteasome and TAP dependent. Nat Immunol 6:287–294

Trombetta ES, Mellman I (2005) Cell biology of antigen processing in vitro and in vivo. Annu Rev Immunol 23:975–1028

Turnquist HR, Raimondi G, Zahorchak AF, Fischer RT, Wang Z, Thomson AW (2007) Rapamycin-conditioned dendritic cells are poor stimulators of allogeneic $CD4^+$ T cells, but enrich for antigen-specific $Foxp3^+$ T regulatory cells and promote organ transplant tolerance. J Immunol 178:7018–7031

Walker DH, Popov VL, Crocquet-Valdes PA, Welsh CJ, Feng HM (1997) Cytokine-induced, nitric oxide-dependent, intracellular antirickettsial activity of mouse endothelial cells. Lab Invest 76:129–138

Weiss S, Bogen B (1989) B-lymphoma cells process and present their endogenous immunoglobulin to major histocompatibility complex-restricted T cells. Proc Natl Acad Sci USA 86:282–286

Weiss S, Bogen B (1991) MHC class II-restricted presentation of intracellular antigen. Cell 64:767–776

Zarour HM, Kirkwood JM, Kierstead LS, Herr W, Brusic V, Slingluff CL Jr, Sidney J, Sette A, Storkus WJ (2000) Melan-A/MART-1(51-73) represents an immunogenic HLA-DR4-restricted epitope recognized by melanoma-reactive $CD4^+$ T cells. Proc Natl Acad Sci USA 97:400–405

Zhou D, Li P, Lott JM, Hislop A, Canaday DH, Brutkiewicz RR, Blum JS (2005) Lamp-2a facilitates MHC class II presentation of cytoplasmic antigens. Immunity 22:571–581

Role of Autophagy and Autophagy Genes in Inflammatory Bowel Disease

Ken Cadwell, Thaddeus S. Stappenbeck, and Herbert W. Virgin

Contents

1 Introduction 142
2 Inflammatory Bowel Disease 143
2.1 Clinical Aspects of Inflammatory Bowel Disease in Humans 143
2.2 Murine Models of Colitis 144
2.3 Murine Models for Disease of the Small Intestine 145
2.4 The Role of Bacteria in Inflammatory Bowel Disease 146
3 Autophagy and Crohn's Disease 148
3.1 Identification of *ATG16L1* and *IRGM1* as Crohn's Disease Susceptibility Genes 148
3.2 The Role of Atg16L1 in Autophagy 149
3.3 The Role of IRGM1 in Autophagy 151
4 The Role of Autophagy in Intestinal Biology 152
4.1 Two Mouse Models to Examine Atg16L1 Function In Vivo 152
4.2 Loss of Atg16L1 Leads to an Increased Inflammatory Response and Increased Expression of the Proinflammatory Cytokine IL-1β in Macrophages 153
4.3 Mice with Reduced Atg16L1 Expression Reproduce Aspects of Crohn's Disease Pathology, and Reveal a Cellular Target for the Role of Atg16L1 in Crohn's Disease 155
5 Conclusion and Future Perspectives 159
References 161

Abstract Polymorphisms associated with two genes in the autophagy pathway, *ATG16L1* and *IRGM1*, have been implicated in susceptibility to Crohn's disease, an idiopathic inflammatory disease typically involving the gastrointestinal tract. The intestinal mucosa is a site of careful immune regulation where the epithelium and immune cells encounter pathogens as well as a robust and diverse population of indigenous microbes that are predominately bacteria. Since the role of autophagy in immunity is broad and expanding, it is unclear which downstream

K. Cadwell, T.S. Stappenbeck, and H.W. Virgin(✉)
Department of Pathology and Immunology, Washington University School of Medicine, 660 South Euclid Avenue, St. Louis, MO, 63110, USA
e-mail: virgin@wustl.edu

B. Levine et al. (eds.), *Autophagy in Infection and Immunity*, Current Topics in Microbiology and Immunology 335,
DOI 10.1007/ 978-3-642-00302-8_7,

functions of autophagy and which cell types are the key factors in Crohn's disease susceptibility. This chapter reviews the recent literature on the roles of *ATG16L1* and *IRGM1* in the autophagy pathway, inflammation, antimicrobial immunity, and the biology of the intestine, and discusses how these genes may contribute to Crohn's disease pathogenesis.

1 Introduction

Human genetic studies have implicated genes associated with the autophagy pathway in several diseases, but it is not always clear *how* autophagy contributes to pathogenesis (Levine and Kroemer 2008). Advances in the field of population genetics have identified two genes in the autophagy pathway, *ATG16L1* and *IRGM1*, as genetic determinants of susceptibility to Crohn's disease, a common inflammatory disease most often involving the small intestine (The Wellcome Trust Case Control Consortium 2007; Rioux et al. 2007; Hampe et al. 2007; Parkes et al. 2007; Barrett et al. 2008). Although the pathogenesis of Crohn's disease is as yet poorly understood, a large body of research has implicated specific genetic and environmental factors that contribute to either onset or severity of the disease (Levine and Kroemer 2008; Pineton de et al. 2008; Ferguson et al. 2007). The major feature of Crohn's disease pathology is exuberant inflammation in the gastrointestinal tract (Neuman 2007). A longstanding hypothesis in this field is that the inflammation in Crohn's disease occurs as an inappropriate response to enteric pathogens or to a shift in the composition of the normal bacterial flora of the small intestine. Accumulating evidence suggests that intracellular bacteria can be surrounded by autophagosomes and targeted to the lysosome for destruction, a subject that is reviewed extensively in other chapters in this volume and in recent review articles (Deretic 2006; Levine and Deretic 2007; Levine and Kroemer 2008). In addition, immune responses that are dependent on autophagy genes but do not require the formation of autophagosomes have also been described (Sanjuan et al. 2007; Jounai et al. 2007; Pyo et al. 2005; Zhao et al. 2008). These considerations have led to a popular model for the role of *ATG16L1* and *IRGM1* in Crohn's disease in which autophagy-mediated control of bacteria is compromised, leading to an aberrant inflammatory response to bacteria (Xavier et al. 2008). However, recent examination of Atg16L1 function in vivo using mouse genetics has challenged this model by revealing unanticipated roles of the autophagy pathway in inflammation and in the biology of epithelial Paneth cells that play a fundamentally important role in the innate immune response in the intestine (Saitoh et al. 2008; Cadwell et al. 2008a). These and other recent findings concerning the roles of *ATG16L1* and *IRGM1* in autophagy and their relation to Crohn's disease will be discussed.

2 Inflammatory Bowel Disease

2.1 Clinical Aspects of Inflammatory Bowel Disease in Humans

Crohn's disease and ulcerative colitis are the two major forms of inflammatory bowel disease. Both are associated with high morbidity and their pathogenesis has been the subject of many excellent review articles (Xavier and Podolsky 2007; Cho 2008; Ferguson et al. 2007; Pineton de et al. 2008; Neuman 2007). The incidence of Crohn's disease has risen in several regions of the globe, suggesting a strong environmental contribution to disease in addition to an undisputed heritable component (Xavier and Podolsky 2007; Ferguson et al. 2007). The incidence of inflammatory bowel disease is highly variable and can be as high as one in 250 people in certain countries including the United States and the United Kingdom (Ferguson et al. 2007). Both Crohn's disease and ulcerative colitis are characterized by chronic and relapsing inflammation of the intestinal mucosa and typically cause clinical symptoms of abdominal pain, severe diarrhea, vomiting, and weight loss. Many features distinguish Crohn's disease from ulcerative colitis. Crohn's disease most commonly occurs in the distal small intestine, but can occur anywhere in the gastrointestinal tract. Ulcerative colitis typically involves the distal rectum and extends proximally. In Crohn's disease, the inflammation is transmural and discontinuous compared to the diffuse superficial mucosal-based inflammation observed in ulcerative colitis. Other features that are more often detected in Crohn's disease than in ulcerative colitis are the involvement of the distal small intestine (the ileum) and perianal region, stenosis, the presence of fistulae and granulomas, and recurrence after surgery.

In addition to these pathological features within the gastrointestinal tract, Crohn's disease and ulcerative colitis are commonly associated with systemic, extra-intestinal features. Inflammation in the eye, both uveitis and episcleritis, are not uncommon in Crohn's disease patients. Arthritis is another complication and can result in chronic pain in large and small joints as well as the spine. There are also examples of blood diseases, such as hypercoagulability, and skin diseases, such as erythema nodosum due to inflammation in the subcutaneous tissue, in Crohn's disease patients. The biliary tract can also be involved in the form of primary sclerosing cholangitis.

The onset of Crohn's disease is typically between the ages of 15 and 30 but can occur at any age. Surgical removal of affected areas is a common treatment for severe Crohn's disease. Lifestyle changes such as dietary restrictions are often recommended to patients. Acute symptoms are alleviated with steroids, but prolonged use can lead to significant side effects. Targeting the TNF-α immune pathway has also been an effective strategy for reducing inflammatory symptoms. Treatment of Crohn's disease can sometimes result in remission, but frequently such remissions are not maintained.

A better fundamental understanding of the mechanisms of pathogenesis of this complex disease may identify specific pathways that can be exploited for improved diagnosis and treatment. It is in this context that recent large genetic studies of disease susceptibility have led to the discovery of important genetic risk factors for Crohn's disease, including some genes associated with pathways such as autophagy that were not previously known to play any role in inflammatory bowel disease. Over 30 genetic loci, including *ATG16L1* and *IRGM1*, have been linked to Crohn's disease susceptibility (Barrett et al. 2008), but the specific relationship between the function of these many genes and the diverse pathologies observed in Crohn's disease is unknown. These genetic studies, and studies driven by these new findings, have the potential to revolutionize our understanding of Crohn's disease.

2.2 Murine Models of Colitis

As a basis for understanding the potential role of autophagy genes in inflammatory bowel disease, we will first review studies of inflammatory bowel disease pathogenesis and causation performed in animals prior to recent analyses of mice lacking or expressing low levels of autophagy proteins. It is clear that a fundamental understanding of the functions of the intestinal epithelium and immune system, as well as the intestinal response to injury, is a prerequisite for comprehending potentially aberrant inflammatory responses in diseases such as Crohn's disease or ulcerative colitis.

To study inducible colonic inflammation in the mouse, many investigators take advantage of the intestinal damaging agent dextran sodium sulfate (DSS). This model has been important for defining how intestinal microbes might contribute to colonic inflammation. The exact mechanisms by which DSS damages the intestinal mucosa are still not completely clear, though it is recognized that prolonged DSS administration leads to increased permeability of the intestinal wall in the colon and changes in colonic epithelial barrier function (Kitajima et al. 1999). Thus, colonic microbes that are normally excluded by the epithelial barrier can potentially gain access to deeper tissues. This may result in direct damage to cells by microbes or their secreted products. Alternatively, the presence of factors derived from the microbes might evoke an inflammatory response that damages the tissue. The ease of use of DSS, via its administration in drinking water, makes it an efficient and popular model of colitis.

Since genetic background can influence the degree of physical damage, inflammation, and efficiency of recovery, one can study the role of individual genes in the colonic injury response using this model (Xavier and Podolsky 2007). For example, mice containing a mutation in the Crohn's disease susceptibility gene *NOD2* have been reported to be hypersensitive to damage caused by DSS (Maeda et al. 2005). However, there are important limitations to the use of the DSS model that must be considered. DSS does not mimic any known chemical exposure in humans, and thus there is no clear parallel between murine exposure to DSS and any known environmental trigger for human inflammatory bowel disease. Moreover, DSS causes colonic inflammation and has not been reported to have as dramatic an effect on the small intestine, a concern for investigators studying Crohn's disease, which commonly involves the distal ileum (see below).

Studies focusing on the *NOD2* gene (also known as *CARD15*), the first of the genetic predisposition loci identified for Crohn's disease, are typical of the difficulties confronting investigators using murine models to study inflammatory bowel disease. Three genetic variants within the coding region of *NOD2*, L1007fsinsC, G908R, and R702W, have been associated with disease risk in European and American populations, and heterozygosity, homozygosity, or compound heterozygosity (carrying a different susceptibility variant on each chromosome) for these mutations can strongly influence susceptibility (Hugot et al. 2001; Ogura et al. 2001; Hampe et al. 2001; Ahmad et al. 2002). *NOD2* encodes a cytoplasmic receptor that recognizes the bacterial product muramyl dipeptide (MDP) found in peptidoglycan (Inohara et al. 2003; Girardin et al. 2003). All three of the above mutations interfere with the ability of NOD2 to activate NF-κB in response to MDP (Inohara et al. 2003). Based on these observations, it has been hypothesized that impairment of NOD2 may lead to decreased recognition and control of intestinal bacteria, and consequently a heightened inflammatory response that can trigger or contribute to the severity of Crohn's disease. To test this model, several groups have generated *Nod2* mutant mice (Pauleau and Murray 2003; Kobayashi et al. 2005; Maeda et al. 2005). In all cases, untreated *Nod2* mutant mice fail to display any of the pathological hallmarks associated with Crohn's disease. However, other results from these different studies are conflicting. In the *Nod2* knock-in model in which the endogenous *Nod2* gene is replaced by a frameshift mutant analogous to the L1007fsinsC susceptibility mutation, treatment with DSS leads to increased inflammation and ulceration in the colon compared to similarly treated control mice (Maeda et al. 2005). In contrast, *Nod2* knockout mice display impaired control of *Listeria monocytogenes* after oral inoculation, but have a normal response to DSS (Kobayashi et al. 2005). The cellular role of Nod2 is controversial as well. Depending on the exact conditions, impairment of Nod2 can lead to either enhanced or diminished production of inflammatory cytokines in response to bacterial products (Pauleau and Murray 2003; Kobayashi et al. 2005; Maeda et al. 2005; Watanabe et al. 2008; Leber et al. 2008). The role of *Nod2* mutations in the production of antimicrobial peptides such as cryptdins by Paneth cells is also controversial (Ouellette 2006; Wehkamp et al. 2004; Simms et al. 2008). Thus, the role of *NOD2* in Crohn's disease pathogenesis is still an open question. However, the important fact remains that a polymorphism in a gene encoding a cytoplasmic sensor for a bacterial product is linked to the disease.

2.3 Murine Models for Disease of the Small Intestine

Important for the Crohn's disease field, only three murine genetic models display an inflammatory response in the distal small intestine. This is in striking contrast to the many models available for studying colitis (Xavier and Podolsky 2007; Strober et al. 2002). In the TNFΔAU model, the AU-rich region within the 3′UTR of TNF-α is deleted, leading to increased stability of the mRNA for this inflammatory cytokine (Kontoyiannis et al. 1999). TNF-α is one of the major inflammatory cytokines implicated in Crohn's disease pathogenesis, and it is an effective therapeutic target in human Crohn's disease (Targan et al. 1997). Predictably, these mice develop chronic systemic inflammation that includes the small intestine. This model is designed to examine the effects of TNF-α overexpression rather than the events leading to the TNF-α response. The SAMP1/Yit mouse has also been used as a Crohn's disease model. SAMP1/Yit mice were generated by selecting mice that spontaneously develop skin lesions (Matsumoto et al. 1998). These mice develop spontaneous ileitis similar to that found in Crohn's disease (Matsumoto et al. 1998; Kosiewicz et al. 2001). However, the specific gene(s) responsible for this remarkable phenotype are unknown and, as yet, this model has not been extensively analyzed at the molecular and cellular level.

A potentially valuable model for spontaneous inflammation in the small intestine has recently been developed via conditional deletion of the *Xbp1* gene in intestinal epithelium (Kaser et al. 2008). *Xbp1* is essential for the proper transcription of a set of genes involved in endoplasmic reticulum (ER) function and maintenance (Acosta-Alvear et al. 2007). This model is especially attractive, since polymorphisms in human *XBP1* are present in some patients with Crohn's disease and ulcerative colitis (Kaser et al. 2008). Conditional deletion of *Xbp1* in the intestinal epithelium leads to spontaneous inflammation of variable intensity and pathology in the small intestines of some adult mice (Kaser et al. 2008). Intriguingly, the spontaneous inflammation occurs in only a certain percentage of mice, suggesting that there is an environmental trigger even in this mouse model. Although the observed inflammation is impressive, *Xbp1* mutant mice do not replicate Crohn's disease exactly since the histology is more consistent with surface ulceration than transmural lesions.

One striking consequence of *Xbp1* deletion is the rapid loss of Paneth cells via apoptosis. Paneth cells are specialized epithelial cells present at the base of the intestinal crypts that function in part by secretion of antimicrobial peptides and other proteins such as lysozyme that alter the intestinal environment (Porter et al. 2002). Since Paneth cells rely heavily on the secretory pathway, these specialized cells may be particularly sensitive to ER stress. Although this loss of Paneth cells may lead to a change in the intestinal microbial community, two Paneth cell-deficient mouse models already exist, and mice from these two lines fail to display spontaneous inflammation (Bry et al. 1994; Bastide et al. 2007). Therefore, the loss of Paneth cells in *Xbp1* mutant mice may not be the sole explanation for the inflammation observed in these mice. It will be interesting to compare Paneth cell-deficient and *Xbp1* mutant mice directly to investigate these issues.

2.4 The Role of Bacteria in Inflammatory Bowel Disease

Both the epidemiology and the literature presented above raise the obvious question of whether a specific bacterial pathogen causes Crohn's disease. Many groups have attempted to implicate a pathogen through indirect association studies using statistical data (Pineton de et al. 2008). These studies most often attempt to identify the etiological agent long after patients have been diagnosed with the disease. It is therefore possible that the pathogen has already been cleared or is limited to a low level of replication at the time of the investigation. This concern, as well as the heterogeneity of the disease, may account for the failure, to date, to identify a pathogen that causes Crohn's disease (Pineton de et al. 2008).

Instead of invoking a single bacterial species as a cause for Crohn's disease, an important alternative hypothesis has been that Crohn's disease is induced by a shift in the commensal intestinal bacterial population, known as dysbiosis. Dysbiosis could potentially be harmful in at least two ways. First, bacteria (or parasites and fungi) that are potentially pathogenic might normally exist at low levels but become predominant and pathogenic during dysbiosis. These species may directly damage the intestinal wall or may be recognized as harmful, and as a result incur an inflammatory response. The best noninflammatory bowel disease clinical example of this paradigm is pseudomembranous colitis where *C. difficile* colonization is elevated following treatment with antibiotics. This inappropriate elevation of *C. difficile* can lead to destruction of the surface epithelium, which then leads to inflammation. Importantly, and in contrast to Crohn's disease (see below), pseudomembranous colitis is responsive to antibiotic therapy. Second, bacterial species that are lost or diminished during dysbiosis could have beneficial functions that contribute to normal homeostasis in the intestine. The ability of commensal bacteria to regulate CD4$^+$ T lymphocyte differentiation and function is an important example of how a subset of the microbiota (Ivanov et al. 2008) or even a bacterial product from a specific species (Mazmanian et al. 2005; Mazmanian et al. 2008) can serve anti-inflammatory functions. Although current techniques may allow the detection of a dramatic alteration in certain bacterial populations, it is difficult to detect a subtle change since the normal bacterial flora of the intestine is incompletely characterized and varies significantly between individuals (McKenna et al. 2008). Proving that dysbiosis is causative may in part require reintroduction of specific bacteria into a gnotobiotic host within a controlled environment (Rawls et al. 2006).

A mechanism of pathogenesis that could contribute to Crohn's disease pathogenesis in the setting of dysbiosis, or even in the normal intestine, is a breakdown in innate or immune tolerance to intestinal contents. In this setting, normally harmless bacteria may evoke an inappropriate immune reaction. For instance, this could occur when lymphocytes that recognize harmless bacteria are improperly activated or fail to be eliminated during development. These models are attractive because an intricate crosstalk between the host and the gut "microbiome" is well established (Turnbaugh et al. 2007). In *TGFβRII/IL-10R2* double mutant mice, a mouse model that reproduces fulminant ulcerative colitis, broad-spectrum antibiotics completely inhibit disease (Kang et al. 2008). This dramatic reversal of inflammation by altering the intestinal bacterial population clearly justifies serious consideration of hyperresponsiveness to normal bacteria as a potential contributor to inflammatory bowel disease. Further support for a role of dysbiosis is provided through the finding that both pathogen- and chemical-induced inflammation of the colon in mice leads to a dramatic shift in the indigenous bacterial population towards Enterobacteriaceae (Lupp et al. 2007).

While there is a clear rationale for considering that the intestinal microbiome or shifts in the microbiome may contribute to Crohn's disease, antibiotics have had limited success in the long-term treatment of Crohn's disease (Sandborn and Feagan 2003; Kozuch and Hanauer 2008). This contrasts with the greater efficacy of steroid treatment or blockade of TNF-α. Part of the difficulty in therapeutics may lie in the fact that symptoms of Crohn's disease vary considerably between patients. For instance, inflammation can spread to the colon in many individuals, which may complicate the diagnosis of Crohn's disease. The antibiotics metronidazole and ciprofloxacin have been reported to benefit patients with colonic Crohn's disease (Sutherland et al. 1991; Blichfeldt et al. 1978; Steinhart et al. 2002; Kozuch and Hanauer 2008). However, clinical trials examining the efficacy of these drugs in treating ileal Crohn's disease (Crohn's disease of the small intestine) have yielded negative results (Sandborn and Feagan 2003; Kozuch and Hanauer 2008). Likewise, clinical trials examining the efficacy of probiotics have also been disappointing (Heilpern and Szilagyi 2008). Antibiotics targeting the commensal bacterial flora likely relieve some of the secondary clinical manifestations of Crohn's disease but do not necessarily address the underlying cause. One possibility is that the wrong antibiotics have been used for these studies. Alternatively, bacteria may initiate a process that becomes independent of their presence later in the disease. Since antibiotics change the composition of the microflora rather than reduce the total number of bacteria, it is equally possible that any bacteria can contribute to pathogenesis, and so current antibiotics will remain ineffective. A last possibility is that bacteria can contribute to Crohn's disease severity in some way, but they are not fundamental to disease etiology.

Taken together, the current literature suggests that dysbiosis is not sufficient to cause intestinal inflammation, although it may be a key step in the process or an important outcome. Gnotobiotic model organism studies (Rawls et al. 2006) allow for colonization of the intestine with specific commensal bacteria. Host tissue analysis using this approach may reveal the specific bacterial communities and pathways that initiate or contribute to intestinal inflammation. In this regard, one must question the utility of using *Salmonella* and *Listeria* species as surrogates for understanding specific aspects of Crohn's disease pathogenesis. Neither of these bacterial species is known to influence disease onset or severity in humans. Although these bacterial models may be useful in "proof of principle" experiments for the analysis of intestinal immunity, their physiological relevance to Crohn's disease is unclear. Regardless of the ultimate answer as to what role bacteria play in Crohn's disease pathogenesis, alterations of intestinal homeostasis by factors other than bacterial infection must be considered.

3 Autophagy and Crohn's Disease

3.1 Identification of ATG16L1 and IRGM1 as Crohn's Disease Susceptibility Genes

The genetic loci for both *ATG16L1* and *IRGM1* were linked to Crohn's disease by genome-wide association studies in which genetic variations between Crohn's disease patients and controls were analyzed using gene chips containing ~10,000 known single nucleotide polymorphisms (SNPs) across the human genome (The Wellcome Trust Case Control Consortium 2007; Rioux et al. 2007; Hampe et al. 2007; Parkes et al. 2007; McCarroll et al. 2008; Barrett et al. 2008). For *ATG16L1*, the genetic linkage is stronger for patients that primarily have ileal disease, which is not necessarily true for the other susceptibility loci. This genetic linkage is attributed to a single SNP (rs2241880) that leads to a threonine to alanine amino acid substitution in a region of the Atg16L1 protein without a known function. The rs2241880 SNP is very common and can be found in as many as 50% of healthy individuals depending on the population examined (The Wellcome Trust Case Control Consortium 2007; Rioux et al. 2007; Hampe et al. 2007). Homozygosity for the risk allele of *ATG16L1* alone confers a twofold increase in risk of ileal Crohn's disease, while homozygosity for the risk allele of *ATG16L1*, *NOD2*, and a third susceptibility locus, *IBD5*, can confer a combined 20-fold increase in risk for Crohn's disease in the same population examined (Prescott et al. 2007). This finding exemplifies the intricate interplay between the various susceptibility loci, and indicates that future studies need to examine how these genes relate to each other in their function, especially in the context of the intestine. Notwithstanding these genetic factors, the environment clearly contributes to Crohn's disease (Jess et al. 2005).

In contrast to *ATG16L1*, the association between *IRGM1* and Crohn's disease is attributed to several noncoding SNPs (The Wellcome Trust Case Control Consortium 2007; Parkes et al. 2007) associated with a 20 kb deletion 2.7 kb upstream of the *IRGM1* transcriptional start site (McCarroll et al. 2008). Significantly different levels of *IRGM1* transcripts can be detected when comparing cells with sequence variations at this locus, indicating that the deleted sequence has a role in transcriptional regulation (McCarroll et al. 2008). However, whether the deletion leads to an increase or a decrease in transcription is highly dependent on the cell line analyzed, underlying the importance of studies with primary cells to complement these foundational genetic studies. It is worth mentioning that, unlike the *ATG16L1* polymorphism which is strictly associated with Crohn's disease (Fisher et al. 2008), the *IRGM1* deletion shows increased frequency in both patients with Crohn's disease and ulcerative colitis (McCarroll et al. 2008), a pathologically distinct form of inflammatory bowel disease. In the North American population that was studied, 10% of healthy individuals carried this particular polymorphism of the *IRGM1* locus compared to 15% and 14% of patients with Crohn's disease and ulcerative colitis, respectively (McCarroll et al. 2008). It will be interesting to see if the combination of the *ATG16L1* and *IRGM1* risk alleles act synergistically and confer a greater risk than one locus alone. Given that more than 30 loci have been linked to Crohn's disease susceptibility (Barrett et al. 2008), it will be essential to determine which loci act via a role in autophagy or other *IRGM1* or *ATG16L1*-dependent processes.

3.2 The Role of Atg16L1 in Autophagy

ATG16L1 (or *Atg16-like 1*) is one of two mammalian homologs of the yeast autophagy gene *Atg16*. The Atg16 protein was first identified in *S. cerevisiae* as a binding partner of the Atg5–Atg12 conjugate (Mizushima et al., 1999). The binding of Atg16 to Atg5–Atg12 is essential for autophagy (Matsushita et al. 2007). Atg16 has an N-terminal Atg5 binding domain and a coiled-coil domain that mediates homomultimerization. Atg16 forms a large and essential multimeric complex with the Atg5–Atg12 conjugate (Mizushima et al. 1999; Kuma et al. 2002; Matsushita et al. 2007). The mammalian Atg16L1 protein was also identified through its interaction with Atg5–Atg12 and, although much larger than yeast Atg16, contains an N-terminal Atg5 binding domain and a coiled-coil domain homologous to the yeast protein (Mizushima et al. 2003). In addition to these evolutionarily conserved regions, both mouse Atg16L1 and human Atg16L1 contain a C-terminal region with seven WD repeats that is not present in yeast Atg16 (Mizushima et al. 2003). WD repeats are protein interaction domains found in functionally diverse proteins, suggesting that there may be undiscovered binding partners of Atg16L1 that interact with this region.

While the purpose of the WD repeats remains obscure, there has been great progress in understanding the biochemical function of Atg16L1. Several observations indicate that Atg16L1 is important for the localization of the nascent autophagosomal machinery. Atg16L1 is found at the isolation membrane during autophagy (Mizushima et al. 2003). In addition to binding itself, the coiled-coil domain also has affinity for Rab33B, a small GTPase involved in membrane trafficking (Itoh et al. 2008). Remarkably, forced targeting of Atg16L1 to the plasma membrane redirects the LC3 conjugation reaction to the plasma membrane (Fujita et al. 2008). Importantly, Atg16L1 deficiency leads to greatly diminished autophagy activity, both in vitro and in vivo. This has been shown in cells derived from mice carrying two distinct mutations that lessen or eliminate the expression of Atg16L1 (Saitoh et al. 2008; Cadwell et al. 2008a).

An important remaining question regarding the role of Atg16L1 in autophagosome formation is how it is targeted to the nascent autophagosome. In yeast, the isolation membrane is generated from an organelle termed the pre-autophagosomal structure (PAS), and only one autophagosome is generated at a time (Reggiori and Klionsky 2005). In contrast, mammalian cells do not have a well-defined PAS and the origin of the autophagosome is still controversial (Axe et al. 2008). During the starvation response, an indiscriminate engulfment of cytoplasm may be sufficient to maintain cell viability. However, other situations may require the engulfment of specific cytoplasmic constituents such as mitochondria, bacteria, parasites, viruses, intracellular membranes, or proteins. Perhaps this requirement for specificity is greater in higher-order organisms that do not have the cytoplasm-to-vacuole trafficking (CVT) pathway that is important for selective autophagy in yeast (Reggiori and Klionsky 2005). It is tempting to speculate that the C-terminal domain of Atg16L1 evolved to mediate one or more specifically targeted forms of autophagy that are utilized in cells of higher-order eukaryotes. The T300A substitution encoded by the Crohn's disease susceptibility polymorphism is found within this extended C-terminal region and may therefore influence the function of the domain containing the WD repeats. A link between the T300A substitution and a defect in autophagy-mediated clearance of bacteria has been suggested by studies examining the localization of the Atg8/LC3 protein to intracellular *Salmonella* (Rioux et al. 2007; Kuballa et al. 2008). There is also evidence that the T300A substitution alters the stability of Atg16L1 protein (Kuballa et al. 2008), and thus it will be critical to determine if the polymorphism associated with Crohn's disease affects general autophagy or a specific aspect of autophagy.

Importantly, new data (Saitoh et al. 2008; Cadwell et al. 2008a) described below show that Atg16L1 can have effects that are highly cell type-specific, for example in either macrophages or Paneth cells. These new findings indicate that studies of the mechanisms responsible for the role of the T300A allele in Crohn's disease pathogenesis will need to be performed in cell systems relevant to these primary cells. This is especially important given the emerging understanding that autophagy can play different roles in different primary cell types.

In comparison to *ATG16L1*, very little is known about *ATG16L2*, the other mammalian homolog of yeast *ATG16*. Although the amino acid sequence of Atg16L2 displays homology with Atg16L1, including a predicted coiled-coil domain and WD repeats, it does not fully compensate for the absence of Atg16L1 in autophagy, since mouse cells lacking sufficient Atg16L1 are autophagy deficient (Saitoh et al. 2008; Cadwell et al. 2008a). It is not currently known if *ATG16L2* has a role in either autophagy or Crohn's disease pathogenesis. Nevertheless, there is one indication that it may have an important role in infectious diseases. A small interfering RNA (siRNA) screen to discover host factors required by HIV-1 replication in HeLa cells identified several autophagy transcripts as well as *ATG16L2* as being necessary (Brass et al. 2008). Since the role often proposed for autophagy in the generation of intracellular membranes required for RNA virus replication remains controversial (Zhao et al. 2007), a genetic model to study Atg16L2 function would be invaluable.

3.3 *The Role of IRGM1 in Autophagy*

Immunity-related GTPase family M member 1 (IRGM1) is a member of the p47 GTPase family, a group of proteins that are most well known for regulating the growth of intracellular pathogens (Taylor et al. 2004). Many studies now indicate a key role for this and related p47 GTPases in autophagy and autophagy-dependent processes (see the chapter by Deretic et al. in this volume). Studies of the murine ortholog of human IRGM1, called Irgm1 or LRG47 (Singh et al. 2006), have shown that this protein is an important downstream effector for interferon-γ (IFN-γ) (Bekpen et al. 2005), a cytokine with a well-established role in protective immunity against intracellular pathogens including *L. monocytogenes*, *Toxoplasma gondii (T. gondii)*, and *Mycobacterium tuberculosis* (Zhao et al. 2008; Flynn et al. 1993). Building on the observation that IFN-γ induces autophagy in certain cell lines (Inbal et al. 2002; Gutierrez et al. 2004), IFN-γ was subsequently shown to increase colocalization of the autophagy machinery and *M. tuberculosis* (Gutierrez et al. 2004). Moreover, IFN-γ-induced autophagy in a murine macrophage cell line has been reported to be dependent on Irgm1/LRG47 (Singh et al. 2006).

Further support for a key role of p47 GTPases in the regulation of autophagy comes from studies suggesting that Irgm1/LRG47 is involved in preventing IFN-γ-induced autophagy-associated cell death in murine T lymphocytes (Feng et al. 2008). In another study, the amount of IRGM1 expression correlated with the degree of colocalization between *Salmonella typhimurium* and LC3 in HeLa cells (McCarroll et al. 2008). These studies suggest that, in certain cell lines, some of these p47 GTPase family members may lie upstream of autophagy or autophagy protein-dependent processes.

Interestingly, recent data demonstrate a different role for the autophagy protein Atg5 in the regulation of p47 GTPase-dependent control of the intracellular parasite *T. gondii* (Zhao et al. 2008). This work places Atg5 upstream in the regulation of GTPase function by showing that Atg5 is required for the recruitment of the p47 GTPase IIGP1 to the parasitophorous vacuole within which *T. gondii* resides and replicates in unactivated macrophages. IFN-γ-induced expression of IIGP1 and recruitment of this GTPase to the parasitophorous vacuole membrane is part of the process responsible for the destruction of the vacuole membrane and exposure of the parasite to the cytoplasm. This ultimately leads to the death and clearance of the parasite from macrophages activated by the combination of IFN-γ and LPS. In addition to a role for Atg5 in recruitment of IIGP1, Atg5 is essential for the clearance of the parasite from activated macrophages (Zhao et al. 2008). However, careful ultrastructural evaluation of parasites undergoing IFN-γ-induced destruction of the parasitophorous vacuole failed to reveal the involvement of classical autophagosomes enveloping the parasite. Lastly, analysis of mice lacking Atg5 expression in macrophages revealed a critical role for Atg5 in the resistance of the animal to challenge from *T. gondii*, providing the first proof in mammals that an autophagy protein is essential for resistance to an intracellular pathogen in vivo. Together, these data demonstrate that there are intimate and, as yet, incompletely understood relationships between autophagy, autophagy proteins, and p47 GTPases involved in innate and acquired immunity.

4 The Role of Autophagy in Intestinal Biology

4.1 Two Mouse Models to Examine Atg16L1 Function In Vivo

The human genetic data and our expanding understanding of the roles of *ATG16L1* and *IRGM1* support a model in which autophagy contributes to Crohn's disease susceptibility, but also leave open critical questions, including what intestinal or immune cell-types have altered function when either *ATG16L1* or *IRGM1* are altered, and what functional abnormalities are present in such cells. Given the strong associations between bacteria and intestinal inflammation, are the effects of autophagy genes in Crohn's disease related only to clearance of intracellular bacteria, or are there additional alterations in cells with aberrant Atg16L1 or IRGM1 expression/function that might contribute to the pathogenesis of Crohn's disease? Two distinct mouse models for the study of Atg16L1 have recently been developed and analyzed (Saitoh et al. 2008; Cadwell et al. 2008a), leading to the answers to some of these important questions and the generation of new questions regarding the function of autophagy proteins in the cell biology and function of two cells known to be critically important for immunity, the macrophage and the Paneth cell.

To take advantage of mouse genetics, researchers must overcome the essential role of autophagy in the survival of newborn mice. In one model of Atg16L1 function, the coiled-coil domain was deleted leading to a nonfunctional Atg16L1 protein (Saitoh et al. 2008) (*Atg16L1* knockout or KO mice herein). Similar to *Atg5* and *Atg7* KO mice (Kuma et al. 2004; Komatsu et al. 2005), *Atg16L1* KO mice die within 24 h after birth. It is uncertain why autophagy-deficient mice do not survive this period, but a contributor to death may be an inability to overcome neonatal starvation (Kuma et al. 2004). To study the impact of Atg16L1 deficiency on cells of the immune system, the investigators generated fetal liver chimeric mice. The mouse fetal liver contains hematopoietic stem cells (HSCs) that are capable of multilineage reconstitution of recipient mice. In their system, fetal liver cells from *Atg16L1* KO or control mice were transferred into lethally irradiated mice, allowing reconstitution of the hematopoietic system of these mice with donor *Atg16L1*-deficient HSCs.

In a second model, investigators generated two mouse lines with hypomorphic (reduced) Atg16L1 protein expression rather than a deletion of the gene (Cadwell et al. 2008a). The two lines were established from ES cells harboring distinct gene trap mutations within the *Atg16L1* locus. Gene trap vectors introduce a false splice acceptor into an intron that can inhibit expression of intact mRNA, frequently resulting in decreased expression of the protein (Stryke et al. 2003). *Atg16L1* hypomorph ($Atg16L1^{HM}$) mice are born at Mendelian ratios and survive to adulthood, thus making them a unique model for studying autophagy. An important consideration for this model is that there is Atg16L1 present in sufficient amounts to allow the mice to survive, indicating that cells from these mice will not be fully deficient in either autophagy or any other function that requires Atg16L1. Studies in both the *Atg16L1* KO and $Atg16L1^{HM}$ models confirm that mammalian Atg16L1 is a bona fide autophagy protein in vivo (Saitoh et al. 2008; Cadwell et al. 2008a). *Atg16L1*-deficient macrophages and embryonic fibroblasts fail to efficiently induce autophagy upon stimulation. Furthermore, thymocytes and the intestinal epithelium harvested from $Atg16L1^{HM}$ mice show defects in autophagy in the absence of starvation or other autophagy-inducing conditions, indicating that Atg16L1 is important for basal autophagy under physiological conditions.

4.2 Loss of Atg16L1 Leads to an Increased Inflammatory Response and Increased Expression of the Proinflammatory Cytokine IL-1β in Macrophages

Similar to studies with certain *Nod2* mutant mice (see above), challenging fetal liver chimeric mice with DSS shows that Atg16L1 expression in hematopoietic cells is necessary for the normal response to intestinal injury. Compared to controls, *Atg16L1* KO chimeric mice have increased mortality and weight loss in response to DSS, indicative of an exaggerated inflammatory response (Saitoh et al. 2008). Indeed, these mice have severe ulceration and infiltration of lymphocytes in the colon. This increased inflammatory response is correlated with increased serum levels of the proinflammatory cytokines IL-1β, IL-6, and IL-18. A role for IL-1β in DSS-induced disease susceptibility was demonstrated using systemic administration of an interfering antibody specific for IL1-β. Analysis of the pathology in the small intestine has not been reported in these mice.

The intersection between Toll-like receptor (TLR) and the autophagy pathway may account for this aberrant injury response in the colon. In addition to the direct destruction of pathogens by the autophagosome, or nonautophagosomal pathways involving autophagy proteins such as GTPase trafficking (Zhao et al. 2008), autophagy is proposed to contribute to innate immunity through TLR activity. Like Nod2, TLRs are important pattern recognition receptors that bind conserved structures derived from various pathogens (Rakoff-Nahoum et al. 2004). Autophagy is essential for the delivery of viral RNA to TLR7 in plasmacytoid dendritic cells, a process that is necessary for the cytokine response to certain viruses (Lee et al. 2007). Also, autophagy genes are necessary for the TLR-mediated delivery of phagocytosed beads carrying TLR ligands from the phagosomes to the lysosome in macrophages (Sanjuan et al. 2007). Surprisingly, the double-membrane vesicles of macroautophagy are absent in this latter process, suggesting an autophagosome-independent function for autophagy genes in the events following phagocytosis. This is reminiscent of the lack of classical autophagosomes in the Atg5-dependent killing of *T. gondii* in activated macrophages described above (Zhao et al. 2008). TLRs are clearly necessary for intestinal homeostasis, since at least a subset of TLRs protect mice from DSS-induced injury of the colon and mortality (Rakoff-Nahoum et al. 2004), an observation that is conspicuously similar to those made in mice reconstituted with HSCs from *Atg16L1* KO mice.

In contrast to previous reports from studies in cell lines (Xu et al. 2007; Delgado et al. 2008), TLR ligands did not induce autophagy in primary macrophages (Saitoh et al. 2008). This underlines the importance of studies in nontransformed primary cells for defining the role of autophagy in immunity and pathogenesis (Virgin 2008; Zhao et al. 2007, 2008; Lee et al. 2007). Thus, not all TLRs are strong inducers of autophagy in macrophages in physiological settings. Instead, the autophagy pathway plays a novel and critical role in regulating the TLR-induced inflammatory response (Saitoh et al. 2008). The absence of Atg16L1, or inhibition of the autophagy pathway by other means, leads to an uncontrolled excess production of the proinflammatory cytokines IL-1β and IL-18 in response to TLR4 activation. This represents a gain of function in autophagy-deficient primary macrophages. Interestingly, exposure of *Atg16L1*-deficient macrophages to several noninvasive bacteria found in the gut elicits this same aberrant inflammatory response, while the invasive bacterium *S. typhimurium* does not, again emphasizing the fundamental importance of studies to define which bacteria, if any, are causative in Crohn's disease. Although the mechanism of this gain of function response in autophagy-deficient primary macrophages awaits elucidation, it appears to be associated with reactive oxygen species (ROS), respiratory byproducts known to accumulate in the absence of autophagy (Zhang et al. 2008). These studies show that, although autophagosomes and autophagy proteins independent of autophagosomes are important in the clearance of intracellular pathogens, the autophagy pathway has another previously unappreciated role in innate immunity.

How does this novel role for autophagy relate to Crohn's disease? One can imagine a scenario where damage to the intestine leads to the exposure of cells such as macrophages to commensal bacteria, an event that is mimicked in mice treated with DSS. In the absence of Atg16L1 function, this may result in uncontrolled proinflammatory cytokine production.

4.3 Mice with Reduced Atg16L1 Expression Reproduce Aspects of Crohn's Disease Pathology, and Reveal a Cellular Target for the Role of Atg16L1 in Crohn's Disease

While the *Atg16L1* KO model is ideal for studying the role of Atg16L1 in hematopoietic lineages, perinatal lethality of these mice does not allow examination of the role of Atg16L1 in the small intestine in adult mice. The newly developed *Atg16L1^{HM}* mice are therefore a unique tool since, instead of relying on a lineage-specific deletion of the gene, there is a systemic reduction of the protein (Cadwell et al. 2008a). As it is unlikely that Atg16L1 function is completely abolished in humans carrying the Crohn's disease risk allele (Kuballa et al. 2008), this model may be more reflective of the disease predisposition state. Indeed, the autophagy defect in the intestinal epithelium of these mice is not as severe as mice with a complete deletion of *Atg5* in this tissue (Cadwell et al. 2008a).

Histological examination of the small intestine from *Atg16L1^{HM}* mice led to the identification of a previously unknown human Crohn's disease pathology: striking morphological, functional, and transcriptional abnormalities in the Paneth cell. Interestingly, like macrophages from *Atg16L1* KO mice (Saitoh et al. 2008), Paneth cells from *Atg16L1*-deficient mice display gain-of-function properties in that they overproduce transcripts associated with PPAR signaling, acute phase reactants, adipocytokine signaling, and lipid metabolism. The proteins encoded by many of these transcripts are directly implicated in inflammation. However, the increased expression of IL-1β in macrophages from *Atg16L1* KO mice is due to post-translational changes in IL-1β, while increases in the cytokine leptin noted in Paneth cells from *Atg16L1^{HM}* mice are due to altered transcriptional regulation.

Paneth cells are considered important for innate immunity since they efficiently generate granules containing antimicrobial peptides, lysozyme, and other proteins that are secreted into the intestinal lumen (Porter et al. 2002). The effect of *Atg16L1* deficiency on these cells is not the same as the absence of Paneth cells described in the *Xbp1* model, although it is interesting that both models have a primary defect in Paneth cells, suggesting a particular sensitivity of these cells to deficiency in autophagy or the unfolded protein response. First, the Paneth cells in *Atg16L1*^{HM} mice have no alterations in cell cycle or cell death while *Xbp1*-deficient Paneth cells die by apoptosis (Kaser et al. 2008). Second, rather than an absence of granules reported in the *Xbp1* model, the Paneth cells in *Atg16L1*^{HM} mice show a variety of granule abnormalities, including reduction in number, heterogeneity in size, and localization to the basal membrane, instead of the normal apical distribution of granules (Fig. 1). In a normal mouse, Paneth granule contents are exocytosed by fusion with the apical membrane. Instead, in the *Atg16L1*^{HM} mice, intact granules can be detected in the lumen as aggregates.

It is difficult to imagine how reduced expression of an autophagy protein accounts for these dramatic cell type-specific effects. Although it is clear that Atg16L1 is a bona fide and essential autophagy protein in mammals, it is possible that Atg16L1 has roles outside of autophagy. To determine if abnormal Paneth cell granule exocytosis is related to a deficiency in autophagy, the essential autophagy gene *Atg5* was deleted specifically in intestinal epithelial cells using Cre-loxP technology (Cadwell et al. 2008a). In this model, mice with a floxed *Atg5* allele are bred with mice expressing the Cre recombinase driven by the villin promoter. Since villin is expressed in all cells of the crypt-villus axis (Madison et al. 2002), *Atg5* is deleted in the Paneth cell and surrounding cell types within the intestinal epithelium. Deletion of *Atg5* in the intestinal epithelium leads to aberrant granule morphology, again in ileal Paneth cells. Although one cannot rule out the possibility that the deletion of *Atg5* in another intestinal epithelial cell type such as the enterocyte is responsible in *trans* for the Paneth cell-specific abnormalities, this is consistent with a critically important role for autophagy proteins in Paneth cells. Similar data has recently been obtained for deletion of *Atg7*, another gene essential for autophagy, using villin-Cre (Cadwell et al. 2008b).

The deletion of *Atg5* or *Atg7* or decreases in Atg16L1 expression all result in abnormalities in Paneth cells, but does this also implicate autophagy/autophagosomes in control of the transcription of inflammatory genes and/or in normal granule exocytosis? There is a role for autophagy in organelle turnover, and the *Xbp1* model indicates that Paneth cells are particularly sensitive to changes in ER homeostasis. Indeed, *Atg16L1*^{HM} Paneth cells display degenerating mitochondria, a lack of normal ER, and the presence of abnormal vesicles throughout the cytoplasm (Cadwell et al. 2008a). However, these observations must be taken in context with the emerging body of literature suggesting that the autophagy proteins are involved in processes that may be independent of double-membrane-bound vesicles. In addition to phagosome–lysosome fusion, autophagy proteins are also implicated in type I interferon signaling (Jounai et al. 2007), recruitment of p47 GTPases to intracellular membranes (Zhao et al. 2008), FADD-mediated cell death and mitochondrial apoptosis (Pyo et al. 2005; Yousefi et al. 2006), and regulation of IL-1β production, as mentioned above. Therefore, future experiments will be required to address the role of autophagy as a cellular process in Paneth cell biology.

Regardless of the exact mechanism by which these Paneth cell-specific abnormalities arise, the *Atg16L1*^{HM} model has direct clinical implications. A retrospective examination of human Crohn's disease resection specimens showed that 100% of the patients that are homozygous for the risk allele of *ATG16L1* have the same morphological abnormalities in Paneth cells observed in *Atg16L1*-deficient mice (Fig. 2). For this reason, the *Atg16L1*^{HM} model is truly opportune since it allowed

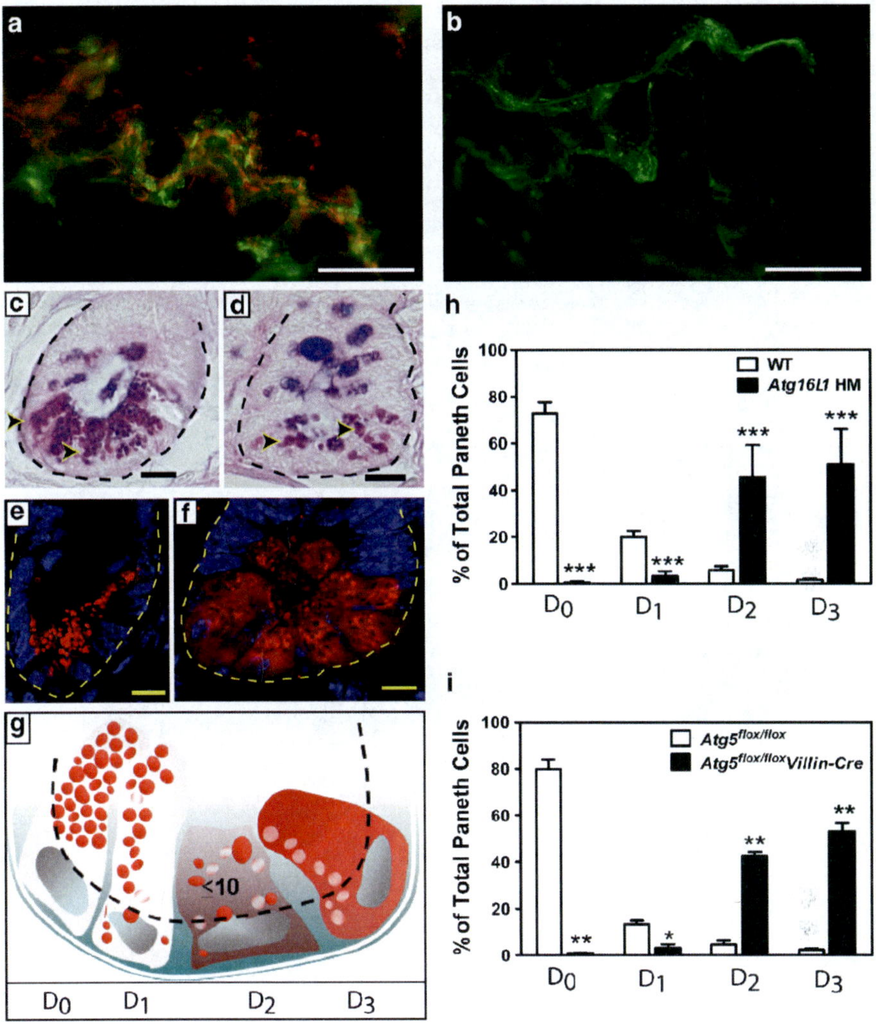

Fig. 1a–i Mutation of *Atg16L1* or *Atg5* leads to disruption of the Paneth cell granule exocytosis pathway. **a**, **b** Whole-mount images taken immediately above the ileal mucosal surface from WT (**a**) and $Atg16L1^{HM}$ (**b**) littermate mice stained with *Helix pomatia* lectin that labels mucus (*green*) and antisera directed against lysozyme (*red*). **c**, **d** Ileal sections from WT (**c**) and $Atg16L1^{HM}$ (**d**) mice were stained with PAS/alcian blue to detect Paneth granules by light microscopy (*dashed line* denotes the crypt unit and *arrowheads* indicate Paneth cells). **e**, **f** Representative images of indirect immunofluorescence of sections stained for lysozyme (*red*) in WT (**e**) and $Atg16L1^{HM}$ (**f**) mouse ileal crypts. **g** Paneth cells display one of four patterns of lysozyme expression (represented in *red*; *white* represents areas that exclude lysozyme): normal (*D0*), disordered (*D1*), depleted (*D2*) and diffuse (*D3*). **h**, **i** Number of Paneth cells from $Atg16L1^{HM}$ (**h**) and $Atg5^{flox/flox}$ *villin-Cre* (**i**) mice displaying each pattern of lysozyme expression. Scale bars: **a**, **b** 200 mm; **c**–**f** 10 mm. **P*, 0.05; ***P*, 0.01; ****P*, 0.001. *P* values were calculated using two-tailed Student's t-test. Error bars, s.e.m. Figure from Cadwell et al. (2008a) reprinted with the permission of Macmillan Publishers Ltd.

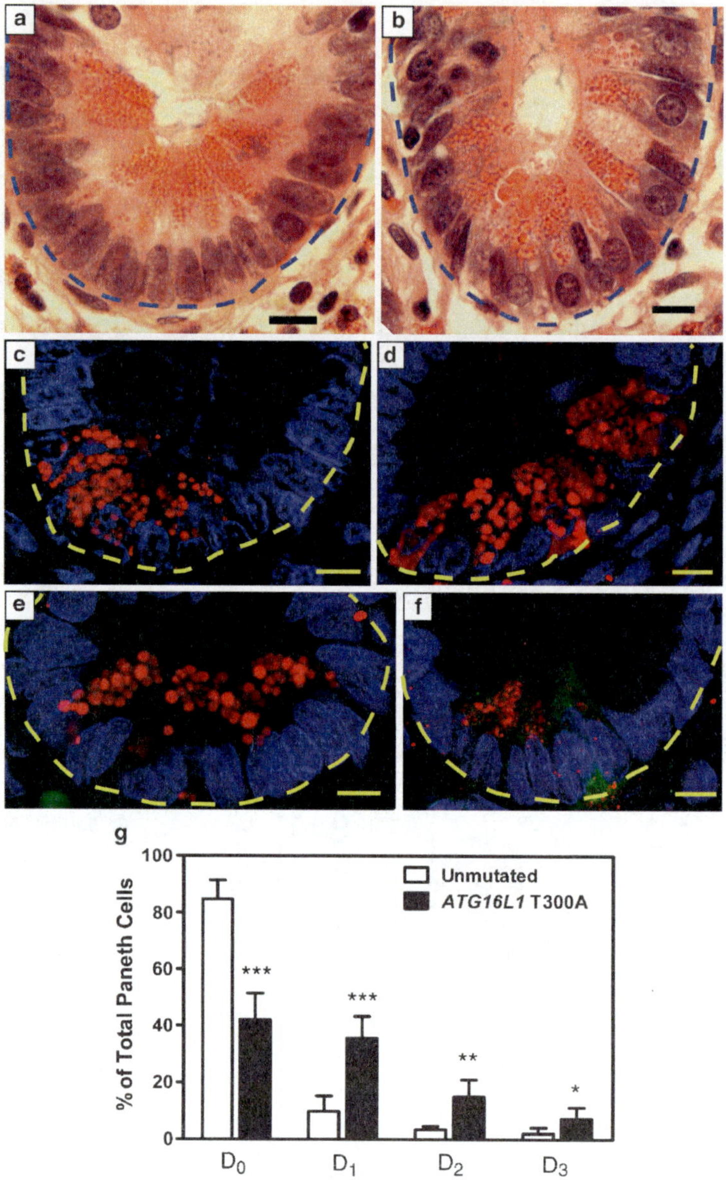

Fig. 2a–g Crohn's disease patients homozygous for the disease risk allele of *ATG16L1* display Paneth cell abnormalities similar to *Atg16L1*^{HM} mice. **a, b** Hematoxylin-and-eosin-stained sections of uninvolved areas from ileo–colic resections from patients with Crohn's disease homozygous for the safe (**a**) or risk (**b**) allele of *ATG16L1* (*blue dashed line* denotes crypt unit). **c–f** Immunofluorescence images of Paneth cells from control patients (**c, e**) and patients with the risk allele (**d, f**) stained for lysozyme (*red*; **c, d**) and double-labeled additionally for leptin (*green*; **e, f**; *yellow dashed line* denotes the crypt unit). **g** Aberrant lysozyme expression was quantified using the same criteria as that used for mouse sections defined in Fig. 3 of Cadwell et al. (2008a). Leptin-positive D3 (see Fig. 1) cells were quantified in patient samples homozygous for the risk allele or homozygous for the safe allele. Scale bars: **a–f** 10 mm. *P, 0.05; **P, 0.01; ***P, 0.001. P values were calculated using two-tailed Student's t-test. Error bars, s.e.m. Figure from Cadwell et al. (2008a) reprinted with the permission of Macmillan Publishers Ltd.

the identification of a previously unknown pathology associated with Crohn's disease patients carrying a specific Crohn's disease risk allele. Given that this study was performed only on specimens from Crohn's disease patients, it will be interesting to determine if healthy individuals carrying the risk allele of *ATG16L1* also have these abnormalities. If they do not, then perhaps an environmental trigger induces this particular pathology only when *ATG16L1* function is deficient. Additionally, since genotyping patients can be cost prohibitive and Crohn's disease is frequently difficult to diagnose, aberrant Paneth cells may be a useful diagnostic tool if it is specific to Crohn's disease and not ulcerative colitis.

There is debate as to whether NOD2 signaling may be important in the secretion or production of antimicrobial peptides by Paneth cells (Wehkamp et al. 2004; Wehkamp et al. 2005; Simms et al. 2008; Voss et al. 2006; Lala et al. 2003; Ogura et al. 2003). These studies did not take into account *ATG16L1* polymorphisms. Therefore, it will be a worthwhile endeavor to check for Paneth cell abnormalities in samples from patients with other Crohn's disease risk alleles in the absence of the *ATG16L1* susceptibility allele.

Now that the Paneth cell defect is defined, it will be necessary to use the $Atg16L1^{HM}$ mice to understand how this specific feature relates to the other pathological hallmarks that are already associated with Crohn's disease. Part of the answer may lie in another feature of the $Atg16L1^{HM}$ mice that is also common in Crohn's disease patients. Among the inflammatory signals overexpressed by Atg16L1-deficient Paneth cells are two adipocytokines, leptin and adiponectin (Cadwell et al. 2008a). Crohn's disease is known to be associated with increased expression of both leptin and adiponectin, immunoregulatory cytokines that have been recognized as products of adipocytes (Barbier et al. 2003; Yamamoto et al. 2005). This overproduction of adipocytokines has previously been attributed to the "creeping fat" commonly observed as a pathologic hallmark of Crohn's disease (Schaffler and Herfarth 2005). The finding that transcripts encoding these inflammatory molecules are overexpressed in $Atg16L1^{HM}$ Paneth cells led to the discovery that Paneth cells from Crohn's disease patients homozygous for the *ATG16L1* risk polymorphism also express leptin protein (Fig. 2) (Cadwell et al. 2008a). This observation was most striking in the most morphologically aberrant Paneth cells in humans. This finding raises the possibility that leptin mRNA is transcribed in excess in individuals with the *ATG16L1* risk polymorphism, similar to $Atg16L1^{HM}$ mice. It will be important to extend these observations through further characterization, such as determining the role of these cytokines in intestinal injury responses.

5 Conclusion and Future Perspectives

Macrophages and Paneth cells, both of which are critical for mucosal immunity in the intestine, are two of the newest cell types in which autophagy genes have been associated with unique cell type-specific roles (Fig. 3). Other chapters in this volume discuss the essential roles of the autophagy pathway in the biology of additional cell types implicated in intestinal inflammation, including lymphocytes and

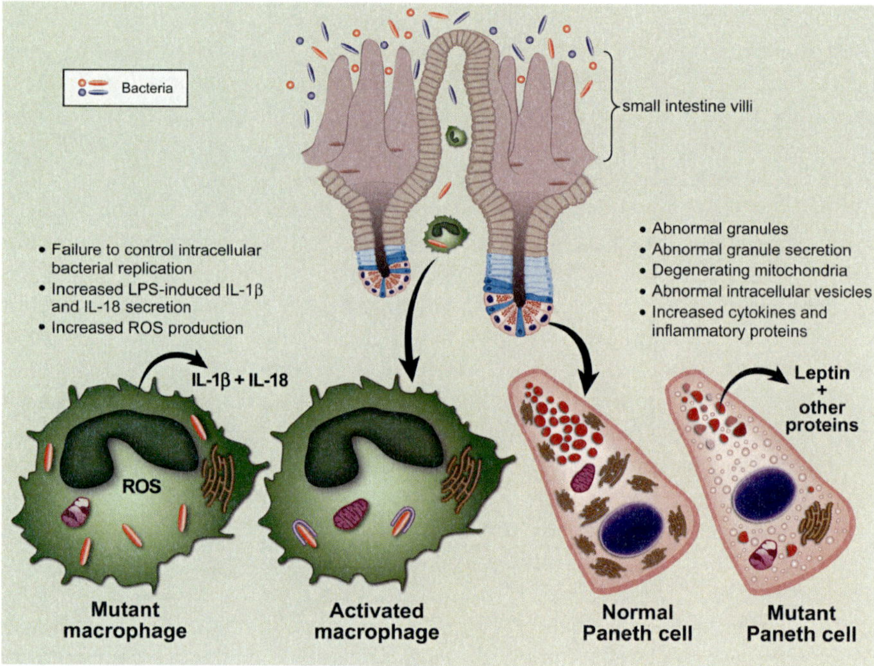

Fig. 3 Mutation of autophagy genes leads to striking abnormalities in key cell types implicated in intestinal inflammation. Macrophages serve a critical role in mucosal immunity by secreting cytokines and other regulatory molecules that shape the immune environment. Macrophages can also phagocytose and destroy bacteria that inappropriately penetrate the epithelial wall. In addition to a failure to control intracellular bacterial replication through classical autophagy (reviewed in other chapters), autophagy mutant macrophages produce increased amounts of reactive oxygen species (ROS) and the proinflammatory cytokines IL-1β and IL-18 (Saitoh et al. 2008). Paneth cells are highly specialized intestinal epithelial cells located at the crypt base. Paneth cells package antimicrobial factors in granules and, upon stimulation, secrete the granule contents into the intestinal lumen. In autophagy mutant Paneth cells, organelle degeneration is accompanied by granule morphology and secretion abnormalities (Cadwell et al. 2008a, b). Moreover, mutant Paneth cells display a gain-of-function overproduction of inflammatory proteins. Importantly, these changes are also observed in human Crohn's disease patients that are homozygous for the *ATG16L1* risk polymorphism

dendritic cells. Thus, a major challenge facing disease research is to discriminate those functions of autophagy that are critical for inflammatory bowel disease pathogenesis from functions of autophagy that may be important in other circumstances. Although the role of autophagy genes in the Paneth cell is the only one validated with Crohn's disease specimens to date, it is reasonable to speculate that autophagy contributes to Crohn's disease pathogenesis through multiple cell type-specific mechanisms that are not mutually exclusive.

Rarely are specific pathological hallmarks of a complicated human disease associated with a specific genetic polymorphism observed in a model organism. Perhaps this is not surprising for a disease like Crohn's disease in which many genetic polymorphisms, and likely environmental factors, act in concert to produce a complex pathologic picture that may vary from person to person. It is fascinating that the two *Atg16L1*-deficient and the two *Nod2*-deficient mutant mouse models all display aberrant production of inflammatory signals, although the mechanisms responsible may be quite distinct. With our current knowledge, there is no obvious unifying mechanism by which the autophagosome, or autophagosome-mediated clearance of intracellular bacteria, could be involved in all of the observed cell type-specific effects of Atg16L1 deficiency in macrophages and Paneth cells. Overproduction of IL-1β and IL-18 in response to LPS was not accompanied by an increase in autophagosome formation (Saitoh et al. 2008). Also, Paneth cell abnormalities in $Atg16L1^{HM}$ mice arise in the absence of exogenous bacterial challenge, and these mice do not exhibit increased sensitivity to oral *L. monocytogenes* challenge (Cadwell et al. 2008a), a clear distinction from the *Nod2* KO mice. Although autophagosomes could be involved in recycling organelles in Paneth cells, autophagosome formation has not previously been associated with transcriptional changes, such as those observed in *Atg16L1*-deficient Paneth cells. The observation that *Atg16L1* or *Atg5*-deficient thymocytes display none of the transcriptional features associated with the aberrant Paneth cells further supports the need for in vivo and cell type-specific studies to define the mechanisms responsible for these seemingly disparate consequences of autophagy protein deficiency. While investigators examine the role of the autophagy pathway in complicated diseases such as Crohn's, these studies indicate that there is yet much to learn about the autophagy proteins themselves.

References

Acosta-Alvear D, Zhou Y, Blais A, Tsikitis M, Lents NH, Arias C, Lennon CJ, Kluger Y, Dynlacht BD (2007) *XBP1* controls diverse cell type- and condition-specific transcriptional regulatory networks. Mol Cell 27:53–66

Ahmad T, Armuzzi A, Bunce M, Mulcahy-Hawes K, Marshall SE, Orchard TR, Crawshaw J, Large O, de SA, Cook JT, Barnardo M, Cullen S, Welsh KI, Jewell DP (2002) The molecular classification of the clinical manifestations of Crohn's disease. Gastro 122:854–866

Axe EL, Walker SA, Manifava M, Chandra P, Roderick HL, Habermann A, Griffiths G, Ktistakis NT (2008) Autophagosome formation from membrane compartments enriched in phosphatidylinositol 3-phosphate and dynamically connected to the endoplasmic reticulum. J Cell Biol 182:685–701

Barbier M, Vidal H, Desreumaux P, Dubuquoy L, Bourreille A, Colombel JF, Cherbut C, Galmiche JP (2003) Overexpression of leptin mRNA in mesenteric adipose tissue in inflammatory bowel diseases. Gastroenterol Clin Biol 27:987–991

Barrett JC, Hansoul S, Nicolae DL, Cho JH, Duerr RH, Rioux JD, Brant SR, Silverberg MS, Taylor KD, Barmada MM, Bitton A, Dassopoulos T, Datta LW, Green T, Griffiths AM, Kistner EO, Murtha MT, Regueiro MD, Rotter JI, Schumm LP, Steinhart AH, Targan SR, Xavier RJ, Libioulle C, Sandor C, Lathrop M, Belaiche J, Dewit O, Gut I, Heath S, Laukens D, Mni M, Rutgeerts P, Van GA, Zelenika D, Franchimont D, Hugot JP, de VM, Vermeire S, Louis E, Cardon LR, Anderson CA, Drummond H, Nimmo E, Ahmad T, Prescott NJ, Onnie CM,

Fisher SA, Marchini J, Ghori J, Bumpstead S, Gwilliam R, Tremelling M, Deloukas P, Mansfield J, Jewell D, Satsangi J, Mathew CG, Parkes M, Georges M, Daly MJ (2008) Genome-wide association defines more than 30 distinct susceptibility loci for Crohn's disease. Nat Genet 40:955–962

Bastide P, Darido C, Pannequin J, Kist R, Robine S, Marty-Double C, Bibeau F, Scherer G, Joubert D, Hollande F, Blache P, Jay P (2007) *Sox9* regulates cell proliferation and is required for Paneth cell differentiation in the intestinal epithelium 1. J Cell Biol 178:635–648

Bekpen C, Hunn JP, Rohde C, Parvanova I, Guethlein L, Dunn DM, Glowalla E, Leptin M, Howard JC (2005) The interferon-inducible p47 (IRG) GTPases in vertebrates: loss of the cell autonomous resistance mechanism in the human lineage. Genome Biol 6:R92

Blichfeldt P, Blomhoff JP, Myhre E, Gjone E (1978) Metronidazole in Crohn's disease. A double blind cross-over clinical trial. Scand J Gastroenterol 13:123–127

Brass AL, Dykxhoorn DM, Benita Y, Yan N, Engelman A, Xavier RJ, Lieberman J, Elledge SJ (2008) Identification of host proteins required for HIV infection through a functional genomic screen. Science 319:921–926

Bry L, Falk P, Huttner K, Ouellette A, Midtvedt T, Gordon JI (1994) Paneth cell differentiation in the developing intestine of normal and transgenic mice. Proc Natl Acad Sci USA 91:10335–10339

Cadwell K, Liu JY, Brown SL, Miyoshi H, Loh J, Lennerz JK, Kishi C, Kc W, Carrero JA, Hunt S, Stone CD, Brunt EM, Xavier RJ, Sleckman BP, Li E, Mizushima N, Stappenbeck TS, Virgin HW (2008a) A key role for autophagy and the autophagy gene *Atg16l1* in mouse and human intestinal Paneth cells. Nature 456:259–263

Cadwell K, Patel KK, Komatsu M, Virgin HW 4th, Stappenbeck TS (2009) A common role for *Atg16L1*, *Atg5* and *Atg7* in small intestinal Paneth cells and Crohn disease. Autophagy 5(2): 250–252.

Cho JH (2008) The genetics and immunopathogenesis of inflammatory bowel disease. Nat Rev Immunol 8:458–466

Delgado MA, Elmaoued RA, Davis AS, Kyei G, Deretic V (2008) Toll-like receptors control autophagy. EMBO J 27:1110–1121

Deretic V (2006) Autophagy as an immune defense mechanism. Curr Opin Immunol 18:375–382

Feng CG, Zheng L, Jankovic D, Bafica A, Cannons JL, Watford WT, Chaussabel D, Hieny S, Caspar P, Schwartzberg PL, Lenardo MJ, Sher A (2008) The immunity-related GTPase Irgm1 promotes the expansion of activated $CD4^+$ T cell populations by preventing interferon-gamma-induced cell death. Nat Immunol 9:1279–1287

Ferguson LR, Shelling AN, Browning BL, Huebner C, Petermann I (2007) Genes, diet and inflammatory bowel disease. Mutat Res 622:70–83

Fisher SA, Tremelling M, Anderson CA, Gwilliam R, Bumpstead S, Prescott NJ, Nimmo ER, Massey D, Berzuini C, Johnson C, Barrett JC, Cummings FR, Drummond H, Lees CW, Onnie CM, Hanson CE, Blaszczyk K, Inouye M, Ewels P, Ravindrarajah R, Keniry A, Hunt S, Carter M, Watkins N, Ouwehand W, Lewis CM, Cardon L, Lobo A, Forbes A, Sanderson J, Jewell DP, Mansfield JC, Deloukas P, Mathew CG, Parkes M, Satsangi J (2008) Genetic determinants of ulcerative colitis include the *ECM1* locus and five loci implicated in Crohn's disease. Nat Genet 40:710–712

Flynn JL, Chan J, Triebold KJ, Dalton DK, Stewart TA, Bloom BR (1993) An essential role for interferon gamma in resistance to *Mycobacterium tuberculosis* infection. J Exp Med 178:2249–2254

Fujita N, Itoh T, Fukuda M, Noda T, Yoshimori T (2008) The Atg16L complex specifies the site of LC3 lipidation for membrane biogenesis in autophagy. Mol Biol Cell 19:2092–2100

Girardin SE, Travassos LH, Herve M, Blanot D, Boneca IG, Philpott DJ, Sansonetti PJ, Mengin-Lecreulx D (2003) Peptidoglycan molecular requirements allowing detection by Nod1 and Nod2. J Biol Chem 278:41702–41708

Gutierrez MG, Master SS, Singh SB, Taylor GA, Colombo MI, Deretic V (2004) Autophagy is a defense mechanism inhibiting BCG and *Mycobacterium tuberculosis* survival in infected macrophages. Cell 119:753–766

Hampe J, Cuthbert A, Croucher PJ, Mirza MM, Mascheretti S, Fisher S, Frenzel H, King K, Hasselmeyer A, Macpherson AJ, Bridger S, van DS, Forbes A, Nikolaus S, Lennard-Jones JE, Foelsch UR, Krawczak M, Lewis C, Schreiber S, Mathew CG (2001) Association between insertion mutation in *NOD2* gene and Crohn's disease in German and British populations. Lancet 357:1925–1928

Hampe J, Franke A, Rosenstiel P, Till A, Teuber M, Huse K, Albrecht M, Mayr G, De La Vega FM, Briggs J, Gunther S, Prescott NJ, Onnie CM, Hasler R, Sipos B, Folsch UR, Lengauer T, Platzer M, Mathew CG, Krawczak M, Schreiber S (2007) A genome-wide association scan of nonsynonymous SNPs identifies a susceptibility variant for Crohn disease in *ATG16L1*. Nat Genet 39:207–211

Heilpern D, Szilagyi A (2008) Manipulation of intestinal microbial flora for therapeutic benefit in inflammatory bowel diseases: review of clinical trials of probiotics, pre-biotics and synbiotics. Rev Recent Clin Trials 3:167–184

Hugot JP, Chamaillard M, Zouali H, Lesage S, Cezard JP, Belaiche J, Almer S, Tysk C, O'Morain CA, Gassull M, Binder V, Finkel Y, Cortot A, Modigliani R, Laurent-Puig P, Gower-Rousseau C, Macry J, Colombel JF, Sahbatou M, Thomas G (2001) Association of *NOD2* leucine-rich repeat variants with susceptibility to Crohn's disease. Nature 411:599–603

Inbal B, Bialik S, Sabanay I, Shani G, Kimchi A (2002) DAP kinase and DRP-1 mediate membrane blebbing and the formation of autophagic vesicles during programmed cell death. J Cell Biol 157:455–468

Inohara N, Ogura Y, Fontalba A, Gutierrez O, Pons F, Crespo J, Fukase K, Inamura S, Kusumoto S, Hashimoto M, Foster SJ, Moran AP, Fernandez-Luna JL, Nunez G (2003) Host recognition of bacterial muramyl dipeptide mediated through *NOD2*. Implications for Crohn's disease. J Biol Chem 278:5509–5512

Itoh T, Fujita N, Kanno E, Yamamoto A, Yoshimori T, Fukuda M (2008) Golgi-resident small GTPase Rab33B interacts with Atg16L and modulates autophagosome formation. Mol Biol Cell 19:2916–2925

Ivanov II, Frutos RL, Manel N, Yoshinaga K, Rifkin DB, Sartor RB, Finlay BB, Littman DR (2008) Specific microbiota direct the differentiation of IL-17-producing T-helper cells in the mucosa of the small intestine. Cell Host Microbe 4:337–349

Jess T, Riis L, Jespersgaard C, Hougs L, Andersen PS, Orholm MK, Binder V, Munkholm P (2005) Disease concordance, zygosity, and *NOD2/CARD15* status: follow-up of a population-based cohort of Danish twins with inflammatory bowel disease. Am J Gastroenterol 100:2486–2492

Jounai N, Takeshita F, Kobiyama K, Sawano A, Miyawaki A, Xin KQ, Ishii KJ, Kawai T, Akira S, Suzuki K, Okuda K (2007) The Atg5 Atg12 conjugate associates with innate antiviral immune responses. Proc Natl Acad Sci USA 104:14050–14055

Kang SS, Bloom SM, Norian LA, Geske MJ, Flavell RA, Stappenbeck TS, Allen PM (2008) An antibiotic-responsive mouse model of fulminant ulcerative colitis. PLoS Med 5:e41

Kaser A, Lee AH, Franke A, Glickman JN, Zeissig S, Tilg H, Nieuwenhuis EE, Higgins DE, Schreiber S, Glimcher LH, Blumberg RS (2008) *XBP1* links ER stress to intestinal inflammation and confers genetic risk for human inflammatory bowel disease. Cell 134:743–756

Kitajima S, Takuma S, Morimoto M (1999) Changes in colonic mucosal permeability in mouse colitis induced with dextran sulfate sodium. Exp Anim 48:137–143

Kobayashi KS, Chamaillard M, Ogura Y, Henegariu O, Inohara N, Nunez G, Flavell RA (2005) Nod2-dependent regulation of innate and adaptive immunity in the intestinal tract. Science 307:731–734

Komatsu M, Waguri S, Ueno T, Iwata J, Murata S, Tanida I, Ezaki J, Mizushima N, Ohsumi Y, Uchiyama Y, Kominami E, Tanaka K, Chiba T (2005) Impairment of starvation-induced and constitutive autophagy in Atg7-deficient mice. J Cell Biol 169:425–434

Kontoyiannis D, Pasparakis M, Pizarro TT, Cominelli F, Kollias G (1999) Impaired on/off regulation of TNF biosynthesis in mice lacking TNF AU-rich elements: implications for joint and gut-associated immunopathologies. Immunity 10:387–398

Kosiewicz MM, Nast CC, Krishnan A, Rivera-Nieves J, Moskaluk CA, Matsumoto S, Kozaiwa K, Cominelli F (2001) Th1-type responses mediate spontaneous ileitis in a novel murine model of Crohn's disease. J Clin Invest 107:695–702

Kozuch PL, Hanauer SB (2008) Treatment of inflammatory bowel disease: a review of medical therapy. World J Gastroenterol 14:354–377

Kuballa P, Huett A, Rioux JD, Daly MJ, Xavier RJ (2008) Impaired autophagy of an intracellular pathogen induced by a Crohn's disease associated *ATG16L1* variant. PLoS ONE 3:e3391

Kuma A, Mizushima N, Ishihara N, Ohsumi Y (2002) Formation of the approximately 350-kDa Apg12-Apg5.Apg16 multimeric complex, mediated by Apg16 oligomerization, is essential for autophagy in yeast. J Biol Chem 277:18619–18625

Kuma A, Hatano M, Matsui M, Yamamoto A, Nakaya H, Yoshimori T, Ohsumi Y, Tokuhisa T, Mizushima N (2004) The role of autophagy during the early neonatal starvation period. Nature 432:1032–1036

Lala S, Ogura Y, Osborne C, Hor SY, Bromfield A, Davies S, Ogunbiyi O, Nunez G, Keshav S (2003) Crohn's disease and the *NOD2* gene: a role for paneth cells. Gastro 125:47–57

Leber JH, Crimmins GT, Raghavan S, Meyer-Morse NP, Cox JS, Portnoy DA (2008) Distinct TLR- and NLR-mediated transcriptional responses to an intracellular pathogen. PLoS Pathog 4:e6

Lee HK, Lund JM, Ramanathan B, Mizushima N, Iwasaki A (2007) Autophagy-dependent viral recognition by plasmacytoid dendritic cells. Science 315:1398–1401

Levine B, Deretic V (2007) Unveiling the roles of autophagy in innate and adaptive immunity. Nat Rev Immunol 7:767–777

Levine B, Kroemer G (2008) Autophagy in the pathogenesis of disease. Cell 132:27–42

Lupp C, Robertson ML, Wickham ME, Sekirov I, Champion OL, Gaynor EC, Finlay BB (2007) Host-mediated inflammation disrupts the intestinal microbiota and promotes the overgrowth of Enterobacteriaceae. Cell Host Microbe 2:204

Madison BB, Dunbar L, Qiao XT, Braunstein K, Braunstein E, Gumucio DL (2002) *Cis* elements of the villin gene control expression in restricted domains of the vertical (crypt) and horizontal (duodenum, cecum) axes of the intestine. J Biol Chem 277:33275–33283

Maeda S, Hsu LC, Liu H, Bankston LA, Iimura M, Kagnoff MF, Eckmann L, Karin M (2005) *Nod2* mutation in Crohn's disease potentiates NF-kappaB activity and IL-1beta processing. Science 307:734–738

Matsumoto S, Okabe Y, Setoyama H, Takayama K, Ohtsuka J, Funahashi H, Imaoka A, Okada Y, Umesaki Y (1998) Inflammatory bowel disease-like enteritis and caecitis in a senescence accelerated mouse P1/Yit strain. Gut 43:71–78

Matsushita M, Suzuki NN, Obara K, Fujioka Y, Ohsumi Y, Inagaki F (2007) Structure of Atg5. Atg16, a complex essential for autophagy. J Biol Chem 282:6763–6772

Mazmanian SK, Liu CH, Tzianabos AO, Kasper DL (2005) An immunomodulatory molecule of symbiotic bacteria directs maturation of the host immune system. Cell 122:107–118

Mazmanian SK, Round JL, Kasper DL (2008) A microbial symbiosis factor prevents intestinal inflammatory disease. Nature 453:620–625

McCarroll SA, Huett A, Kuballa P, Chilewski SD, Landry A, Goyette P, Zody MC, Hall JL, Brant SR, Cho JH, Duerr RH, Silverberg MS, Taylor KD, Rioux JD, Altshuler D, Daly MJ, Xavier RJ (2008) Deletion polymorphism upstream of IRGM associated with altered IRGM expression and Crohn's disease. Nat Genet 40:1107–1112

McKenna P, Hoffmann C, Minkah N, Aye PP, Lackner A, Liu Z, Lozupone CA, Hamady M, Knight R, Bushman FD (2008) The macaque gut microbiome in health, lentiviral infection, and chronic enterocolitis. PLoS Pathog 4:e20

Mizushima N, Noda T, Ohsumi Y (1999) Apg16p is required for the function of the Apg12p-Apg5p conjugate in the yeast autophagy pathway. EMBO J 18:3888–3896

Mizushima N, Kuma A, Kobayashi Y, Yamamoto A, Matsubae M, Takao T, Natsume T, Ohsumi Y, Yoshimori T (2003) Mouse Apg16L, a novel WD-repeat protein, targets to the autophagic isolation membrane with the Apg12-Apg5 conjugate. J Cell Sci 116:1679–1688

Neuman MG (2007) Immune dysfunction in inflammatory bowel disease. Transl Res 149:173–186

Ogura Y, Bonen DK, Inohara N, Nicolae DL, Chen FF, Ramos R, Britton H, Moran T, Karaliuskas R, Duerr RH, Achkar JP, Brant SR, Bayless TM, Kirschner BS, Hanauer SB,

Nunez G, Cho JH (2001) A frameshift mutation in *NOD2* associated with susceptibility to Crohn's disease. Nature 411:603–606

Ogura Y, Lala S, Xin W, Smith E, Dowds TA, Chen FF, Zimmermann E, Tretiakova M, Cho JH, Hart J, Greenson JK, Keshav S, Nunez G (2003) Expression of NOD2 in Paneth cells: a possible link to Crohn's ileitis. Gut 52:1591–1597

Ouellette AJ (2006) Paneth cell alpha-defensin synthesis and function. Curr Top Microbiol Immunol 306:1–25

Parkes M, Barrett JC, Prescott NJ, Tremelling M, Anderson CA, Fisher SA, Roberts RG, Nimmo ER, Cummings FR, Soars D, Drummond H, Lees CW, Khawaja SA, Bagnall R, Burke DA, Todhunter CE, Ahmad T, Onnie CM, McArdle W, Strachan D, Bethel G, Bryan C, Lewis CM, Deloukas P, Forbes A, Sanderson J, Jewell DP, Satsangi J, Mansfield JC, Cardon L, Mathew CG (2007) Sequence variants in the autophagy gene *IRGM* and multiple other replicating loci contribute to Crohn's disease susceptibility. Nat Genet 39:830–832

Pauleau AL, Murray PJ (2003) Role of nod2 in the response of macrophages to Toll-like receptor agonists. Mol Cell Biol 23:7531–7539

Pineton de CG, Colombel JF, Poulain D, rfeuille-Michaud A (2008) Pathogenic agents in inflammatory bowel diseases. Curr Opin Gastroenterol 24:440–447

Porter EM, Bevins CL, Ghosh D, Ganz T (2002) The multifaceted Paneth cell. Cell Mol Life Sci 59:156–170

Prescott NJ, Fisher SA, Franke A, Hampe J, Onnie CM, Soars D, Bagnall R, Mirza MM, Sanderson J, Forbes A, Mansfield JC, Lewis CM, Schreiber S, Mathew CG (2007) A nonsynonymous SNP in *ATG16L1* predisposes to ileal Crohn's disease and is independent of *CARD15* and *IBD5*. Gastro 132:1665–1671

Pyo JO, Jang MH, Kwon YK, Lee HJ, Jun JI, Woo HN, Cho DH, Choi B, Lee H, Kim JH, Mizushima N, Oshumi Y, Jung YK (2005) Essential roles of Atg5 and FADD in autophagic cell death: dissection of autophagic cell death into vacuole formation and cell death. J Biol Chem 280:20722–20729

Rakoff-Nahoum S, Paglino J, Eslami-Varzaneh F, Edberg S, Medzhitov R (2004) Recognition of commensal microflora by toll-like receptors is required for intestinal homeostasis. Cell 118:229–241

Rawls JF, Mahowald MA, Ley RE, Gordon JI (2006) Reciprocal gut microbiota transplants from zebrafish and mice to germ-free recipients reveal host habitat selection. Cell 127:423–433

Reggiori F, Klionsky DJ (2005) Autophagosomes: biogenesis from scratch? Curr Opin Cell Biol 17:415–422

Rioux JD, Xavier RJ, Taylor KD, Silverberg MS, Goyette P, Huett A, Green T, Kuballa P, Barmada MM, Datta LW, Shugart YY, Griffiths AM, Targan SR, Ippoliti AF, Bernard EJ, Mei L, Nicolae DL, Regueiro M, Schumm LP, Steinhart AH, Rotter JI, Duerr RH, Cho JH, Daly MJ, Brant SR (2007) Genome-wide association study identifies new susceptibility loci for Crohn disease and implicates autophagy in disease pathogenesis. Nat Genet 39:596–604

Saitoh T, Fujita N, Jang MH, Uematsu S, Yang BG, Satoh T, Omori H, Noda T, Yamamoto N, Komatsu M, Tanaka K, Kawai T, Tsujimura T, Takeuchi O, Yoshimori T, Akira S (2008) Loss of the autophagy protein Atg16L1 enhances endotoxin-induced IL-1beta production. Nature 456:264–268

Sandborn WJ Feagan BG (2003) Review article: mild to moderate Crohn's disease—defining the basis for a new treatment algorithm. Aliment Pharmacol Ther 18:263–277

Sanjuan MA, Dillon CP, Tait SW, Moshiach S, Dorsey F, Connell S, Komatsu M, Tanaka K, Cleveland JL, Withoff S, Green DR (2007) Toll-like receptor signaling in macrophages links the autophagy pathway to phagocytosis. Nature 450:1253–1257

Schaffler A, Herfarth H (2005) Creeping fat in Crohn's disease: travelling in a creeper lane of research?. Gut 54:742–744

Simms LA, Doecke JD, Walsh MD, Huang N, Fowler EV, Radford-Smith GL (2008) Reduced alpha-defensin expression is associated with inflammation and not *NOD2* mutation status in ileal Crohn's disease. Gut 57:903–910

Singh SB, Davis AS, Taylor GA, Deretic V (2006) Human IRGM induces autophagy to eliminate intracellular mycobacteria. Science 313:1438–1441

Steinhart AH, Feagan BG, Wong CJ, Vandervoort M, Mikolainis S, Croitoru K, Seidman E, Leddin DJ, Bitton A, Drouin E, Cohen A, Greenberg GR (2002) Combined budesonide and antibiotic therapy for active Crohn's disease: a randomized controlled trial. Gastro 123:33–40

Strober W, Fuss IJ, Blumberg RS (2002) The immunology of mucosal models of inflammation. Annu Rev Immunol 20:495–549

Stryke D, Kawamoto M, Huang CC, Johns SJ, King LA, Harper CA, Meng EC, Lee RE, Yee A, L'Italien L, Chuang PT, Young SG, Skarnes WC, Babbitt PC, Ferrin TE (2003) BayGenomics: a resource of insertional mutations in mouse embryonic stem cells. Nucleic Acids Res 31:278–281

Sutherland L, Singleton J, Sessions J, Hanauer S, Krawitt E, Rankin G, Summers R, Mekhjian H, Greenberger N, Kelly M (1991) Double blind, placebo controlled trial of metronidazole in Crohn's disease. Gut 32:1071–1075

Targan SR, Hanauer SB, Van Deventer SJ, Mayer L, Present DH, Braakman T, DeWoody KL, Schaible TF, Rutgeerts PJ (1997) A short-term study of chimeric monoclonal antibody cA2 to tumor necrosis factor alpha for Crohn's disease. Crohn's disease cA2 study group. N Engl J Med 337:1029–1035

Taylor GA, Feng CG, Sher A (2004) p47 GTPases: regulators of immunity to intracellular pathogens. Nat Rev Immunol 4:100–109

The Wellcome Trust Case Control Consortium (2007) Genome-wide association study of 14,000 cases of seven common diseases and 3,000 shared controls. Nature 447:661–678

Turnbaugh PJ, Ley RE, Hamady M, Fraser-Liggett CM, Knight R, Gordon JI (2007) The human microbiome project. Nature 449:804–810

Virgin H (2008) A 'fly-by' killing with a primordial cellular weapon. Nat Immunol 9:827–829

Voss E, Wehkamp J, Wehkamp K, Stange EF, Schroder JM, Harder J (2006) NOD2/CARD15 mediates induction of the antimicrobial peptide human beta-defensin-2. J Biol Chem 281:2005–2011

Watanabe T, Asano N, Murray PJ, Ozato K, Tailor P, Fuss IJ, Kitani A, Strober W (2008) Muramyl dipeptide activation of nucleotide-binding oligomerization domain 2 protects mice from experimental colitis. J Clin Invest 118:545–559

Wehkamp J, Harder J, Weichenthal M, Schwab M, Schaffeler E, Schlee M, Herrlinger KR, Stallmach A, Noack F, Fritz P, Schroder JM, Bevins CL, Fellermann K, Stange EF (2004) *NOD2* (*CARD15*) mutations in Crohn's disease are associated with diminished mucosal alpha-defensin expression. Gut 53:1658–1664

Wehkamp J, Salzman NH, Porter E, Nuding S, Weichenthal M, Petras RE, Shen B, Schaeffeler E, Schwab M, Linzmeier R, Feathers RW, Chu H, Lima H Jr, Fellermann K, Ganz T, Stange EF, Bevins CL (2005) Reduced Paneth cell alpha-defensins in ileal Crohn's disease. Proc Natl Acad Sci U S A 102:18129–18134

Xavier RJ, Podolsky DK (2007) Unravelling the pathogenesis of inflammatory bowel disease. Nature 448:427–434

Xavier RJ, Huett A, Rioux JD (2008) Autophagy as an important process in gut homeostasis and Crohn's disease pathogenesis. Gut 57:717–720

Xu Y, Jagannath C, Liu XD, Sharafkhaneh A, Kolodziejska KE, Eissa NT (2007) Toll-like receptor 4 is a sensor for autophagy associated with innate immunity. Immunity 27:135–144

Yamamoto K, Kiyohara T, Murayama Y, Kihara S, Okamoto Y, Funahashi T, Ito T, Nezu R, Tsutsui S, Miyagawa JI, Tamura S, Matsuzawa Y, Shimomura I, Shinomura Y (2005) Production of adiponectin, an anti-inflammatory protein, in mesenteric adipose tissue in Crohn's disease. Gut 54:789–796

Yousefi S, Perozzo R, Schmid I, Ziemiecki A, Schaffner T, Scapozza L, Brunner T, Simon HU (2006) Calpain-mediated cleavage of Atg5 switches autophagy to apoptosis. Nat Cell Biol 8:1124–1132.

Zhang H, Bosch-Marce M, Shimoda LA, Tan YS, Baek JH, Wesley JB, Gonzalez FJ, Semenza GL (2008) Mitochondrial autophagy is an HIF-1-dependent adaptive metabolic response to hypoxia. J Biol Chem 283:10892–10903

Zhao Z, Thackray LB, Miller BC, Lynn TM, Becker MM, Ward E, Mizushima NN, Denison MR, Virgin HW (2007) Coronavirus replication does not require the autophagy gene *ATG5*. Autophagy 3:581–585

Zhao Z, Fux B, Goodwin M, Dunay IR, Strong D, Miller BC, Cadwell K, Delgado MA, Ponpuak M, Green KG, Schmidt RE, Mizushima N, Deretic V, Sibley LD, Virgin HW (2008) Autophagosome-independent essential function for the autophagy protein Atg5 in cellular immunity to intracellular pathogens. Cell Host Microbe 4:458–469

Autophagy in Immunity Against *Mycobacterium tuberculosis*: a Model System to Dissect Immunological Roles of Autophagy

Vojo Deretic, Monica Delgado, Isabelle Vergne, Sharon Master, Sergio De Haro, Marisa Ponpuak, and Sudha Singh

Contents

1 Introduction: Autophagy as an Antimicrobial Defense Mechanism Against Bacteria, Protozoan Parasites and Viruses 170
2 *M. tuberculosis* Parasitizes Host Macrophages 172
3 Autophagy Eliminates Intracellular *M. tuberculosis* 173
4 Unique Properties of Autolysosomes in Microbial Killing 173
5 Autophagy is an Effector of Th1/Th2 Polarization 174
6 Immunity-Related GTPases (IRGs) Regulate Autophagy in Antimicrobial Defense and Inflammation 176
7 Autophagy is an Effector of Pattern Recognition Receptor Signaling 178
8 Conclusions and a Model 182
References 183

Abstract The recognition of autophagy as an immune mechanism has been affirmed in recent years. One of the model systems that has helped in the development of our current understanding of how autophagy and more traditional immunity systems cooperate in defense against intracellular pathogens is macrophage infection with *Mycobacterium tuberculosis*. *M. tuberculosis* is a highly significant human pathogen that latently infects billions of people and causes active disease in millions of patients worldwide. The ability of the tubercle bacillus to persist in human populations rests upon its macrophage parasitism. One of the initial reports on the ability of autophagy to act as a cell-autonomous innate immunity mechanism capable of eliminating intracellular bacteria was on *M. tuberculosis*. This model system has further contributed to the recognition of multiple connections between conventional immune regulators and autophagy. In this chapter, we will review

V. Deretic (✉), M. Delgado, I. Vergne, S. Master, S. De Haro,
M. Ponpuak, and S. Singh
Department of Molecular Genetics and Microbiology, University of New Mexico Health Sciences Center, 915 Camino de Salud NE, Albuquerque, NM, 87131, USA
e-mail: vderetic@salud.unm.edu

B. Levine et al. (eds.), *Autophagy in Infection and Immunity*,
Current Topics in Microbiology and Immunology 335,
DOI 10.1007/ 978-3-642-00302-8_8,

how these studies have helped to establish the following principles: (1) autophagy functions as an innate defense mechanism against intracellular microbes; (2) autophagy is under the control of pattern recognition receptors (PRR) such as Toll-like receptors (TLR), and it acts as one of the immunological output effectors of PRR and TLR signaling; (3) autophagy is one of the effector functions associated with the immunity-regulated GTPases, which were initially characterized as molecules involved in cell-autonomous defense, but whose mechanism of function was unknown until recently; (4) autophagy is an immune effector of Th1/Th2 T cell response polarization—autophagy is activated by Th1 cytokines (which act in defense against intracellular pathogens) and is inhibited by Th2 cytokines (which make cells accessible to intracellular pathogens). Collectively, the studies employing the *M. tuberculosis* autophagy model system have contributed to the development of a more comprehensive view of autophagy as an immunological process. This work and related studies by others have led us to propose a model of how autophagy, an ancient innate immunity defense, became integrated over the course of evolution with other immune mechanisms of ever-increasing complexity.

1 Introduction: Autophagy as an Antimicrobial Defense Mechanism Against Bacteria, Protozoan Parasites and Viruses

The role of autophagy as a cell-autonomous antimicrobial defense has been suspected for a long time, but remained difficult to define until a recent burst of studies capitalizing on the genetic definition of autophagic machinery (Deretic 2005; Levine and Deretic 2007; Schmid and Munz 2007). These studies (Andrade et al. 2006; Birmingham et al. 2006; Checroun et al. 2006; Cullinane et al. 2008; Gutierrez et al. 2004; Liang et al. 1998; Ling et al. 2006; Liu et al. 2005; Nakagawa et al. 2004; Ogawa et al. 2005; Orvedahl et al. 2007; Py et al. 2007; Singh et al. 2006; Talloczy et al. 2002; Yano et al. 2008) have demonstrated that autophagy can function as a cell's defense against bacteria, protozoa and viral pathogens. At a first approximation, autophagy can eliminate intracellular pathogens in a process akin to the sequestration and degradation of large macromolecular aggregates or surplus and dysfunctional intracellular organelles. This is in keeping with one of the primary functions of autophagy as a cytoplasmic clean-up process. A subset of the recent studies have demonstrated a role of autophagy in the elimination of microorganisms such as *M. tuberculosis* that reside within phagosomes (Gutierrez et al. 2004), intracellular pathogens that escape from the phagosome into the cytosol such as *Shigella* (Ogawa et al. 2005), and extracellular pathogens when they erode into the interior of the host cell, as demonstrated for Group A *Streptococci* (Nakagawa et al. 2004). A further testament to the cell-protection role of autophagy is provided by the evolutionary adaptations of pathogens that have developed countermeasures to defend themselves against autophagy (Jackson et al. 2005). For example, HSV-1 interferes with autophagy using a specific gene product ICP34.5 (Orvedahl et al. 2007). *Shigella* normally evades autophagy but falls pray upon the loss of one of its intracellular motility regulators (Ogawa et al. 2005), while *Listeria* inhibits autophagic maturation using its pore-forming toxins to counter lumenal acidification (Birmingham et al. 2008).

Once the primary function of autophagy as a cell-autonomous defense against invading microbes had been established, the next important step was to examine whether links existed between the well-established innate and adaptive immune systems and autophagy. This was important to establish, since autophagy has been mostly viewed as a system that is involved in feeding cells during starvation and acts as a cell death/survival pathway. The role of autophagy in the control of intracellular *M. tuberculosis* has grown into a model system for making such connections. For example, the first publication showing that autophagy can eliminate *M. tuberculosis* also demonstrated that the cardinal Th1 cytokine IFN-γ can induce autophagy to eliminate an intracellular microbe (Gutierrez et al. 2004). This paved the way for a demonstration that autophagy is a previously unappreciated effector of Th1/Th2 polarization (Harris et al. 2007). It turned out that, in contrast to the protective function of IFN-γ via autophagy, the Th2 cytokines IL-4 and IL-13 inhibited autophagy and counteracted IFN-γ-induced autophagy, thus sparing intracellular mycobacteria (Harris et al. 2007). The opposing roles of Th1 and Th2 cytokines in dictating the macrophage's ability to control intracellular bacteria such as *M. tuberculosis* can now be attributed, at least in part, to autophagy-activating effects of Th1 cytokines and autophagy-repressing effects of Th2 cytokines.

Additional important connections between conventional immune systems and autophagy have followed with the use of the *M. tuberculosis* macrophage system, including the following examples. (1) *Autophagy as an effector of Toll-like receptors (TLRs), pattern recognition receptors (PRRs), and pathogen-associated molecular pattern (PAMP) signaling*. Two different groups have shown that activation of innate immunity using TLR ligands can stimulate the autophagic elimination of *M. tuberculosis* (Delgado et al. 2008; Xu et al. 2007). These and additional studies have shown that, in general, stimulating pattern recognition PRRs with PAMP can activate autophagy (Delgado et al. 2008; Sanjuan et al. 2007), and that this matters in vivo (Virgin 2008; Yano et al. 2008). (2) *Immunity-related GTPases (IRGs) and autophagy*. IRGs are now recognized as being regulators of cell-autonomous defense systems downstream of IFN-γ activation (Martens and Howard 2006; Taylor et al. 2004), but their mechanism of action—how they carry out their antimicrobial function—was still to be elucidated (Howard 2008). Since one of the murine IRGs (Irgm1, also known as LRG47) protected against *M. tuberculosis* (MacMicking et al. 2003), and IFN-γ induced both Irgm1 expression and autophagy, we tested whether expression of Irgm1 alone can induce autophagy (Gutierrez et al. 2004). This turned out to be the case (Gutierrez et al. 2004). A similar relationship, albeit with a somewhat altered form vis-à-vis IFN-γ, also held up when the sole human IRG protein, IRGM, was tested. Thus, the model system for the autophagy-based control of intracellular *M. tuberculosis* has helped to establish several key connections between conventional regulatory immunity processes and autophagy.

2 *M. tuberculosis* Parasitizes Host Macrophages

M. tuberculosis asymptomatically infects over a billion people and causes millions of new active disease cases annually, with a 25-30% mortality rate worldwide (Dye et al. 1999). Of particular importance is the strong link between the incidence of active tuberculosis and AIDS (Goldfeld et al. 2008; Nunn et al. 2005). HIV and *M. tuberculosis* are intimately associated on the global health stage, and one-third to two-thirds of all AIDS patients around the world are coinfected with *M. tuberculosis* (Reid et al. 2006). In HIV-infected individuals, overt tuberculosis can occur before the $CD4^+$ counts drop to levels that allow other less potent pathogens to survive and other symptoms of AIDS to show (Reid et al. 2006), and thus tuberculosis can often serve as a sentinel disease for underlying HIV infection. Worldwide, the treatment of HIV has impacted on tuberculosis control, whereas tuberculosis is a frequent cause of death in those with AIDS (Nahid and Daley 2006). Tuberculosis and AIDS, most alarmingly in combination, have been recognized as global health emergencies (http://www.who.int/en/).

One of the best characterized virulence determinants of *M. tuberculosis* and a key feature of its pathogenesis is the ability of the tubercle bacillus to infect and survive in macrophages by blocking the maturation of its phagosome into a degradative organelle called the phagolysosome (Armstrong and Hart 1971). This paradigm is often referred to as the inhibition of phagosome–lysosome fusion, phagosomal maturation block, or inhibition of phagolysosome biogenesis (Armstrong and Hart 1971). By preventing phagosomal maturation, *M. tuberculosis* avoids the bactericidal (Pieters 2008) and, with some exceptions (Majlessi et al. 2007), antigen-processing environment of the phagolysosome (Ramachandra et al. 2005; Torres et al. 2006) immediately upon phagocytosis, and possibly for prolonged periods of time. The suppression of phagosome maturation by *M. tuberculosis* is a critical process within the infectious cycle that allows the pathogen to establish a foothold. The confinement of *M. tuberculosis* within the phagosome, which is typical of its survival in macrophages, may extend to at least some parts of the protective granuloma. The status quo is maintained in these dynamic structures for extended periods of time, including during latency (Manabe and Bishai 2000; Russell 2007). Latency is the "preferred" state of *M. tuberculosis*, which asymptomatically persists in billons of people (Manabe and Bishai 2000). Most infected individuals are not ill and remain asymptomatic so long as a large number of immunological mediators needed for control, in particular IFN-γ and TNF-α (Flynn and Chan 2001) and T cells, continue to cooperate effectively. The latter include MHC II-restricted $CD4^+$ T cells, MHC I-restricted $CD8^+$ cells (Ottenhoff et al. 2008) and "unconventional" T cells (CD1a-c-restricted T cells, γδ-restricted T cells, invariant TCR CD1d-restricted NKT cells, variant/diverse TCR CD1d-restricted NKT cells, Treg cells, Th17 cells, and nonclassical HLA-E restricted $CD8^+$ T cells) (Behar and Boom 2008; Lewinsohn et al. 2000; Moody et al. 2004). This complex immunological control can break down with age, nutritional/environmental changes, HIV infection, or upon immunosuppressive therapy (Saunders and Britton 2007).

3 Autophagy Eliminates Intracellular *M. tuberculosis*

There are very few known macrophage microbicidal mechanisms that are capable of killing *M. tuberculosis*. Conventional antibacterial effectors such as reactive oxygen and nitrogen intermediates, acidification of the phagosome, and degradation of microbes in the phagolysosome do not readily inhibit *M. tuberculosis*, a property that has been known for decades (Armstrong and Hart 1975). This has been reaffirmed by modern molecular analyzes, as the known *M. tuberculosis* determinants of persistence in macrophages include: (1) specific inhibitors of phagosome-lysosome fusion (Vergne et al. 2004); (2) resistance to acidic conditions of the lysosome (Vandal et al. 2008); (3) disruption of recruitment of iNOS to the vicinity of the phagosome (Davis et al. 2007), and (4) protection against oxidants by the components of the lipid-rich envelope (Yuan et al. 1995).

In contrast to the notorious resistance of *M. tuberculosis* to other microbicidal effectors, our studies have shown that autophagy can efficiently eliminate intracellular *M. tuberculosis* (Gutierrez et al. 2004). These initial observations have been confirmed by several groups in different contexts, including a report showing that autolysosomes contain ubiquitin fragments that act as mycobactericidal peptides (Alonso et al. 2007), a study linking TLR stimulation and autophagy (Xu et al. 2007), a screen for novel autophagy inducers (Floto et al. 2007), and a recent study (Biswas et al. 2008) showing that the previously reported ATP stimulation of P2X7-receptor leading to the elimination of intracellular mycobacteria (Lammas et al. 1997) is achieved through autophagy (Biswas et al. 2008).

4 Unique Properties of Autolysosomes in Microbial Killing

The autophagy–*M. tuberculosis* model system has also provided insight into some special properties that are possibly unique to autophagic killing of intracellular microbes. There are two points that need to be made before describing published experimental findings. Firstly, while one may assume that having any lysosomal environment is microbicidal enough to eliminate any microbe, this is not always the case. For example, when *M. tuberculosis* is forced (by antibody opsonization) into a standard phagolysosome it still does not get killed. This has been known since 1975 (Armstrong and Hart 1975). Secondly, autophagy captures normal cytosolic components and processes/digests them in autolysosomes, thus creating a different lysosomal composition compared to the conventional lysosome. Taking these two points together, and the reports that autophagy can kill mycobacteria (Xu et al. 2007; Floto et al. 2007; Biswas et al. 2008; Lammas et al. 1997), it is possible that novel microbicidal components are present in the autolysosome.

Initial progress in this direction has been made using *M. tuberculosis* in a study in which Alonso et al. (Alonso et al. 2007) showed that ubiquitin fragments generated by ubiquitin digestion in autophagosomes endow lysosomes with higher mycobactericidal capacities. This is in keeping with the earlier reports that ubiquitin fragments generally possess antimicrobial properties (Kieffer et al. 2003), although how they were delivered to pathogens was not known. The experimental evidence indicates that autophagy, after capturing cytosolic ubiquitin or ubiquitinated proteins or their aggregates, proteolytically generates ubiquitin fragments during maturation into autolysosomes, and that these fragments are delivered (directly or indirectly) to intracellular *M. tuberculosis* to enhance its killing. Whether ubiquitin fragments are produced upon the induction of autophagy beforehand or are produced in the same compartment in which *M. tuberculosis* resides in cells activated for autophagy is not known at present. Furthermore, additional cytosolic components may contribute to the special microbicidal properties of autolysosomes. For instance, cytoplasmic proteins such as ubiquicidin (a unique ribosomal protein, S30, with homology to ubiquitin) and the ribosomal polypeptides S19 and L30 all display antimicrobial activity (Howell et al. 2003). Since ribosomes are common autophagic substrates (Kraft et al. 2008), some of these polypeptides may be present in autophagosomes. Intriguingly, microbicidal ribosomal peptides have been isolated from colonic epithelial cells based on their antimicrobial activity against enteric bacteria, and given the recently uncovered links between autophagy and inflammatory bowel disease (specifically Crohn's disease) (Burton et al. 2007; Hampe et al. 2007; Massey and Parkes 2007; Parkes et al. 2007; Rioux et al. 2007), these potential links and other unique aspects of autolysosomes warrant further investigations.

5 Autophagy is an Effector of Th1/Th2 Polarization

It is now known that autophagy is regulated by immunologically relevant cytokines and ligands (Andrade et al. 2006; Arico et al. 2001; Djavaheri-Mergny et al. 2006; Gutierrez et al. 2004; Harris et al. 2007; Inbal et al. 2002; Li et al. 2006; Paludan et al. 2005; Petiot et al. 2000; Pyo et al. 2005; Schlottmann et al. 2008). This knowledge has its early roots in nonimmunological studies where cytokines were used simply as convenient agonists or antagonists to induce or repress autophagy (Arico et al.

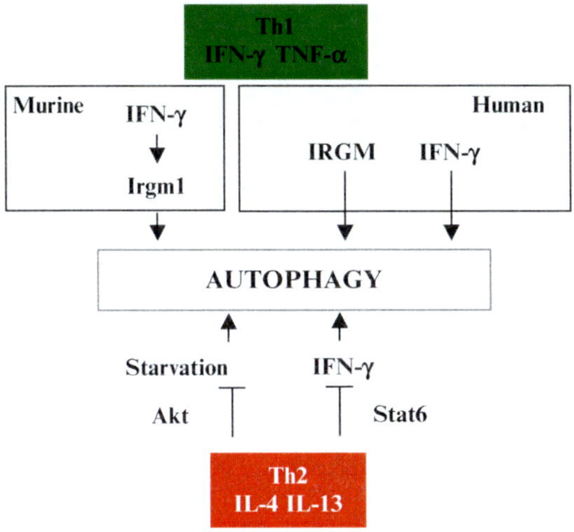

Fig. 1 Th1/Th2 polarization and autophagy regulation: Th1 cytokines activate and Th2 cytokines inhibit autophagy. Mouse cells: IFN-γ activates expression of immunity-related GTPase Irgm1 (LRG-47) to induce autophagy. Human cells: the IRG factor IRGM is expressed independently of IFN-γ, but is required for IFN-γ-induced autophagy. IL-4 and IL-13 inhibit autophagy. Inhibition of starvation-induced autophagy by IL-4 and IL-13 depends on Akt/PKB; inhibition of IFN-γ autophagy by IL-4 and IL-13 depends on Stat-6. Th1 cytokines (in particular IFN-γ and TNF-α) are critical for protection against tuberculosis. Th2 cytokines have been shown to be permissive in mycobacterial diseases

2001; Inbal et al. 2002; Petiot et al. 2000). With this as a starting point, we have used the autophagy–*M. tuberculosis* model system to show that key Th1/Th2-polarization cytokines, IFN-γ, IL-4 and IL-13, affect autophagy in immunologically relevant contexts (Fig. 1) (Gutierrez et al. 2004; Harris et al. 2007; Singh et al. 2006). In addition to these cytokines, TNF-α has been shown to activate autophagy under conditions where NF-κB is inhibited (Djavaheri-Mergny et al. 2006). Other cell-mediated immunity regulatory systems can induce autophagy, such as CD40L-CD40 stimulation in the context of protection against the parasite *Toxoplasma gondii* (Andrade et al. 2006), apparently in association with TNFα secreted downstream of CD40–TRAF6 stimulation (Sabauste et al. 2007). The key Th1 cytokine IFN-γ, which is strongly associated with protective immunity against *M. tuberculosis* (Fortin et al. 2007), stimulates autophagy (Gutierrez et al. 2004; Inbal et al. 2002; Pyo et al. 2005). How IFN-γ signaling induces autophagy is yet to be fully defined. In the mouse, one pathway includes STAT-1-dependent expression of IRG (Gutierrez et al. 2004; Singh et al. 2006). In murine cells, IFN-γ induces autophagy in an Irgm1 (also known as LRG47)-dependent manner. However, the human equivalent, IRGM, is required for autophagy but its expression is not controlled by IFN-γ, thus indicating that additional signaling pathways are involved.

In contrast to Th1 cytokines, which induce autophagy, the Th2 cytokines IL-4 and IL-13 are antagonists of autophagy (Harris et al. 2007). This is in part based on the activation of the Akt-Tor cascade by IL-4 and IL-13, and in part on the STAT-6 pathway (Harris et al. 2007). Treatment of macrophages with IL-4 and IL-13 inhibits starvation- or IFN-γ-induced autophagic delivery of mycobacteria into degradative compartments and counteracts mycobacterial killing in infected macrophages stimulated for autophagy (Harris et al. 2007).

IL-4 and IL-13 signal through a shared receptor, IL-4Rα, which complexes with the γ-common chain in the case of IL-4 or with IL-13Rα1 in the case of IL-13 as a ligand (Nelms et al. 1999). IL-13 can also signal through a high-affinity receptor, IL-13Rα2. Once the IL-4 and IL-13 receptors are engaged, this results in not only the activation of the STAT-6 pathway (well appreciated in immunological studies) but also signaling via the insulin receptor substrate (IRS)-1 and 2 (Nelms et al. 1999). The signaling via IRS stimulates the Akt pathway, which provides the basis for IL-4 and IL-13 inhibiting autophagy induced by starvation (Harris et al. 2007). However, a different signaling pathway, independent of Akt and dependent on STAT-6, is required to suppress IFN-γ-induced autophagy. The inhibitory action of IL-4 and IL-13 translates into the inhibition of autophagic control of intracellular *M. tuberculosis* (Harris et al. 2007). Collectively, the induction of autophagy by IFN-γ, the autophagic control of *M. tuberculosis* following activation with Th1 cytokines, the inhibition by IL-4 and IL-13 of autophagy, and the suppression of autophagic killing of *M. tuberculosis* by Th2 cytokines indicate that autophagy is an effector of Th1/Th2 polarization. This in turn can help explain why Th1 cytokines are protective against and Th2 cytokines are permissive to intracellular pathogens. Significantly, the overriding suppressive effects of IL-4 and IL-13 on the induction of autophagy by IFN-γ can explain why Th1/Th2 polarization does not need to be sharply defined (and indeed this rarely happens) in infection sites. It turns out that presence of Th2 cytokines may override IFN-γ when cytokine responses are mixed.

6 Immunity-Related GTPases (IRGs) Regulate Autophagy in Antimicrobial Defense and Inflammation

The work on *M. tuberculosis* led to the initial connection between IRGs and autophagy in murine cells (Gutierrez et al. 2004), and has been recently expanded to the control of *M. tuberculosis* in human cells (Singh et al. 2006). IFN-γ is a major correlate of immunity against tuberculosis, but the exact nature of IFN-γ antimycobacterial action remained an elusive issue, as neither reactive oxygen nor reactive nitrogen intermediates could explain its potent antimycobacterial action (MacMicking et al. 2003). It has been demonstrated that IFN-γ acts through an IRG family member, Irgm1 (LRG-47), to control *M. tuberculosis* in murine macrophages and in the mouse model of tuberculosis (MacMicking et al. 2003), and that autophagy is the result of Irgm1 action (Gutierrez et al. 2004).

Thus, in the case of *M. tuberculosis* it appears that at least some dots have been connected, between IFN-γ, IRG, autophagy, and intracellular control of the tubercle bacilli.

The IRG factors (Fig. 2) were initially recognized as being mediators of efficient cell-autonomous defense (at the time of an unknown nature) against intracellular pathogens. The mode of action for these systems remained a mystery until autophagy was connected to at least one member (Irgm1/LRG-47) of the IRG family in the mouse (Gutierrez et al. 2004) and to the only member (IRGM) of this family in humans (Singh et al. 2006). The role of autophagy downstream of murine IRG has also been demonstrated in control of *T. gondii*, where another murine IRG protein, Irgm3 (also known as IGTP) has been implicated in autophagy induction and association with autophagosomes (Ling et al. 2006). The additional proposed and still contested roles for IRGs (Howard 2008) include the enhancement of phagosomal acidification (MacMicking et al. 2003), the disintegration of the parasitophorous vacuole membrane (Martens et al. 2005) and the aberrant function of hematopoietic stem cells during infection (Feng et al. 2008). Significantly,

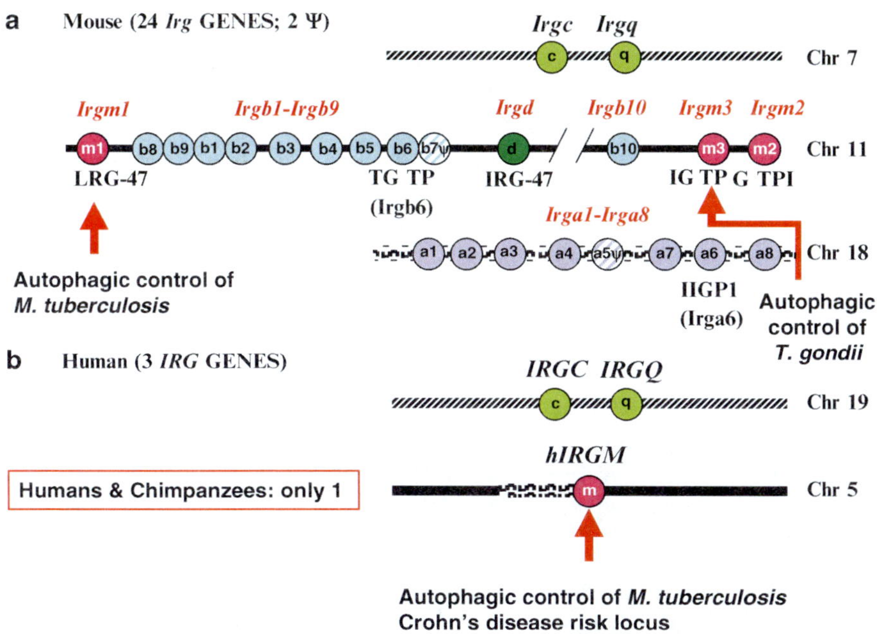

Fig. 2 Immunity-related GTPases (IRGs): genomic organization and roles in autophagy. The overall genomic organization of IRG genes. The mouse and humans are at the opposite ends of the spectrum in terms of the number of IRG genes. The murine Irg loci with dual names (underneath the chromosomes) have been implicated in defense against intracellular pathogens. *Red arrows* indicate those IRG that have been implicated in the autophagic control of intracellular pathogens, *Mycobacterium tuberculosis* and *Toxoplasma gondii*. Modified with permission from Singh et al. (2006)

the majority of the above mechanisms are compatible with autophagy, can be potentially explained by autophagy, or may involve autophagy as a component.

The IRG family (Bekpen et al. 2005) in the mouse is represented by a total of 23 *Irg* genes (Fig. 2). There are 19 interferon-controlled complete *Irg* genes. *Irgm1*, *Irgm2*, *Irgm3*, *Irgb1*, *Irgb3*, *Irgb5*, *Irg6*, *Irgb8*, *Irgb 9*, and *Irgb10* are on mouse chromosome 11. The *Irgb7ψ* within the *Irgb* cluster is a pseudogene. The *Irg* cluster on the mouse chromosome 18 has *Irga1*, *Irga2*, *Irga3*, *Irga4*, *Irga6*, *Irga7*, and *Irga8* genes, and *Irga5ψ* as a pseudogene. The expression of the murine *Irg* genes depends on activation with interferon via a combination of IRES and GAS promoter elements, with the exception of *Irgb5* and *Irgb9*, for which only IRES sites have been recognized in bioinformatics approaches. The two *Irg* paralogs on mouse chromosome 7 represent an interferon-independent gene (*Irgc*) or an incomplete, quasi-GTPase (*Irgq*) gene. Based on this, Irgc and Irgq have been a priori (although experimental evidence is lacking) excluded as potential immune regulators or effectors (Bekpen et al. 2005). Figure 2 also features the pregenomic era names of the murine *Irgs*, from the research period when several of them were individually characterized.

The prolific nature of the *Irg* loci in the mouse genome starkly contrasts with the dearth of *IRG* genes in the human genome. In humans, there are only three *IRG* paralogs, with *IRGM* thus far being the only *IRG* gene considered and functionally characterized in human cells (Singh et al. 2006). *IRGC* and *IRGQ* on human chromosome 19, which are syntenic with the mouse *Irgc* and *Irgq*, have not been (by analogy to the murine genes) considered thus far in immunity. The sole immunologically implicated human *IRG* gene, *IRGM*, on human chromosome 5 is surrounded by chromosomal regions syntenic with the murine *Irgm* and *Irga* chromosomal loci, suggesting its genetic correspondence to the immunologically defined murine *IRG*. Moreover, human IRGM has been functionally characterized as playing a role in immune processes (Chaturvedi et al. 2008; Singh et al. 2006). Although IRGM represents a much-shortened (N-terminally and C-terminally truncated) version relative to the murine Irgm proteins, it exceeds the size of Ras and Rab proteins. Curiously, human IRGM is not regulated by IFN-γ, and is instead constitutively expressed from the long terminal repeat (LTR) of a human endogenous retrovirus repetitive element, ERV9. However, IRGM is still required for full autophagy activation in cells stimulated with IFN-γ, starvation, or rapamycin (Fig. 1). The exact mechanisms by which IRGM in human cells and Irgm1 (LRG-47) in murine cells promote autophagy are not known and are presently being investigated.

7 Autophagy is an Effector of Pattern Recognition Receptor Signaling

The autophagy–*M. tuberculosis* system has also helped to connect immunologically relevant autophagy with innate immunity receptor signaling. The innate immunity receptors and downstream effectors are responsible for early detection and initial elimination of invading microbes plus the modulation of adaptive immunity that subsequently develops (Ishii et al. 2008; Medzhitov 2007). The innate immunity receptors (Fig. 3), collectively referred to as PRRs, encompass three major classes: TLRs, retinoic acid-inducible gene I (RIG-I)-like helicase receptors (RLRs), and nucleotide-binding and oligomerization domain (NOD)-like receptors (NLRs) (Takeuchi and Akira 2008). PRRs recognize PAMP and induce a number of proinflammatory cytokines as a well-known, conventional output. A new addition to the repertoire of PRR stimulation outputs is the recently described induction of autophagy downstream of TLR stimulation (Delgado et al. 2008; Lee et al. 2007; Sanjuan et al. 2007; Xu et al. 2007) (Fig. 3). Two out of the four initial reports used *M. tuberculosis* killing and mycobacterial phagosome maturation as a physiological output to demonstrate that autophagy induced downstream of TLR is physiologically and immunologically relevant (Delgado et al. 2008; Xu et al. 2007).

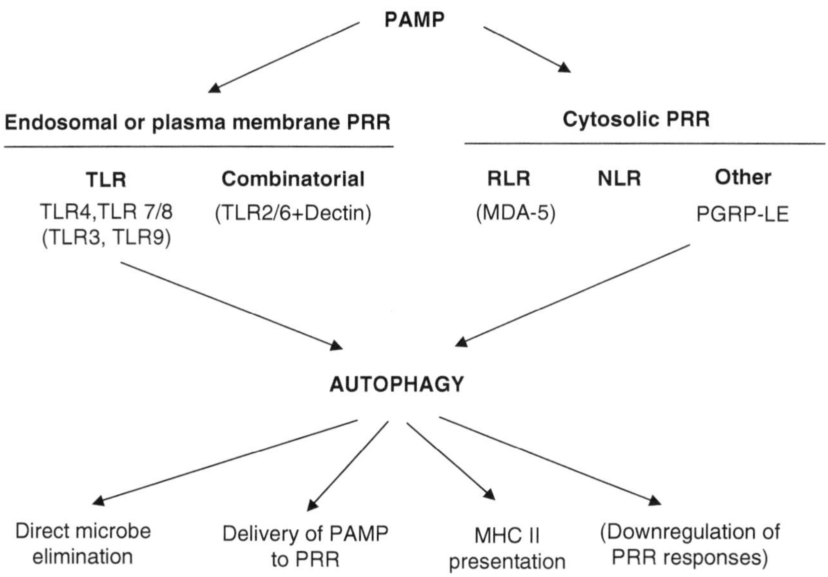

Fig. 3 Autophagy is a downstream effector of PRR signaling. *PAMP*, pathogen associated molecular patterns (microbial products, e.g., LPS/endotoxin, peptidoglycan constituents, viral replication intermediates etc.); *PRR*, pattern recognition receptors (*TLR*, Toll-like receptors; *RLR*, RIG-I-like receptors; *NLR*, Nod-like receptors); *PGRP-LE*, a *Drosophila* cytosolic PRR; *Dectin*, fungal cell wall receptor). Autophagy downstream of PRR can result in direct microbial elimination (e.g., killing of *M. tuberculosis*) (Delgado et al. 2008; Xu et al. 2007), and increases MHC II-restricted presentation of microbial cytosolic antigens (e.g., EBNA1 protein of EBV) (Paludan et al. 2005). Autophagy has also been shown to deliver TLR ligands to endosomal TLR (e.g., VSV nucleic acids to TLR7), and to play a role in downregulating PRR signaling (e.g., Atg5–Atg12 complex inhibits RIG-I-IPS-1 signaling) (Jounai et al. 2007). *Parentheses* indicate PRR or processes where the indicated function has not been firmly established

TLR1, TLR2, TLR4, TLR5, and TLR6 partition primarily to the cell surface and recognize bacterial components, whereas TLR3, TLR7, TLR8, and TLR9 are primarily located in the endosomal compartments and recognize viral products (Lee and Kim 2007). TLRs recruit a different combination of four TIR domain-containing adaptor molecules: myeloid differentiation primary response protein 88 (MyD88), employed by all TLRs except TLR3; TIR domain-containing adaptor protein (TIRAP) or MyD88 adaptor-like (MAL), used by TLR2 and TLR4 as a bridge to recruit MyD88; TIR domain-containing adaptor-inducing interferon-β (TRIF) or TIR domain-containing adaptor molecule 1 (TICAM-1), employed by TLR3 and TLR4; and TRIF-related adaptor molecule (TRAM) or TICAM-2, employed only by TLR4 for interactions with TRIF (Kawai et al. 2004; Lee et al. 2007; O'Neill and Bowie 2007). A group of TLRs signal exclusively through MyD88 (TLR1, TLR2, TLR5, TLR6, TLR7, TLR8 and TLR9). TLR3 signals exclusively via TRIF. TLR4 signals with both MyD88 and TRIF. The duality in the context of TLR4 has been underscored by the discovery that TLR4 acts in a sequential manner, with the MyD88 pathway engaged on the plasma membrane and TRIF engaged upon endocytosis in the early endosome (Kagan et al. 2008). The majority of these signaling cascades activate NF-κB and AP-1, leading to the production of inflammatory cytokines and chemokines, which in turn recruit and activate innate immune cells such as monocytes, neutrophils, and natural killer cells (Lee and Kim 2007). TLR3, TLR4, TLR7, TLR8 and TLR9 furthermore activate IRF3 or IRF7, leading to the production of IFN-α and IFN-β (type I IFN) (Lee and Kim 2007). Type I IFN can induce an antiviral state in most cells (Lee and Kim 2007). In addition to the above well-appreciated PRR outputs, it has recently been demonstrated that the stimulation of a number of TLRs with their cognate ligands activates autophagy as a defense mechanism that is capable of directly eliminating intracellular pathogens (Delgado et al. 2008; Sanjuan et al. 2007; Xu et al. 2007).

The TLR4 agonist LPS/endotoxin induces autophagy (Delgado et al. 2008; Sanjuan et al. 2007; Xu et al. 2007). An increase in double-membrane vacuoles monitored by electron microscopy (EM) was reported in RAW cells after LPS treatment (Xu et al. 2007). The punctate distribution of LC3, a sign of autophagosome formation, was also observed by immunofluorescence in human alveolar macrophages upon LPS stimulation (Xu et al. 2007). Induction of autophagy with LPS was reported by Xu et al. to be dependent on TLR4, TRIF, RIP1, and p38 MAPK, but to be independent of MyD88 (Xu et al. 2007). The same group reported that LPS treatment induced localization of *M. tuberculosis* in autophagosomes. TLR4 was also found by others to induce autophagy (Delgado et al. 2008; Sanjuan et al. 2007).

Two different ligands for mouse TLR7 induce autophagy in RAW, J774 macrophage-like cells, and (to a lesser extent) in primary murine bone marrow-derived macrophages (Delgado et al. 2008). Single-stranded (ss) RNA induced LC3 puncta formation, LC3-I-to-LC3-II conversion, and formation of late-stage autophagosomal profiles (autolysosomes) detected by EM (Delgado et al. 2008). LC3-II conversion was detected as early as 30 min following ssRNA addition, in the presence of Bafilomycin A1 (to preserve LC3-II by inhibiting autophagic flux)

(Delgado et al. 2008). Imiquimod, an artificial TLR7 agonist, induced LC3 puncta formation (Delgado et al. 2008; Sanjuan et al. 2007) and increased the proteolysis of long-lived proteins in RAW macrophages (Delgado et al. 2008). Both TLR7 ligands increased GFP-LC3 puncta in murine macrophages from transgenic GFP-LC3 mice (Delgado et al. 2008). The autophagy induced by TLR7 ligands depends on Beclin 1, TLR7 and MyD88, as shown by siRNA knockdown experiments (Delgado et al. 2008). Autophagy elicited by TLR7 agonists can induce the killing of heterologous targets (intracellular *M. tuberculosis*) (Delgado et al. 2008), in a similar manner to autophagy induced by starvation, rapamycin, or overexpression of LRG47 (Irgm1) (Alonso et al. 2007; Gutierrez et al. 2004; Singh et al. 2006). The TLR7-induced killing of *M. tuberculosis* depended on MyD88, Beclin 1 and Atg5 (Delgado et al. 2008). It is important to note that *M. tuberculosis* is not known to stimulate TLR7, and yet induction of autophagy by an artificially added TLR7 agonist resulted in the elimination of intracellular *M. tuberculosis*, indicating that PRR-induced autophagy may not discriminate among microbial targets, and that this could be exploited in the future for therapeutic purposes.

Importantly, PRR-induced autophagy has been detected under conditions where cells were infected with a virus corresponding to a natural infection. Infection of HeLa cells with HIV, a TLR7/TLR8-activating virus, stimulated autophagy (Delgado et al. 2008). However, in pDCs there was no detectable autophagic increase upon infection with vesicular stomatitis virus (Lee et al. 2007). This may be due to an already high baseline level of autophagy for dendritic cells, as reported by Schmid et al. (2007). Thus, the induction of autophagy may be cell-type dependent. Nevertheless, autophagy even plays a role in TLR signaling in pDCs, as it serves to deliver cytosolic viral ligands to endosomally localized TLR7 (Fig. 3) (Lee et al. 2007), as described in detail in the chapter by Tal and Iwasaki in this volume.

There are mixed reports regarding TLR9 and CpG in the induction of autophagy (Delgado et al. 2008; Sanjuan et al. 2007), and more work is needed to address these discrepancies. Moreover, some TLRs require combinatorial stimulation, as TLR2 agonists (Pam3CSK4 or by Pam2CSK4) alone did not induce autophagy (Delgado et al. 2008), but zymosan engaged TLR2/TLR6 and Dectin-1, inducing autophagic markers (Delgado et al. 2008; Sanjuan et al. 2007). These issues and downstream signaling pathways leading to autophagy induction remain to be studied in detail. Another TLR ligand, dsRNA, can induce autophagy (Delgado et al. 2008), which is indicative of TLR3 engagement. However, an exclusive role for TLR3 was not demonstrated in these experiments. Since poly(I:C) can also activate MDA-5 (Takeuchi and Akira 2008), perhaps an RLR was engaged in those experiments, although MDA-5 has not been studied in the context of autophagy (Fig. 3).

Other PRRs have been shown to play a role in inducing autophagy. For example, Yano et al. (2008) have shown in *Drosophila* that a cytosolic PRR, PGRP-LE, which recognizes diaminopimelic acid-type peptidoglycan, induces autophagy which then protects the fruit fly from *Listeria monocytogenes* infection. At present it is not known whether NLRs play an analogous role in autophagy induction.

In addition to the activation of autophagy downstream of PRR stimulation, autophagy proteins seem to affect PRR signaling. The Atg5–Atg12 conjugate, which is a key regulator of autophagy, directly associates with one of the RLRs, RIG-I, and its downstream partner IPS-1 through the CARD domains. This association has been reported to negatively regulate the signaling mediated by IPS-1 and to suppress type I IFN production (Jounai et al. 2007). The biological meaning of Atg5–Atg12 negative regulation of RIG-I–IPS-1 signaling was interpreted by Jounai et al. as viral interference via autophagy with type I IFN production. It is perhaps interesting to also consider an alternative possibility that autophagic proteins may inhibit proinflammatory PRR signaling at a stage after induction in order to prevent potentially deleterious excessive stimulation of proinflammatory cytokines.

8 Conclusions and a Model

Figure 4 collates the presently known layers of conventional immunity regulators superimposed on the role of autophagy in immunity. We believe that autophagy may have been one of the very earliest eukaryotic cell defenses against pathogens. For example, phagocytosis or uptake of bacteria by eukaryotic cells (amoebae feeding on bacteria still occurs today) may sometimes, depending on the pray, result in bacterial escape from the phagosome, as in the case of *Rickettsia* spp., *Shigella*, and *Listeria*. Autophagy may help cells survive such events. Incidentally, a *Rickettsia*-like α-protobacterium is believed to be the mitochondrial ancestor, and present-day mitochondria remain one of the classical substrates for autophagy (mitophagy) (Lemasters 2005; Lyamzaev et al. 2008; Sandoval et al. 2008; Schweers et al. 2007; Twig et al. 2008). During subsequent evolution, layers of immune regulation have been added to control autophagy, as we have covered extensively in this chapter. These layers involve PRR (specifically TLR and perhaps others), a step that links early recognition of microbial products to the induction of autophagy, which can eliminate intracellular pathogens such as *M. tuberculosis* (Delgado et al. 2008; Sanjuan et al. 2007; Xu et al. 2007; Yano et al. 2008). In vertebrates, IRG (p47 GTPases) (Martens and Howard 2006), along with other proposed functions (Howard 2008), became coupled to autophagy as a defense against highly evolved intracellular pathogens such as *M. tuberculosis* and *Toxoplasma* (Gutierrez et al. 2004; Ling et al. 2006; Singh et al. 2006). In its most evolutionarily advanced stage, cytokine networks have also gained control over autophagy in order to optimize the autophagic response to pathogens. This is reflected in autophagy being upregulated by Th1 cytokines and downregulated by Th2 cytokines, in keeping with the protective role of Th1 cytokines against intracellular bacteria and protozoan parasites, and the permissive role of Th2 cytokines towards intracellular pathogens. Th2 cytokines are protective against metazoan parasites (e.g., helminthes), but interfere with the protective Th1 responses. Hence, autophagy appears to be one of the antimicrobial effectors that can explain the immunological consequences of the Th1/Th2 polarization of T cell responses (Gutierrez et al. 2004; Harris et al. 2007).

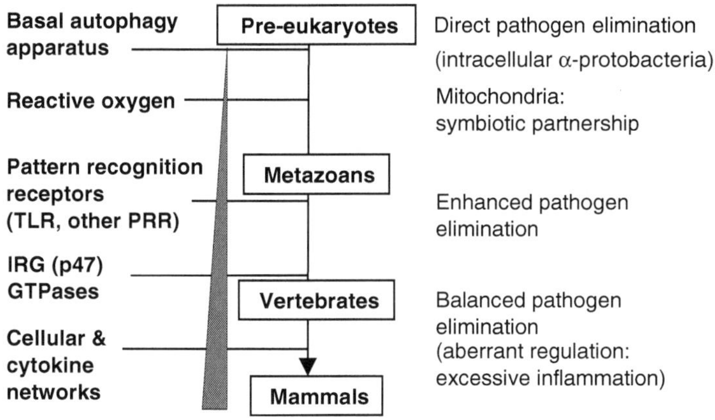

Fig. 4 Autophagy and immune regulatory systems: an evolutionary timeline. Autophagy may be one of the very earliest eukaryotic cell-autonomous defenses against pathogens. Feeding by protozoa (e.g., amoebae) on bacteria via phagocytosis has been preserved to the present day. In metazoans, phagocytic uptake of bacteria occurs by the cells of the reticulo-endothelial system as an innate immune defense. Even mammalian cells that do not readily phagocytose microbes can be induced by bacterial pathogens for phagocytosis. Sometimes, bacteria escape from the phagosome, as in the case of *Rickettsia* spp., *Shigellae* spp., and *Listeria* spp., and they kill the host cell. A *Rickettsia*-like α-protobacterium is believed to be the mitochondrial ancestor, and present-day mitochondria remain one of the classical substrates for autophagy (mitophagy) (Lemasters 2005; Lyamzaev et al. 2008; Sandoval et al. 2008; Schweers et al. 2007; Twig et al. 2008). During evolution, immune systems with increasing levels of complexity have been integrated with autophagy as an integral part of comprehensive innate and adaptive immune defense networks. (i) PRR (specifically TLR and perhaps others) link the early recognition of microbial products to the induction of autophagy, which can eliminate intracellular pathogens such as *M. tuberculosis* (Delgado et al. 2008; Sanjuan et al. 2007; Xu et al. 2007; Yano et al. 2008). (ii) IRG (p47 GTPases) (Martens and Howard 2006) were coupled in vertebrates to autophagy as a defense against highly evolved intracellular pathogens such as *M. tuberculosis* and *Toxoplasma* (Gutierrez et al. 2004; Ling et al. 2006; Singh et al. 2006). (iii) Cytokine networks and Th1/Th2 polarization also control autophagy in order to optimize autophagic response. Autophagy is upregulated by Th1 cytokines and downregulated by Th2 cytokines. This partially explains the protective role of Th1 cytokines and the permissive role of Th2 cytokines vis-à-vis intracellular pathogens, with mycobacterial infections being a prime example (Gutierrez et al. 2004; Harris et al. 2007)

References

Alonso S, Pethe K, Russell DG, Purdy GE (2007) Lysosomal killing of mycobacterium mediated by ubiquitin-derived peptides is enhanced by autophagy. Proc Natl Acad Sci USA 104:6031–6036

Andrade RM, Wessendarp M, Gubbels MJ, Striepen B, Subauste CS (2006) CD40 induces macrophage anti-*Toxoplasma gondii* activity by triggering autophagy-dependent fusion of pathogen-containing vacuoles and lysosomes. J Clin Invest 116:2366–2377

Arico S, Petiot A, Bauvy C, Dubbelhuis PF, Meijer AJ, Codogno P, Ogier-Denis E (2001) The tumor suppressor PTEN positively regulates macroautophagy by inhibiting the phosphatidylinositol 3-kinase/protein kinase B pathway. J Biol Chem 276:35243–35246

Armstrong JA, Hart PDA (1971) Response of cultured macrophages to *Mycobacterium tuberculosis*, with observations on fusion of lysosomes with phagosomes. J Exp Med 134:713–740

Armstrong JA, Hart PD (1975) Phagosome-lysosome interactions in cultured macrophages infected with virulent tubercle bacilli. reversal of the usual nonfusion pattern and observations of bacterial survival. J Exp Med 142:1–16

Behar S, Boom W (2008) Unconventional T cells. In: Kaufmann S, Britton W (eds) Handbook of tuberculosis: immunology and cell biology. Wiley-VCH, Weinheim, pp 157–183

Bekpen C, Hunn JP, Rohde C, Parvanova I, Guethlein L, Dunn DM, Glowalla E, Leptin M, Howard JC (2005) The interferon-inducible p47 (IRG) GTPases in vertebrates: loss of the cell autonomous resistance mechanism in the human lineage. Genome Biol 6:R92

Birmingham CL, Canardien V, Kaniuk NA, Steinberg BE, Higgins DE, Brumell JH (2008) Listeriolysin O allows *Listeria monocytogenes* replication in macrophage vacuoles. Nature 451:350–354

Birmingham CL, Smith AC, Bakowski MA, Yoshimori T, Brumell JH (2006) Autophagy controls *Salmonella* infection in response to damage to the *Salmonella*-containing vacuole. J Biol Chem 28(6):11374–11383

Biswas D, Qureshi OS, Lee WY, Croudace JE, Mura M, Lammas DA (2008) ATP-induced autophagy is associated with rapid killing of intracellular mycobacteria within human monocytes/macrophages. BMC Immunol 9:35

Burton PR et al. (The Wellcome Trust Case Control Consortium) (2007) Genome-wide association study of 14,000 cases of seven common diseases and 3,000 shared controls. Nature 447:661–678

Chaturvedi A, Dorward D, Pierce SK (2008) The B cell receptor governs the subcellular location of Toll-like receptor 9 leading to hyperresponses to DNA-containing antigens. Immunity 28:799–809

Checroun C, Wehrly TD, Fischer ER, Hayes SF, Celli J (2006) Autophagy-mediated reentry of *Francisella tularensis* into the endocytic compartment after cytoplasmic replication. Proc Natl Acad Sci USA 103:14578–14583

Cullinane M, Gong L, Li X, Lazar-Adler N, Tra T, Wolvetang E, Prescott M, Boyce JD, Devenish RJ, Adler B (2008) Stimulation of autophagy suppresses the intracellular survival of *Burkholderia pseudomallei* in mammalian cell lines. Autophagy 4:744–753

Davis AS, Vergne I, Master SS, Kyei GB, Chua J, Deretic V (2007) Mechanism of inducible nitric oxide synthase exclusion from mycobacterial phagosomes. PLoS Pathog 3:e186

Delgado MA, Elmaoued RA, Davis AS, Kyei G, Deretic V (2008) Toll-like receptors control autophagy. EMBO J 27:1110–1121

Deretic V (2005) Autophagy in innate and adaptive immunity. Trends Immunol 26:523–528

Djavaheri-Mergny M, Amelotti M, Mathieu J, Besancon F, Bauvy C, Souquere S, Pierron G, Codogno P (2006) NF-kappaB activation represses tumor necrosis factor-alpha-induced autophagy. J Biol Chem 281:30373–30382

Dye C, Scheele S, Dolin P, Pathania V, Raviglione MC (1999) Consensus statement. Global burden of tuberculosis: estimated incidence, prevalence, and mortality by country. WHO Global Surveillance and Monitoring Project. JAMA 282:677–686

Feng CG, Weksberg DC, Taylor GA, Sher A, Goodell MA (2008) The p47 GTPase Lrg-47 (Irgm1) links host defense and hematopoietic stem cell proliferation. Cell Stem Cell 2:83–89

Floto RA, Sarkar S, Perlstein EO, Kampmann B, Schreiber SL, Rubinsztein DC (2007) Small molecule enhancers of rapamycin-induced TOR inhibition promote autophagy, reduce toxicity in Huntington's disease models and enhance killing of mycobacteria by macrophages. Autophagy 3:620–622

Flynn JL, Chan J (2001) Immunology of tuberculosis. Annu Rev Immunol 19:93–129

Fortin A, Abel L, Casanova JL, Gros P (2007) Host genetics of mycobacterial diseases in mice and men: forward genetic studies of BCG-osis and tuberculosis. Annu Rev Genomics Hum Genet 8:163–192

Goldfeld A, Ranjbar S, Tsitsikov E (2008) Tuberculosis/human immunodeficiency virus coinfection and the host immune response. In: Kaufmann S, Britton W (eds) Handbook of tuberculosis: immunology and cell biology. Wiley-VCH, Weinheim, pp 347–368

Gutierrez MG, Master SS, Singh SB, Taylor GA, Colombo MI, Deretic V (2004) Autophagy is a defense mechanism inhibiting BCG and *Mycobacterium tuberculosis* survival in infected macrophages. Cell 119:753–766

Hampe J, Franke A, Rosenstiel P, Till A, Teuber M, Huse K, Albrecht M, Mayr G, De La Vega FM, Briggs J, Gunther S, Prescott NJ, Onnie CM, Hasler R, Sipos B, Folsch UR, Lengauer T, Platzer M, Mathew CG, Krawczak M, Schreiber S (2007) A genome-wide association scan of nonsynonymous SNPs identifies a susceptibility variant for Crohn disease in *ATG16L1*. Nat Genet 39:207–211

Harris J, De Haro SA, Master SS, Keane J, Roberts EA, Delgado M, Deretic V (2007) T helper 2 cytokines inhibit autophagic control of intracellular *Mycobacterium tuberculosis*. Immunity 27:505–517

Howard J (2008) The IRG proteins: a function in search of a mechanism. Immunobiology 213:367–375

Howell SJ, Wilk D, Yadav SP, Bevins CL (2003) Antimicrobial polypeptides of the human colonic epithelium. Peptides 24:1763–1770

Inbal B, Bialik S, Sabanay I, Shani G, Kimchi A (2002) DAP kinase and DRP-1 mediate membrane blebbing and the formation of autophagic vesicles during programmed cell death. J Cell Biol 157:455–468

Ishii KJ, Koyama S, Nakagawa A, Coban C, Akira S (2008) Host innate immune receptors and beyond: making sense of microbial infections. Cell Host Microbe 3:352–363

Jackson WT, Giddings TH, Jr, Taylor MP, Mulinyawe S, Rabinovitch M, Kopito RR, Kirkegaard K (2005) Subversion of cellular autophagosomal machinery by RNA viruses. PLoS Biol 3:e156

Jounai N, Takeshita F, Kobiyama K, Sawano A, Miyawaki A, Xin KQ, Ishii KJ, Kawai T, Akira S, Suzuki K, Okuda K (2007) The Atg5-Atg12 conjugate associates with innate antiviral immune responses. Proc Natl Acad Sci USA 104:14050–14055

Kagan JC, Su T, Horng T, Chow A, Akira S, Medzhitov R (2008) TRAM couples endocytosis of Toll-like receptor 4 to the induction of interferon-beta. Nat Immunol 9:361–368

Kawai T, Sato S, Ishii KJ, Coban C, Hemmi H, Yamamoto M, Terai K, Matsuda M, Inoue J, Uematsu S, Takeuchi O, Akira S (2004) Interferon-alpha induction through Toll-like receptors involves a direct interaction of IRF7 with MyD88 and TRAF6. Nat Immunol 5:1061–1068

Kieffer AE, Goumon Y, Ruh O, Chasserot-Golaz S, Nullans G, Gasnier C, Aunis D, Metz-Boutigue MH (2003) The N- and C-terminal fragments of ubiquitin are important for the antimicrobial activities. FASEB J 17:776–778

Kraft C, Deplazes A, Sohrmann M, Peter M (2008) Mature ribosomes are selectively degraded upon starvation by an autophagy pathway requiring the Ubp3p/Bre5p ubiquitin protease. Nat Cell Biol 10:602–610

Lammas DA, Stober C, Harvey CJ, Kendrick N, Panchalingam S, Kumararatne DS (1997) ATP-induced killing of mycobacteria by human macrophages is mediated by purinergic P2Z(P2X7) receptors. Immunity 7:433–444

Lee HK, Lund JM, Ramanathan B, Mizushima N, Iwasaki A (2007) Autophagy-dependent viral recognition by plasmacytoid dendritic cells. Science 315:1398–1401

Lee MS, Kim YJ (2007) Signaling pathways downstream of pattern-recognition receptors and their cross talk. Annu Rev Biochem 76:447–480

Lemasters JJ (2005) Selective mitochondrial autophagy, or mitophagy, as a targeted defense against oxidative stress, mitochondrial dysfunction, and aging. Rejuvenation Res 8:3–5

Levine B, Deretic V (2007) Unveiling the roles of autophagy in innate and adaptive immunity. Nat Rev Immunol 7:767–777

Lewinsohn DM, Briden AL, Reed SG, Grabstein KH, Alderson MR (2000) *Mycobacterium tuberculosis*-reactive $CD8^+$ T lymphocytes: the relative contribution of classical versus non-classical HLA restriction. J Immunol 165:925–930

Li C, Capan E, Zhao Y, Zhao J, Stolz D, Watkins SC, Jin S, Lu B (2006) Autophagy is induced in $CD4^+$ T cells and important for the growth factor-withdrawal cell death. J Immunol 177:5163–5168

Liang XH, Kleeman LK, Jiang HH, Gordon G, Goldman JE, Berry G, Herman B, Levine B (1998) Protection against fatal Sindbis virus encephalitis by Beclin, a novel Bcl-2-interacting protein. J Virol 72:8586–8596

Ling YM, Shaw MH, Ayala C, Coppens I, Taylor GA, Ferguson DJ, Yap GS (2006) Vacuolar and plasma membrane stripping and autophagic elimination of *Toxoplasma gondii* in primed effector macrophages. J Exp Med 203:2063–2071

Liu Y, Schiff M, Czymmek K, Talloczy Z, Levine B, Dinesh-Kumar SP (2005) Autophagy regulates programmed cell death during the plant innate immune response. Cell 121:567–577

Lyamzaev KG, Nepryakhina OK, Saprunova VB, Bakeeva LE, Pletjushkina OY, Chernyak BV, Skulachev VP (2008) Novel mechanism of elimination of malfunctioning mitochondria (mitoptosis): formation of mitoptotic bodies and extrusion of mitochondrial material from the cell. Biochim Biophys Acta 1777:817–825

MacMicking JD, Taylor GA, McKinney JD (2003) Immune control of tuberculosis by IFN-gamma-inducible LRG-47. Science 302:654–659

Majlessi L, Combaluzier B, Albrecht I, Garcia JE, Nouze C, Pieters J, Leclerc C (2007) Inhibition of phagosome maturation by mycobacteria does not interfere with presentation of mycobacterial antigens by MHC molecules. J Immunol 179:1825–1833

Manabe YC, Bishai WR (2000) Latent *Mycobacterium tuberculosis*-persistence, patience, and winning by waiting. Nat Med 6:1327–1329

Martens S, Howard J (2006) The interferon-inducible GTPases. Annu Rev Cell Dev Biol 22:559–589

Martens S, Parvanova I, Zerrahn J, Griffiths G, Schell G, Reichmann G, Howard JC (2005) Disruption of *Toxoplasma gondii* parasitophorous vacuoles by the mouse p47-resistance GTPases. PLoS Pathog 1:e24

Massey D, Parkes M (2007) Common pathways in Crohn's disease and other inflammatory diseases revealed by genomics. Gut 56:1489–1492

Medzhitov R (2007) Recognition of microorganisms and activation of the immune response. Nature 449:819–826

Moody DB, Young DC, Cheng TY, Rosat JP, Roura-Mir C, O'Connor PB, Zajonc DM, Walz A, Miller MJ, Levery SB, Wilson IA, Costello CE, Brenner MB (2004) T cell activation by lipopeptide antigens. Science 303:527–531

Nahid P, Daley CL (2006) Prevention of tuberculosis in HIV-infected patients. Curr Opin Infect Dis 19:189–193

Nakagawa I, Amano A, Mizushima N, Yamamoto A, Yamaguchi H, Kamimoto T, Nara A, Funao J, Nakata M, Tsuda K, Hamada S, Yoshimori T (2004) Autophagy defends cells against invading group A *Streptococcus*. Science 306:1037–1040

Nelms K, Keegan AD, Zamorano J, Ryan JJ, Paul WE (1999) The IL-4 receptor: signaling mechanisms and biologic functions. Annu Rev Immunol 17:701–738

Nunn P, Williams B, Floyd K, Dye C, Elzinga G, Raviglione M (2005) Tuberculosis control in the era of HIV. Nat Rev Immunol 5:819–826

O'Neill LA, Bowie AG (2007) The family of five: TIR-domain-containing adaptors in Toll-like receptor signalling. Nat Rev Immunol 7:353–364

Oawa M, Yoshimori T, Suzuki T, Sagara H, Mizushima N, Sasakawa C (2005) Escape of intracellular *Shigella* from autophagy. Science 307:727–731

Orvedahl A, Alexander D, Tallóczy Z, Sun Q, Wei Y, Zhang W, Burns D, Leib D, Levine B (2007) HSV-1 ICP34.5 confers neurovirulence by targeting the Beclin 1 autophagy protein. Cell Host Microbe 1:23–35

Ottenhoff H, Lewinsohn D, Lewinsohn D (2008) Human CD4 and CD8 cell responses to *Mycobacterium tuberculosis*: antigen specificity, function, implications and applications. In: Kaufmann S, Britton D (eds) Handbook of tuberculosis: immunology and cell biology. Wiley-VCH, Weinheim, pp 119–155

Paludan C, Schmid D, Landthaler M, Vockerodt M, Kube D, Tuschl T, Munz C (2005) Endogenous MHC class II processing of a viral nuclear antigen after autophagy. Science 307:593–596

Parkes M, Barrett JC, Prescott NJ, Tremelling M, Anderson CA, Fisher SA, Roberts RG, Nimmo ER, Cummings FR, Soars D, Drummond H, Lees CW, Khawaja SA, Bagnall R, Burke DA, Todhunter CE, Ahmad T, Onnie CM, McArdle W, Strachan D, Bethel G, Bryan C, Lewis CM, Deloukas P, Forbes A, Sanderson J, Jewell DP, Satsangi J, Mansfield JC, Cardon L, Mathew CG (2007) Sequence variants in the autophagy gene *IRGM* and multiple other replicating loci contribute to Crohn's disease susceptibility. Nat Genet 39:830–832

Petiot A, Ogier-Denis E, Blommaart EF, Meijer AJ, Codogno P (2000) Distinct classes of phosphatidylinositol 3′-kinases are involved in signaling pathways that control macroautophagy in HT-29 cells. J Biol Chem 275:992–998

Pieters J (2008) *Mycobacterium tuberculosis* and the macrophage: maintaining a balance. Cell Host Microbe 3:399–407

Py BF, Lipinski MM, Yuan J (2007) Autophagy limits *Listeria monocytogenes* intracellular growth in the early phase of primary infection. Autophagy 3:117–125

Pyo JO, Jang MH, Kwon YK, Lee HJ, Jun JI, Woo HN, Cho DH, Choi B, Lee H, Kim JH, Mizushima N, Oshumi Y, Jung YK (2005) Essential roles of Atg5 and FADD in autophagic cell death: dissection of autophagic cell death into vacuole formation and cell death. J Biol Chem 280:20722–20729

Ramachandra L, Smialek JL, Shank SS, Convery M, Boom WH, Harding CV (2005) Phagosomal processing of *Mycobacterium tuberculosis* antigen 85B is modulated independently of mycobacterial viability and phagosome maturation. Infect Immun 73:1097–1105

Reid A, Scano F, Getahun H, Williams B, Dye C, Nunn P, De Cock KM, Hankins C, Miller B, Castro KG, Raviglione MC (2006) Towards universal access to HIV prevention, treatment, care, and support: the role of tuberculosis/HIV collaboration. Lancet Infect Dis 6:483–495

Rioux JD, Xavier RJ, Taylor KD, Silverberg MS, Goyette P, Huett A, Green T, Kuballa P, Barmada MM, Datta LW, Shugart YY, Griffiths AM, Targan SR, Ippoliti AF, Bernard EJ, Mei L, Nicolae DL, Regueiro M, Schumm LP, Steinhart AH, Rotter JI, Duerr RH, Cho JH, Daly MJ, Brant SR (2007) Genome-wide association study identifies new susceptibility loci for Crohn disease and implicates autophagy in disease pathogenesis. Nat Genet 39: 596–604

Russell DG (2007) Who puts the tubercle in tuberculosis? Nat Rev Microbiol 5:39–47

Sabauste C, Andrade R, Wessendarp M (2007) CD40-TRAF6 and autophagy-dependent anti-microbial activity in macrophages. Autophagy 3:245–248

Sandoval H, Thiagarajan P, Dasgupta SK, Schumacher A, Prchal JT, Chen M, Wang J (2008) Essential role for Nix in autophagic maturation of erythroid cells. Nature 454:232–235

Sanjuan MA, Dillon CP, Tait SW, Moshiach S, Dorsey F, Connell S, Komatsu M, Tanaka K, Cleveland JL, Withoff S, Green DR (2007) Toll-like receptor signalling in macrophages links the autophagy pathway to phagocytosis. Nature 450:1253–1257

Saunders BM, Britton WJ (2007) Life and death in the granuloma: immunopathology of tuberculosis. Immunol Cell Biol 85:103–111

Schlottmann S, Buback F, Stahl B, Meierhenrich R, Walter P, Georgieff M, Senftleben U (2008) Prolonged classical NF-kappaB activation prevents autophagy upon *E. coli* stimulation in vitro: a potential resolving mechanism of inflammation. Mediators Inflamm 2008:725–854

Schmid D Munz C (2007) Innate and adaptive immunity through autophagy. Immunity 27:11–21

Schmid D, Pypaert M, Munz C (2007) Antigen-loading compartments for major histocompatibility complex class II molecules continuously receive input from autophagosomes. Immunity 26:79–92

Schweers RL, Zhang J, Randall MS, Loyd MR, Li W, Dorsey FC, Kundu M, Opferman JT, Cleveland JL, Miller JL, Ney PA (2007) NIX is required for programmed mitochondrial clearance during reticulocyte maturation. Proc Natl Acad Sci USA 104:19500–19505

Singh SB, Davis AS, Taylor GA, Deretic V (2006) Human IRGM induces autophagy to eliminate intracellular mycobacteria. Science 313:1438–1441

Takeuchi O, Akira S (2008) MDA5/RIG-I and virus recognition. Curr Opin Immunol 20:17–22

Talloczy Z, Jiang W, Virgin HW IV, Leib DA, Scheuner D, Kaufman RJ, Eskelinen EL, Levine B (2002) Regulation of starvation- and virus-induced autophagy by the eIF2alpha kinase signaling pathway. Proc Natl Acad Sci USA 99:190–195

Taylor GA, Feng CG, Sher A (2004) p47 GTPases: regulators of immunity to intracellular pathogens. Nat Rev Immunol 4:100–109

Torres M, Ramachandra L, Rojas RE, Bobadilla K, Thomas J, Canaday DH, Harding CV, Boom WH (2006) Role of phagosomes and major histocompatibility complex class II (MHC-II) compartment in MHC-II antigen processing of *Mycobacterium tuberculosis* in human macrophages. Infect Immun 74:1621–1630

Twig G, Elorza A, Molina AJ, Mohamed H, Wikstrom JD, Walzer G, Stiles L, Haigh SE, Katz S, Las G, Alroy J, Wu M, Py BF, Yuan J, Deeney JT, Corkey BE, Shirihai OS (2008) Fission and selective fusion govern mitochondrial segregation and elimination by autophagy. Embo J 27:433–446

Vandal OH, Pierini LM, Schnappinger D, Nathan CF, Ehrt S (2008) A membrane protein preserves intrabacterial pH in intraphagosomal *Mycobacterium tuberculosis*. Nat Med 14:849–854

Vergne I, Chua J, Singh S, Deretic V (2004) Cell biology of *Mycobacterium tuberculosis* phagosome. Annu Rev Cell Dev Biol 20:367–394

Virgin H (2008) A 'fly-by' killing with a primordial cellular weapon. Nat Immunol 9: 827–829

Xu Y, Jagannath C, Liu XD, Sharafkhaneh A, Kolodziejska KE, Eissa NT (2007) Toll-like receptor 4 is a sensor for autophagy associated with innate immunity. Immunity 27:135–144

Yano T, Mita S, Ohmori H, Oshima Y, Fujimoto Y, Ueda R, Takada H, Goldman WE, Fukase K, Silverman N, Yoshimori T, Kurata S (2008) Autophagic control of *Listeria* through intracellular innate immune recognition in *Drosophila*. Nat Immunol 9:908–916

Yuan Y, Lee RE, Besra GS, Belisle JT, Barry CE, 3rd (1995) Identification of a gene involved in the biosynthesis of cyclopropanated mycolic acids in *Mycobacterium tuberculosis*. Proc Natl Acad Sci USA 92:6630–6634

Autophagy in Immunity Against Intracellular Bacteria

Ju Huang and John H. Brumell

Contents

1 Autophagy as an Immune Defense Mechanism Against Bacteria 191
2 Autophagy of *Salmonella enterica* Serovar Typhimurium 191
  2.1 *Salmonella* in Disease 191
  2.2 Lifestyle of Intracellular *S. typhimurium* in Mammalian Cells 192
  2.3 Virulence of *S. typhimurium*: T3SS and Effectors 194
  2.4 Autophagy in Defense Against *S. typhimurium* 195
  2.5 Multiple Fates of Intracellular *S. typhimurium* 196
3 Autophagy of *Listeria monocytogenes* 196
  3.1 *L. monocytogenes* in Disease 196
  3.2 Lifestyle of *L. monocytogenes* in Mammalian Cells 197
  3.3 Virulence Factors of *L. monocytogenes* 197
  3.4 Autophagy *of L. monocytogenes* 199
  3.5 Spacious *Listeria*-Containing Phagosomes 201
  3.6 Multiple Fates of Different *L. monocytogenes* Populations 203
4 Autophagy of *Shigella flexneri* and Other Intracellular Bacteria 203
5 Bacterial Autophagy in Inflammatory Bowel Disease 205
6 Summary 208
References 209

Abstract Autophagy is an innate immune defense mechanism against various intracellular bacterial pathogens, such as *Salmonella enterica* serovar Typhimurium (*S. typhimurium*), *Listeria monocytogenes* and *Shigella flexneri*. *S. typhimurium* uses type three secretion systems (T3SSs) to invade mammalian cells and replicate in *Salmonella*-containing vacuoles (SCVs). A small population of intracellular

---

J. Huang and J.H. Brumell
Cell Biology Program, Hospital for Sick Children, Toronto, ON, Canada M5G 1X8

J.H. Brumell (✉)
Department of Molecular Genetics and Institute of Medical Science,
University of Toronto, Toronto, ON, Canada M5S 1A8
e-mail: john.brumell@sickkids.ca

B. Levine et al. (eds.), *Autophagy in Infection and Immunity*,
Current Topics in Microbiology and Immunology 335,
DOI 10.1007/ 978-3-642-00302-8_9, © Springer-Verlag Berlin Heidelberg 2009

S. typhimurium is targeted by autophagy shortly after infection. Evidence suggests that these bacteria are present within SCVs that have been damaged by high levels of T3SS activity. Autophagy limits the growth of *S. typhimurium* in host cells. Therefore, autophagy can be considered to protect the cytosol of eukaryotic cells from bacterial colonization. *L. monocytogenes* secretes the pore-forming cytolysin listeriolysin O (LLO) to disrupt the phagosome and escape into the cytosol, where it acquires actin-based motility. Autophagy can target *L. monocytogenes* in the cytosol under specific experimental conditions. However, *L. monocytogenes* utilizes several virulence factors to evade being killed by the autophagy system. A newly appreciated population of *L. monocytogenes* undergoes slow growth in specialized vacuoles termed spacious *Listeria*-containing phagosomes (SLAPs), the formation of which requires bacterial LLO and host autophagy. In the cytosol, *S. flexneri* can also be a target for autophagy in the absence of a T3SS effector, IcsB, that normally impairs the interaction between Atg5 and wild-type bacteria. Therefore, autophagy can recognize intracellular bacteria in a variety of ways, leading to different fates for these bacteria in host cells. The inefficient autophagy of enteric bacteria in genetically compromised individuals may contribute to the pathogenesis of Crohn's disease.

Abbreviations

CD	Crohn's disease
EEA1	Early endosome antigen 1
ER	Endoplasmic reticulum
GEF	Guanine nucleotide exchange factor
IRG	Immunity-related GTPase
LAMP-1	Lysosome-associated membrane protein 1
LLO	Listeriolysin O
3-MA	3-Methyladenine
M6PR	Mannose-6-phosphate receptor
MEF	Mouse embryonic fibroblast
NLR	NOD-like receptor
NOD	Nucleotide-binding oligomerization domain
p.i.	Postinfection
PLC	Phospholipase C
SCID	Severe combined immunodeficiency
SCV	*Salmonella*-containing vacuole
SLAP	Spacious *Listeria*-containing phagosome
SNP	Single nucleotide polymorphism
SPI	*Salmonella* pathogenicity island
T3SS	Type three secretion system
TLR	Toll-like receptor

1 Autophagy as an Immune Defense Mechanism Against Bacteria

Some pathogenic bacteria have adopted an intracellular lifestyle during infection of their host. The fate of these bacteria in host cells is determined by the balance between bacterial virulence factors and host defense mechanisms. One of the major antibacterial pathways is phagocytosis, which traps the bacterium in an endocytic membranous compartment (termed the phagosome) and delivers it to the lysosome for degradation (Fang 2004; Tosi 2005). However, intracellular bacteria have developed strategies to avoid phagosomal killing. For example, some bacteria are capable of impairing or delaying fusion of the phagosome with the lysosome, while others can disrupt the phagosome membrane and escape into the cytosol (Sinai and Joiner 1997; Cossart and Sansonetti 2004). Autophagy is another intracellular immune mechanism that can target intracellular bacteria, either within phagosomes (e.g., *Mycobacterium tuberculosis*, see detailed review in the chapter by Deretic et al. in this volume), in damaged vacuoles (e.g., *Salmonella enterica* serovar Typhimurium), or in the cytosol (e.g., Group A *Streptococcus*, see detailed review in the chapter by Yoshimori and Amano in this volume), and kill them through the autophagosome–lysosome pathway (Gutierrez et al. 2004; Nakagawa et al. 2004; Huang and Klionsky 2007). Interestingly, some bacteria can avoid recognition by autophagy when they are present in the cytosol, such as *Listeria monocytogenes* and *Shigella flexneri* (Ogawa et al. 2005; Ogawa and Sasakawa 2006; Birmingham et al. 2007; Py et al. 2007; Birmingham et al. 2008b). The battle between bacteria and host defenses results in heterogeneous fates of intracellular bacteria.

This chapter focuses on the autophagy of two bacterial species, *S. enterica* and *L. monocytogenes*. We will give an overview of how these bacteria manipulate host cells, how they interact with autophagy, and the possible role of bacterial autophagy in inflammatory bowel diseases.

2 Autophagy of *Salmonella enterica* Serovar Typhimurium

2.1 *Salmonella in Disease*

S. enterica are Gram-negative pathogenic bacteria that cause a variety of diseases in mammals. *S. enterica* serovar Typhi infects humans and leads to typhoid fever, whereas *S. enteric* serovar Typhimurium (*S. typhimurium*) can infect a broad range of hosts and is a significant cause of gastroenteritis in humans (Finlay and Brumell 2000; Haraga et al. 2008). *S. typhimurium* infection of genetically susceptible mice, however, manifests a typhoid fever-like systemic disease, making these bacteria a useful model to study the pathogenesis of this disease (Finlay and Brumell 2000). *S. typhimurium* normally enters the human digestive tract through oral ingestion of contaminated food. Once the bacteria reach the intestine, they are able to disrupt the mucous layer and penetrate the epithelium. This leads to the recruitment of immune cells to the intestinal lumen, which release cytokines, chemokines and cause local inflammation (gastroenteritis) (Ohl and Miller 2001; Haraga et al. 2008). In the systemic disease mouse model, the bacteria can gain access to and survive in macrophages after breaching the epithelial layer, then disseminate to other locations, such as Peyer's patches, the liver, spleen, and bone marrow, causing systemic illness (Alpuche-Aranda et al. 1994; Finlay and Brumell 2000; Ohl and Miller 2001).

Although mouse models of infection are available, there are many technical difficulties in studying the molecular mechanisms of *S. typhimurium* pathogenicity in vivo. Therefore, in vitro methods using cultured mammalian cells provide a powerful tool for investigating the interactions between bacteria and host cells on a molecular level. By these means, the cellular functions of *S. typhimurium* virulence factors as well as the countering host immune defense mechanisms are being revealed (Finlay and Brumell 2000; Haraga et al. 2008).

2.2 *Lifestyle of Intracellular S. typhimurium in Mammalian Cells*

In vitro studies have shown that *S. typhimurium* invasion of epithelial cells is a "force feeding" process, in contrast to the receptor-mediated endocytosis processes normally exploited by other pathogenic bacteria (Brumell et al. 1999). During invasion, *S. typhimurium* uses a syringe-needle structure, called a type three secretion system (T3SS), and injects a set of virulence proteins into the host cell cytosol. These bacterial proteins (referred to as effectors) rearrange the host cytoskeleton to induce plasma membrane ruffles and promote the engulfment of the bacterium into the cell (Brumell et al. 1999) (Fig. 1). This process is very rapid and results in the bacteria being sealed in a spacious membrane-bound compartment, termed the *Salmonella*-containing vacuole (SCV), which resembles the phagosome in macrophages. SCVs undergo a maturation process, characterized by the sequential acquisition of early and late endosomal proteins (Brumell and Grinstein 2004). The early endosome antigen-1 (EEA1) and Rab5 GTPase associate with the SCV transiently (5–15 min postinfection, p.i.), to be replaced by late endosomal markers, such as the Rab7 GTPase and lysosomal-associated membrane protein 1(LAMP-1) after 30–60 min p.i. (Brumell and Grinstein 2004).

SCVs are diverted from lysosomal fusion, as evidenced by a lack of accumulation of mannose-6-phosphate receptor (M6PR) and cathepsin D on the SCV during the first 3 h p.i. (Knodler and Steele-Mortimer 2003; Brumell and Grinstein 2004). This enables the bacteria to avoid degradation in the harsh lysosomal lumen. By 6–8 h p.i., the membrane of the SCV extends and forms long tubular structures, termed *Salmonella*-induced filaments (Sifs), which are associated with rapid bacterial replication within the late SCV (Garcia-del Portillo et al. 1993; Brumell et al. 2002a; Birmingham et al. 2005) (Fig. 1). The exact function of Sifs is not yet clear, although many effectors required for full virulence in animal models are associated

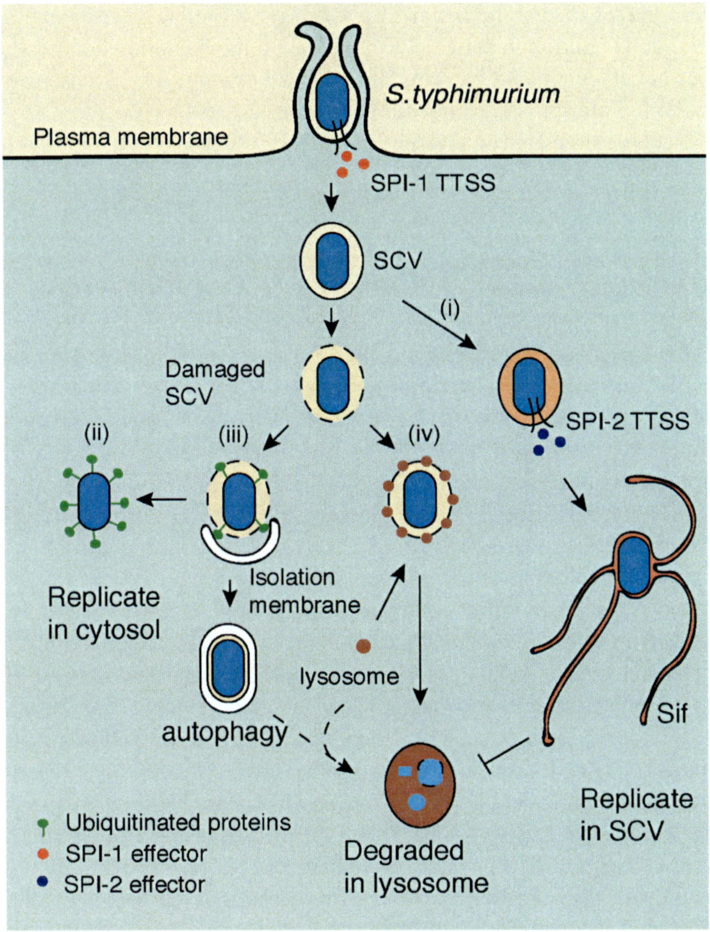

Fig. 1a–d Multiple populations of intracellular *S. typhimurium* in epithelial cells in vitro. *S. typhimurium* utilizes the SPI-1 T3SS to invade host cells. The majority of bacteria remain in SCVs and replicate in Sif-associated mature SCVs (**a**). A portion of bacteria disrupt the SCV and enter the cytosol where they associate with ubiquitinated proteins and replicate (**b**). Autophagy targets a population of bacteria within damaged SCVs and possibly eliminates them through fusion with lysosomes (**c**). Damaged SCVs are also repaired by a special lysosome repair mechanism, eventually killing the bacteria inside (**d**)

with Sif formation in vitro (Birmingham et al. 2005). Bacterial deletion mutants lacking the effector SifA, which is essential for Sif formation, disrupt the SCV and are released into cytosol during late infection times (Beuzon et al. 2000). Exposure of bacteria to the cytosol causes restricted growth in macrophages, although it results in enhanced replication in epithelial cells (Beuzon et al. 2000; Brumell et al. 2001; Brumell et al. 2002b). This suggests that a cell type-specific killing mechanism is present in the cytosol of macrophages that is lacking in epithelial cells.

In macrophages, *S. typhimurium* SCVs follow a similar route of maturation as in epithelial cells (Rathman et al. 1997). However, the bacteria can also induce different types of cell death after entering the cell. In macrophages, caspase-1-dependent cell death, called pyroptosis, can occur early after infection (less than 1 h), whereas autophagic cell death has been observed after 5 h p.i. (Hersh et al. 1999; Hernandez et al. 2003; Fink and Cookson 2007).

2.3 Virulence of *S. typhimurium*: T3SS and Effectors

Many virulence factors of *S. typhimurium* are encoded in *Salmonella* pathogenic islands (SPIs) on the bacterial chromosome. *SPI-1* and *SPI-2* encode two different T3SSs, which are utilized by the bacterium to transport their respective effectors directly into host cells (Galan 2001). These effectors can mimic host protein functions and manipulate host cellular events. The two T3SSs of *S. typhimurium* are expressed independently but their effectors may function cooperatively during the infection of epithelial cells (Galan 2001; Haraga et al. 2008). The *SPI-1* T3SS mediates the initial invasion process, including rearrangement of the actin cytoskeleton to form membrane ruffles. At least six known effectors (SopE, SopE2, SopB, SptP, SipA, SipC) are involved in this process (Galan 2001; Haraga et al. 2008). For example, SopE and SopE2 act as guanine nucleotide exchange factors (GEFs) and activate host Rho family GTPases to regulate the actin cytoskeleton (Galan 2001; Haraga et al. 2008). Another *SPI-1* T3SS effector, SipB, was shown to induce pyroptosis as well as autophagic cell death in macrophages, likely through association with caspase 1 or by disrupting the mitochondria (Hernandez et al. 2003; Hueffer and Galan 2004).

During late infection (>4 h p.i.), *SPI-1* T3SS expression is downregulated and expression of the *SPI-2* T3SS system is upregulated (Eriksson et al. 2003). Effectors secreted from this latter system are directly released into the host cytosol across the SCV membrane, and manipulate SCV trafficking to promote intracellular bacterial growth (Chen et al. 1996). *SPI-2* T3SS mutants show restricted proliferation in cultured murine macrophages and significantly reduced virulence in mouse models of infection (Cirillo et al. 1996; Shea et al. 1996; Hensel et al. 1998). Over 20 effectors are secreted by this system, though most of their functions are unknown (Haraga et al. 2008). Among them, the most extensively studied effector is SifA, which is required for the formation of Sifs and to maintain the integrity of the SCV (Beuzon 2000, Brumell 2001).

Although the two T3SSs appear to function sequentially, crosstalk does exist. Two effectors, Slrp and SspH1, were found to be secreted by both systems, and certain *SPI-2* genes are necessary for the transcription of *SPI-1* genes (Deiwick et al. 1998; Miao et al. 1999; Tsolis et al. 1999). In addition, the expression of several *SPI-1* effectors persists long after invasion (Drecktrah et al. 2005; Brawn et al. 2007; Giacomodonato et al. 2007). The existence of two T3SSs and a large number of effectors reflects the complex interaction of this bacterium during its fight for survival in host cells.

2.4 Autophagy in Defense Against *S. typhimurium*

During in vitro infection of cultured cells, intracellular *S. typhimurium* modulates the SCV into a replicative niche. However, not all bacteria follow this route. We and others have found that a small but significant population (~10–20%) disrupt SCVs and escape into the cytosol early after invasion (Perrin et al. 2004; Birmingham and Brumell 2006; Birmingham et al. 2006). Once exposed to the cytosol, this population of bacteria are recognized by the ubiquitin system and become associated with polyubiquitinated proteins (Perrin et al. 2004) (Fig. 1). These bacteria undergo replication during the first 6 h p.i. (Perrin et al. 2004). We further found that a population of intracellular bacteria is targeted by autophagy in epithelial cells and fibroblasts (Birmingham et al. 2006) (Fig. 1). The autophagosome marker LC3 is recruited to the intracellular bacteria early after invasion, reaching maximal levels at 1 h p.i. when ~20% of all bacteria are sequestered by autophagosomes (Birmingham et al. 2006). This event is transient, as there is a rapid decline in the percentage of $LC3^+$ bacteria after 1 h of infection. Translocation of LC3 to the bacterium requires the core autophagy machinery since it does not occur in *Atg5* knockout cells (Birmingham et al. 2006). Targeting of intracellular *S. typhimurium* maintains bacteria in SCVs and protects the cytosol from bacterial colonization (Birmingham et al. 2006).

The signal(s) that trigger autophagy of *S. typhimurium* are not yet clear. Immunofluorescence studies show mosaic-style staining of LC3 and SCV markers around the bacterium, implying that the SCV is damaged and autophagosomes are possibly recruited to repair the gaps in the vacuolar membranes (Birmingham and Brumell 2006; Birmingham et al. 2006). Indeed, some of the bacteria (~50%) targeted by autophagy colocalized with polyubiquitinated proteins, indicating that they have been exposed to the cytosol. It was reported that wild-type *S. typhimurium* can form pores in early SCV membranes and this requires the *SPI-1* T3SS apparatus. Mutants lacking this system do not have pore-forming activity and are not capable of inducing autophagy (Roy et al. 2004; Birmingham et al. 2006). Inhibiting bacterial protein synthesis immediately after invasion (10 min) also abolished the recruitment of LC3 to bacteria at 1 h (Birmingham et al. 2006). These results suggest that *SPI-1* T3SS-regulated SCV injury is a possible signal that triggers autophagy of *S. typhimurium*. Damaged organelles, such as abnormal ER and mitochondria, have been reported to be specifically targeted by autophagy (Priault et al. 2005; Ogata et al. 2006; Yorimitsu et al. 2006). Therefore, the targeting of damaged SCVs may reflect an ancestral "housekeeping" function of autophagy.

S. typhimurium also induces autophagy in macrophages, resulting in autophagic cell death (Hernandez et al. 2003). Instead of specific targeting of the intracellular bacteria, like that observed in epithelial cells, numerous multimembranous autophagosomes accumulate in the cytosol of bacteria-infected macrophages (Hernandez et al. 2003). This is caused by the *SPI-1* T3SS effector SipB, which localizes to mitochondria and disrupts normal mitochondrial morphology (Hernandez et al. 2003). Expression of SipB alone in different cell lines also induces the formation of autophagosomes

(Hernandez et al. 2003). As yet, it is difficult to determine whether the bacteria-induced autophagic cell death is a host defense response or a bacterial virulence strategy.

2.5 Multiple Fates of Intracellular *S. typhimurium*

In vitro studies reveal the existence of several populations of intracellular *S. typhimurium* in both phagocytic and nonphagocytic cell types (Fig. 1). Heterogeneity in both the bacterial innoculum and the cell lines used for infection likely impact on these outcomes. For example, it is known that rapidly growing bacterial cultures contain fast- and slow-growing bacteria, as well as some dead bacteria that may be internalized passively by host cells during in vitro infection. During infection of epithelial cells, the majority of intracellular bacteria (~70–80%) reside in SCVs and undergo replication by limiting fusion with lysosomes (Brumell and Grinstein 2004; Birmingham et al. 2005). Early postinvasion, a small fraction (~20%) of the bacteria damage the SCV membrane and become exposed to cytosol, where they become associated with ubiquitinated proteins (Birmingham et al. 2006). The consequence of this process remains unclear, but may be associated with antigen presentation in other cell types (Perrin et al. 2004). SCV damage activates a lysosome repair system that results in the killing of the bacteria (Roy et al. 2004). SCV injury also causes a population of bacteria to be targeted by autophagy and restricted in replication (Birmingham and Brumell 2006; Birmingham et al. 2006). Half of this population of bacteria is also labeled with ubiquitin (Birmingham et al. 2006). However, whether there is a link between ubiquitination and autophagy of bacteria is not known.

3 Autophagy of *Listeria monocytogenes*

3.1 *L. monocytogenes* in Disease

L. monocytogenes is a Gram-positive bacterium that infects a wide range of hosts, causing listeriosis in humans (Pamer 2004; Lecuit 2007; Swaminathan and Gerner-Smidt 2007). Immunocompromised individuals, pregnant women and infants are particularly susceptible to this pathogen. Oral ingestion of contaminated food is the common method of importing the bacteria into the intestinal tract, where they penetrate through the epithelial barrier and cause enteritis. Following access to the bloodstream, the bacteria quickly spread to the liver and spleen, where they colonize macrophages (Lecuit 2007). During severe infection, the bacteria can penetrate the blood–brain barrier and infect the central nervous system, resulting in encephalitis and meningitis (Lecuit 2007). *L. monocytogenes* is also able to cross the placental barrier in pregnant women, infecting the fetus and causing abortions or stillbirths (Lecuit 2007). The abilities of the bacterium to actively invade epithelial cells and macrophages, replicate in these cells, and spread to neighboring cells are managed by complex bacterial virulence mechanisms. Upon infection, different levels of host immune responses are activated. Phagocytes and cytokines of the innate immune system play essential roles in controlling early infections. IFN-γ in particular has shown the ability to activate macrophages to restrict the replication of intracellular bacteria (Pamer 2004; Zenewicz and Shen 2007). Adaptive immune responses, especially $CD8^+$ T cells, are required for the complete clearance of *L. monocytogenes* (Pamer 2004; Zenewicz and Shen 2007).

3.2 Lifestyle of L. monocytogenes in Mammalian Cells

In vitro studies have established a model of *L. monocytogenes* lifestyle in both phagocytic and nonphagocytic cells at the molecular level. The bacterium specifically binds to host cell membrane adhesion molecules to promote a "zippering" style of internalization into the host cell (Hamon et al. 2006; Ireton 2007) (Fig. 2). Initially, the bacteria are taken up into phagosomes, but within 30 min they rapidly lyse the phagosomal membrane and escape into the cytosol (Cossart and Sansonetti 2004; Dussurget et al. 2004). Once in the cytosol, *L. monocytogenes* recruit host actin to their surface, resulting in the formation of an actin comet tail which propels the bacterium throughout the host cell (Dussurget et al. 2004; Gouin et al. 2005) (Fig. 2). This intracellular motility allows the bacteria to escape targeting by cytosolic defense mechanisms and rapidly replicate in the cytosol (discussed below). The motile bacteria are also able to form protrusions from the cell membrane and push inward into neighboring cells. The bacteria then escape into the cytosol of the neighboring cell to perpetuate the infectious cycle (Dussurget et al. 2004; Gouin et al. 2005).

3.3 Virulence Factors of L. monocytogenes

Expression of most of the key virulence factors of *L. monocytogenes* is controlled by the PrfA transcriptional activator (Dussurget et al. 2004; Scortti et al. 2007). PrfA regulates the expression of at least nine genes (*inlA, inlB, inlC, plcA, plcB, hly, mpl, actA, and hpt*) that contribute to virulence (Dussurget et al. 2004; Scortti et al. 2007). The adhesins internalin A (InlA) and InlB recognize host cell membrane receptors and mediate bacterial entry (Hamon et al. 2006; Ireton 2007). InlA binds to E-cadherin, a cell–cell adhesion molecule, and in turn causes actin polymerization at the entry site that leads to membrane extension to enclose the bacterium (Dussurget et al. 2004; Ireton 2007). InlB recognizes the receptor tyrosine kinase

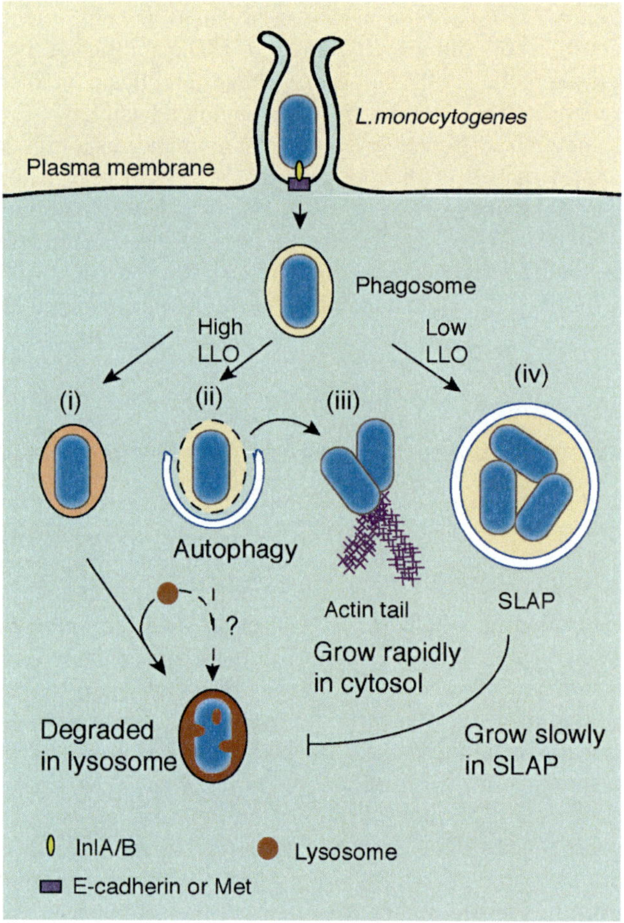

Fig. 2a–d Different intracellular fates of *L. monocytogenes* in host cells. Through binding of the bacterial proteins InlA or InlB (or other adhesins) with host cell surface receptors, bacteria are taken up into cells. Shortly after entry, ~60% of *L. monocytogenes* remain in phagosomes and are delivered to and degraded in lysosomes (**a**). However, a population of bacteria can lyse the phagosome membrane by secreting LLO and escape into the cytosol. In phagosomes damaged by LLO, bacteria can be recognized by autophagy (**b**). Phospholipases may allow bacteria to escape from autophagosomes. In the cytosol, bacteria can evade autophagy through the expression of ActA, which allows bacteria to undergo actin-based motility and spread from cell to cell (**c**). A small portion of bacteria remain in modified phagosomes (SLAPs) due to inefficient LLO activity and grow slowly in this compartment (**d**)

c-Met on the surfaces of hepatocytes, endothelial cells and other non-epithelial cells, leading to the activation of downstream signaling pathways of c-Met and the regulation of actin polymerization (Hamon et al. 2006; Ireton 2007). InlA is required for bacteria to cross the intestinal epithelium, whereas InlB mediates invasion of the spleen and liver (Ireton 2007). Internalin F was recently shown to promote the binding of *L. monocytogenes* to host cells and bacterial invasion, though its receptor is not yet known (Kirchner and Higgins 2008).

The *hly* gene encodes listeriolysin O (LLO), a pore-forming cytolysin (Kayal and Charbit 2006; Schnupf and Portnoy 2007). Once bacteria are internalized into phagosomes, LLO is secreted and inserts into host cell membrane through binding to cholesterol (Kayal and Charbit 2006). Polymerization of LLO monomers results in pore formation in the membrane, which is critical to the escape of the bacterium into cytosol after invasion and after cell-to-cell spread (Kayal and Charbit 2006; Seveau et al. 2007). An *hly* deletion strain of *L. monocytogenes* is not able to enter the cytosol and is eventually degraded in the phagolysosome (Alvarez-Dominguez et al. 1997). It has been shown that LLO is also involved in cellular signaling pathways, such as the activation of NF-κB and the production of proinflammatory cytokines (Hamon et al. 2006). Recently LLO was found to regulate the autophagic recognition of intracellular *L. monocytogenes* and the persistent growth of this bacterium within vacuoles (Birmingham et al. 2008a; Birmingham et al. 2008b). Coordinating with the pore-forming function of LLO, two phospholipases C (PLCs), phosphatidylinositol (PI)-PLC and phosphatidylcholine (PC)-PLC, also participate in phagosomal membrane lysis. These enzymes hydrolyze a broad range of phospholipids, the products of which are involved in certain cellular signaling pathways, such as calcium-dependent signaling (Dussurget et al. 2004). However, the exact mechanism by which PI-PLC and PC-PLC promote phagosomal escape is not clear.

Once *L. monocytogenes* escapes into the cytosol, it undergoes actin-based motility. This is mediated by the bacterial virulence protein ActA (Gouin et al. 2005). The function of ActA is to recruit the actin-nucleating complex Arp2/3 and stimulate actin polymerization on the surface of the bacterium (Gouin et al. 2005). ActA is unevenly distributed on the bacterial surface, with a high concentration at one pole that causes the formation of a long actin tail. The *actA* deletion strain is immotile and deficient in cell-to-cell spread (Cossart and Sansonetti 2004; Gouin et al. 2005).

3.4 Autophagy of *L. monocytogenes*

The first evidence that autophagy can target intracellular *L. monocytogenes* was presented by Rich et al. (2003). They found that 95% of *actA* mutant bacteria enter the cytosol after 3 h of infection in the macrophage cell line J774 (Rich et al. 2003). These bacteria are not able to move to neighboring cells and thus remain in the cytosol. However, when bacterial protein synthesis is inhibited by treatment with the bacteriostatic antibiotic chloramphenicol, the cytosolic bacteria become wrapped in double-membrane autophagosomes (Rich et al. 2003). This event can be inhibited by autophagy inhibitors, and the bacteria-containing autophagosomes finally acquire the lysosomal protein LAMP-1, suggesting that the bacteria are targeted by the autophagy pathway for degradation in the autophagolysosomes (Rich et al. 2003).

Recently, we and others have found that under normal conditions, the wild-type *L. monocytogenes* is also a target of autophagy (Birmingham et al. 2007; Py et al. 2007; Birmingham et al. 2008b) (Fig. 2). In the macrophage cell line RAW 264.7, recruitment of LC3 to *L. monocytogenes* is observed early after infection, peaking at 1 h p.i. with a population of 37% of intracellular bacteria colocalizing with this autophagy maker (Birmingham et al. 2007). This level of LC3 recruitment does not occur during infection by an *hly* deletion strain, suggesting the requirement of LLO for the induction of autophagy (Birmingham et al. 2007). Although the level of membrane-conjugated LC3 increases throughout the 8 h infection time, indicating increased autophagic activity, the colocalization between bacteria and LC3 decreased after 1 h and dropped to a basal level of approximately 13% by 4 h p.i. (Birmingham et al. 2007). Intracellular bacteria replicate rapidly after 2 h of infection (Birmingham et al. 2007). These observations suggest that the bacteria can evade autophagy when they are present in the cytosol. Interestingly, most bacteria that acquire LC3 are not associated with actin (Birmingham et al. 2007). Further experiments showed that ActA expression during the first 3 h of infection, which allowed the initiation of actin-based motility, was sufficient for autophagy evasion (Birmingham et al. 2007). Furthermore, we found that mutant bacteria lacking PI-PLC or PC-PLC expression sustained a high percentage (20–35%) of colocalization with LC3 up to 8 h p.i. (Birmingham et al. 2007). These results suggest that actin polymerization as well as the bacterial phospholipases mediate bacterial autophagy evasion (Fig. 2). Virulence factors such as PC- and PI-PLC probably disrupt the inner membrane of the autophagosomes because the wild-type bacteria were observed only in single-membrane vacuoles, whereas the PrfA mutant that expresses constitutively induced LLO (and thus autophagy can be induced) was often viewed in double-membrane vacuoles (Birmingham et al. 2007).

The signal(s) that initiate autophagy of *L. monocytogenes* remain unclear. The requirement of the pore-forming toxin LLO for the induction of autophagy implies that the damaged vacuole is a possible signal early during infection, similar to the autophagy of *S. typhimurium*. *L. monocytogenes* is probably targeted within a perforated vacuole before complete escape into the cytosol. A recent study of *Listeria* infection in *Drosophila melanogaster* demonstrated that a peptidoglycan-recognition protein (PGRP) member, PGRP-LE, which acts as an intracellular pattern-recognition receptor, is crucial for autophagy targeting of the bacteria (Yano et al. 2008). The recognition of the specific peptidoglycan in the *L. monocytogenes* cell wall by PGRP-LE results in the induction of bacterial autophagy and the suppression of bacterial growth in *Drosophila* (Yano et al. 2008). During infection of *Drosophila* S2 cells, which are macrophage-like cells and do not express detectable levels of PGRP-LE, intracellular bacteria were not recognized by autophagy. When PGRP-LE was transfected into and expressed into S2 cells, cytosolic bacteria were targeted by autophagosomes (Yano et al. 2008). This study suggests that intracellular innate immune recognition of *L. monocytogenes* is a signal that triggers autophagy of the bacteria in flies. It remains unknown whether the same mechanism exists in mammals.

The ability of *L. monocytogenes* to evade autophagy in macrophages at late infection times reflects its adapted strategy to colonize the cytosol, which is critical for dissemination and disease. The exact mechanisms by which ActA, PI-PLC, and PC-PLC mediate bacterial evasion of autophagy are unknown. For early targeting of the bacteria, it is possible that phospholipases disrupt the autophagosome membrane or manipulate signaling pathways that induce autophagy. For evasion in the cytosol, it is possible that actin-based motility allows the bacteria to inhibit the interaction between bacteria and autophagic components, literally allowing them to "outrun" autophagy. An alternative possibility is that ActA acts as a "molecular shield" for the same purpose, in a manner similar to *S. flexneri* (discussed below).

3.5 Spacious Listeria-Containing Phagosomes

It has long been believed that intracellular *L. monocytogenes* either undergo phagocytosis and are killed by the phagosome–lysosomal pathway or escape into cytosol and rapidly replicate in this compartment (Pamer 2004; Gouin et al. 2005; Seveau et al. 2007). However, during persistent *L. monocytogenes* infection of severe combined immunodeficiency (SCID) mice, the majority of bacteria are localized within large vacuolar compartments in macrophages in the liver (Bhardwaj et al. 1998; Birmingham et al. 2008a) (Fig. 3). Recently we found that *L. monocytogenes* can occupy vacuolar compartments in vitro (Birmingham et al. 2008a). In macrophages at 4 h p.i., a small population of *L. monocytogenes* (~13%) is localized in spacious vacuoles that are morphologically similar to those formed in the SCID mouse model of persistent infection (Birmingham et al. 2008a). These compartments were termed spacious *Listeria*-containing phagosomes (SLAPs) (Birmingham et al. 2008a) (Fig. 3). SLAPs are labeled with LAMP-1 and the vacuolar type proton ATPase (v-ATPase; "proton pump") but not with the lysosomal hydrolase cathepsin D (Birmingham et al. 2008a). They do not accumulate the acidotropic dye Lysotracker and have a neutral pH, which suggests impaired fusion with lysosomes (Birmingham et al. 2008a).

It is believed that SLAPs form from primary phagosomes diverted from the lysosomal degradation pathway. SLAP formation requires the bacterial toxin LLO as well as the host autophagy machinery (Birmingham et al. 2008a). Efficient activity of LLO leads to the disruption of phagosomes and the escape of the bacteria into cytosol (Birmingham et al. 2008a). Innate immune defenses, such as cathepsin D, have been reported to impair the LLO activity and limit bacterial escape from phagosomes (del Cerro-Vadillo et al. 2006). Inefficient LLO activity that is not sufficient for bacterial escape appears to promote SLAP formation. LLO acts to disrupt the proton gradient across the SLAP membranes, thereby blocking fusion with lysosomes (Henry et al. 2006; Birmingham et al. 2008a). An *hly* deletion strain expressing an isopropyl β-D-1-thiogalactopyranoside-inducible LLO (iLLO), which only has one-third of the hemolytic activity of the wild-type strain, cannot escape into

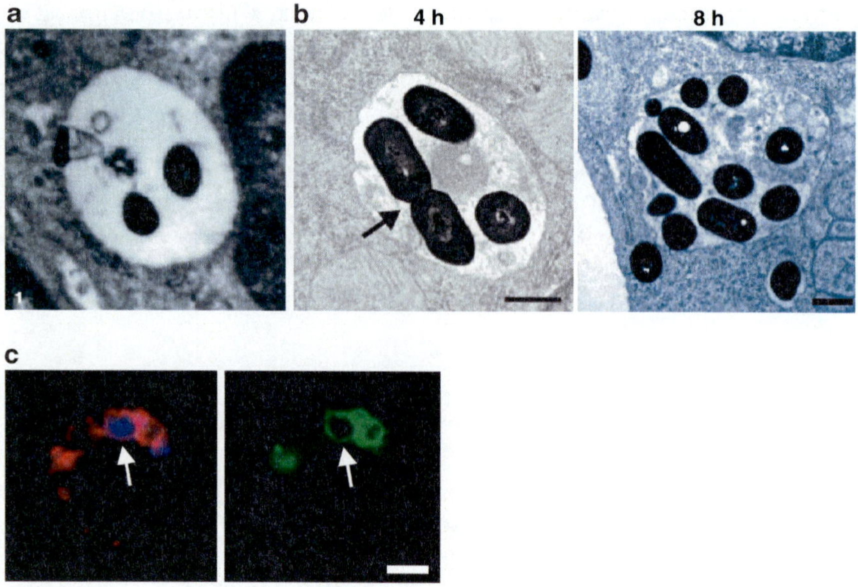

Fig. 3a–c *L. monocytogenes* colonizes SLAPs both in vivo and in vitro. **a** Liver granulomas from infected SCID mice were analyzed by transmission electron microscopy (TEM). An example of a SLAP containing multiple bacteria is shown. **b** Murine macrophages (RAW 264.7) were infected for 4 and 8 h and then subjected to TEM analysis. Examples of SLAPs are shown. Bacteria replicate slowly in these compartments. Scale bar: 0.5 μm. **c** GFP-LC3-transfected macrophages were infected with *L. monocytogenes* for 4 h and immunostained with antibodies against LAMP-1 (*red*) and the bacteria (*blue*). Colocalization of GFP-LC3 with LAMP-1-labeled SLAPs was detected by fluorescence microscopy. Scale bar: 5 μm. Images are reproduced from Birmingham et al. (2008a)

cytosol and undergoes slow growth within SLAPs (Birmingham et al. 2008a). Host autophagy is also required for SLAP formation. The autophagosome marker LC3 localizes to SLAPs and autophagy components are required for SLAP formation during iLLO infection (Birmingham et al. 2008a) (Fig. 3). SLAPs are different from bacteria-containing autophagosomes observed at 1 h p.i. The latter are tight-fitting autophagosomes, whereas one SLAP contains multiple bacteria in a spacious vacuole (Birmingham et al. 2007; Birmingham et al. 2008a). Bacteria in SLAPs undergo active, albeit slow, replication compared to those present in the cytosol (Birmingham et al. 2008a). In *Atg5*-knockout MEFs, most *hly* mutant bacteria expressing iLLO replicate rapidly in the cytosol, suggesting the requirement of autophagy in maintaining iLLO bacteria in SLAPs (Birmingham et al. 2008a). Therefore, SLAPs reflect a stalemate state between pathogen virulence and host innate immune factors. The mechanisms of SLAP formation may also apply in vivo, since the SLAPs observed in SCID mice during persistent infection are almost indistinguishable from those formed in cultured macrophages (Birmingham et al. 2008a).

3.6 Multiple Fates of Different L. monocytogenes Populations

Based on current studies, several populations of *L. monocytogenes* have been identified in infected cells (Fig. 2). De Chastellier and Berche (de Chastellier and Berche 1994) have carefully characterized the different populations in murine macrophages. They found that at 1 h p.i., ~65% of all *L. monocytogenes* are in acidic phagolysosomes and are degraded. Approximately 20% of the bacteria escape into cytosol and replicate, whereas ~15% reside in immature phagosomes (de Chastellier and Berche 1994). We have since found that, at this time point, a population of bacteria are targeted by autophagy and exist in single-membrane compartments (Birmingham et al. 2007). Hence, these bacteria are not distinguishable from those within phagosomes and were not detected by electron microscopy. The fate of bacteria targeted by autophagy is not clear. We found that bacterial growth is not affected in *Atg5*-knockout MEFs after 3 h p.i., although a delayed replication of *L. monocytogenes* has been shown in *Atg5*-knockout MEFs compared with wild-type cells (Birmingham et al. 2007; Py et al. 2007). In *Drosophila*, autophagy targets the cytosolic bacteria and restricts their growth. Knockdown of Atg5 expression in hemocytes from flies resulted in enhanced bacterial survival (Yano et al. 2008).

In the cytosol, bacteria can initiate actin-based motility and begin to replicate quickly. At 4 h p.i., the cytosolic population constitutes ~50% of all intracellular bacteria and the population with damaged morphology within mature phagolysosomes decreases to ~30% (de Chastellier and Berche 1994). The remaining 20% of the bacteria remain intact in phagosomes. Indeed, this population of bacteria may include those in SLAPs, as ~13% bacteria colocalized with LC3 at this time and had normal morphology (Birmingham et al. 2008a).

4 Autophagy of *Shigella flexneri* and Other Intracellular Bacteria

Besides *S. typhimurium*, *L. monocytogenes*, *M. tuberculosis* (introduced in this chapter and discussed in detail in the chapter by Deretic et al. in this volume) and Group A *Streptococcus* (introduced in the chapter by Yoshimori and Amano in this volume), other bacteria have also been reported to be targets of autophagy, including *Rickettsia conorii* and *S. flexneri*. Little is known about the autophagy of *Rickettsia*. Only one report has shown that in vitro incubation of these bacteria with guinea pig polymorphonuclear leukocytes induces the formation of autophagosomes containing the bacteria (Rikihisa 1984). These autophagosomes have lysosomal features, suggesting they are possibly involved in bacterial clearance.

S. flexneri is a Gram-negative bacterial pathogen that causes shigellosis in humans. It can penetrate the intestinal epithelium and infect macrophages, resulting in the production of proinflammatory cytokines and the recruitment of neutrophils to the infection site, which causes inflammation (Ogawa and Sasakawa 2006).

Similar to *S. typhimurium* invasion, *S. flexneri* also utilizes a T3SS to secrete effectors into the host cell to trigger plasma membrane ruffling and promote bacterial internalization (Cossart and Sansonetti 2004; Ogawa and Sasakawa 2006) (Fig. 4). *S. flexneri* is able to disrupt the resulting vacuole and escape into the cytosol, where it recruits actin via the outer membrane protein VirG (also called IcsA) (Gouin et al. 2005; Ogawa and Sasakawa 2006). VirG is concentrated at one pole of the bacterium and activates the nucleation of actin filaments in a similar manner to ActA during *L. monocytogenes* infection (Gouin et al. 2005; Ogawa and Sasakawa 2006). In the cytosol, *S. flexneri* is targeted by the autophagy machinery in the absence of IcsB, a T3SS effector important for bacterial proliferation in the cytosol (Ogawa et al. 2005) (Fig. 4). At 2 h p.i., ~30% of *icsB* mutant bacteria colocalize

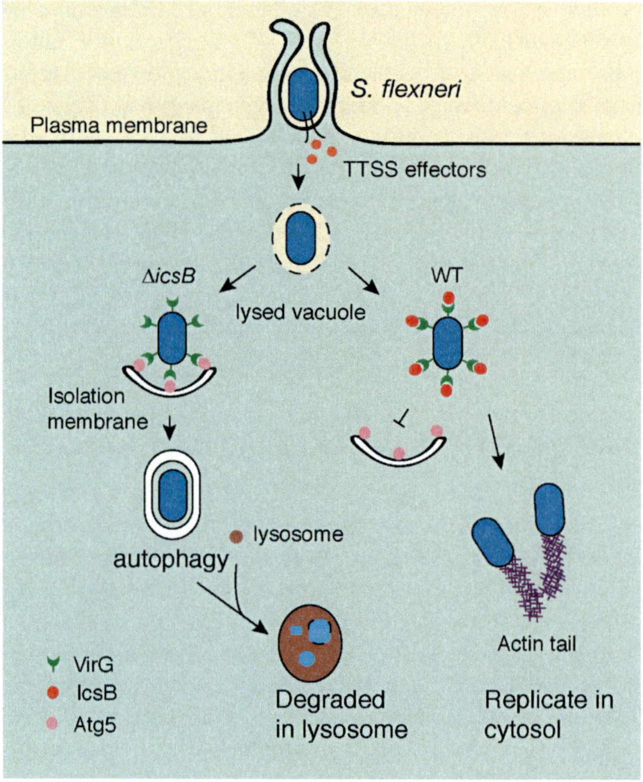

Fig. 4 Different intracellular fates of *S. flexneri* in host cells. *S. flexneri* invade nonphagocytic cells by T3SS-induced membrane ruffling. Most of the bacteria lyse the vacuoles to enter the cytosol. Wild-type (WT) bacteria secrete the effector protein IcsB, which binds the bacterial surface protein VirG and blocks its interaction with the autophagy protein Atg5. This allows bacteria to evade autophagy and replicate in the cytosol. δ*icsB* bacteria are targeted by autophagy through the interaction of VirG and Atg5 and are eventually degraded in lysosomes

with LC3, and this population increases to ~50% by 6 h p.i. (Ogawa et al. 2005). In contrast, only ~10% of wild-type bacteria label with LC3 throughout the 6 h infection period. During infection with *icsB* mutant, VirG on the bacterial surface physically interacts with the host protein Atg5, an essential autophagy component, and thus likely initiates autophagosome formation around the bacterium (Ogawa et al. 2005). However, this interaction is blocked by competitive binding of IcsB to VirG, which enables the wild-type bacterium to evade autophagy (Ogawa et al. 2005). In *Atg5*-knockout MEFs, *icsB* mutant *S. flexneri* replicates the same as the wild type, suggesting that the restricted growth of *icsB* mutant in normal cells is the result of its inability to evade autophagy recognition (Ogawa et al. 2005).

In primary macrophages, *S. flexneri* infection induces pyroptosis and production of active IL-1β (Suzuki et al. 2007). The bacterial T3SS system as well as the cytosolic bacterial sensor, nucleotide-binding oligomerization domain (NOD)-like receptor (NLR) Ipaf, are responsible for the activation of caspase-1 and pyroptosis (Suzuki et al. 2007). Interestingly, in the absence of Ipaf and caspase-1, autophagy of bacteria is increased during *S. flexneri* infection, implying an inhibition of autophagy by caspase 1 (Suzuki et al. 2007). At 30 min p.i., up to 80% of the intracellular bacteria associate with LC3 in caspase-1- and Ipaf-deficient macrophages, whereas only 20% of bacteria are targeted by autophagy in wild-type cells (Suzuki et al. 2007). T3SS-deficient *S. flexneri* do not induce autophagy in either cell type (Suzuki et al. 2007). Interestingly, *virG* mutant bacteria show the same ability to induce autophagy as wild-type bacteria, suggesting that VirG is not involved in *S. flexneri*-induced autophagy in primary macrophages, although it is necessary in epithelial cells (Ogawa et al. 2005; Suzuki et al. 2007). Autophagy during *S. flexneri* infection presents a cellular protective mechanism, since inhibition of autophagy results in more cell death (Suzuki et al. 2007).

5 Bacterial Autophagy in Inflammatory Bowel Disease

Inflammatory bowel disease is a chronic disorder caused by abnormal inflammation in the intestinal tract. Ulcerative colitis and Crohn's disease (CD) are the two major forms of inflammatory bowel disease (Podolsky 2002; Mizoguchi and Mizoguchi 2008). Ulcerative colitis is most often restricted to the colon, while CD involves patchy inflammation throughout the gastrointestinal tract (Xavier and Podolsky 2007). The pathogenesis of these disorders is now believed to involve multiple factors, including environmental conditions, genetic composition of the individual, and bacterial infection (Podolsky 2002; Xavier and Podolsky 2007). Readers are directed to several excellent reviews on the pathogenesis of inflammatory bowel disease by Podolsky (2002), Xavier and Podolsky (2007) and Mizoguchi and Mizoguchi (2008). Here we will focus on the current understanding of autophagy in CD, especially bacteria-related autophagy.

In CD patients, the most commonly affected areas are the ileum and colon. The infiltration of neutrophils into the epithelium and the formation of granulomas are hallmarks of affected regions (Xavier and Podolsky 2007). These responses are believed to be caused by abnormal innate and adaptive immune responses, such as impaired epithelial barriers, a failed response to mucus penetration, defective bacterial sensing, and augmented secretion of inflammatory cytokines (Cobrin and Abreu 2005; Vignal et al. 2007; Xavier and Podolsky 2007; Mizoguchi and Mizoguchi 2008). Bacteria are believed to be a major risk factor for CD since the colon is rich in commensal bacteria and abnormal control of these "good bacteria" may cause excessive immune responses, leading to inflammation (Podolsky 2002; Tlaskalova-Hogenova et al. 2004; Abubakar et al. 2008; Sartor 2008). In addition, some pathogenic bacteria, including invasive *Escherichia coli* and *Mycobacterium avium paratuberculosis*, have been linked to CD (Baumgart et al. 2007).

The genetic makeup of the individual is thought to predispose to CD, and more than 30 genes have now been associated with this disease (Hugot et al. 2001; Duerr et al. 2006; Consortium 2007; Hampe et al. 2007; Parkes et al. 2007; Prescott et al. 2007; Rioux et al. 2007). The first-identified and best-studied CD-associated gene is *NOD2/CARD5*, which encodes an intracellular receptor for pathogen recognition (Hugot et al. 2001). Stimulation of NOD2 can initiate various immune responses, such as the secretion of cytokines, chemokines and antimicrobial peptides, and the killing of bacteria (Le Bourhis et al. 2007). CD-associated *NOD2* mutations exhibit reduced recognition of pathogens (through sensing peptidoglycan moieties) and increased susceptibility to bacterial infection (Maeda et al. 2005; Vignal et al. 2007). Another strong CD-associated gene is *IL23R*. This gene encodes a subunit of the receptor for interleukin 23, which is a proinflammatory cytokine that activates organ-specific inflammation (Duerr et al. 2006).

Recent genome-wide association studies have revealed new CD-associated single-nucleotide polymorphisms (SNPs) in the genes *ATG16L1* and *IRGM*, suggesting a role of autophagy in the disease (Hampe et al. 2007; Massey and Parkes 2007; Parkes et al. 2007; Glas et al. 2008). ATG16L1 is the mammalian homolog of yeast *ATG16* and is an essential component of autophagy (Kuma et al. 2002; Mizushima et al. 2003). It forms a complex with ATG5 and ATG12 and is involved in targeting LC3 to the membrane for lipidation (Mizushima et al. 2003; Fujita et al. 2008). ATG16L1 contains an N-terminal coiled-coil region, which is involved in binding to ATG5 and self-oligomerization, and C-terminal WD repeats of unknown function (Mizushima et al. 2003). The CD-associated SNP (T300A) is located close to the WD-repeat region. Knockdown of the endogenous expression of ATG16L1 in epithelial cells results in deficient autophagy of *S. typhimurium* (Rioux et al. 2007). When cells only express the ATG16L1 (T300A) variant in contrast to the T300T variant, autophagy of bacteria is deficient, but basal autophagy and the ability of the ATG16L1 (T300A) variant to form dimers or interact with ATG5 are not affected, suggesting the association of bacterial handling with CD (Kuballa et al. 2008).

The function of ATG16L1 in vivo was recently characterized by two different groups simultaneously (Cadwell et al. 2008; Saitoh et al. 2008). Mice that are hypomorphic for ATG16L1 expression ($ATG16L1^{HM}$) express less than half of the protein level of ATG16L1 than wild-type mice, and have impaired autophagy activity (Cadwell et al. 2008). Abnormal Paneth cells, which are specialized intestinal epithelial cells secreting antimicrobial granule contents into mucus to control the intestinal environment, were observed in the ilea of these mice (Cadwell et al. 2008). Aberrant granule distribution, disrupted granule secretion and elevated transcription of genes involved in injury responses in Paneth cells are major phenotypes resulting from deficient ATG16L1 expression (Cadwell et al. 2008). Deletion of another autophagy gene, Atg5, in mouse intestinal epithelium also caused the same abnormalities in Paneth cells, suggesting that these effects are autophagy specific (Cadwell et al. 2008). Interestingly, similar phenotypes were also observed in tissue sections from human CD patients carrying the ATG16L1 risk allele (Cadwell et al. 2008).

In a separate study, *Atg16L1* mutant mice that lack the coiled-coil domain in *Atg16L1* were generated by Akira and colleagues (Saitoh et al. 2008). *Atg16L1*-deficient macrophages were found to display elevated production of the inflammatory cytokines IL-1β and IL-18 when stimulated by lipopolysaccharide, a major endotoxin found in Gram-negative bacteria (Saitoh et al. 2008). In a dextran sulfate sodium (DSS)-induced experimental model of colitis, mice lacking functional Atg16L1 in hematopoietic cells had more severe inflammation in their colons than wild-type mice (Saitoh et al. 2008). Higher production of IL-1β and IL-18 was also observed in DSS-treated *Atg16L1* mutant mice (Saitoh et al. 2008). These studies revealed novel roles of autophagy in the maintenance of normal Paneth cell function as well as the control of inflammation, providing potential mechanisms for how *ATG16L1* is involved in the pathogenesis of CD.

IRGM, the human homolog of mouse Irgm1, is a member of the murine immunity-related GTPase (IRG) family (Taylor 2007). IRGs are upregulated by IFN-γ and are involved in the control of intracellular pathogens (Taylor 2007). Similar to mouse Irgm1, human IRGM was found to be involved in bacterial autophagy in macrophages (Gutierrez et al. 2004; Singh et al. 2006). However, the expression of IRGM is constitutive in human cell lines and not responsive to IFN-γ (Singh et al. 2006). Overexpression of IRGM upregulates autophagy and enhances the clearance of intracellular *M. tuberculosis* in vitro (Singh et al. 2006). *IRGM* has been identified as a susceptibility gene for CD by genome-wide association studies (Consortium 2007; Parkes et al. 2007; Raelson et al. 2007; Roberts et al. 2008). The correlation of IRGM with CD seems to be disease-type and population specific. It only associates with ileal CD but not other types of CD, such as ileocolonic or colonic CD in New Zealand Caucasians (Roberts et al. 2008). Moreover, the association of *IRGM* with CD in Canadian children was found to be not significant (Amre et al. 2009). The role of IRGM in CD is not yet known, but it might involve its role in the autophagic clearance of bacteria. The two CD-associated SNPs immediately flanking *IRGM* are probably located at the gene-regulatory region and may affect protein expression (Parkes et al. 2007). A 20-kb deletion polymorphism immediately upstream of *IRGM*, which has a high linkage disequilibrium with the CD-associated SNP, showed different expression levels in different tissue cell lines compared to the non-disease control allele (McCarroll et al. 2008). Knockdown endogenous expression of IRGM results in impaired autophagy of intracellular *S. typhimurium*, while elevating its expression causes increased targeting of *S. typhimurium* by autophagy (McCarroll et al. 2008).

Recent studies have revealed that autophagy can be regulated by extra- and intracellular bacterial recognition. Toll-like receptors (TLRs), which recognize a wide range of extracellular bacterial or viral components, were found to be able to induce autophagy upon ligand binding (Xu et al. 2007; Delgado et al. 2008) (see the detailed discussions in the chapters by Tal and Iwasaki and Deretic et al. in this volume). Also, TLR-induced autophagy promotes phagosome maturation (Sanjuan et al. 2007). These results link autophagy with classical TLR-mediated antimicrobial immune mechanisms. Intracellular pathogen-recognition receptors, such as NLRs, are also linked to autophagy. For example, *S. flexneri* can induce Ipaf (a member of NLR)-dependent pyroptosis in macrophages, which was found to inhibit autophagy (Suzuki et al. 2007). As mentioned above, in the absence of Ipaf, autophagy is induced by *S. flexneri* infection, and inhibition of autophagy resulted in increased cell death (Suzuki et al. 2007). The Ipaf/caspase-1 pathway is involved in the processing of IL-1β and IL-18, which are important inflammatory cytokines that are believed to play roles in CD (Sansonetti et al. 2000; Maeda et al. 2005). Deficient autophagy causes more cell death during *S. flexneri* infection, and this may lead to an excessive production of the cytokines. The NLR Naip5 also regulates caspase-1 activation and autophagy in *Legionella pneumophila*-infected cells (Swanson and Molofsky 2005). In *Drosophila*, the intracellular receptor PGRP-LE is critical to the induction of autophagy to target cytosolic *L. monocytogenes* (Yano et al. 2008). Together, these results reveal an important link between bacterial sensing and autophagy, and suggest that defects in both can contribute to the development of inflammatory bowel diseases.

Besides innate immunity, autophagy is also involved in adaptive immune responses. For example, it is known that autophagy plays a role in presenting cytosolic antigens on major histocompatibility complex (MHC) class II molecules (Dengjel et al. 2005; Paludan et al. 2005; Munz 2006; Schmid et al. 2006) (see the chapter by Gannagé and Münz in this volume). Components of autophagy are also critical for T and B lymphocyte proliferation and death (Pua et al. 2007; Miller et al. 2008) (see the chapter by Pua and He in this volume). The involvement of autophagy in both innate and adaptive immune systems makes it an important player in the regulation of the pathogenesis of inflammatory bowel disease.

6 Summary

Autophagy can target different intracellular bacteria in host cells in various ways. *S. typhimurium* is recognized by autophagy in damaged SCVs. Autophagy targets *L. monocytogenes* in damaged phagosomes, but is evaded by this bacterium when present in the cytosol. *S. flexneri* can also escape from autophagy in the cytosol. The different levels of interaction between bacteria and autophagy give rise to multiple populations of intracellular bacteria in host cells. When studying bacterial infection in vitro, it is important to keep the heterogeneity of these different populations in mind, as manipulations of either host cell or pathogen can lead to alterations in each population. As a component of human immunity, autophagy may play a role in the pathogenesis of certain inflammatory bowel diseases. Further study of the mechanisms by which autophagy targets intracellular bacteria and the virulence factors that pathogens use to evade autophagy are warranted.

Acknowledgments John H. Brumell, Ph.D., holds an Investigators in Pathogenesis of Infectious Disease Award from the Burroughs Wellcome Fund. Ju Huang was supported by a Restracomp Fellowship from the Hospital for Sick Children and is the recipient of a CAG/CIHR/CCFC post-doctoral fellowship administered by the Canadian Association of Gastroenterology. We thank Dr. Cheryl Birmingham for discussion and critical reading of the manuscript.

References

Abubakar I, Myhill D, Aliyu SH, Hunter PR (2008) Detection of *Mycobacterium avium subspecies paratuberculosis* from patients with Crohn's disease using nucleic acid-based techniques: a systematic review and meta-analysis. Inflamm Bowel Dis 14:401–410

Alpuche-Aranda CM, Racoosin EL, Swanson JA, Miller SI (1994) Salmonella stimulate macrophage macropinocytosis and persist within spacious phagosomes. J Exp Med 179:601–608

Alvarez-Dominguez C, Roberts R, Stahl PD (1997) Internalized *Listeria monocytogenes* modulates intracellular trafficking and delays maturation of the phagosome. J Cell Sci 110 (Pt 6):731–743

Amre DK, Mack DR, Morgan K, Krupoves A, Costea I, Lambrette P, Grimard G, Dong J, Feguery H, Bucionis V, Deslandres C, Levy E, Seidman EG (2009) Autophagy gene *ATG16L1* but not *IRGM* is associated with Crohn's disease in Canadian children. Inflamm Bowel Dis 15:501–507

Baumgart M, Dogan B, Rishniw M, Weitzman G, Bosworth B, Yantiss R, Orsi RH, Wiedmann M, McDonough P, Kim SG, Berg D, Schukken Y, Scherl E, Simpson KW (2007) Culture independent analysis of ileal mucosa reveals a selective increase in invasive *Escherichia coli* of novel phylogeny relative to depletion of *Clostridiales* in Crohn's disease involving the ileum. Isme J 1:403–418

Beuzon CR, Meresse S, Unsworth KE, Ruiz-Albert J, Garvis S, Waterman SR, Ryder TA, Boucrot E, Holden DW (2000) *Salmonella* maintains the integrity of its intracellular vacuole through the action of SifA. Embo J 19:3235–3249

Bhardwaj V, Kanagawa O, Swanson PE, Unanue ER (1998) Chronic *Listeria* infection in SCID mice: requirements for the carrier state and the dual role of T cells in transferring protection or suppression. J Immunol 160:376–384

Birmingham CL, Brumell JH (2006) Autophagy recognizes intracellular *Salmonella enterica* serovar Typhimurium in damaged vacuoles. Autophagy 2:156–158

Birmingham CL, Jiang X, Ohlson MB, Miller SI, Brumell JH (2005) *Salmonella*-induced filament formation is a dynamic phenotype induced by rapidly replicating *Salmonella enterica* serovar typhimurium in epithelial cells. Infect Immunol 73:1204–1208

Birmingham CL, Smith AC, Bakowski MA, Yoshimori T, Brumell JH (2006) Autophagy controls *Salmonella* infection in response to damage to the *Salmonella*-containing vacuole. J Biol Chem 281:11374–11383

Birmingham CL, Canadien V, Gouin E, Troy EB, Yoshimori T, Cossart P, Higgins DE, Brumell JH (2007) *Listeria monocytogenes* evades killing by autophagy during colonization of host cells. Autophagy 3:442–451

Birmingham CL, Canadien V, Kaniuk NA, Steinberg BE, Higgins DE, Brumell JH (2008a) Listeriolysin O allows *Listeria monocytogenes* replication in macrophage vacuoles. Nature 451:350–354

Birmingham CL, Higgins DE, Brumell JH (2008b) Avoiding death by autophagy: interactions of *Listeria monocytogenes* with the macrophage autophagy system. Autophagy 4:368–371

Brawn LC, Hayward RD, Koronakis V (2007) *Salmonella* SPI1 effector SipA persists after entry and cooperates with a SPI2 effector to regulate phagosome maturation and intracellular replication. Cell Host Microbe 1:63–75

Brumell JH, Grinstein S (2004) *Salmonella* redirects phagosomal maturation. Curr Opin Microbiol 7:78–84

Brumell JH, Steele-Mortimer O, Finlay BB (1999) Bacterial invasion: force feeding by *Salmonella*. Curr Biol 9:R277–280

Brumell JH, Rosenberger CM, Gotto GT, Marcus SL, Finlay BB (2001) SifA permits survival and replication of *Salmonella typhimurium* in murine macrophages. Cell Microbiol 3:75–84

Brumell JH, Goosney DL, Finlay BB (2002a) SifA, a type III secreted effector of *Salmonella typhimurium*, directs *Salmonella*-induced filament (Sif) formation along microtubules. Traffic 3:407–415

Brumell JH, Tang P, Zaharik ML, Finlay BB (2002b) Disruption of the *Salmonella*-containing vacuole leads to increased replication of *Salmonella enterica* serovar typhimurium in the cytosol of epithelial cells. Infect Immun 70:3264–3270

Cadwell K, Liu JY, Brown SL, Miyoshi H, Loh J, Lennerz JK, Kishi C, Kc W, Carrero JA, Hunt S, Stone CD, Brunt EM, Xavier RJ, Sleckman BP, Li E, Mizushima N, Stappenbeck TS, Virgin HW (2008) A key role for autophagy and the autophagy gene *Atg16l1* in mouse and human intestinal Paneth cells. Nature 456:259–263

Chen LM, Hobbie S, Galan JE (1996) Requirement of CDC42 for *Salmonella*-induced cytoskeletal and nuclear responses. Science 274:2115–2118

Cirillo DM, Heffernan EJ, Wu L, Harwood J, Fierer J, Guiney DG (1996) Identification of a domain in Rck, a product of the *Salmonella typhimurium* virulence plasmid, required for both serum resistance and cell invasion. Infect Immun 64:2019–2023

Cobrin GM, Abreu MT (2005) Defects in mucosal immunity leading to Crohn's disease. Immunol Rev 206:277–295

Consortium WTCC (2007) Genome-wide association study of 14,000 cases of seven common diseases and 3,000 shared controls. Nature 447:661–678

Cossart P, Sansonetti PJ (2004) Bacterial invasion: the paradigms of enteroinvasive pathogens. Science 304:242–248

de Chastellier C, Berche P (1994) Fate of *Listeria monocytogenes* in murine macrophages: evidence for simultaneous killing and survival of intracellular bacteria. Infect Immun 62:543–553

Deiwick J, Nikolaus T, Shea JE, Gleeson C, Holden DW, Hensel M (1998) Mutations in *Salmonella* pathogenicity island 2 (*SPI2*) genes affecting transcription of *SPI1* genes and resistance to antimicrobial agents. J Bacteriol 180:4775–4780

del Cerro-Vadillo E, Madrazo-Toca F, Carrasco-Marin E, Fernandez-Prieto L, Beck C, Leyva-Cobian F, Saftig P, Alvarez-Dominguez C (2006) Cutting edge: a novel nonoxidative phagosomal mechanism exerted by cathepsin-D controls *Listeria monocytogenes* intracellular growth. J Immunol 176:1321–1325

Delgado MA, Elmaoued RA, Davis AS, Kyei G, Deretic V (2008) Toll-like receptors control autophagy. EMBO J 27:1110–1121

Dengjel J, Schoor O, Fischer R, Reich M, Kraus M, Muller M, Kreymborg K, Altenberend F, Brandenburg J, Kalbacher H, Brock R, Driessen C, Rammensee HG, Stevanovic S (2005) Autophagy promotes MHC class II presentation of peptides from intracellular source proteins. Proc Natl Acad Sci USA 102:7922–7927

Drecktrah D, Knodler LA, Galbraith K, Steele-Mortimer O (2005) The *Salmonella* SPI1 effector SopB stimulates nitric oxide production long after invasion. Cell Microbiol 7:105–113

Duerr RH, Taylor KD, Brant SR, Rioux JD, Silverberg MS, Daly MJ, Steinhart AH, Abraham C, Regueiro M, Griffiths A, Dassopoulos T, Bitton A, Yang H, Targan S, Datta LW, Kistner EO, Schumm LP, Lee AT, Gregersen PK, Barmada MM, Rotter JI, Nicolae DL, Cho JH (2006) A

genome-wide association study identifies *IL23R* as an inflammatory bowel disease gene. Science 314:1461–1463
Dussurget O, Pizarro-Cerda J, Cossart P (2004) Molecular determinants of *Listeria monocytogenes* virulence. Annu Rev Microbiol 58:587–610
Eriksson S, Lucchini S, Thompson A, Rhen M, Hinton JC (2003) Unravelling the biology of macrophage infection by gene expression profiling of intracellular *Salmonella enterica*. Mol Microbiol 47:103–118
Fang FC (2004) Antimicrobial reactive oxygen and nitrogen species: concepts and controversies. Nat Rev Microbiol 2:820–832
Fink SL, Cookson BT (2007) Pyroptosis and host cell death responses during *Salmonell*a infection. Cell Microbiol 9:2562–2570
Finlay BB, Brumell JH (2000) *Salmonella* interactions with host cells: in vitro to in vivo. Philos Trans R Soc Lond B Biol Sci 355:623–631
Fujita N, Itoh T, Omori H, Fukuda M, Noda T, Yoshimori T (2008) The Atg16L complex specifies the site of LC3 lipidation for membrane biogenesis in autophagy. Mol Biol Cell 19:2092–2100
Galan JE (2001) *Salmonella* interactions with host cells: type III secretion at work. Annu Rev Cell Dev Biol 17:53–86
Garcia-del Portillo F, Zwick MB, Leung KY, Finlay BB (1993) *Salmonella* induces the formation of filamentous structures containing lysosomal membrane glycoproteins in epithelial cells. Proc Natl Acad Sci USA 90:10544–10548
Giacomodonato MN, Uzzau S, Bacciu D, Caccuri R, Sarnacki SH, Rubino S, Cerquetti MC (2007) SipA, SopA, SopB, SopD and SopE2 effector proteins of *Salmonella enterica* serovar Typhimurium are synthesized at late stages of infection in mice. Microbiology 153:1221–1228
Glas J, Konrad A, Schmechel S, Dambacher J, Seiderer J, Schroff F, Wetzke M, Roeske D, Torok HP, Tonenchi L, Pfennig S, Haller D, Griga T, Klein W, Epplen JT, Folwaczny C, Lohse P, Goke B, Ochsenkuhn T, Mussack T, Folwaczny M, Muller-Myhsok B, Brand S (2008) The *ATG16L1* gene variants rs2241879 and rs2241880 (T300A) are strongly associated with susceptibility to Crohn's disease in the German population. Am J Gastroenterol 103:682–691
Gouin E, Welch MD, Cossart P (2005) Actin-based motility of intracellular pathogens. Curr Opin Microbiol 8:35–45
Gutierrez MG, Master SS, Singh SB, Taylor GA, Colombo MI, Deretic V (2004) Autophagy is a defense mechanism inhibiting BCG and *Mycobacterium tuberculosis* survival in infected macrophages. Cell 119:753–766
Hamon M, Bierne H, Cossart P (2006) *Listeria monocytogenes*: a multifaceted model. Nat Rev Microbiol 4:423–434
Hampe J, Franke A, Rosenstiel P, Till A, Teuber M, Huse K, Albrecht M, Mayr G, De La Vega FM, Briggs J, Gunther S, Prescott NJ, Onnie CM, Hasler R, Sipos B, Folsch UR, Lengauer T, Platzer M, Mathew CG, Krawczak M, Schreiber S (2007) A genome-wide association scan of nonsynonymous SNPs identifies a susceptibility variant for Crohn disease in *ATG16L1*. Nat Genet 39:207–211
Haraga A, Ohlson MB, Miller SI (2008) Salmonellae interplay with host cells. Nat Rev Microbiol 6:53–66
Henry R, Shaughnessy L, Loessner MJ, Alberti-Segui C, Higgins DE, Swanson JA (2006) Cytolysin-dependent delay of vacuole maturation in macrophages infected with *Listeria* monocytogenes. Cell Microbiol 8:107–119
Hensel M, Shea JE, Waterman SR, Mundy R, Nikolaus T, Banks G, Vazquez-Torres A, Gleeson C, Fang FC, Holden DW (1998) Genes encoding putative effector proteins of the type III secretion system of *Salmonella* pathogenicity island 2 are required for bacterial virulence and proliferation in macrophages. Mol Microbiol 30:163–174
Hernandez LD, Pypaert M, Flavell RA, Galan JE (2003) A *Salmonella* protein causes macrophage cell death by inducing autophagy. J Cell Biol 163:1123–1131

Hersh D, Monack DM, Smith MR, Ghori N, Falkow S, Zychlinsky A (1999) The *Salmonella* invasin SipB induces macrophage apoptosis by binding to caspase-1. Proc Natl Acad Sci USA 96:2396–2401

Huang J, Klionsky DJ (2007) Autophagy and human disease. Cell Cycle 6:1837–1849

Hueffer K, Galan JE (2004) *Salmonella*-induced macrophage death: multiple mechanisms, different outcomes. Cell Microbiol 6:1019–1025

Hugot JP, Chamaillard M, Zouali H, Lesage S, Cezard JP, Belaiche J, Almer S, Tysk C, O'Morain CA, Gassull M, Binder V, Finkel Y, Cortot A, Modigliani R, Laurent-Puig P, Gower-Rousseau C, Macry J, Colombel JF, Sahbatou M, Thomas G (2001) Association of NOD2 leucine-rich repeat variants with susceptibility to Crohn's disease. Nature 411:599–603

Ireton K (2007) Entry of the bacterial pathogen *Listeria monocytogenes* into mammalian cells. Cell Microbiol 9:1365–1375

Kayal S, Charbit A (2006) Listeriolysin O: a key protein of *Listeria monocytogenes* with multiple functions. FEMS Microbiol Rev 30:514–529

Kirchner M, Higgins DE (2008) Inhibition of ROCK activity allows InlF-mediated invasion and increased virulence of *Listeria monocytogenes*. Mol Microbiol 68:749–767

Knodler LA, Steele-Mortimer O (2003) Taking possession: biogenesis of the *Salmonella*-containing vacuole. Traffic 4:587–599

Kuballa P, Huett A, Rioux JD, Daly MJ, Xavier RJ (2008) Impaired autophagy of an intracellular pathogen induced by a Crohn's disease associated *ATG16L1* variant. PLoS ONE 3:e3391

Kuma A, Mizushima N, Ishihara N, Ohsumi Y (2002) Formation of the approximately 350-kDa Apg12-Apg5.Apg16 multimeric complex, mediated by Apg16 oligomerization, is essential for autophagy in yeast. J Biol Chem 277:18619–18625

Le Bourhis L, Benko S, Girardin SE (2007) Nod1 and Nod2 in innate immunity and human inflammatory disorders. Biochem Soc Trans 35:1479–1484

Lecuit M (2007) Human listeriosis and animal models. Microbes Infect 9:1216–1225

Maeda S, Hsu LC, Liu H, Bankston LA, Iimura M, Kagnoff MF, Eckmann L, Karin M (2005) *Nod2* mutation in Crohn's disease potentiates NF-kappaB activity and IL-1beta processing. Science 307:734–738

Massey DC, Parkes M (2007) Genome-wide association scanning highlights two autophagy genes, *ATG16L1* and *IRGM*, as being significantly associated with Crohn's disease. Autophagy 3:649–651

McCarroll SA, Huett A, Kuballa P, Chilewski SD, Landry A, Goyette P, Zody MC, Hall JL, Brant SR, Cho JH, Duerr RH, Silverberg MS, Taylor KD, Rioux JD, Altshuler D, Daly MJ, Xavier, RJ (2008) Deletion polymorphism upstream of IRGM associated with altered IRGM expression and Crohn's disease. Nat Genet 40:1107–1112

Miao EA, Scherer CA, Tsolis RM, Kingsley RA, Adams LG, Baumler AJ, Miller SI (1999) *Salmonella typhimurium* leucine-rich repeat proteins are targeted to the SPI1 and SPI2 type III secretion systems. Mol Microbiol 34:850–864

Miller BC, Zhao Z, Stephenson LM, Cadwell K, Pua HH, Lee HK, Mizushima NN, Iwasaki A, He YW, Swat W, Virgin HWT (2008) The autophagy gene *Atg5* plays an essential role in B lymphocyte development. Autophagy 4:309–314

Mizoguchi A, Mizoguchi E (2008) Inflammatory bowel disease, past, present and future: lessons from animal models. J Gastroenterol 43:1–17

Mizushima N, Kuma A, Kobayashi Y, Yamamoto A, Matsubae M, Takao T, Natsume T, Ohsumi Y, Yoshimori T (2003) Mouse Apg16L, a novel WD-repeat protein, targets to the autophagic isolation membrane with the Apg12-Apg5 conjugate. J Cell Sci 116:1679–1688

Munz C (2006) Autophagy and antigen presentation. Cell Microbiol 8:891–898

Nakagawa I, Amano A, Mizushima N, Yamamoto A, Yamaguchi H, Kamimoto T, Nara A, Funao J, Nakata M, Tsuda K, Hamada S, Yoshimori T (2004) Autophagy defends cells against invading group A *Streptococcus*. Science 306:1037–1040

Ogata M, Hino S, Saito A, Morikawa K, Kondo S, Kanemoto S, Murakami T, Taniguchi M, Tanii I, Yoshinaga K, Shiosaka S, Hammarback JA, Urano F, Imaizumi K (2006) Autophagy is activated for cell survival after endoplasmic reticulum stress. Mol Cell Biol 26:9220–9231

Ogawa M, Sasakawa C (2006) Intracellular survival of *Shigella*. Cell Microbiol 8:177–184
Ogawa M, Yoshimori T, Suzuki T, Sagara H, Mizushima N, Sasakawa C (2005) Escape of intracellular *Shigella* from autophagy. Science 307:727–731
Ohl ME, Miller SI (2001) *Salmonella*: a model for bacterial pathogenesis. Annu Rev Med 52:259–274
Paludan C, Schmid D, Landthaler M, Vockerodt M, Kube D, Tuschl T, Munz C (2005) Endogenous MHC class II processing of a viral nuclear antigen after autophagy. Science 307:593–596
Pamer EG (2004) Immune responses to *Listeria monocytogenes*. Nat Rev Immunol 4:812–823
Parkes M, Barrett JC, Prescott NJ, Tremelling M, Anderson CA, Fisher SA, Roberts RG, Nimmo ER, Cummings FR, Soars D, Drummond H, Lees CW, Khawaja SA, Bagnall R, Burke DA, Todhunter CE, Ahmad T, Onnie CM, McArdle W, Strachan D, Bethel G, Bryan C, Lewis CM, Deloukas P, Forbes A, Sanderson J, Jewell DP, Satsangi J, Mansfield JC, Cardon L, Mathew CG (2007) Sequence variants in the autophagy gene *IRGM* and multiple other replicating loci contribute to Crohn's disease susceptibility. Nat Genet 39:830–832
Perrin AJ, Jiang X, Birmingham CL, So NS, Brumell JH (2004) Recognition of bacteria in the cytosol of mammalian cells by the ubiquitin system. Curr Biol 14:806–811
Podolsky DK (2002) Inflammatory bowel disease. N Engl J Med 347:417–429
Prescott NJ, Fisher SA, Franke A, Hampe J, Onnie CM, Soars D, Bagnall R, Mirza MM, Sanderson J, Forbes A, Mansfield JC, Lewis CM, Schreiber S, Mathew CG (2007) A nonsynonymous SNP in *ATG16L1* predisposes to ileal Crohn's disease and is independent of *CARD15* and *IBD5*. Gastroenterology 132:1665–1671
Priault M, Salin B, Schaeffer J, Vallette FM, di Rago JP, Martinou JC (2005) Impairing the bioenergetic status and the biogenesis of mitochondria triggers mitophagy in yeast. Cell Death Differ 12:1613–1621
Pua HH, Dzhagalov I, Chuck M, Mizushima N, He YW (2007) A critical role for the autophagy gene *Atg5* in T cell survival and proliferation. J Exp Med 204:25–31
Py BF, Lipinski MM, Yuan J (2007) Autophagy limits *Listeria monocytogenes* intracellular growth in the early phase of primary infection. Autophagy 3:117–125
Raelson JV, Little RD, Ruether A, Fournier H, Paquin B, Van Eerdewegh P, Bradley WE, Croteau P, Nguyen-Huu Q, Segal J, Debrus S, Allard R, Rosenstiel P, Franke A, Jacobs G, Nikolaus S, Vidal JM, Szego P, Laplante N, Clark HF, Paulussen RJ, Hooper JW, Keith TP, Belouchi A, Schreiber S (2007) Genome-wide association study for Crohn's disease in the Quebec Founder Population identifies multiple validated disease loci. Proc Natl Acad Sci USA 104:14747–14752
Rathman M, Barker LP, Falkow S (1997) The unique trafficking pattern of *Salmonella typhimurium*-containing phagosomes in murine macrophages is independent of the mechanism of bacterial entry. Infect Immun 65:1475–1485
Rich KA, Burkett C, Webster P (2003) Cytoplasmic bacteria can be targets for autophagy. Cell Microbiol 5:455–468
Rikihisa Y (1984) Glycogen autophagosomes in polymorphonuclear leukocytes induced by Rickettsiae. Anat Rec 208:319–327
Rioux JD, Xavier RJ, Taylor KD, Silverberg MS, Goyette P, Huett A, Green T, Kuballa P, Barmada MM, Datta LW, Shugart YY, Griffiths AM, Targan SR, Ippoliti AF, Bernard EJ, Mei L, Nicolae DL, Regueiro M, Schumm LP, Steinhart AH, Rotter JI, Duerr RH, Cho JH, Daly MJ, Brant SR (2007) Genome-wide association study identifies new susceptibility loci for Crohn disease and implicates autophagy in disease pathogenesis. Nat Genet 39:596–604
Roberts RL, Hollis-Moffatt JE, Gearry RB, Kennedy MA, Barclay ML, Merriman TR (2008) Confirmation of association of *IRGM* and *NCF4* with ileal Crohn's disease in a population-based cohort. Genes Immun 9:561–565
Roy D, Liston DR, Idone VJ, Di A, Nelson DJ, Pujol C, Bliska JB, Chakrabarti S, Andrews NW (2004) A process for controlling intracellular bacterial infections induced by membrane injury. Science 304:1515–1518
Saitoh T, Fujita N, Jang MH, Uematsu S, Yang BG, Satoh T, Omori H, Noda T, Yamamoto N, Komatsu M, Tanaka K, Kawai T, Tsujimura T, Takeuchi O, Yoshimori T, Akira S (2008) Loss

of the autophagy protein Atg16L1 enhances endotoxin-induced IL-1beta production. Nature 456:264–268

Sanjuan MA, Dillon CP, Tait SW, Moshiach S, Dorsey F, Connell S, Komatsu M, Tanaka K, Cleveland JL, Withoff S, Green DR (2007) Toll-like receptor signalling in macrophages links the autophagy pathway to phagocytosis. Nature 450:1253–1257

Sansonetti PJ, Phalipon A, Arondel J, Thirumalai K, Banerjee S, Akira S, Takeda K, Zychlinsky A (2000) Caspase-1 activation of IL-1beta and IL-18 are essential for *Shigella flexneri*-induced inflammation. Immunity 12:581–590

Sartor RB (2008) Microbial influences in inflammatory bowel diseases. Gastroenterology 134:577–594

Schmid D, Dengjel J, Schoor O, Stevanovic S, Munz C (2006) Autophagy in innate and adaptive immunity against intracellular pathogens. J Mol Med 84:194–202

Schnupf P, Portnoy DA (2007) Listeriolysin O: a phagosome-specific lysin. Microbes Infect 9:1176–1187

Scortti M, Monzo HJ, Lacharme-Lora L, Lewis DA, Vazquez-Boland JA (2007) The PrfA virulence regulon. Microbes Infect 9:1196–1207

Seveau S, Pizarro-Cerda J, Cossart P (2007) Molecular mechanisms exploited by *Listeria monocytogenes* during host cell invasion. Microbes Infect 9:1167–1175

Shea JE, Hensel M, Gleeson C, Holden DW (1996) Identification of a virulence locus encoding a second type III secretion system in *Salmonella typhimurium*. Proc Natl Acad Sci USA 93:2593–2597

Sinai AP, Joiner KA (1997) Safe haven: the cell biology of nonfusogenic pathogen vacuoles. Annu Rev Microbiol 51:415–462

Singh SB, Davis AS, Taylor GA, Deretic V (2006) Human IRGM induces autophagy to eliminate intracellular mycobacteria. Science 313:1438–1441

Suzuki T, Franchi L, Toma C, Ashida H, Ogawa M, Yoshikawa Y, Mimuro H, Inohara N, Sasakawa C, Nunez G (2007) Differential regulation of caspase-1 activation, pyroptosis, and autophagy via Ipaf and ASC in *Shigella*-infected macrophages. PLoS Pathog 3:e111

Swaminathan B, Gerner-Smidt P (2007) The epidemiology of human listeriosis. Microbes Infect 9:1236–1243

Swanson MS, Molofsky AB (2005) Autophagy and inflammatory cell death, partners of innate immunity. Autophagy 1:174–176

Taylor GA (2007) IRG proteins: key mediators of interferon-regulated host resistance to intracellular pathogens. Cell Microbiol 9:1099–1107

Tlaskalova-Hogenova H, Stepankova R, Hudcovic T, Tuckova L, Cukrowska B, Lodinova-Zadnikova R, Kozakova H, Rossmann P, Bartova J, Sokol D, Funda DP, Borovska D, Rehakova Z, Sinkora J, Hofman J, Drastich P, Kokesova A (2004) Commensal bacteria (normal microflora), mucosal immunity and chronic inflammatory and autoimmune diseases. Immunol Lett 93:97–108

Tosi MF (2005) Innate immune responses to infection. J Allergy Clin Immunol 116:241–249; quiz 250

Tsolis RM, Townsend SM, Miao EA, Miller SI, Ficht TA, Adams LG, Baumler AJ (1999) Identification of a putative *Salmonella enterica* serotype Typhimurium host range factor with homology to IpaH and YopM by signature-tagged mutagenesis. Infect Immun 67:6385–6393

Vignal C, Singer E, Peyrin-Biroulet L, Desreumaux P, Chamaillard M (2007) How *NOD2* mutations predispose to Crohn's disease?. Microbes Infect 9:658–663

Xavier RJ, Podolsky DK (2007) Unravelling the pathogenesis of inflammatory bowel disease. Nature 448:427–434

Xu Y, Jagannath C, Liu XD, Sharafkhaneh A, Kolodziejska KE, Eissa NT (2007) Toll-like receptor 4 is a sensor for autophagy associated with innate immunity. Immunity 27:135–144

Yano T, Mita S, Ohmori H, Oshima Y, Fujimoto Y, Ueda R, Takada H, Goldman WE, Fukase K, Silverman N, Yoshimori T, Kurata S (2008) Autophagic control of *Listeria* through intracellular innate immune recognition in *Drosophila*. Nat Immunol 9:908–916
Yorimitsu T, Nair U, Yang Z, Klionsky DJ (2006) Endoplasmic reticulum stress triggers autophagy. J Biol Chem 281:30299–30304
Zenewicz LA, Shen H (2007) Innate and adaptive immune responses to *Listeria monocytogenes*: a short overview. Microbes Infect 9:1208–1215

Group A *Streptococcus*: A Loser in the Battle with Autophagy

Tamotsu Yoshimori and Atsuo Amano

Contents

1 Invasion of Host Cells by Group A *Streptococcus* ... 218
2 Escape of GAS from Endosomes ... 218
3 Capture of GAS by Autophagy ... 219
4 Winner or Loser? ... 220
5 Differences from Canonical Autophagy ... 222
6 Autophagy in *Staphylococcus aureus* Infection ... 224
7 Concluding Remarks ... 225
References ... 225

Abstract Autophagy is an intracellular bulk degradation/recycling system for turning over cellular constituents. In one of the more remarkable findings among the recent developments in the field of autophagy, it was found that autophagy can also eliminate bacteria that invade host cells. The first evidence of this phenomenon came from an analysis of group A *Streptococcus* (GAS), the etiological agent underlying diverse human diseases. This bacterium is often internalized into nonphagocytic cells via the endocytic pathway, and then escapes from endosomes into the cytoplasm by secreting streptolysin O. The bacteria that escape into the cytoplasm induce and are captured by a unique membranous structure, which shares characteristics and molecular machinery with canonical autophagosomes but has some distinctions, including its large size. Subsequent fusion with lysosomes causes the death of most intracellular GAS. These findings have opened up a new field of innate immunity: the intracellular immune system against pathogens that penetrate the first defense of the endocytic pathway. The role of autophagy in *Staphylococcus aureus* infection is also discussed.

T. Yoshimori (✉)
Department of Cell Regulation, Research Institute for Microbial Diseases,
Osaka University, Suita-Osaka 565-0871, Japan
e-mail: tamyoshi@biken.osaka-u.ac.jp

A. Amano
Department of Oral Frontier Biology, Osaka University Graduate School of Dentistry,
Suita-Osaka 565-0871, Japan

B. Levine et al. (eds.), *Autophagy in Infection and Immunity*,
Current Topics in Microbiology and Immunology 335,
DOI 10.1007/ 978-3-642-00302-8_10,

1 Invasion of Host Cells by Group A *Streptococcus*

The Gram-positive bacterium *Streptococcus pyogenes* (group A *Streptococcus*; GAS) is a ubiquitous and versatile pathogen that causes a wide range of human infections, from clinically uncomplicated conditions, including pharyngitis (strep throat) and impetigo, to life-threatening invasive disorders such as necrotizing fasciitis (flesh-eating disease) and streptococcal toxic shock syndrome (Smith et al. 2005). In addition, GAS organisms cause postinfectious immune sequelae, including glomerulonephritis, and acute rheumatic fever—a leading cause of cardiovascular morbidity and mortality, especially in developing countries throughout the world (Carapetis et al. 2005).

GAS reportedly adheres to and invades various types of cultured human epithelial cells, and its adhesive components, known as invasins, promote bacterial invasion of human cells (Hynes 2004). Indeed, infection generally occurs in the peritonsillar region of the throat or on the surface of the skin, where the bacteria grow extracellularly or intracellularly. Subsequently, intracellular organisms are thought to activate the focal adhesion complex and induce cytoskeletal rearrangements, thereby disabling cellular functions such as adhesion, migration, and proliferation (Tomasini-Johansson et al. 2001, Cywes and Wessels 2001). In addition, intracellular GAS induces proinflammatory cytokine production (Nakagawa et al. 2004a) and often causes apoptosis (Nakagawa et al. 2001). Intracellular infection likely provides a nutritionally rich shelter, offering protection from components of the host immune system, including phagocytes and humoral antibodies, as well as from certain antibiotics (Molinari et al. 2000).

In many cases, microorganisms internalized by host cells are efficiently eliminated by host defense mechanisms with phagosomes (Vieira et al. 2002). However, intracellular GAS was suggested to survive either within the phagosomes of professional phagocytic cells, such as polymorphonuclear leukocytes, or within the endosomes of nonphagocytic cells, such as those found in pharyngeal epithelium (Cunningham 2000). Clinical observation indicated the possibility that virulent strains of GAS were more resistant to phagocytosis and killing by human neutrophils, and caused severe invasive disease involving bacterial dissemination from normally sterile sites of colonization into deeper tissues, as well as the bloodstream and lymphatics, and various organs (Sumby et al. 2006). Therefore, it was suspected that some virulent GAS clones, such as JRS4, were able to escape from phagosomes/endosomes to the cytoplasm, where they multiplied and possibly disseminated into neighboring cells by eliciting actin polymerization, such as in the case of *Listeria* spp. (Molinari et al. 2000). However, it remained unclear how intracellular GAS evaded phagosomes/endosomes, and whether the bacterium is, in fact, safe inside cells.

2 Escape of GAS from Endosomes

To clarify the fate of intracellular GAS, we incubated HeLa cells with the JRS4 strain for 1 h and then washed and cultured the cells for various times in the presence of antibiotics to kill any remaining extracellular bacteria. At first, GAS appeared in early endosomes labeled by the FYVE domain of early endosomal autoantigen 1 (EEA1), a marker for this compartment (Nakagawa et al. 2004b). This colocalization then gradually decreased over time. GAS secretes streptolysin O (SLO), a member of a conserved family of cholesterol-dependent pore-forming cytolysins. We found that JRS4 Δ*SLO*, an isogenic SLO-deficient mutant of strain JRS4, did not show decreased endosomal localization; instead, it remained within FYVE-positive endosomes for a longer time (Nakagawa et al. 2004b). Since *Listeria monocytogenes* is known to escape from endocytic compartments to the cytoplasm by secreting listeriolysin O (LLO), a member of the same cytolysin family (see the chapter by Huang and Brumell in this volume), it is very likely that GAS escapes from endosomes into the cytoplasm via a SLO-dependent mechanism to avoid lysosomal degradation, the final destination of the endocytic pathway (see Fig. 3). How SLO and LLO break down the endosomal membrane is unclear, but at high concentrations, these proteins may form pores on the membrane to destroy the integrity of the lipid bilayer.

3 Capture of GAS by Autophagy

Remarkably, escape to the cytoplasm does not result in freedom for GAS; instead, it is trapped by autophagy. We observed that the bacterial cluster was surrounded by GFP-LC3, a standard marker of autophagosomes (Nakagawa et al. 2004b). These compartments are unexpectedly large, sometimes over 10 µm in diameter (Fig. 1, right), whereas the sizes of the autophagosomes hitherto observed are limited to 0.5-1.0 µm in diameter in a variety of species and tissues. Thus, we termed these special autophagosomes GcAVs (GAS-containing autophagosome-like vacuoles). To exclude the possibility that GFP-LC3 directly binds to the bacterial surface, we performed both conventional and immunoelectron microscopy, and found double-membrane cisternae surrounding a GAS cluster; this is a characteristic, albeit an

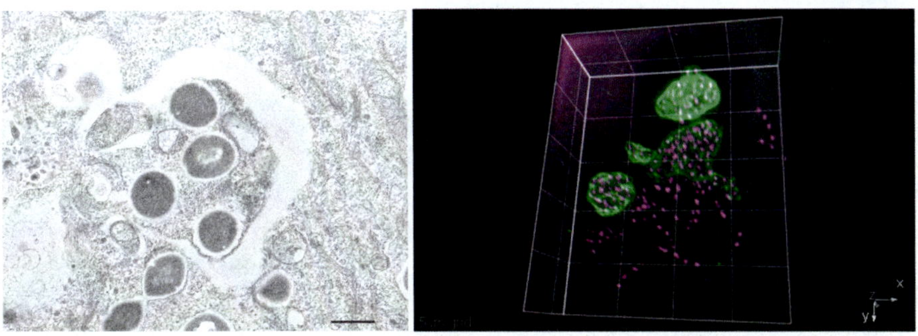

Fig. 1 GcAVs. *Left panel.* The electron micrograph shows the double membrane-bound cisterna (so-called isolation membrane/phagophore) surrounding several GAS cells in the cytoplasm to form a GcAV. *Bar*, 0.5 µm. *Right panel.* A three-dimensional image of a large GcAV in HeLa cells expressing enhanced GFP-LC3 (*green*) at 3 h post infection with GAS. The bacteria were visualized by staining bacterial DNA with propidium iodide. *Grid*, 5 µm. Reproduced with permission. Amano A and Yoshimori T, Autophagy in Immunity and Infection; a novel immune effector, edited by Vojo Deretic, p143, 2006, copyright Wiley-VCH Verlag GmbH & Co. KGaA

extremely large one, of autophagosomes and isolation membranes (pre-autophagosomes) (Fig. 1, left). We also observed single-membrane-bound compartments containing degraded GAS and cytosol, which again resemble—but are larger than—autolysosomes (autophagosomes fused with lysosomes). Finally, we confirmed that GFP-LC3 was localized to the GcAV membrane by immunoelectron microscopy. Therefore, despite their large size, GcAVs are a type of autophagic membrane structure. The sizes of the GcAVs increase in a time-dependent manner after infection, and at least 20–30 bacterial cells were associated with each single GcAV.

We previously demonstrated that LC3 exists in two molecular forms; 18 kDa LC3-I is cytosolic, whereas 16 kDa LC3-II (lipidated LC3) binds to autophagosomes (Kabeya et al. 2000). In general, LC3-II levels correlate with the number of autophagosomes. We found that the amount of LC3-II in GAS-infected cells significantly increased. In addition, Atg5 is essential for autophagosome formation. This protein binds to isolation membranes and then departs from the membrane upon maturation to an autophagosome (Mizushima et al. 2001). To determine whether Atg5 is required for GcAV formation, we infected *Atg5*$^{-/-}$ mouse embryonic stem (ES) cells or mouse embryonic fibroblasts (MEFs) with GAS; GcAVs were never detected in these experiments. Electron microscopy revealed GAS in the cytoplasm without any surrounding membrane (this also confirmed the escape of GAS from endosomes). These results indicate that the basic autophagic machinery, including LC3 and Atg5, is involved in GcAV formation.

4 Winner or Loser?

The appearance of GcAVs within 30 min of infection was followed by a time-dependent increase in the number and size of GcAVs. About 80% of intracellular GAS was sequestered by GcAVs at 4 h post-infection. To determine whether sequestered bacteria would survive after capture in GcAVs, we conducted colony formation assays using the intracellular bacteria (Nakagawa et al. 2004b) (Fig. 2). In wild-type MEFs at 4 h postinfection, only 20% of intracellular GAS lived. This percentage correlates with that of GAS not trapped by GcAVs. On the other hand, there was significantly increased survival of GAS in *Atg5*$^{-/-}$ MEFs was significantly suppressed; 60% of intracellular GAS survived at 5 h post-infection. Tannic acid (TA) is a cell-impermeable fixative that, at low concentrations, prevents fusion of secretory vesicles with the plasma membrane, while it does not affect intracellular membrane dynamics. Therefore, we expected that treatment of cells with moderate levels of TA would prevent escape of GAS through the plasma membrane into the culture medium. Indeed, under these conditions, the distinction between wild-type and *atg5*$^{-/-}$ MEFs was clearer: in TA-treated *Atg5*$^{-/-}$ MEFs, the number of living GAS increased at 2 h postinfection and was maintained at the same high level for 5 h postinfection; in contrast, the number of surviving GAS decreased in TA-treated wild-type MEFs with a similar rate and kinetics as in untreated wild-type cells. Based on these results, we concluded that intracellular GAS is effectively killed by the Atg5-dependent autophagic machinery; however, in cells lacking the autophagic

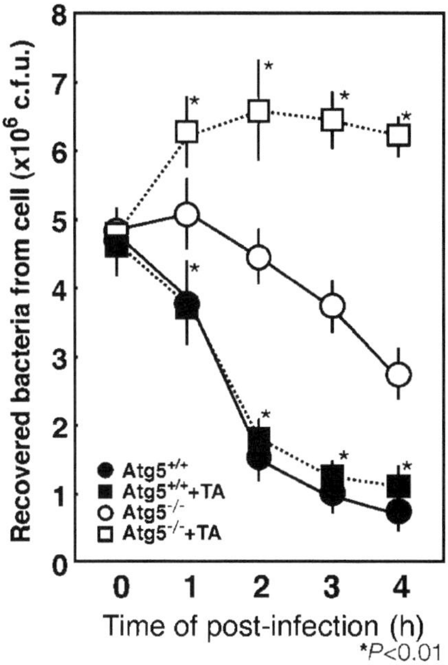

Fig. 2 Most intracellular GAS are killed by autophagy. The viability of intracellular GAS in wild-type MEFs (*closed symbols*) and *Atg5*$^{-/-}$ MEFs (*open symbols*) was measured in the presence (*squares*) or absence (*circles*) of tannic acid (TA) at a final concentrartion of 0.5%. Reproduced with permission from Nakagawa et al. 2004b

machinery, GAS is not killed, but rather multiplies and can even be released from host cells. Thus, autophagy plays a protective role against GAS infection.

We also pursued the mechanism by which GAS are killed by GcAVs (Nakagawa et al. 2004b). GcAVs formed in GAS-infected HeLa cells eventually colocalized with lysosome-associated membrane protein 1 (LAMP-1), a late endosome/lysosome marker. Over 90% of GcAVs were LAMP-1 positive at 4 h postinfection, indicating efficient fusion of GcAVs and lytic compartments. Addition of the lysosomal protease inhibitors leupeptin and E64d to the culture medium greatly suppressed the decrease in living intracellular GAS in wild-type cells. This treatment, however, did not affect the number of live GAS in *Atg5*$^{-/-}$ cells. Thus, intracellular GAS is killed by lysosomal proteases following fusion of GcAVs with lysosomes.

One question arising from the elucidation of the itinerary of intracellular GAS is why GAS cannot break GcAVs by secreting SLO. Although there is no answer at the moment, one possibility is that the content of cholesterol, the target of SLO, in the GcAV membrane is lower than that in endosomes. At present, the lipid composition of autophagosomes is unknown, since it is very difficult to isolate the compartment with high purity. Interestingly, we have often observed that GcAVs are surrounded by another isolation membrane (unpublished data). SLO may take longer to break GcAVs than endosomes because of low cholesterol levels, and during this time, subsequent autophagic sequestration could seal holes in the initial GcAV. Such serial membrane formation may also cause the integration of several GcAVs into larger structures. This sealing hypothesis for a bacteria-containing membrane-bound compartment is reminiscent of spacious *Listeria*-containing phagosome (SLAPs) formation (see the chapter by Huang and Brumell in this volume), although GAS and *L. monocytogenes* have opposite fates.

5 Differences from Canonical Autophagy

The characterization of the host cell response to intracellular GAS described above strongly indicates that the autophagic machinery, which is conserved from yeast to human and is ubiquitous among most cell types, acts as an innate immune system against the bacteria (Fig. 3). However, autophagy against GAS differs in several respects from basal or starvation-induced autophagy (Table 1).Usually, autophagy is suppressed to a basal level in cells under nutrient-rich conditions. GAS infection strongly induces autophagy even in the presence of sufficient nutrients. Very few GcAVs were formed in cells infected with JRS4 Δ*SLO*, suggesting that the emergence of the bacteria into the cytoplasm or the destruction of endosomes triggers autophagic induction (Nakagawa et al. 2004b). In any case, it is likely that some surveillance system exists in the cytoplasm. Moreover, in principle, autophagy is a nonselective process sequestering a random portion of the cytoplasm. GcAVs, however, seem to specifically surround GAS clusters, although small fractions of cytosol around the bacteria are also engulfed. The target molecule for this selectivity and how it is recognized remain enigmatic.

The lifetime of GcAVs is significantly longer than that of canonical autophagosomes. Autophagosomes form for 5–10 min and fuse with lysosomes within several 10 min after formation. Atg5 localizes to the membrane only during its formation, LC3 on the inner membrane is degraded upon fusion with lysosomes, and LC3 on the outer membrane leaves autolysosomes following cleavage by Atg4. Thus, both markers should label these compartments for a few hours. However, these markers remain on GcAVs for several hours or days. It is possible that this is due to the continuous formation of isolation membranes around the same GAS cluster, as discussed above. Such a phenomenon has not been observed in canonical autophagy, suggesting that an interaction between the bacteria and host cell continues even after GcAV trapping.

Is the molecular machinery underlying autophagy against GAS completely conserved with canonical autophagy? The answer is no: we recently found that GcAV formation requires a certain small GTPase that is not essential for canonical autophagosome formation (unpublished data). Presumably, autophagy targeting GAS is a subtype of autophagy that is specialized for defending cells against invading pathogens. During evolution, host cells have established specific types of autophagy armed with additional molecular mechanisms allowing selective targeting and the formation of a large capturing compartment. Furthermore, future studies are needed to examine whether other Atg proteins in addition to Atg5 and LC3 are also involved in the process; while they are essential to canonical autophagy, some of them may be unnecessary for autophagy against GAS.

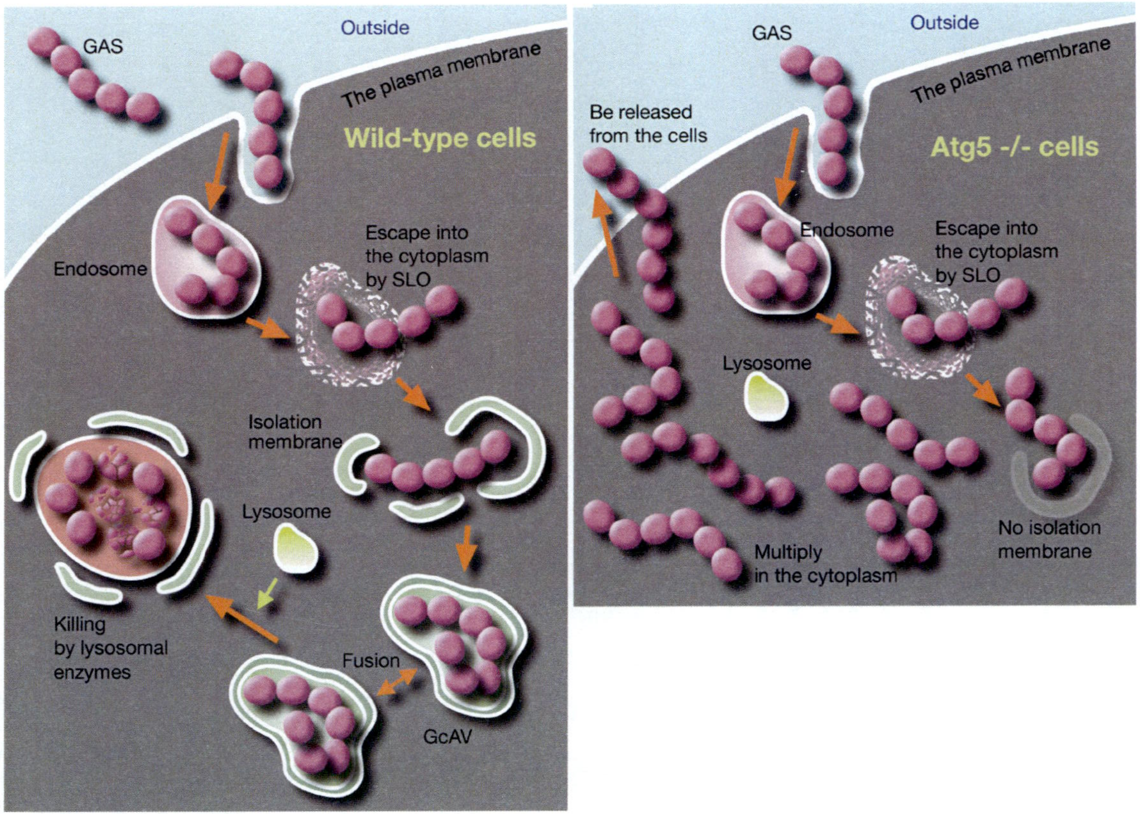

Fig. 3 A schematic illustration of the itinerary of GAS invading host nonphagocytic cells with (*left panel*, wild-type cells) or without (*right panel*, *Atg5*$^{-/-}$ cells) autophagic activity. Reproduced with permission. Amano A and Yoshimori T, Autophagy in Immunity and Infection; a novel immune effector, edited by Vojo Deretic, p148, 2006, copyright Wiley-VCH Verlag GmbH & Co. KGaA

Table 1 The differences between GcAVs and canonical autophagosomes

	GcAVs	Canonical autophagosomes
Induction	Formed only when GAS appears in the cytoplasm	Induced by starvation, rapamycin, etc. Formed at a basal level in the absence of the inducer
Selectivity	Seems to selectively surround GAS	Principally nonselective
Size	Several μm to over 10 μm	0.5–10 μm
Lifetime	Over 4 h	Several 10 min

6 Autophagy in *Staphylococcus aureus* Infection

A substantial increase in resistance to antimicrobial agents among bacterial pathogens has been found, particularly in Gram-positive bacteria; this compromises traditional therapies (Finch 2006). *Staphylococcus aureus*, one of the most ubiquitous Gram-positive pathogens, is a major cause of infections in both hospitals and care centers, and has exhibited significant resistance to methicillin (methicillin-resistant *S. aureus*, MRSA) (Kollef and Micek 2006). MRSA is generally considered to be a nosocomial pathogen that is associated with higher levels of morbidity and mortality as compared to other diseases caused by pathogens susceptible to methicillin. The prevalence of MRSA is increasing significantly in many parts of the world, with resistance rates of greater than 50% in the United States (Wisplinghoff et al. 2004) and some European countries (EARSS Management Team 2006).

Like GAS, *S. aureus* was previously thought to be an extracellular bacterium; however, the adhesion and invasion of nonprofessional phagocytic cells by *S. aureus* has been implicated in the pathogenesis of invasive and metastatic infections (Sinha and Herrmann 2005). Furthermore, studies have shown the effective autophagic elimination of both GAS and *S. aureus*, as the latter was sequestered in LC3 compartments and degraded by autolysosomes (personal communication from Nakagawa et al. 2006). Notably, among the various strains of *S. aureus*, some, including MRSA strains, have a marked resistance to autophagic elimination, as they are trapped by autophagosomes but thereafter escape from vacuoles into the cytoplasm. While it has been shown that the pathogenicity of MRSA is related to its resistance to antibiotics, resistance to autophagic degradation may also contribute.

Nearly simultaneously, another study reported that *S. aureus* replicated within autophagosomes and subsequently escaped into the cytoplasm (Schnaith et al. 2007). In infected HeLa cells, *S. aureus* were swiftly sequestered by autophagosomes; however, these compartments did not become acidified or acquire the lysosomal marker LAMP-2, indicating an arrest of autophagosome maturation and lack of fusion with lysosomes. Finally, *S. aureus* was found to escape from the autophagosomes into the cytoplasm, resulting in caspase-independent host cell death. In contrast, a *S. aureus* strain lacking the accessory gene regulator (*agr*) system failed to escape from autophagosomes and was eventually degraded.

The *agr* system, which is controlled by quorum sensing, tightly regulates the expression of various *S. aureus* virulence factors (Novick 2003). Although the *S. aureus*-derived factors that mediate escape from autophagosomes have not been identified, they seem to be under the control of the *agr* system. In addition, some, but not all, *S. aureus* strains were able to induce host cell death after invasion; this correlated with the virulence of the particular strain tested (Krut et al. 2003) and suggests that the ability to escape from autophagy is a critical virulence determinant. Therefore, methicillin resistance may be related to the *agr* system. Additional studies are needed to elucidate the specific mechanism used by *S. aureus* to defeat autophagy.

7 Concluding Remarks

Finding that autophagy efficiently kills GAS revealed a role for autophagy in intracellular immunity against invading pathogens. This expands the biological function of autophagy beyond its original role in recycling. It is clear that there are a variety of interactions between autophagy and diverse microorganisms. Although many microorganisms can evade or even subvert autophagic defense, it is clear that not all invaders are able to avoid the activity of autophagy.

The autophagic killing of GAS is exceptionally efficient. GAS infection strongly induces autophagy. Starvation or treatment with rapamycin, an autophagy-inducing drug, has no additive effect on GcAV formation, suggesting that GAS infection triggers the process maximally (unpublished observation). However, it should be noted that 20% of intracellular GAS survives autophagy. GAS apparently loses to autophagy in the first round, but the survivors may play some role in GAS pathology. In addition, attenuation of autophagic activity in skin or throat epithelial cells would allow GAS to penetrate into deeper tissues and cause severe infectious disease. Future work is required to elucidate these points as well as the molecular mechanisms of the formation and maintenance of GcAVs.

References

Amano A, Nakagawa I, Yoshimori T (2006) Autophagy in innate immunity against intracellular bacteria. J Biochem 140:161–166

Carapetis JR, Steer AC, Mulholland EK, Weber M (2005) The global burden of group A streptococcal diseases. Lancet Infect Dis 5:685–694

Cunningham MW (2000) Pathogenesis of group A streptococcal infections. Clin Microbiol Rev 13:470–511

Cywes C, Wessels MR (2001) Group A *Streptococcus* tissue invasion by CD44-mediated cell signalling. Nature 414:648–652

European Antimicrobial Resistance Surveillance System (EARSS)Management Team (2006) Antimicrobial resistance in Europe. In: EARSS annual report 2005 (eds) EARSS, Bilthoven, pp 41–65

Finch R (2006) Gram-positive infections: lessons learnt and novel solutions. Clin Microbiol Infect 12 (Suppl 8): 3–8

Hynes W (2004) Virulence factors of the group A streptococci and genes that regulate their expression. Front Biosci 9:3399–3433

Kabeya Y, Mizushima N, Ueno T, Yamamoto A, Kirisako T, Noda T, Kominami E, Ohsumi Y, Yoshimori T (2000) LC3, a mammalian homolog of yeast Apg8p, is localized in autophagosome membranes after processing. EMBO J 19:5720–5728

Kollef MH, Micek ST (2006) Methicillin-resistant *Staphylococcus aureus*: a new community-acquired pathogen? Curr Opin Infect Dis 19:161–168

Krut O, Utermöhlen O, Schlossherr X, Krönke M (2003) Strain-specific association of cytotoxic activity and virulence of clinical *Staphylococcus aureus* isolates. Infect Immun 71:2716–2723

Mizushima N, Yamamoto A, Hatano M, Kobayashi Y, Kabeya Y, Tokuhisa T, Ohsumi Y, Yoshimori T (2001) Dissection of autophagosome formation using *Apg5*-deficient mouse embryonic stem cells. J Cell Biol 152:657–667

Molinari G, Rohde M, Guzman CA, Chhatwal GS (2000) Two distinct pathways for the invasion of *Streptococcus pyogenes* in non-phagocytic cells. Cell Microbiol 2:145–154

Nakagawa I, Nakata M, Kawabata S, Hamada S (2001) Cytochrome c-mediated caspase-9 activation triggers apoptosis in *Streptococcus pyogenes*-infected epithelial cells. Cell Microbiol 3:395–405

Nakagawa I, Nakata M, Kawabata S, Hamada S (2004a) Transcriptome analysis and gene expression profiles of early apoptosis-related genes in *Streptococcus pyogenes*-infected epithelial cells. Cell Microbiol 6:939–952

Nakagawa I, Amano A, Mizushima N, Yamamoto A, Yamaguchi H, Kamimoto T, Nara A, Funao J, Nakata M, Tsuda K, Hamada S, Yoshimori T (2004b) Autophagy defenses cells against invading group A *Streptococcus*. Science 306:1037–1040

Novick RP (2003) Autoinduction and signal transduction in the regulation of staphylococcal virulence. Mol Microbiol 48:1429–1449

Schnaith A, Kashkar H, Leggio SA, Addicks K, Krönke M, Krut O (2007) *Staphylococcus aureus* subvert autophagy for induction of caspase-independent host cell death. J Biol Chem 282:2695–2706

Sinha B, Herrmann M (2005) Mechanism and consequences of invasion of endothelial cells by *Staphylococcus aureus*. Thromb Haemostasis 94:266–277

Smith A, Lamagni TL, Oliver I, Efstratiou A, George RC, Stuart JM (2005) Invasive group A streptococcal disease: should close contacts routinely receive antibiotic prophylaxis? Lancet Infect Dis 5:494–500

Sumby P, Whitney AR, Graviss EA, DeLeo FR, Musser JM (2006) Genome-wide analysis of group A streptococci reveals a mutation that modulates global phenotype and disease specificity. PLoS Pathog 2:41–49

Tomasini-Johansson BR, Kaufman NR, Ensenberger MG, Ozeri V, Hanski E, Mosher DF (2001) A 49-residue peptide from adhesin F1 of *Streptococcus pyogenes* inhibits fibronectin matrix assembly. J Biol Chem 276:23430–23439

Vieira OV, Botelho RJ, Grinstein S (2002) Phagosome maturation: aging gracefully. Biochem J 366:689–704

Wisplinghoff H, Bischoff T, Tallent SM, Seifert H, Wenzel RP, Edmond MB (2004) Nosocomial bloodstream infections in US hospitals: analysis of 24,179 cases from a prospective nationwide surveillance study. Clin Infect Dis 39:309–317

Autophagy Subversion by Bacteria

Emanuel Campoy and María I. Colombo

Contents

1 Coxiella burnetii 228
  1.1 Bacterial Morphology and Antigenic Variation 228
  1.2 *Coxiella burnetii* Type IV Secretion System 229
  1.3 The *C. burnetii* Replicative Niche 230
  1.4 The Autophagic Pathway Favors *C. burnetii* Infection 230
2 Legionella pneumophila 232
  2.1 *L. pneumophila* Generates a Customized Bacterial Compartment 232
  2.2 The Type IV Secretion System Encoded by the *dot/icm* Gene Complex 233
  2.3 *L. pneumophila* Exploits the Host Autophagic Pathway 233
3 Staphylococcus aureus 235
  3.1 *S. aureus* Small Cell Variants 235
  3.2 The Global Regulator Systems *agr* and *sar* 235
  3.3 Intracellular Survival of *S. aureus*: Escaping from the Phagosomal Compartment 236
  3.4 *S. aureus* Subverts the Autophagy Pathway 237
4 Porphyromonas gingivalis 238
  4.1 The Intracellular Life of *P. gingivalis* and the Autophagy Pathway 239
  4.2 The Role of Gingipains 240
5 Anaplasma phagocytophilum 241
  5.1 *A. phagocytophilum* Interacts with the Autophagic Pathway 241
  5.2 The Autophagic Pathway Favors *A. phagocytophilum* Infection 242
6 Brucella abortus 242
  6.1 The Intracellular Niche of *B. abortus* 242
7 Concluding Remarks 245
References 246

Abstract Autophagy is an important cell survival process during nitrogen starvation conditions, and it also plays a housekeeping role, removing superfluous or aged organelles. Autophagy has also been linked to host cell control of several intracellular

E. Campoy and M.I. Colombo (✉)
Laboratorio de Biología Celular y Molecular, Instituto de Histología y Embriología (IHEM)-CONICET, Facultad de Ciencias Médicas, Universidad Nacional de Cuyo, Mendoza, Argentina
e-mail: mcolombo@fcm.uncu.edu.ar

B. Levine et al. (eds.), *Autophagy in Infection and Immunity*,
Current Topics in Microbiology and Immunology 335,
DOI 10.1007/ 978-3-642-00302-8_11,

microorganisms. However, since it is an important host defense mechanism, some pathogens have also evolved strategies to exploit or subvert autophagy. Thus, certain pathogens harness autophagy, leading to persistent infection and pathogenesis. In this chapter we highlight our current understanding of those bacterial pathogens that transit through the autophagic pathway, efficiently replicating and surviving within the host cell. In addition, we discuss present knowledge of how autophagy modulation affects the infectious capacities and life cycles of several intracellular pathogens.

1 *Coxiella burnetii*

The obligate intracellular bacterium *C. burnetii* is the causal agent of Q fever in humans. The acute form of the disease can range from asymptomatic infection to atypical pneumonia, granulomatous hepatitis, or a self-limited febrile illness. The manifestations of chronic Q fever are endocarditis, hepatitis, osteomyelitis or infected aortic aneurysms (Heinzen et al. 1999). *C. burnetii* disseminates and replicates in a wide variety of tissues but the initial targets are the alveolar macrophages. *C. burnetii* can survive for long periods in natural environments as it is highly resistant to elevated temperature, desiccation, UV light and various chemical disinfectants. In general, both humans and animals become infected by inhaling contaminated aerosols. Due to its low infectious dose and high environmental resistance, the Centers for Disease Control and Prevention (CDC) consider this pathogen a Class B agent of bioterrorism (Madariaga et al. 2003).

1.1 Bacterial Morphology and Antigenic Variation

C. burnetii is an obligate intracellular microorganism that cannot be grown in axenic media. It is a small pleomorphic rod (0.2-0.4 μm wide, 0.4–1.0 μm long) with a membrane similar to that of Gram-negative bacteria. Based on phylogenetic analyses, *C. burnetii* is considered a γ-proteobacteria. *Coxiella* 16S rRNA sequence analyses indicate that this bacterium has a specific relationship to the genus *Legionella* (Weisburg et al. 1989; Zusman et al. 2003). This relationship is supported by the strong homologies found between a number of *Coxiella* and *Legionella* genes.

Two antigenic variations have been described in *C. burnetii*: (1) the highly virulent phase I Nine Mile, which is found in infected hosts and insect vectors and has a full-length lipopolysaccharide (LPS) on the cell surface; (2) the less virulent phase II Nine Mile bacteria, which can be obtained by multiple passages through chicken embryos and has a truncated LPS. The uptake of phase II bacteria by monocytes is mediated by both $\alpha v\beta 3$ integrin and complement receptor 3 (CR3), whereas only the $\alpha v\beta 3$ integrin mediates the uptake of phase I. This differential internalization mechanism has been implicated in a change in the intracellular transport in the host cell (Capo et al. 1999).

Coxiella has a biphasic developmental cycle. Two main morphological cell types have been described: the small cell variant (SCV), which is the infectious form, and the large cell variant (LCV), with replicative characteristics (Coleman et al. 2004). At the ultrastructural level, the LCVs appear as less electron-dense bacteria, with nucleoid-dense filaments radiating from the central region towards the bacterial periphery. The SCVs have a compact rod-shaped appearance, with a dense central region corresponding to condensed nucleoid filaments surrounded by ribonucleoproteins (Heinzen et al. 1996). The SCV is considered the metabolically inactive, nonreplicative but environmentally stable cell variant, and it has a thicker peptidoglycan layer, whereas the LCV is the metabolically active, replicative form of the bacteria (Heinzen et al. 1999). Both forms can coexist in the same parasitophorous vacuole where the bacteria replicate. After replication, the bacteria are released from the host cell as a result of cell lysis or possibly exocytosis, and these released bacteria can then infect other cells.

1.2 Coxiella burnetii Type IV Secretion System

Several pathogens have been found to utilize specialized Type IV secretion systems (T4SSs) to survive intracellularly (Brouqui et al. 1993). The T4SS has homology to plasmid transfer systems but it has been modified as a protein export apparatus and it is present in several intracellular pathogens such as *L. pneumophila*, *C. burnetii*, *B. abortus* and *H. pylori*. A group of 24 genes called *dot/icm* (defective in organelle trafficking/intracellular multiplication) were described for the first time in *L. pneumophila* (Howe et al. 2003; Zamboni and Rabinovitch 2003). Although recent studies have identified many Dot/Icm-translocated effector proteins, most of their functions remain to be determined (Seshadri and Samuel 2001; Varghees et al. 2002). Since *C. burnetii* is an obligate intracellular pathogen, genetic manipulations are difficult to perform and so there is very limited information available about its pathogenic systems. Genomic sequence data indicate that *C. burnetii* has 21 genes with similar components to the *Legionella* Dot/Icm T4SS. Interestingly, it has been shown that overexpression of some of these genes can restore the growth of *Legionella* mutants deficient for *dot/icm* gene products, indicating that both T4SSs are functionally related (Zusman et al. 2003; Zamboni and Rabinovitch 2003).

In summary, T4SS are believed to be key elements in determining the intracellular fate and the replication capability of these pathogens via the secretion of effector molecules. However, actual proteins exported by these types of systems have not been fully identified.

1.3 The C. burnetii Replicative Niche

C. burnetii generates large and spacious parasitophorous vacuoles (PV) with lysosomal features in which the bacteria replicate (Heinzen et al. 1999). The in vitro development of the PV can be carried out in numerous cell types, including macrophage- and fibroblast-like cells. This custom-made compartment largely depends on the function of the *Coxiella* T4SS since cumulative evidence indicates that this bacterium actively participates in the biogenesis of its vacuole. Although *C. burnetii* has perfectly adapted to survive and replicate in the harsh environment of the acidic, phagolysosomal-like vacuoles, it seems that the bacterium initially prefers a slightly less acidic compartment. Indeed, it has been shown that phase I *C. burnetii* delays fusion with lysosomes, at least during the first 2 h of infection (Howe et al. 2003). Similarly, using phase II *Coxiella*, we have observed that the colocalization between *C. burnetii* phagosomes and cathepsin D is low during the first 20-40 min after internalization (Romano et al. 2007). Note that this delay is actively controlled by the bacterium, since it is hampered by chloramphenicol treatment, indicating that bacterial protein synthesis is crucial in this process.

The PVs that form in *C. burnetii*-infected cells also display some hallmark features of an autophagosomal compartment (Beron et al. 2002; Gutierrez et al. 2005). Indeed, the nascent *Coxiella*-containing phagosomes formed after bacterial internalization fuse rapidly with autophagic compartments labeled by the protein LC3 (Romano et al. 2007). Interestingly, this association is actively modulated by the bacterium, again implicating the requirement of the *Coxiella* T4SS in this interaction via the injection of effector molecules.

One of the key questions is: how is the large PV generated, which in many cases nearly occupies the entire cell? Also, which intracellular compartments supply membranes to generate these large vacuoles? Compelling evidence indicates that *C. burnetii*-containing vacuoles fuse homo- and heterotypically with other vesicles of the endophagocytic system (Veras et al. 1994). Therefore, these fusogenic properties are important for the assembly of the spacious *Coxiella* PV. Furthermore, as indicated above, the *Coxiella* phagosomes also fuse with compartments of the autophagic pathway that likely contribute by supplying membrane. In addition, recent evidence from our laboratory indicates that the early secretory pathway via the delivery of Rab1-labeled vesicles also seems to be critical in the development of the large *Coxiella* vacuole (Campoy and Colombo, manuscript in preparation). Indeed, Rab1 depletion by siRNA results in the formation of altered *Coxiella* PVs and deficient bacterial replication, suggesting that, similar to *L. pneumophila*, the early secretory pathway is also involved in the biogenesis of the *Coxiella* replicative niche.

1.4 The Autophagic Pathway Favors C. burnetii Infection

As mentioned above, the *Coxiella* replicative vacuole is decorated by the autophagic protein LC3 (Gutierrez et al. 2005). Furthermore, the autophagosome marker monodansylcadaverine (MDC) accumulates in both the PV and in smaller vesicles in close proximity to the *Coxiella* vacuole (Biederbick et al. 1995; Munafo and Colombo 2001). In addition, the PV is also decorated by Rab7 and Rab24, members of the Rab GTPase family, which are also involved in autophagy-related processes. In addition, we have evidence that Beclin 1, a protein that forms a complex with Class III phosphatidylinositol 3-kinase (PI3K) (i.e., Vps34) that is essential for the initial events of autophagy, is also recruited to the PV membrane (Vázquez and Colombo, Cell Death and Differentiation). Interestingly, our results indicate that Beclin 1 is present on the *Coxiella* vacuole even at late times postinfection, suggesting that autophagy is active during the entire course of infection. These results clearly indicate that the limiting membrane of the *Coxiella* PV is decorated by proteins that are components of the autophagic machinery, confirming the connection with this pathway.

C. burnetii exhibits a growth cycle characterized by a lag phase, which lasts for approximately two days postinfection (p.i.) (Coleman et al. 2004). In infected Chinese hamster ovary cells, this lag period is characterized by the presence of small *Coxiella* PV at 24 h p.i, whereas at 48 h the bacteria are found in spacious vacuoles and the cells contain only a few bacteria. During this lag period, the differentiation process between SCV and LCV is believed to occur. Subsequently, the *Coxiella* PV is almost completely filled with bacteria by 96 h p.i., indicating a very fast replication rate, which continues over the next four days (Coleman et al. 2004).

The role of autophagy in *Coxiella* vacuole development has been assessed under different experimental conditions (Gutierrez et al. 2005). First, the bacterial response to activation or inhibition of the autophagic pathway has been analyzed using different pharmacologic agents. Wortmannin (WM) and 3-methyladenine (3-MA) inhibit autophagy by blocking PI3K activity (Blommaart et al. 1997). Both compounds hamper *Coxiella* vacuole formation, suggesting that autophagy has a critical role in both PV development and bacterial replication (Beron et al. 2002). In contrast, when cells are subjected to autophagy induction (i.e., amino acid deprivation) prior to infection, the percentage of infected cells and both the size and development of vacuoles increase (Gutierrez et al. 2005). As expected, this enhancing effect can be blocked by WM or 3-MA (Beron et al. 2002). Similarly, treatment of the cells with rapamycin, an inhibitor of the kinase TOR (a critical kinase involved in suppression of the autophagic pathway) (Cutler et al. 1999; reviewed by van Sluijters et al. 2000; Klionsky and Emr 2000), markedly increases the percentage of cells containing large PVs at 12 h p.i. Both starvation and rapamycin significantly enhance *C. burnetii* replication and viability, as determined by an infectious focus-forming units (FFU) assay (Gutierrez et al.2005), indicating that autophagy promotes the development and growth of this bacterium. Second, the infection process has also been analyzed by the overexpression of autophagic proteins like LC3, Rab24, and Beclin 1 that are known to be recruited to the PV. Interestingly, overexpression of these proteins accelerates the formation of large *Coxiella*-replicative vacuoles (Gutierrez et al. 2005). In contrast, overexpression of mutated forms of these proteins delays vacuole development, although by 48 h this effect is reversed, indicating that *Coxiella* is capable of bypassing the delay caused by the altered proteins (Gutierrez et al. 2005). Thus, overexpression of proteins involved in the autophagic pathway accelerates the development of the compartment in which *C. burnetii* replicates, suggesting that components of the cellular machinery involved in autophagy are subverted to promote *C. burnetii* replication and differentiation.

It is worth noting that, early after internalization, *C. burnetii* increases cellular autophagy levels, possibly by injecting factors across the phagosomal membrane, via the T4SS. Thus, the injected proteins may stimulate the autophagic pathway in the host cell and the newly generated autophagosomes may fuse with the parasitophorous vacuoles. It may be beneficial for the bacteria to reside in an autophagic-like vacuole, since this compartment represents a continuous source of metabolites for bacterial growth and development (Gutierrez et al. 2005). The accessibility to nutrients may promote differentiation from the SCV to the LCV form, favoring bacterial replication. Thus, *C. burnetii* may be diverted to the autophagic pathway to enable differentiation into a variant form that is able to tolerate the harsh environment of the lysosomal compartment (Swanson and Fernandez-Moreira 2002). In summary, *Coxiella* subverts the autophagy pathway and thereby generates a large replicative niche that creates a more permissive environment for bacterial replication.

2 *Legionella pneumophila*

L. pneumophila is a Gram-negative bacterial pathogen that in humans causes Legionnaires' disease. Following inhalation, this facultative intracellular pathogen (Krech et al. 1980) replicates within macrophages and monocytes, leading to a potentially lethal pneumonia (Marra and Shuman 1992). Subsequently, the microorganism also infects alveolar epithelial cells, and then hematogenous bacterial spread to different organs can occur.

2.1 *L. pneumophila Generates a Customized Bacterial Compartment*

The ability of *L. pneumophila* to manipulate host cell processes is an important hallmark of its pathogenesis that allows the microorganism to create a customized replicative compartment within the host cell. Following internalization into target cells, the bacterium is engulfed in a phagosome that avoids the interaction with the lysosomal pathway. As early as 15 min after uptake, the *Legionella*-containing compartment (LCC) lacks plasma membrane markers as well as endocytic markers (i.e., Rab5) (Clemens and Horwitz 1996), while mitochondria and small ER-derived vesicles are in close apposition to the LCC (Horwitz 1983). The vacuoles bearing *L. pneumophila* fuse with vesicles derived from the early secretory pathway and most of the *Legionella* vacuoles are surrounded by double membranes studded with ribosomes (Swanson and Isberg 1995; Kagan and Roy 2002). Indeed, 1 h after infection, a substantial number of LCCs stain positive for the protein Rab1 (a small GTPase involved in ER-Golgi transport) and for the v-SNARE Sec22b (Derre and Isberg 2004). Endoplasmic reticulum (ER) proteins like BIP and calnexin, as well as the recombinant ER marker KDEL-YFP, also colocalize with the *Legionella* vacuoles. Within this ER-derived compartment, *L. pneumophila* differentiates into a replicative form. Interestingly, after a pause of several hours, the compartment harboring *Legionella* becomes acidic and is labeled by lysosomal markers, indicating that the LCC finally fuses with lysosomes (Sturgill-Koszycki and Swanson 2000; Swanson and Sturgill-Koszycki 2000). After the replication stage, the cell is lysed and neighboring cells are infected.

2.2 The Type IV Secretion System Encoded by the *dot/icm* Gene Complex

The initial avoidance of fusion with lysosomes, the generation of a tailored replication compartment, and the final lysis of the host cell are events actively modulated by the type IV secretion system encoded by the *dot/icm* gene complex (Zamboni and Rabinovitch 2003). The *dot* genes identified from dot mutants (defective for organelle trafficking) encode a large putative membrane complex that functions as a secretion system to inject effector molecules into the host cell (Vogel et al. 1998; Roy and Tilney 2002). It has been shown that the dotA product is required to regulate the traffic of the LCC in order to prevent fusion with the lysosomal compartment at the initial stage. Indeed, phagosomes containing dotA mutants show, as early as 5 min postinfection, rapid accumulation of the lysosomal glycoprotein LAMP-1 and the GTPase Rab7, whereas the majority of wild-type *L. pneumophila* phagosomes do not acquire these proteins (Roy and Tilney 2002). Evidence has also been presented indicating that factor(s) released via the type IV secretion system are responsible for inducing an autophagic response in macrophages. Interestingly, autophagy is activated by soluble products present in sterile cultures, indicating that phagocytosis of the bacteria is not required (Amer and Swanson 2005).

2.3 *L. pneumophila* Exploits the Host Autophagic Pathway

Cumulative evidence indicates that interaction with the autophagic pathway is a key event in the development of the *L. pneumophila* replication compartment (Amer and Swanson 2005). At the ultrastructural level, the replication vacuole resembles a nascent autophagosome (i.e., limited by a double membrane). Furthermore, the autophagic protein, Atg7, involved in the formation of the isolation membrane (Mizushima et al. 1998) transiently associates with the pathogen-containing vacuoles.

In contrast, no association is observed with phagosomes containing the *dotA* mutant, indicating that the bacterium uses the T4SS for recognition by the autophagy machinery. Subsequently, Atg8/LC3 and the autophagic marker MDC label the LCC. Furthermore, autophagic-like vacuoles that harbor *Legionella* mature more slowly than autophagosomes stimulated by starvation or rapamycin treatment. Thus, existing data support the idea that *L. pneumophila* delivers virulence factors via the T4SS to retard the maturation of the vacuole, postponing the arrival of lysosomal enzymes until *Legionella* develops into a more resistant form.

As expected for an autophagic process, the acquisition of autophagic markers by LCCs can be prevented by treatment with the autophagy inhibitor 3-MA. Interestingly, when autophagy is blocked by 3-MA, a large proportion of the bacteria are killed by 2 h after infection (Amer and Swanson 2005). Thus, macrophages deliver *Legionella* phagosomes to the endo/lysosomal pathway when autophagosome formation is inhibited by 3-MA or when the production of ER-derived vesicles is decreased by brefeldin A, preventing bacterial replication (Watarai et al. 2001; Kagan and Roy 2002) (Kagan et al. 2004). These results suggest that the pathogen intersects the autophagy pathway to evade endo/lysosomal degradation. Thus, *Legionella* trafficking in macrophages reveals a reciprocal relationship between the delivery of phagosomes directly to the lysosomal pathway and their capture by autophagy. Note that *L. pneumophila* can replicate in different autophagy null mutants of *Dictyostelium discoideum*, indicating that a functional autophagy pathway does not seem to be required for intracellular bacterial replication in this amoeba (Otto et al. 2004).

Upon autophagy induction in macrophages, the number of intracellular *L. pneumophila* surrounded by ER markers and bacterial growth increases. In contrast, blocking the early secretory pathway with brefeldin A decreases the colocalization between the autophagic enzyme Atg7 and *Legionella*-containing vacuoles (Amer and Swanson 2005), indicating a dynamic interplay between both the secretory and autophagic pathways. Since autophagy is indeed activated by this pathogen, the cumulative data support the hypothesis that *Legionella* exploits autophagy to favor the development of its replicative niche.

It is interesting to note that the interaction of *Legionella* with lipid rafts at the plasma membrane seems to be required for the localization of this pathogen within autophagic-like structures (Amer and Swanson 2005). Certain components of the lipid rafts such as cholesterol promote both pathogen uptake and autophagosome formation in infected cells. In contrast, cholesterol does not seem to be required for autophagosome formation in starvation or rapamycin-induced autophagy.

The high fusogenic capacity of the LCC is a key feature in the infection and persistence of this intracellular pathogen. The T4SS is a crucial element that actively participates in this process. These characteristics are indispensable to favor the development of the *Legionella*-replication compartment inside the host cell.

3 *Staphylococcus aureus*

The Gram-positive bacterium *Staphylococcus aureus* is a facultative anaerobic cocci responsible for a variety of serious infections. Among the different clinical syndromes caused by *S. aureus*, endocarditis, pneumonia, septic arthritis, and skin and soft tissue infections are the most frequent. This pathogen expresses several factors that combat the first lines of defense against infection exerted by macrophages and neutrophils. For example, *S. aureus* is resistant to defensins, probably because of the composition of its cell wall and cytoplasmic membrane. Also, its polysaccharide-containing capsule is important for bacterial evasion of neutrophil phagocytosis (Luong and Lee 2002). As a consequence, *S. aureus* is able to disseminate from local sites of infection by a hematogenous route, leading to septic and toxic shock (Lowy 1998). These severe complications may lead to multiple organ failure (Marrack and Kappler 1990). Also, *S. aureus* frequently colonizes the lungs of patients with cystic fibrosis, causing persistent infections. Although *S. aureus* has been classically considered to be an extracellular pathogen, there is also evidence that it can invade and colonize a wide variety of mammalian cells (Hudson et al. 1995; Almeida et al. 1996; Bayles et al. 1998; Jarry and Cheung 2006).

3.1 S. aureus Small Cell Variants

Similar to other bacterial pathogens, *S. aureus* has small colony variants (SCVs) with unusual colony morphology (i.e., nonpigmented, nonhemolytic colonies when grown on rabbit blood agar) that grow very slowly. These metabolically inactive SCVs are more resistant to antibiotics, and so these forms may play a critical role in intracellular survival and in the development of persistent staphylococcal infections refractory to antibiotic treatment (reviewed by Proctor et al. 1994; von et al. 2000). Indeed, SCVs can be isolated from chronically infected airways of patients suffering from cystic fibrosis (Besier et al. 2007). Thus, SCVs are sheltered within the host cell, but upon completion of antibiotic therapy, these variants revert to the fast-growing and highly virulent form, leading to cell lysis.

3.2 The Global Regulator Systems agr and sar

In the last few years, much progress has been made in understanding the regulation of staphylococcal pathogenic factors. Two regulatory systems, *agr* and *sar*, control the expression of many of the *S. aureus* virulence genes (Projan et al. 1994). The *sar* locus is required for the expression of the *agr* locus, which encodes a sensing system that utilizes an octapeptide pheromone produced by the organism itself. This *agr* system enhances the production of cell wall-associated binding factors such as collagen- and fibronectin-binding proteins according to the bacterial cell density (Ji et al. 1995). As cells enter the postexponential phase, two promoters in the *agr* system are activated and the secretion of proteins such as toxins, lipases and proteases is enhanced; a concomitant reduction in the expression of cell wall-associated proteins such as adhesins also occurs (Ji et al. 1997).

The ability of *S. aureus* to invade mammalian cells and induce apoptosis is dependent on factors regulated by Agr and Sar. Mutant strains of these global regulators are incapable of inducing apoptotic cell death, indicating that *agr*- and *sar*-dependent factors are essential molecules involved in apoptosis induction (Wesson et al. 1998). Furthermore, it was shown that the stage of growth is crucial for optimal bacterial internalization. Interestingly, *agr* and *sar* mutants have elevated levels of internalization in comparison to those of wild-type bacteria, probably because these mutants express a higher number of surface proteins required for internalization (Projan et al. 1994).

3.3 Intracellular Survival of S. aureus: Escaping from the Phagosomal Compartment

The capacity of *S. aureus* to invade cells and survive intracellularly are critical features of staphylococcal persistence and chronic disease (Kubica et al. 2008). Cell wall-associated molecules are involved in cell adhesion and perhaps also in protection against host cell defenses. The bacterium is engulfed as soon as it is in close proximity to the host cell and it attaches to the host cell surface via the fibronectin- and collagen-binding proteins present on the bacterial cell wall. One of the key features of *S. aureus* infection is the generation of a series of pathogenic factors such as secreted enzymes and toxins. These products are involved in the degradation of tissue components and in the lysis and death of host cells. The alpha toxin, a secreted pore-forming toxin which generates pores at the plasma membrane, is at least partially responsible for *S. aureus*-induced cytotoxicity, leading to cell death via the intrinsic death pathway, independent of death receptor signaling (Bantel et al. 2001).

In pulmonary epithelial cells, *S. aureus* is internalized by a zipper-like mechanism, with the formation of tight pseudopodia engulfing individual bacterium. In these cells, within a few minutes after internalization, the bacteria are located in phagosomes with one or two bacteria per vacuole, a number that increases over time (Kahl et al. 2000). Interestingly, these phagosomes containing live bacteria do not fuse with lysosomes, whereas phagosomes containing heat-killed bacteria do interact with lysosomal compartments (Kahl et al. 2000). Subsequently, the bacteria escape from the phagosomal compartment and actively multiply, most likely within the cytoplasm (Bayles et al. 1998; Jarry and Cheung 2006). Intracellular replication is initiated after a lag period of several hours. An important question is how *S. aureus* escapes from the phagosomal compartment. It has recently been demonstrated in a cystic fibrosis cell line that the escape from the vacuole requires the activity of a gene product controlled by the *agr* regulator system, since an AgrA mutant was unable to escape from the endosomal compartment (Jarry and Cheung 2006). As indicated above, the two-component regulatory system AgrCA controls the expression of multiple virulence genes, including several toxins and numerous secreted enzymes (for a review see (Novick 2003)). One of the major toxins upregulated by the *agr* system is α-hemolysin, a 34 kD secreted protein that is a pore-forming toxin (Gray and Kehoe 1984). Note that overexpression of α-hemolysin under an inducible promoter partially rescues the phenotype of an *S. aureus agr* mutant, indicating that this toxin, perhaps in conjunction with another factor, is required for the lysis of the phagosomal membrane to allow bacterial escape into the cytoplasm.

3.4 *S. aureus* Subverts the Autophagy Pathway

In a recent publication, evidence was presented indicating that *S. aureus* transits via the autophagy pathway to generate a protective niche where the bacterium can survive before escaping into the cytoplasm (Schnaith et al. 2007). The intracellular fate of *S. aureus* was analyzed in HeLa cells by fluorescence microscopy and at the ultrastructural level. At 1–2 h postinfection, the bacteria-containing compartment is in close apposition to multilamellar vesicles, which eventually trap the phagosome. These autophagosomal-like structures are positive for Rab7, a protein present in late endocytic vesicles but also in autophagic vacuoles (Gutierrez et al. 2004; Jager et al. 2004). During the first 3 h, the majority of these compartments colocalize with the autophagic protein GFP-LC3, whereas this colocalization is lost by 5 h, suggesting that the bacteria escape from the autophagosome. Indeed, by 8–12 h postinfection, the bulk of the *S. aureus* is found free in the cytoplasm. Interestingly, the escape of the bacteria into the host cell cytoplasm correlates with signs of cytotoxicity and cell death.

The colocalization of *S. aureus*-containing compartments with LC3, as well as the presence of multilammellar membranes surrounding the *S. aureus* phagosomes, depends on an active *agr* system. Indeed, an *agr*-deficient strain is not enclosed by double-membrane structures and does not colocalize with LC3 at all postinfection times analyzed (Schnaith et al. 2007), indicating that an *agr*-dependent factor(s) is responsible for stimulating autophagy in the host cell. However, when autophagy is pharmacologically activated by rapamycin treatment, the majority of the *agr*-deficient *S. aureus* colocalizes with the protein LC3. Based on studies of colocalization with the lysosomal marker LAMP-2, it appears that there is impaired fusion between this bacterial-autophagosomal compartment and lysosomes. Phagosomes containing wild-type *S. aureus* seldom colocalize with LAMP-2, whereas the *agr*-deficient mutant clearly acquires this marker. Similarly, in contrast to the vacuoles containing wild-type *S. aureus*, the *agr*-deficient bacteria are labeled by Lysotracker, a marker of acidic compartments, indicating that the wild-type bacteria also circumvent acidification. Thus, this pathogen inhibits the maturation of the autophagosomal compartment, avoiding both acidification and fusion with the lysosome.

The number of intracellular pathogens is markedly increased upon treatment with the autophagy inducer rapamycin, whereas wortmannin treatment dramatically reduces the intracellular growth of the bacteria. Bacterial replication is markedly impaired in MEFs deficient for the autophagy protein Atg5, indicating that a functional autophagic pathway is required for proper *S. aureus* replication. On the other hand, consistent with the observation that *agr*-deficient *S. aureus* are targeted to a lysosomal compartment, these mutants are rapidly eliminated by the host cell. In contrast, treatment with rapamycin increases the intracellular load of the *agr*-deficient *S. aureus* strains, indicating that pharmacological induction of autophagy changes the mutant bacteria from a replication-deficient pattern to a replication-competent phenotype.

Regarding the cytotoxic effects of *S. aureus*, it is noteworthy that in HeLa cells *S. aureus* induces cell death independent of caspase activation. The cell lysis and death caused by *S. aureus* are not prevented by overexpression of the apoptosis inhibitor XIAP, but they are suppressed by Bcl-2. It is important to take into account that as well as functioning as an antiapoptotic protein, Bcl-2 also inhibits autophagy by binding to the autophagic protein Beclin 1 (Pattingre et al. 2005). Thus, it has been postulated that in HeLa cells, *S. aureus* leads to an autophagic type of cell death also known as Type II cell death (Amano et al. 2006). Indeed, in cells infected with *S. aureus*, marked vacuolization, a hallmark of autophagic cell death, is observed. Furthermore, autophagy activation by rapamycin enhances the cytotoxic effects of wild-type *S. aureus*. In contrast, no signs of cell death are observed in cells deficient for Atg5, confirming the role of autophagy in *S. aureus*-dependent cell death.

In summary, infection with *S. aureus*, similar to *L. pneumophila*, *P. gingivalis* and *C. burnetii*, seems to take control of the autophagy pathway, actively delaying fusion with a degradative protease-containing compartment. In addition, *S. aureus* infection leads to an autophagic cell death independent of caspase activation.

4 *Porphyromonas gingivalis*

Porphyromonas gingivalis, a Gram-negative bacterial pathogen and natural component of the oral mucosal microbiome, is the causative agent of periodontal disease (i.e., gingivitis), which can lead to periodontal bone loss (Fiehn et al. 1992). This pathogen has also been involved in cardiovascular disease, pulmonary infections and atherosclerosis. Gingival epithelial cells are among the first host cells colonized by *P. gingivalis*. *P. gingivalis* invades and replicates in human endothelial cells, producing an array of potential virulence factors including extracellular proteases. The binding of the bacterial fimbriae to the host cellular β1-integrin receptor is required for invasion (Deshpande et al. 1998; reviewed by Yilmaz 2008) and for activation of endothelial cells, which leads to the development of atherogenesis (Takahashi et al. 2006). Activation of signaling molecules and actin rearrangements are key events upon the interaction of the bacterium with plasma membrane components. It is interesting to note that, as described for *Legionella*, lipid rafts also seem to be involved in *P. gingivalis* internalization. The protein caveolin 1 colocalizes with *P. gingivalis*, and reduction of plasma membrane cholesterol or knockdown of caveolin 1 decreases bacterial uptake (Tamai et al. 2005). However, it is not known if the interaction with lipids rafts is important for the subsequent connection with the autophagy pathway, as has been previously demonstrated for *L. pneumophila* (Amer and Swanson 2005).

4.1 The Intracellular Life of P. gingivalis and the Autophagy Pathway

P. gingivalis is highly invasive and can rapidly infect cultured cells, where it successfully replicates. After uptake, the bacteria reside in a phagosomal compartment, colocalizing with the small GTPase Rab5 (Dorn et al. 2001), a marker of early endocytic compartments. However, these phagosomes are not labeled by the mannose 6-phosphate receptor (M6PR), a late endosomal marker, indicating that the bacteria avoid the maturation of its surrounding phagosome. Cumulative evidence indicates that the *P. gingivalis*-containing compartment is diverted to the autophagy pathway. First, analysis at the ultrastructural level demonstrated that the pathogen is localized in compartments surrounded by several membrane layers resembling autophagosomal structures (Dorn et al. 1999). The bacterium is able to replicate in these double-membrane vacuoles, since images of duplicating bacteria were observed. Subsequently, in a more detailed study, it was described that the autophagic protein HsGsap (i.e., Atg7) is recruited to the phagosome early after internalization (Dorn et al. 2001). The compartment then becomes labeled by the ER resident chaperone BIP and also by the lysosomal membrane protein LAMP-1. However, no lysosomal enzymes are detected, indicating that *P. gingivalis* prevents the formation of autolysosomes. Thus, in human coronary artery endothelial cells (HCAEC), *P. gingivalis* localizes within autophagosomes that interact with ER-derived vesicles, but the bacteria hamper autophagosome-lysosome fusion (reviewed by Belanger et al. 2007).

Compelling evidence indicates that a functional autophagy pathway is required in HCAEC cells for the generation of the intracellular niche of *P. gingivalis* and for the establishment of persistent infection. Cell treatment with classical autophagy inhibitors, WM or 3-MA, diverts the bacteria from an early compartment to a vacuole labeled by M6PR that subsequently acquires the lysosomal enzyme cathepsin L (Dorn et al. 2001). Moreover, WM treatment also interferes with the intersection with ER-derived vesicles, as evidenced by the lack of the acquisition of the protein BIP. As would be expected for diversion towards a lysosomal degradative pathway, bacterial survival is dramatically reduced (Dorn et al. 2001). In agreement with these observations, in 3-MA- or WM-treated cells, bacteria with clear signs of degradation can be visualized at the ultrastructural level at later times after infection.

An important point to consider is that the interaction of *P. gingivalis* with the autophagy pathway appears to depend on the cell type and perhaps also on the bacterial strain used. In contrast to HCAEC cells, in primary gingival epithelial cells (GEC) and KB epithelial cells, the bacterium is found free in the cytoplasm and in the perinuclear area (Belton et al. 1999; Houalet-Jeanne et al. 2001). Interestingly, in GEC cells, the bacteria seem to disseminate directly from cell to cell through actin-based membrane protrusions, bypassing the need for bacterial escape into the extracellular medium (Yilmaz 2008). In addition, a recent publication has shown that in HCAEC cells the majority of the internalized microorganisms seem to escape the autophagic pathway, and instead traffic exclusively through the phagocytic pathway to a lysosomal compartment (Yamatake et al. 2007). However, depending upon the bacterial strain used, considerable colocalization with LC3 can be observed during the first hours after internalization. Moreover, in infected cells, the autophagy pathway is indeed activated, as demonstrated by LC3 puncta formation and the accumulation of LC3-II (Yamatake et al. 2007; reviewed by Kadowaki et al. 2007). Nevertheless, the same authors found important differences depending upon the strain used. Thus, future experiments are needed to clarify the behavior of different *P. gingivalis* strains in different cell types and the signficance of these interactions during in vivo infection of different tissues.

Although, as indicated above, the interaction of *P. gingivalis* with autophagy-like compartments seems to be strain dependent or cell-type specific, it is known that this bacterium must acquire nutrients from host-derived substrates. Therefore, the diversion towards the autophagic pathway may not only provide a way to avoid lysosomal degradation but also a means of acquiring essential nutrients.

4.2 The Role of Gingipains

This periodontal bacterium is known to produce a unique class of cysteine proteinases termed gingipains. It has been proposed that gingipains represent the most significant virulence factor produced by this bacterium. These proteolytic enzymes play a key role in altering host defense mechanisms and also in microorganism survival, since gingipains appear to participate in bacterial resistance against destruction by the host cell (reviewed by Kadowaki et al. 2007). However, it has been proposed that gingipains have no effect in the microenvironment of the phagolysosomal vacuole (Yamatake et al. 2007); thus, the molecular mechanism by which gingipains confer resistance to *P. gingivalis* is still unknown. Even though a bacterial mutant lacking gingipains enhances autophagic responses to a similar degree as wild-type bacteria in infected cells (Yamatake et al. 2007), it remains to be fully determined whether gingipains have any role in modulating autophagy.

5 *Anaplasma phagocytophilum*

Anaplasma phagocytophilum is a Gram-negative obligate intracellular bacterium of the order Rickettsiales (Goodman et al. 1996; Dumler and Bakken 1998; Demma et al. 2005). It is the causative agent of tick-borne fever (TBF), an emerging zoonosis in the United States and in other regions of the world. This infection is characterized by fever, headache, myalgias, thrombocytopenia, and leukopenia. The principal reservoirs are sheep, cattle, and goats. In humans, this pathogen causes granulocytic anaplasmosis. Bacterial invasion of neutrophil granulocytes is the hallmark of the disease, but other cells such as endothelial cells are also infected by this pathogen (Goodman et al. 1996; Munderloh et al. 2004; Herron et al. 2005).

One of the principal characteristics of this obligate intracellular bacterium is its ability to survive inside the harsh environment of the neutrophil by seizing host cell functions and regulating critical antimicrobial activities. For example, it has been demonstrated that neutrophils of sheep infected with TBF have a reduced phagocytic capacity, a situation that may further alter neutrophil function and predispose the host to opportunistic infections and dysregulation of inflammation.

5.1 *A. phagocytophilum* Interacts with the Autophagic Pathway

After phagocytosis, *A. phagocytophilum* resides in a nonacidic membrane-bound vacuole (i.e., inclusion) which avoids fusion with the lysosomal compartment (Webster et al. 1998; Mott et al. 1999). Some inclusions are positive for M6PR, a marker of late endocytic compartments. However, no lysosomal enzymes are detected, indicating that formation of an autolysosome is impaired (Dunn 1990; Dorn et al. 2002; Eskelinen et al. 2002). The interaction between *A. phagocytophilum* and the autophagic pathway has been established. Several hallmarks of autophagosomes, including a double-lipid bilayer membrane visualized by electron microscopy techniques, as well as the presence of autophagic proteins on the inclusion-limiting membrane, have been observed. Using fluorescence microscopy, Hua Niu and collaborators found that, in contrast to uninfected cells in which GFP-LC3 has a predominantly diffuse distribution, GFP-LC3 changes to a punctate pattern during infection (Niu et al. 2008). Consistently, an increase in the amount of the LC3-II form was detected by western blot analysis by three days p.i. At earlier times after infection (i.e., 14 h p.i.), no colocalization of GFP-LC3 with the inclusion is evident. However, by 48 and 72 h p.i., most of the chimeric protein colocalizes with the *A. phagocytophilum* inclusion, which is virtually encased by GFP-LC3. In agreement with these observations, autophagosome formation is undetectable during the *A. phagocytophilum* lag phase (i.e., up to 24 h p.i.), whereas autophagosomes are readily observed during logarithmic growth (from 24 to 72 h p.i.). Thus, the amount of autophagosomes is associated with the increase in bacterial load, suggesting that autophagosome formation is indeed stimulated in response to infection.

5.2 *The Autophagic Pathway Favors A. phagocytophilum Infection*

Since *A. phagocytophilum* has only a limited number of genes for de novo amino acid biosynthesis, it is forced to acquire various components from the host cytoplasm in order to replicate within the vacuole (Hotopp et al. 2006). Thus, residing in an autophagic compartment replete with nutrients may be beneficial for this intracellular pathogen. Indeed, pretreatment of host cells with the autophagy inducer rapamycin increases *A. phagocytophilum* infection approximately twofold (Niu et al. 2008). In contrast, treatment with 3-MA significantly inhibits *A. phagocytophilum* infection, although it does not affect binding or internalization, suggesting that internalized bacteria cannot replicate when autophagy is blocked by this drug.

Present knowledge supports the idea that autophagy facilitates *A. phagocytophilum* growth. During *A. phagocytophilum* infection, the activity of the autophagic pathway correlates with productive *A. phagocytophilum* replication and not with *A. phagocytophilum* destruction. Thus, this pathogen subverts the autophagy pathway, leading to the biogenesis of an early autophagosome-like compartment where the pathogen actively replicates, while avoiding fusion of this compartment with lysosomes.

6 *Brucella abortus*

Brucella abortus is a Gram-negative intracellular bacterium and the causative agent of brucellosis, a zoonotic disease that causes systemic symptoms involving several organs and tissues (Corbel 1980). One of the major sources of infection is the ingestion of unpasteurized diary products, although this pathogen can also infect persons working with infected animals via small injuries or aerosols. In humans, this chronic disease is characterized by fever and weakness. Musculoskeletal system involvement is the most common complication of brucellosis, while life-threatening complications are meningitis and endocarditis (Abrahams and Tylkowski 1985). Brucellosis, an endemic disease, leads to serious economic losses in developing countries. Furthermore, infection with *B. abortus* continues to cause a human health risk despite the progress in eradicating the disease from domestic animals (Street et al. 1975).

6.1 *The Intracellular Niche of B. abortus*

Macrophages and other professional phagocytes, as well as nonprofessional phagocytes such as HeLa or Vero cells are colonized by *B. abortus*. Within these cells, the bacterium survives and efficiently replicates (Jones and Winter 1992; reviewed by Smith et al. 1990; Gorvel and Moreno 2002). *B abortus* is internalized upon binding to receptors present at the plasma membrane, with subsequent activation of members of the Rho family leading to the rearrangement of the actin cytoskeleton. After engulfment, the bacteria traffic to autophagosome-like compartments instead of trafficking via the classical phagocytic pathway (reviewed by Gorvel and Moreno 2002). Indeed, at the ultrastructural level, one of the first lines of evidence suggesting that *Brucella* is diverted towards the autophagic pathway in HeLa cells was the observation of multiple membranes surrounding the bacteria (Pizarro-Cerda et al. 1998). In cells subjected to autophagy induction (i.e., starvation), an increase in bacterial growth is observed, whereas 3-MA or WM treatment yields lower bacterial numbers, indicating that transit via the autophagic pathway plays a key role in *Brucella* intracellular replication.

With a few minutes after internalization, *B. abortus* localizes in a compartment labeled by a transferrin receptor, Rab5, and the protein EEA1, markers of early endocytic compartments. However, at later times after infection, no markers of late endocytic compartments such as Rab7 or M6PR colocalize with the bacterium. Likewise, the lysosomal enzyme cathepsin D is not detected in the *Brucella*-containing compartment (Pizarro-Cerda et al. 1998), clearly indicating that fusion with the lysosomal compartment is impaired. However, evidence has been presented that in macrophages at later times after infection times (i.e., 24 h), a substantial fraction of the live bacteria are targeted to lysosomal compartments. This latter observation suggests that in professional macrophages, *B. abortus* delays this interaction instead of inhibiting fusion with the lysosome (Arenas et al. 2000). Another difference related to the cell type is the colocalization with both autophagic and ER markers. In HeLa cells, the *B. abortus*-containing compartment is labeled with the autophagic marker MDC and with the protein sec61β, a subunit of the ER translocon. In contrast, in macrophages, the bacteria does not colocalize with either an ER-labeling compound or MDC, suggesting that *Brucella* does not seem to be targeted to the autophagic pathway in professional phagocytes (Arenas et al. 2000).

Since only certain ER markers are found in the *Brucella*-containing compartment, it is believed that only specialized regions of the ER contribute to the development of the *Brucella* replicative niche. It has been postulated that in HeLa cells, *B. abortus* transits from an autophagosomal-like compartment towards an ER-derived compartment where the bacterium actively replicates. This final compartment, characterized by a single membrane surrounding the replicating bacteria, acquires other ER markers such as calnexin and protein disulfide isomerase (Pizarro-Cerda and Cossart 2006). Interestingly, treatment of infected cells with brefeldin A induces the localization of Golgi markers to the limiting membrane of the *Brucella* compartment, which is consistent with the Golgi redistribution to the ER induced by this drug. It has been shown that the biogenesis of this replicative compartment depends on the *Brucella* T4SS VirB (Comerci et al. 2001; Celli et al. 2003). In addition, the *Brucella*-containing vacuole interacts with ER exit sites (ERES), and disruption of these ERES by inhibition of the small GTPase Sar1 inhibits *Brucella* replication (Celli et al. 2003).

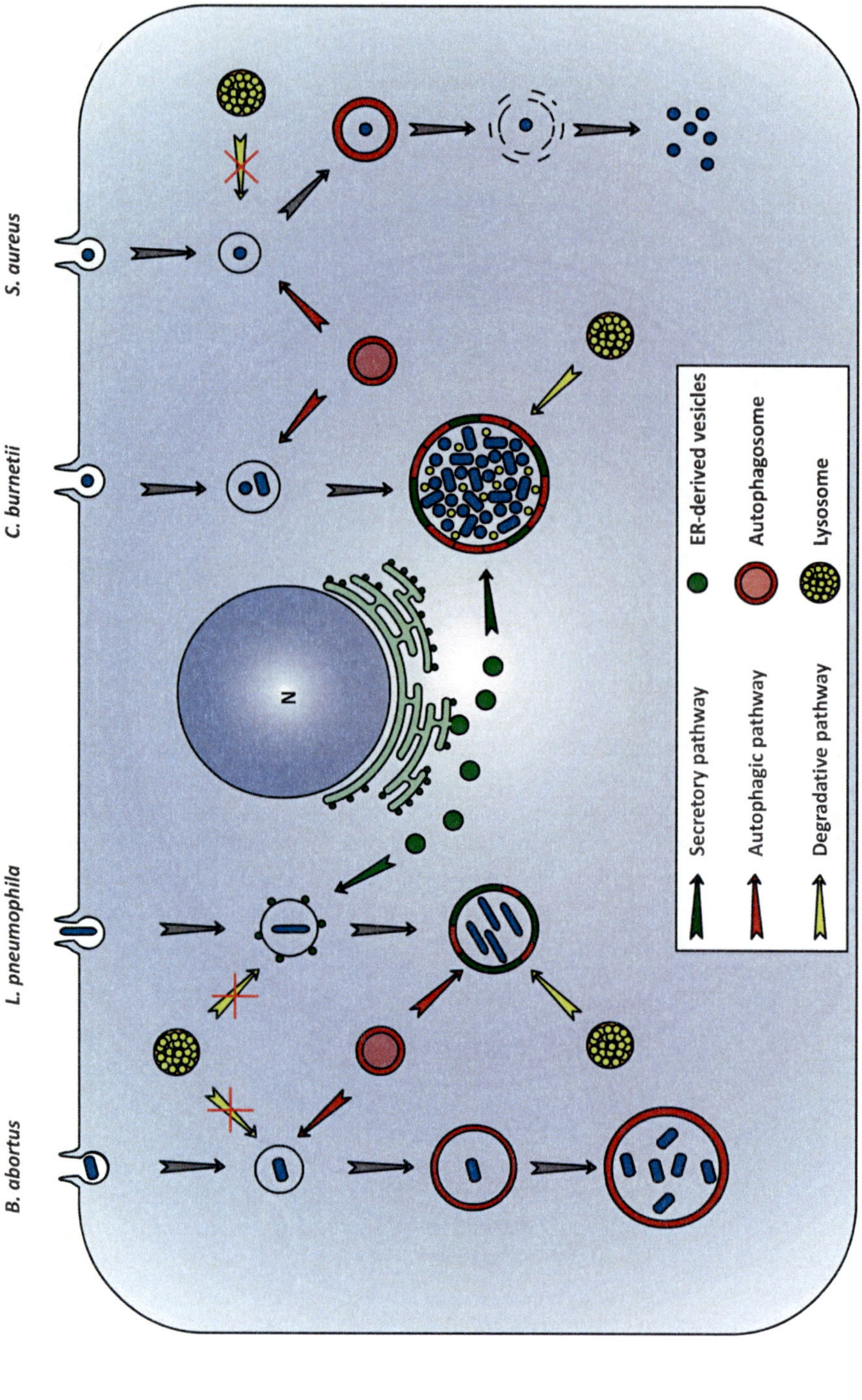
B. abortus
L. pneumophila
C. burnetii
S. aureus
N
Secretory pathway
Autophagic pathway
Degradative pathway
ER-derived vesicles
Autophagosome
Lysosome

Like many other pathogens, *Brucella* has a two-component regulatory system known as BvrS-BvrR (virulence-related sensory and regulatory proteins) (Sola-Landa et al. 1998). It has been shown that the *bvrS* and *bvrR* mutants inefficiently invade HeLa cells and are targeted to a cathepsin D-positive compartment without trafficking via the autophagy pathway. As expected, these mutants are rapidly degraded by lysosomes. The expression of a bvrR plasmid recovers the wild-type phenotype and bacterial targeting to autophagic compartments.

In summary, experiments in HeLa cells clearly show that *Brucella* traffics via the autophagy pathway at some point during infection, and that this event benefits micro-organism survival. However, the autophagosome does not seem to be the replication niche for virulent *B. abortus*, since this compartment is no longer labeled by MDC after prolonged infection. Further experiments are necessary to confirm this and to determine the molecular machinery involved in *B. abortus* intracellular transit.

7 Concluding Remarks

In this chapter, we have summarized the interaction between autophagy and several intracellular bacterial pathogens that colonize host cells causing chronic and persistent diseases. As illustrated in Fig. 1, some of these bacteria generate a replicative niche, subverting the autophagic pathway, but ultimately fuse with a degradative compartment. In many cases, how these bacteria resist this lytic compartment is still far from being understood. Other pathogens, by intersecting the autophagy pathway, either avoid or delay the interaction with the degradative compartment or use this pathway to escape into the cytoplasm. How these bacteria manipulate the autophagic machinery and the identities the bacterial factors that govern individual transport steps of the target cell remain to be elucidated. Several bacterial pathogens take advantage of host cell autophagy to generate a replicative compartment allowing the colonization and successful intracellular replication of the microbe. It is tempting to speculate that the identification of bacterial virulence

Fig. 1 Model showing how different intracellular bacterial pathogens intercept and subvert the autophagic pathway. The model shows different strategies that bacteria use to survive within the host cell by subverting autophagy. After engulfment, *B. abortus* traffics to autophago some-like compartments, impairing fusion with the degradative pathway, and replicating in an ER-derived compartment. Early after internalization, *L. pneumophila* evades lysosomal fusion and interacts with the ER by a mechanism induced by *Legionella*-secreted factors. *Legionella* intercepts the autophagy pathway, and several hours after fusion with vesicles derived from the early secretory pathway, *Legionella*-containing compartments fuse with lysosomes to establish its replication vacuoles. Within a few minutes after infection, *C. burnetii* actively interacts with the autophagic pathway in order to generate its replicative niche within acidic and degradative (i.e., lysosomal-like) compartments. The early secretory pathway also contributes to the development of the *Coxiella*-replicative vacuole. The strategy used by *S. aureus* seems to involve an early fusion event with autophagosomes, avoiding fusion with lysosomes. After this interaction with autophagic compartments, *S. aureus* escapes into and replicates inside the host cell cytoplasm

factors that modulate the host cell molecular machinery may form the basis for novel therapeutic interventions against intracellular pathogens. Furthermore, the knowledge acquired may shed new light on the pathogenesis of several of these bacterial diseases.

Acknowledgments We thank Cristina Vázquez and Cecilia Lerena, members of our laboratory, for helpful suggestions and critical reading of this review. Our work in this area is supported by grants from Agencia Nacional de Promoción Científica y Tecnológica (PICT2004 #20711 and PICT 2005 #38420), CONICET (PIP #5943) and SECTyP (Universidad Nacional de Cuyo).

Note added in proof A very recent publication has demonstrated that *Coxiella* can be grown in axenic media. Please see the following publication: Omsland A, Cockrell DC, Howe D, Fischer ER, Virtaneva K, Sturdevant DE, Porcella SF, Heinzen RA. Host cell-free grwoth of the Q fever bacterium *Coxiella burnetii*. Proc Natl Acad Sci USA. 2009 Mar 17;106(11):4430–4

References

Abrahams MA, Tylkowski CM (1985) *Brucella* osteomyelitis of a closed femur fracture. Clin Orthop Relat Res 194–196

Almeida RA, Matthews KR, Cifrian E, Guidry AJ, Oliver SP (1996) *Staphylococcus aureus* invasion of bovine mammary epithelial cells. J Dairy Sci 79:1021–1026

Amano A, Nakagawa I, Yoshimori T (2006) Autophagy in innate immunity against intracellular bacteria. J Biochem 140:161–166

Amer AO, Swanson MS (2005) Autophagy is an immediate macrophage response to Legionella pneumophila. Cell Microbiol 7:765–778

Arenas GN, Staskevich AS, Aballay A, Mayorga LS (2000) Intracellular trafficking of *Brucella abortus* in J774 macrophages. Infect Immun 68:4255–4263

Bantel H, Sinha B, Domschke W, Peters G, Schulze-Osthoff K, Janicke RU (2001) Alpha-toxin is a mediator of *Staphylococcus aureus*-induced cell death and activates caspases via the intrinsic death pathway independently of death receptor signaling. J Cell Biol 155:637–648

Bayles KW, Wesson CA, Liou LE, Fox LK, Bohach GA, Trumble WR (1998) Intracellular *Staphylococcus aureus* escapes the endosome and induces apoptosis in epithelial cells. Infect Immun 66:336–342

Belanger M, Rodrigues P, Progulske-Fox A (2007) Genetic manipulation of *Porphyromonas gingivalis*. Curr Protoc Microbiol Chapter 13:Unit13C

Belton CM, Izutsu KT, Goodwin PC, Park Y, Lamont RJ (1999) Fluorescence image analysis of the association between *Porphyromonas gingivalis* and gingival epithelial cells. Cell Microbiol 1:215

Besier S, Ludwig A, Ohlsen K, Brade V, Wichelhaus TA (2007) Molecular analysis of the thymidine-auxotrophic small colony variant phenotype of *Staphylococcus aureus*. Int J Med Microbiol 297:217–225

Beron W, Gutierrez MG, Rabinovitch M, Colombo MI (2002) *Coxiella burnetii* localizes in a Rab7-labeled compartment with autophagic characteristics. Infect Immun 70:5816–5821

Biederbick A, Kern HF, Elsasser HP (1995) Monodansylcadaverine (MDC) is a specific in vivo marker for autophagic vacuoles. Eur J Cell Biol 66:3–14

Blommaart EF, Krause U, Schellens JP, Vreeling-Sindelarova H, Meijer AJ (1997) The phosphatidylinositol 3-kinase inhibitors wortmannin and LY294002 inhibit autophagy in isolated rat hepatocytes. Eur J Biochem 243:240–246

Capo C, Lindberg FP, Meconi S, Zaffran Y, Tardei G, Brown EJ, Raoult D, Mege JL (1999) Subversion of monocyte functions by *Coxiella burnetii:* impairment of the cross-talk between alphavbeta3 integrin and CR3. J Immunol 163:6078–6085

Celli J, de CC, Franchini DM, Pizarro-Cerda J, Moreno E, Gorvel JP (2003) *Brucella* evades macrophage killing via VirB-dependent sustained interactions with the endoplasmic reticulum. J Exp Med 198:545–556

Clemens DL, Lee BY, Horwitz MA (2000) Deviant expression of Rab5 on phagosomes containing the intracellular pathogens *Mycobacterium tuberculosis* and *Legionella pneumophila* is associated with altered phagosomal fate. Infect Immun 68:2671–2684

Coleman SA, Fischer ER, Howe D, Mead DJ, Heinzen RA (2004) Temporal analysis of *Coxiella burnetii* morphological differentiation. J Bacteriol 186:7344–7352

Comerci DJ, Martinez-Lorenzo MJ, Sieira R, Gorvel JP, Ugalde RA (2001) Essential role of the VirB machinery in the maturation of the *Brucella abortus*-containing vacuole. Cell Microbiol 3:159–168

Corbel MJ (1980) The effect of synthetic polynucleotides on the intracellular growth of *Brucella abortus* in vitro and in vivo. Br Vet J 136:175–181

Cutler NS, Heitman J, Cardenas ME (1999) TOR kinase homologs function in a signal transduction pathway that is conserved from yeast to mammals. Mol Cell Endocrinol 155:135–142

Demma LJ, Holman RC, McQuiston JH, Krebs JW, Swerdlow DL, Epidemiology of human ehrlichiosis and anaplasmosis in the United States, 2001–2002. Am J Trop Med Hyg. 2005 Aug;73(2):400–9

Derre I, Isberg RR (2004) Macrophages from mice with the restrictive Lgn1 allele exhibit multifactorial resistance to *Legionella pneumophila*. Infect Immun 72:6221–6229

Deshpande RG, Khan MB, Genco CA (1998) Invasion of aortic and heart endothelial cells by *Porphyromonas gingivalis*. Infect Immun 66:5337–5343

Dorn BR, Dunn WA Jr, Progulske-Fox A (1999) Invasion of human coronary artery cells by periodontal pathogens. Infect Immun 67:5792–5798

Dorn BR, Dunn WA Jr, Progulske-Fox A (2002) Bacterial interactions with the autophagic pathway. Cell Microbiol 4:1–10

Dorn BR, Dunn WA, Progulske-Fox A (2001) *Porphyromonas gingivalis* traffics to autophagosomes in human coronary artery endothelial cells, Infect. Immun 69:5698–5708

Dumler JS, Bakken JS (1998) Human ehrlichioses: newly recognized infections transmitted by ticks. Annu Rev Med 49:201–213

Dunn WA (1990) Studies on the mechanisms of autophagy: maturation of the autophagic vacuole. J Cell Biol 110:1935–1945

Eskelinen EL, Prescott AR, Cooper J, Brachmann SM, Wang L, Tang X, Backer JM, Lucocq JM (2002) Inhibition of autophagy in mitotic animal cells. Traffic 3:878–893

Fiehn NE, Klausen B, Evans RT (1992) Periodontal bone loss in *Porphyromonas gingivalis*-infected specific pathogen-free rats after preinoculation with endogenous *Streptococcus sanguis*. J Periodontal Res 27:609–614

Goodman JL, Nelson C, Vitale B, Madigan JE, Dumler JS, Kurtti TJ, Munderloh UG (1996) Direct cultivation of the causative agent of human granulocytic ehrlichiosis. N Engl J Med 334:209–215

Gorvel JP, Moreno E (2002) *Brucella* intracellular life: from invasion to intracellular replication. Vet Microbiol 90:281–297

Gutierrez MG, Vazquez CL, Munafo DB, Zoppino FC, Beron W, Rabinovitch M, Colombo MI (2005) Autophagy induction favours the generation and maturation of the *Coxiella*-replicative vacuoles. Cell Microbiol 7:981–993

Gray GS, Kehoe M (1984) Primary sequence of the alpha-toxin gene from *Staphylococcus aureus* wood 46. Infect Immun 46:615–618

Heinzen RA, Hackstadt T, Samuel JE (1999) Developmental biology of *Coxiella burnetii*. Trends Microbiol 7:149–154

Heinzen RA, Howe D, Mallavia LP, Rockey DD, Hackstadt T (1996) Developmentally regulated synthesis of an unusually small, basic peptide by *Coxiella burnetii*. Mol Microbiol 22:9–19

Herron MJ, Ericson ME, Kurtti TJ, Munderloh UG (2005) The interactions of *Anaplasma phagocytophilum*, endothelial cells, and human neutrophils. Ann N Y Acad Sci 1063:374–382

Hotopp JC, Lin M, Madupu R, Crabtree J, Angiuoli SV, Eisen JA, Seshadri R, Ren Q, Wu M, Utterback TR, Smith S, Lewis M, Khouri H, Zhang C, Niu H, Lin Q, Ohashi N, Zhi N, Nelson W, Brinkac LM, Dodson RJ, Rosovitz MJ, Sundaram J, Daugherty SC, Davidsen T, Durkin AS, Gwinn M, Haft DH, Selengut JD, Sullivan SA, Zafar N, Zhou L, Benahmed F, Forberger H, Halpin R, Mulligan S, Robinson J, White O, Rikihisa Y, Tettelin H (2006) Comparative genomics of emerging human ehrlichiosis agents. PLoS Genet 2:e21

Houalet-Jeanne S, Pellen-Mussi P, Tricot-Doleux S, Apiou J, Bonnaure-Mallet M (2001) Assessment of internalization and viability of *Porphyromonas gingivalis* in KB epithelial cells by confocal microscopy. Infect Immun 69:7146–7151

Horwitz MA (1983) The Legionnaires' disease bacterium (*Legionella pneumophila*) inhibits phagosome-lysosome fusion in human monocytes. J Exp Med 158:2108–2126

Howe D, Melnicakova J, Barak I, Heinzen RA (2003) Maturation of the *Coxiella burnetii* parasitophorous vacuole requires bacterial protein synthesis but not replication. Cell Microbiol 5:469–480

Hudson MC, Ramp WK, Nicholson NC, Williams AS, Nousiainen MT (1995) Internalization of *Staphylococcus aureus* by cultured osteoblasts. Microb Pathog 19:409–419

Jager S, Bucci C, Tanida I, Ueno T, Kominami E, Saftig P, Eskelinen EL (2004) Role for Rab7 in maturation of late autophagic vacuoles. J Cell Sci 117:4837–4848

Jarry TM, Cheung AL (2006) *Staphylococcus aureus* escapes more efficiently from the phagosome of a cystic fibrosis bronchial epithelial cell line than from its normal counterpart. Infect Immun 74:2568–2577

Ji G, Beavis R, Novick RP (1997) Bacterial interference caused by autoinducing peptide variants. Science 276:2027–2030

Ji G, Beavis RC, Novick RP (1995) Cell density control of staphylococcal virulence mediated by an octapeptide pheromone. Proc Natl Acad Sci USA 92:12055–12059

Jones SM, Winter AJ (1992) Survival of virulent and attenuated strains of *Brucella abortus* in normal and gamma interferon-activated murine peritoneal macrophages. Infect Immun 60:3011–3014

Juhas M, Crook DW, Hood DW (2008) Type IV secretion systems: tools of bacterial horizontal gene transfer and virulence, Cell Microbiol. 49:965–976

Kadowaki T, Takii R, Yamatake K, Kawakubo T, Tsukuba T, Yamamoto K (2007) A role for gingipains in cellular responses and bacterial survival in *Porphyromonas gingivalis*-infected cells. Front Biosci 12:4800–4809

Kagan JC, Roy CR (2002) *Legionella* phagosomes intercept vesicular traffic from endoplasmic reticulum exit sites. Nat Cell Biol 4:945–954

Kagan JC, Stein MP, Pypaert M, Roy CR (2004) *Legionella* subvert the functions of Rab1 and Sec22b to create a replicative organelle. J Exp Med 199:1201–1211

Kahl BC, Goulian M, van WW, Herrmann M, Simon SM, Kaplan G, Peters G, Cheung AL (2000) *Staphylococcus aureus* RN6390 replicates and induces apoptosis in a pulmonary epithelial cell line. Infect Immun 68:5385–5392

Klionsky DJ, Emr SD (2000) Autophagy as a regulated pathway of cellular degradation. Science 290:1717–1721

Krech U, Pagon S, Sonnabend W (1980) Microbiological aspects of Legionnaires' disease. Schweiz Med Wochenschr 110:1739–1745

Kubica M, Guzik K, Koziel J, Zarebski M, Richter W, Gajkowska B, Golda A, iag-Gudowska A, Brix K, Shaw L, Foster T, Potempa J (2008) A potential new pathway for *Staphylococcus aureus* dissemination: the silent survival of S. aureus phagocytosed by human monocyte-derived macrophages. PLoS ONE 3:e1409

Lowy FD (1998) *Staphylococcus aureus* infections. N Engl J Med 339:520–532

Luong TT, Lee CY (2002) Overproduction of type 8 capsular polysaccharide augments *Staphylococcus aureus* virulence. Infect Immun 70:3389–3395

Madariaga MG, Rezai K, Trenholme GM, Weinstein RA (2003) Q fever: a biological weapon in your backyard. Lancet Infect Dis 3:709–721

Marra A, Shuman HA (1992) Genetics of *Legionella pneumophila* virulence. Annu Rev Genet 26:51–69

Marrack P, Kappler J (1990) The staphylococcal enterotoxins and their relatives. Science 248:1066

Mizushima N, Noda T, Yoshimori T, Tanaka Y, Ishii T, George MD, Klionsky DJ, Ohsumi M, Ohsumi Y (1998) A protein conjugation system essential for autophagy. Nature 395:395–398

Mott J, Barnewall RE, Rikihisa Y (1999) Human granulocytic ehrlichiosis agent and *Ehrlichia chaffeensis* reside in different cytoplasmic compartments in HL-60 cells. Infect Immun 67:1368–1378

Munafo DB, Colombo MI (2001) A novel assay to study autophagy: regulation of autophagosome vacuole size by amino acid deprivation. J Cell Sci 114:3619–3629

Munderloh UG, Lynch MJ, Herron MJ, Palmer AT, Kurtti TJ, Nelson RD, Goodman JL (2004) Infection of endothelial cells with *Anaplasma marginale* and *A. phagocytophilum*. Vet Microbiol 101:53–64

Niu H, Yamaguchi M, Rikihisa Y (2008) Subversion of cellular autophagy by *Anaplasma phagocytophilum*. Cell Microbiol 10:593–605

Novick RP (2003) Autoinduction and signal transduction in the regulation of staphylococcal virulence. Mol Microbiol 48:1429–1449

Otto GP, Wu MY, Kazgan N, Anderson OR, Kessin RH (2004) *Dictyostelium* macroautophagy mutants vary in the severity of their developmental defects. J Biol Chem 279:15621–15629

Pattingre S, Tassa A, Qu X, Garuti R, Liang XH, Mizushima N, Packer M, Schneider MD, Levine B (2005) Bcl-2 antiapoptotic proteins inhibit Beclin 1-dependent autophagy. Cell 122:927–939

Pizarro-Cerda J, Cossart P (2006) Bacterial adhesion and entry into host cells. Cell 124:715–727

Pizarro-Cerda J, Meresse S, Parton RG, van der GG, Sola-Landa A, Lopez-Goni I, Moreno E, Gorvel JP (1998) *Brucella abortus* transits through the autophagic pathway and replicates in the endoplasmic reticulum of nonprofessional phagocytes. Infect Immun 66:5711–5724

Proctor RA, Balwit JM, Vesga O (1994) Variant subpopulations *of Staphylococcus aureus* as cause of persistent and recurrent infections, Infect Agents Dis 3:302–312

Projan SJ, Brown-Skrobot S, Schlievert PM, Vandenesch F, Novick RP (1994) Glycerol monolaurate inhibits the production of beta-lactamase, toxic shock toxin-1, and other staphylococcal exoproteins by interfering with signal transduction. J Bacteriol 176:4204–4209

Romano PS, Gutierrez MG, Beron W, Rabinovitch M, Colombo MI (2007) The autophagic pathway is actively modulated by phase II *Coxiella burnetii* to efficiently replicate in the host cell. Cell Microbiol 9:891–909

Roy CR, Tilney LG (2002) The road less traveled: transport of *Legionella* to the endoplasmic reticulum. J Cell Biol 158:415–419

Schnaith A, Kashkar H, Leggio SA, Addicks K, Kronke M, Krut O (2007) *Staphylococcus aureus* subvert autophagy for induction of caspase-independent host cell death. J Biol Chem 282:2695–2706

Seshadri R, Samuel JE (2001) Characterization of a stress-induced alternate sigma factor, RpoS, of *Coxiella burnetii* and its expression during the development cycle. Infect Immun 69:4874–4883

Smith R, Adams III,LG, Sowa BA, Ficht TA (1990) Induction of lymphocyte responsiveness by the outer membrane-peptidoglycan complex of rough strains of *Brucella abortus*. Vet Immunol Immunopathol 26:31–48

Sola-Landa A, Pizarro-Cerda J, Grillo MJ, Moreno E, Moriyon I, Blasco JM, Gorvel JP, Lopez-Goni I (1998) A two-component regulatory system playing a critical role in plant pathogens and endosymbionts is present in *Brucella abortus* and controls cell invasion and virulence. Mol Microbiol 29:125–138

Street L, Grant WW Alva JDJr, (1975) Brucellosis in childhood. Pediatrics 55:416–421

Sturgill-Koszycki S, Swanson MS (2000) *Legionella pneumophila* replication vacuoles mature into acidic, endocytic organelles. J Exp Med 192:1261–1272

Swanson MS, Fernandez-Moreira E (2002) A microbial strategy to multiply in macrophages: the pregnant pause. Traffic 3:170–177

Swanson MS, Isberg RR (1995) Association of *Legionella pneumophila* with the macrophage endoplasmic reticulum. Infect Immun 63:3609–3620

Swanson MS, Sturgill-Koszycki I (2000) Exploitation of macrophages as a replication niche by *Legionella pneumophila*. Trends Microbiol 8:47–49

Takahashi Y, Davey M, Yumoto H, Gibson FC III, Genco CA (2006) Fimbria-dependent activation of pro-inflammatory molecules in *Porphyromonas gingivalis* infected human aortic endothelial cells. Cell Microbiol 8:738–757

Tamai R, Asai Y, Ogawa T (2005) Requirement for intercellular adhesion molecule 1 and caveolae in invasion of human oral epithelial cells by *Porphyromonas gingivalis*. Infect Immun 73:6290–6298

van Sluijters DA, Dubbelhuis PF, Blommaart EF, Meijer AJ (2000) Amino-acid-dependent signal transduction. Biochem J 351:Pt 3 545–550

Varghees S, Kiss K, Frans G, Braha O, Samuel JE (2002) Cloning and porin activity of the major outer membrane protein P1 from *Coxiella burnetii*. Infect Immun 70:6741–6750

Veras PS, de CC, Moreau MF, Villiers V, Thibon M, Mattei D, Rabinovitch M (1994) Fusion between large phagocytic vesicles: targeting of yeast and other particulates to phagolysosomes that shelter the bacterium *Coxiella burnetii* or the protozoan *Leishmania amazonensis* in Chinese hamster ovary cells. J Cell Sci 107:(Pt 11) 3065–3076

Vogel JP, Andrews HL, Wong SK, Isberg RR (1998) Conjugative transfer by the virulence system of *Legionella pneumophila*. Science 279:873–876

Watarai M, Derre I, Kirby J, Growney JD, Dietrich WF, Isberg RR (2001) *Legionella pneumophila* is internalized by a macropinocytotic uptake pathway controlled by the Dot/Icm system and the mouse *Lgn1* locus. J Exp Med 194:1081–1096

Webster P, IJdo JW, Chicoine LM, Fikrig E (1998) The agent of human granulocytic ehrlichiosis resides in an endosomal compartment. J Clin Invest 101:1932–1941

Weisburg WG, Dobson ME, Samuel JE, Dasch GA, Mallavia LP, Baca O, Mandelco L, Sechrest JE, Weiss E, Woese CR (1989) Phylogenetic diversity of the Rickettsiae. J Bacteriol 171:4202–4206

Wesson CA, Liou LE, Todd KM, Bohach GA, Trumble WR, Bayles KW (1998) *Staphylococcus aureus Agr* and *Sar* global regulators influence internalization and induction of apoptosis. Infect Immun 66:5238–5243

Yamatake K, Maeda M, Kadowaki T, Takii R, Tsukuba T, Ueno T, Kominami E, Yokota S, Yamamoto K (2007) Role for gingipains in *Porphyromonas gingivalis* traffic to phagolysosomes and survival in human aortic endothelial cells. Infect Immun 75:2090–2100

Yilmaz O (2008) The chronicles of *Porphyromonas gingivalis*: the microbium, the human oral epithelium and their interplay, Microbiology 154:2897–2903.

Zamboni DS, McGrath S, Rabinovitch M, Roy CR (2003) *Coxiella burnetii* express type IV secretion system proteins that function similarly to components of the *Legionella pneumophila* Dot/Icm system, Mol. Microbiol 49:965–976

Zusman T, Yerushalmi G, Segal G (2003) Functional similarities between the *icm/dot* pathogenesis systems of *Coxiella burnetii* and *Legionella pneumophila*. Infect Immun 71:3714–3723

Autophagy in Immunity Against *Toxoplasma gondii*

Carlos S. Subauste

Contents

1 Introduction 252
2 CD40 and *T. gondii* Infection 253
 2.1 CD40 Induces Toxoplasmacidal Activity by Vacuole-Lysosomal Fusion 254
 2.2 CD40 Induces Toxoplasmacidal Activity Through Autophagy 256
3 Relevance of Autophagy in *T. gondii* Infection 259
4 Conclusion 260
References 261

Abstract A decisive outcome during host-pathogen interaction is governed by whether pathogen-containing vacuoles fuse with lysosomes. Fusion with lysosomes typically kills microbes. *Toxoplasma gondii* represents a classical example of an intracellular pathogen that survives within host cells by preventing the endosomal-lysosomal compartments from fusing with the vacuoles that contain the pathogen. Thus, *T. gondii* provides an excellent model to determine if the immune system can target a pathogen for lysosomal degradation. CD40, a major regulator of cell-mediated immunity, activates macrophages to kill *T. gondii* through a process that requires recruitment of autophagosomes around the parasitophorous vacuole, leading to lysosomal degradation of the parasite. These studies demonstrate that cell-mediated immunity can activate autophagy to kill a pathogen. CD40-induced autophagy likely contributes to resistance against *T. gondii*, particularly in neural tissues, the main sites affected by this pathogen.

C.S. Subauste (✉)
Department of Ophthalmology and Visual Sciences, Case Western Reserve University School of Medicine, Cleveland, OH, 44106, USA; Department of Medicine, Division of Infectious Diseases and HIV Medicine, Case Western Reserve University School of Medicine, Cleveland, OH, 44106, USA; Department of Pathology, Case Western Reserve University School of Medicine, Cleveland, OH, 44106, USA
e-mail: carlos.subauste@case.edu

B. Levine et al. (eds.), *Autophagy in Infection and Immunity,*
Current Topics in Microbiology and Immunology 335,
DOI 10.1007/ 978-3-642-00302-8_12,

1 Introduction

*Toxoplasma gondii*is an obligate intracellular protozoan that infects a broad variety of hosts, including humans. *T. gondii* infection is commonly asymptomatic. However, clinical evidence of disease (toxoplasmosis) can occur, especially in immunodeficient patients and in babies infected in utero (Montoya et al. 2005). *T. gondii* exists in three forms: (1) the tachyzoite or invasive form of the parasite (tachyzoites can infect almost any nucleated cell through a process of active invasion) (Carruthers and Boothroyd 2007); (2) the tissue cyst that appears to persist in tissues of the infected host for life; and (3) the oocyst that forms during the sexual cycle, which takes place exclusively in the intestines of acutely infected felines. Humans become infected by eating poorly coked meat that contains tissue cysts, by ingesting food or water contaminated with oocysts, or by materno-fetal transmission (Montoya et al. 2005).

During the acute phase of the infection, the tachyzoites disseminate through the bloodstream. This is followed by the chronic phase of infection, which is characterized by the disappearance of tachyzoites and the formation of tissue cysts (primarily in the central nervous system and skeletal muscle). The development of an immune response is accompanied by the formation of *T. gondii* tissue cysts that characterize the chronic phase of infection.

Seropositivity for antibodies against *T. gondii* infection varies depending on the geographical location and on dietary habits, and in some parts of the world more than 50% of individuals are chronically infected with *T. gondii* (Montoya et al. 2005). *T. gondii* can cause disease in adults that manifests primarily as ocular toxoplasmosis and/or cerebral toxoplasmosis. Toxoplasmic encephalitis occurs in immunocompromised humans such as AIDS patients or organ transplant recipients (Israelski and Remington 1993; Navia et al. 1986). In contrast, ocular toxoplasmosis occurs not only in immunocompromised patients but also in immunocompetent ones. Indeed, ocular toxoplasmosis (toxoplasmic retinochoroiditis) is the most common form of posterior uveitis in otherwise healthy individuals (McCannel et al. 1996). Acute infection in pregnant women can lead to congenital transmission of the infection to the fetus and result in severe neurological and ocular sequelae or abortion (Kravetz and Federman 2005).

Tachyzoites of *T. gondii* survive within mammalian cells by residing in a specialized niche called the parasitophorous vacuole (Lingelbach and Joiner 1998). This structure is formed during the process of active invasion of host cells. As the parasite penetrates these cells, the tachyzoite modifies the membrane that surrounds it by excluding many host cell proteins and by inserting proteins of parasitic origin (Hakansson et al. 2001; Joiner and Roos 2002). The extensively modified membrane of the parasitophorous vacuole allows the organism to survive by permitting the transport of nutrients (Martin et al. 2007). Host endocytic structures are recruited to the parasitophorous vacuole and delivered intact into the vacuolar space (Coppens et al. 2006). However, no release of endosomal contents into the parasitophorous vacuole takes place. Thus, the membrane of the parasitophorous vacuole fulfills two important functions: it provides access to nutrients, including those from the endocytic compartment of the host cell, and it prevents actual fusion of the parasitophorous vacuole with endosomes or lysosomes. The dogma had been that the nonfusogenic nature of the parasitophorous vacuole was irreversible and was established at the time of active invasion (Joiner et al. 1990; Mordue and Sibley 1997). A fundamental question in regard to host-pathogen interactions is whether the immune system can change the nonfusogenicity of the parasitophorous vacuole and target it for lysosomal degradation. This question is important because *T. gondii* cannot withstand the lysosomal environment. We recently reported that cell-mediated immunity accomplishes this task through CD40 (Andrade et al. 2006; Subauste et al. 2007a).

2 CD40 and *T. gondii* Infection

Studies in mice have revealed that IFN-γ is essential for protection against *T. gondii*. IFN-γ-deficient mice die of fulminant infection, even when challenged with an avirulent strain of the parasite (Gazzinelli et al. 1991; Suzuki et al. 1989; Suzuki et al. 1988). Similarly, IFN-γ is required to control the parasite in the brain and eye (Gazzinelli et al. 1994; Gazzinelli et al. 1993; Suzuki et al. 1989). TNF-α, NOS2 and the production of nitric oxide are also critical for resistance against toxoplasmic encephalitis and ocular toxoplasmosis (Deckert-Schluter et al. 1998; Gazzinelli et al. 1994, 1993; Hayashi et al. 1996a, b; Roberts et al. 2000; Scharton-Kersten et al. 1997; Yap et al. 1998). These findings are in keeping with the notion that IFN-γ and TNF-α confer protection against the parasite by synergistically inducing NOS2 expression.

Although these studies provide important information on the nature of protective immunity by reporting a key role of IFN-γ/TNF-α-induced NOS2 in resistance against *T. gondii* in neural tissue, it is likely that there are other components of cellular immunity that are also required for control of the parasite in these sites. This possibility is particularly relevant to humans because of their restricted expression of NOS2 and because patients with a mutation in IFN-γR1 that prevents recruitment of STAT-1 do not develop disease after *T. gondii* infection, while STAT-$1^{-/-}$ mice die acutely after challenge with the parasite (Gavrielescu et al. 2004; Janssen et al. 2002; Lieberman et al. 2004).

CD40 is a member of the TNF receptor superfamily expressed on various cells, including antigen-presenting cells (van Kooten and Banchereau 2000). Its counter receptor, CD154 (CD40 ligand), is expressed primarily on activated T cells (van Kooten and Banchereau 2000). The CD40-CD154 pathway mediates resistance against *T. gondii* in humans and mice. Patients with X-linked Hyper IgM syndrome (X-HIGM, a congenital immunodeficiency caused by a lack of functional CD154) are susceptible to toxoplasmosis (Leiva et al. 1998; Subauste et al. 1999). While the CD40-CD154 pathway restricts the growth of *T. gondii* in peripheral tissues during the acute phase of infection (Subauste and Wessendarp 2006), the role of this pathway is particularly important for controlling the parasite in the brain. $CD154^{-/-}$ mice are susceptible to toxoplasmic encephalitis (Reichmann et al. 2000).

Although the CD40-CD154 pathway promotes Th1-type cytokine response against *T. gondii* in vitro in human cells and in vivo in mice (Subauste and Wessendarp 2000; Subauste et al. 1999; Subauste, unpublished observations), animal studies indicate that this pathway also activates mechanisms of host resistance that act independently of IFN-γ (Reichmann et al. 2000). $CD154^{-/-}$ mice die of toxoplasmic encephalitis despite upregulation of IFN-γ in the brain that is similar to that in infected wild-type mice (Reichmann et al. 2000). This raises the possibility that CD40 utilizes an alternate pathway to promote resistance to *T. gondii* in the brain.

2.1 CD40 Induces Toxoplasmacidal Activity by Vacuole-Lysosomal Fusion

Macrophages are key effectors of resistance against *T. gondii* (Gazzinelli et al. 1993; Robben et al. 2005). The inflammatory infiltrate in toxoplasmosis is characterized not only by macrophages but also by the presence of T cells (Gazzinelli et al. 1994; Schluter et al. 1991). Thus, $CD154^{+}$*T. gondii*-reactive activated $CD4^{+}$ T cells likely provide CD40 signaling to macrophages. Whereas *T. gondii* survives and multiplies in resting macrophages, CD40 stimulation of these cells allows them to acquire toxoplasmacidal activity (Andrade et al. 2003, 2005a, b, 2006). This effect is mediated by not only recombinant CD154 or an agonistic anti-CD40 mAb, but also by CD154 expressed on the membrane of activated $CD4^{+}$ T cells (Andrade et al. 2005a). Of relevance to mechanisms of resistance that act independently of IFN-γ/NOS2, the killing of *T. gondii* tachyzoites induced by CD40 does not require IFN-γ or effector molecules downstream of this cytokine such as NOS2 and immunity-related GTPases (IRGs) (Andrade et al. 2003, 2005a; Subauste and Wessendarp 2006). In addition, the killing of the parasite is not mediated by the oxidative pathway or starvation for tryptophan (Andrade et al. 2005a).

CD40 alters the fate of intracellular tachyzoites because it causes fusion of parasite-containing vacuoles with late endosomes and lysosomes, as assessed by colocalization with the acidotropic dye Lysotracker Red and the late endosomal/lysosomal markers mannose-6-phosphate receptor, Rab7, LAMP-1, LAMP-2, CD63 and cathepsin D (Andrade et al. 2006; Subauste et al. 2007a). A critical question was whether parasitophorous vacuoles, which by definition had been considered nonfusogeneic, fuse with late endosomes/lysosomes. The formation of parasitophorous vacuoles involves the secretion of contents of parasite organelles located in the apex of the organisms (Archbarou et al. 1991; Carruthers and Sibley 1997; Cesbron-Delauw 1994). The use of transgenic parasites that express a fluorescent protein targeted to the dense granules allowed us to demonstrate that vacuoles in which contents of the dense granules are secreted into the vacuolar lumen fuse with late endosomes/lysosomes (Andrade et al. 2006) (Fig. 1). Moreover, preformed parasitophorous vacuoles still fuse with late endosomes/lysosomes even if

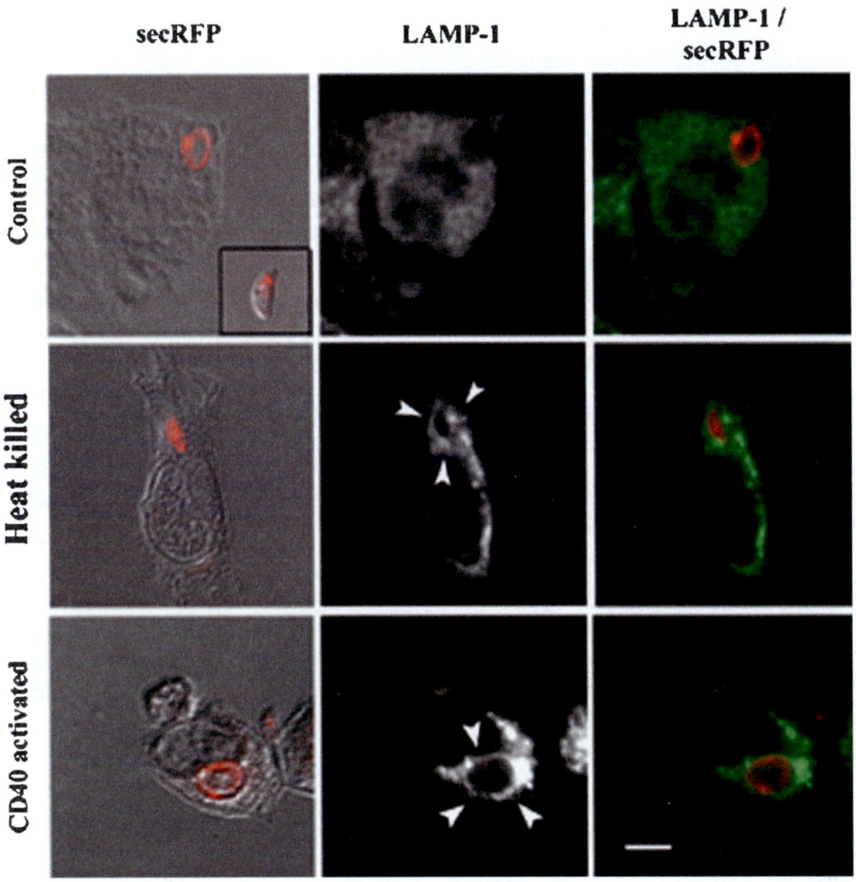

Fig. 1 CD40 stimulation induces convergence of parasitophorous vacuoles and late endosomes/lysosomes. Mouse resident peritoneal macrophages were incubated with control or stimulatory anti-CD40 mAbs and subsequently challenged with transgenic *T. gondii* that express RFP in their dense granules. These parasites secrete RFP into the lumen of parasitophorous vacuoles (*T. gondii*-secRFP). Expression of LAMP-1 was assessed at 8 h after challenge. *Inset* represents an extracellular tachyzoite with accumulation of fluorescence in dense granules. Macrophages were also incubated with heat-killed *T. gondii*-secRFP, and LAMP-1 expression was assessed at 1 h after challenge. *Arrowheads* indicate colocalization of LAMP-1 around *T. gondii*-containing compartments. Scale bar: 5 μm. Reprinted from Andrade et al. (2006) with permission

CD40 is engaged 18 h after infection (Andrade et al. 2006). Therefore, the immune system through CD40 alters a fundamental aspect of the biology of *T. gondii*, the avoidance of lysosomal fusion.

CD40 induces killing of *T. gondii* through vacuole-lysosome fusion because pharmacologic inhibition of lysosomal enzymes, vacuolar ATPase, phosphoinositide-3-kinase (PI3K), and, more importantly, expression of a dominant negative mutant of Rab7 as well as knockdown of Class III PI3K (hVps34) abrogate the killing of *T. gondii* induced by CD40 (Andrade et al. 2006). These studies uncovered a new paradigm in which the interaction between CD154 on T cells and CD40 expressed on macrophages leads to the killing of an intracellular pathogen via the induction of vacuole-lysosomal fusion (Andrade et al. 2006; Subauste et al. 2007a). The fact that in vivo CD40 stimulation induces macrophage toxoplasmacidal activity and reduces the parasite load (Subauste and Wessendarp 2006) strongly suggests that vacuole-lysosome fusion induced by CD40 contributes to host protection.

2.2 CD40 Induces Toxoplasmacidal Activity Through Autophagy

Lysosomal degradation as a mechanism to kill pathogens that normally reside within nonfusogenic vacuoles requires rerouting these vacuoles to the lysosomal compartment. In the case of *T. gondii*, this possibility is further supported by studies of the expression of endosomal and lysosomal markers; late endosomal/lysosomal markers are recruited to the parasitophorous vacuole despite the lack of evidence of recruitment of the early endosomal markers Rab5 and EEA1 (Andrade et al. 2006). This raises the possibility that CD40 reroutes the parasitophorous vacuole to the endosomal/lysosomal compartment using a mechanism that is distinct from the classical pathway of phago-lysosomal fusion. In this regard, autophagy is a process that directs cytoplasmic material and organelles to lysosomes (Levine and Klionsky 2004; Mizushima et al. 2002). This is a ubiquitous and highly conserved process in which an isolation membrane encircles portions of cytosol and organelles, leading to the formation of autophagosomes (Dunn 1994; Levine and Klionsky 2004; Mizushima et al. 2002). Fusion between autophagosomes and endosomes/lysosomes culminates in the formation of autolysosomes and the degradation of their contents (Dunn 1994). Thus, autophagy represents an alternate route to the lysosomal compartment.

Autophagy acts as an innate response that targets pathogens such as *Streptococcus pyogenes*, metabolically arrested *Listeria monocytogenes, Salmonella enterica* and *Francisella tularensis* (Birmingham et al. 2006; Checroun et al. 2006; Nakagawa et al. 2004; Rich et al. 2003). In the absence of the virulence factors IscB and ICP34.5, *Shigella flexneri* and herpes simplex virus (HSV), respectively, are targeted by autophagosomes (Ogawa et al. 2004; Talloczy et al. 2002). Pharmacological or starvation-induced autophagy also kills mycobacteria (Gutierrez et al. 2004; Singh et al. 2006). An important point is whether the immune system activates autophagy to induce the killing of pathogens. IFN-γ induces autophagosome formation in phagosomes containing *M. tuberculosis* (Gutierrez et al. 2004; Singh et al. 2006) and it has been proposed that IFN-γ kills the pathogen through autophagy. In the case of *T. gondii*, CD40 stimulation of macrophages causes recruitment of the highly specific autophagy marker LC3 (Atg8) around the parasitophorous vacuole (Andrade et al. 2006) (Fig. 2). Consistent with the notion that autophagy is responsible for vacuole-lysosomal fusion, recruitment of LC3 around the parasitophorous vacuole precedes the recruitment of LAMP-1 (Andrade et al. 2006). Moreover, knockdown of Beclin 1 revealed that autophagy is required for fusion

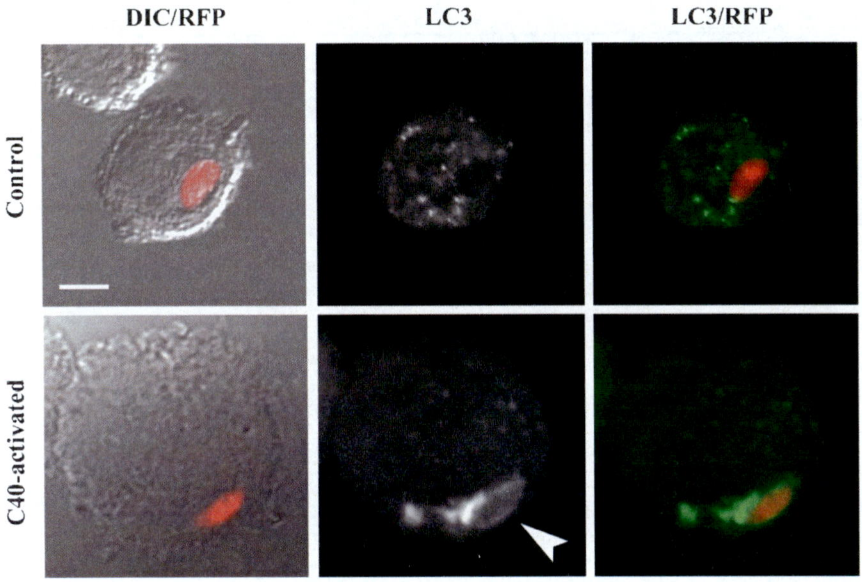

Fig. 2 CD40 stimulation induces vacuole-lysosome fusion and antimicrobial activity through autophagy. Control and CD40-activated RAW 264.7 cells that express a chimera of the extracellular domain of human CD40 and the intracellular tail of mouse CD40 were transfected with LC3-EGFP and infected with *T. gondii* that expresses RFP in their cytoplasm. Monolayers were examined by confocal microscopy at 5 h after challenge. *Arrowhead* indicates accumulation of LC3 around vacuole. Scale bar: 5 μm. Reprinted from Andrade et al. (2006) with permission

of the parasitophorous vacuole with late endosomes/lysosomes and for the killing of *T. gondii* by CD40-activated macrophages (Andrade et al. 2006). Thus, these studies demonstrated that autophagy can be activated by adaptive immunity to reroute *T. gondii* to the lysosomal compartment, resulting in macrophage antimicrobial activity.

More recent studies have begun to identify the signals downstream of CD40 that control autophagy. CD40 signals through the recruitment of adapter proteins since it lacks kinase activity. TNF receptor-associated factors (TRAFs) are major mediators of the effects of CD40 (Bishop et al. 2007). The intracytoplasmic tail of CD40 has two binding sites that directly recruit TRAF2 and TRAF3 and a binding site that directly recruits TRAF6 (Ishida et al. 1996; Lu et al. 2003; Pullen et al. 1998). The TRAF6 binding site plays a dual role in the autophagic killing of *T. gondii*: it enhances autocrine production of TNF-α (Mukundan et al. 2005), and TRAF6 signals downstream of CD40 cooperate with TNF-α to mediate activate autophagy. As a result of this process, autophagosomes are recruited around the parasitophorous vacuole and this is followed by Rab7-dependent fusion with late endosomes/lysosomes and killing of the organism (Andrade et al. 2005b, 2006) (Fig. 3). While TNF-α can modulate autophagy (Djavaheri-Mergny et al. 2006), this cytokine by itself is inadequate for the killing of *T. gondii* by macrophages (Andrade et al. 2005b). Vacuole-lysosomal fusion and killing of the parasite requires synergy

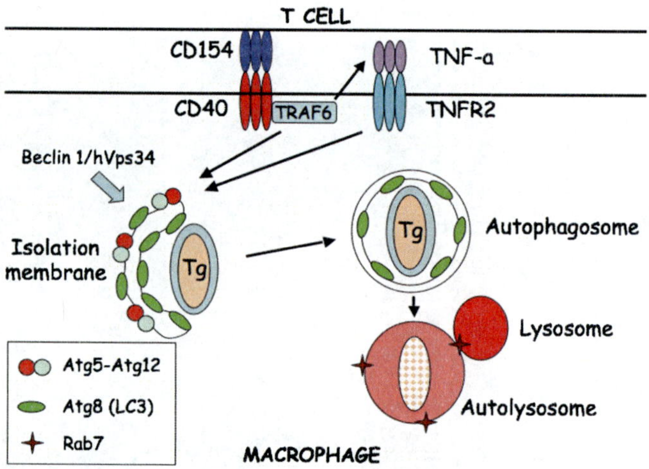

Fig. 3 Autophagy-dependent killing of *T. gondii* induced by CD40. *T. gondii*-specific $CD4^+$ T cell acquires expression of CD154 after interaction with infected macrophages. CD40-CD154 binding results in the recruitment of TRAF6 to the intra-cytoplasmic tail of CD40, which in turn enhances TNF-α production. TRAF6 signals downstream of CD40 and TNF-α signaling synergize to cause recruitment of autophagosomes to the parasitophorous vacuole that contains the pathogen (*Tg*). This process is dependent on Beclin 1 and Class III PI3K. Subsequent recruitment of Rab7 controls fusion with the lysosomes and killing of *T. gondii*

between the CD40-TRAF6 pathway and the TNF-α signaling triggered by CD40 (Andrade et al. 2005b; Subauste et al. 2007a).

IFN-γ has been reported to induce the recruitment of autophagosomes around *T. gondii* (Ling et al. 2006). However, this phenomenon follows the disruption of the parasitophorous vacuole membrane caused by IFN-γ (Ling et al. 2006), a mechanism by which IFN-γ has been reported to kill *T. gondii* (Martens et al. 2005). Thus, the formation of autophagosomes in IFN-γ-treated macrophages is likely not the cause of death of the parasite but rather a response to the presence of altered structures (disrupted parasitophorous membranes and denuded tachyzoites) within the host cell. This finding would be consistent with the known role of autophagy in the removal of altered intracellular structures. Moreover, studies in human and mouse macrophages revealed that pharmacological inhibition of lysosomal enzymes, vacuolar ATPase, PI3K (including the autophagy inhibitor 3-methyladenine), knockdown of hVps34, expression of dominant negative Rab7, and knockdown of Beclin 1 did not affect the killing induced by IFN-γ but ablated that caused by CD40 (Andrade et al. 2006). Interestingly, while autophagy does not mediate the anti-*T. gondii* activity triggered by IFN-γ, the autophagy protein Atg5 is involved in autophagy-independent killing of the parasite in IFN-γ-activated phagocytic cells (Zhao et al. 2008). Atg5 was required for the recruitment of the IFN-γ-inducible IRG IIGP1 (Irga6) to the parasitophorous vacuole, the damage to this structure, and the clearance of the parasite (Zhao et al. 2008) (see the chapter by Cadwell et al. in this volume). The different mechanisms used by CD40 and

IFN-γ to kill *T. gondii* may partially explain the cooperation observed between CD40 and IFN-γ to promote control of the parasite (Andrade et al. 2003).

3 Relevance of Autophagy in *T. gondii* Infection

Despite increasing evidence of the role of autophagy in killing pathogens in vitro, there is still limited evidence that autophagy mediates host protection in vivo. However, recent studies in a mouse model of HSV-1 infection and a *Drosophila* model of *Listeria* infection provide strong evidence to this effect. HSV-1 prevents autophagy by producing the neurovirulence factor ICP34.5 that binds and blocks the effect of Beclin 1 (Orvedahl et al. 2007). A strain of HSV-1 deficient in the Beclin 1 binding domain of ICP34.5 does not cause encephalitis in wild-type mice but causes disease in mice deficient in PKR, a signaling molecule linked to autophagy (Orvedahl et al. 2007). Furthermore, in *Drosophila*, recognition of a bacterial pathogen-associated molecular pattern (PAMP) by a pattern recognition receptor (PRR) was shown to be crucial for autophagy induction during *Listeria monocytogenes* infection, and autophagy prevented intracellular bacterial growth and promoted host survival (Yano et al. 2008).

It is likely that autophagy is also protective against *T. gondii* in vivo, particularly in neural tissue. This is supported by the fact that, despite the upregulation of IFN-γ, the CD40-CD154 pathway is still required for resistance against toxoplasmic encephalitis, together with the demonstration that macrophages (key effectors of protection against the parasite) kill *T. gondii* through autophagy, a process that requires CD40 but is independent of IFN-γ and NOS2. These results support the notion that while IFN-γ/TNF-α-induced NOS2 are needed for resistance against toxoplasmosis, they are insufficient for protection and require CD40-induced autophagic killing of *T. gondii* within macrophages for full control of *T. gondii* (Fig. 4). This paradigm can explain why IFN-γ appears insufficient for control of *T. gondii* in neural tissue (Reichmann et al. 2000; Yap et al. 1998).

CD40-induced autophagy may be particularly relevant to humans infected with *T. gondii*. The immune response against the parasite in mice does not fully mimic immunity against *T. gondii* as it occurs in humans. Although IFN-γ is clearly indispensable for controlling the parasite in mice, IFN-γ-independent mechanisms of resistance against *T. gondii* appear to be more effective in humans. In this regard, children with an autosomal dominant defect in IFN-γR1 that causes a deletion in the STAT-1 binding site do not develop disease when infected with *T. gondii* (Janssen et al. 2002). In marked contrast, STAT-$1^{-/-}$ mice die within one week after infection with the parasite (Gavrielescu et al. 2004; Lieberman et al. 2004). The differences between the immune response in humans and mice also apply to the downstream effectors of IFN-γ, NOS2 and IRGs. Whereas NOS2 is key to resistance against ocular and cerebral toxoplasmosis (Hayashi et al. 1996a; Hayashi et al. 1996b; Scharton-Kersten et al. 1997), NOS2 in humans is more tightly regulated than in rodents, and the production of nitric oxide appears to be weaker in humans

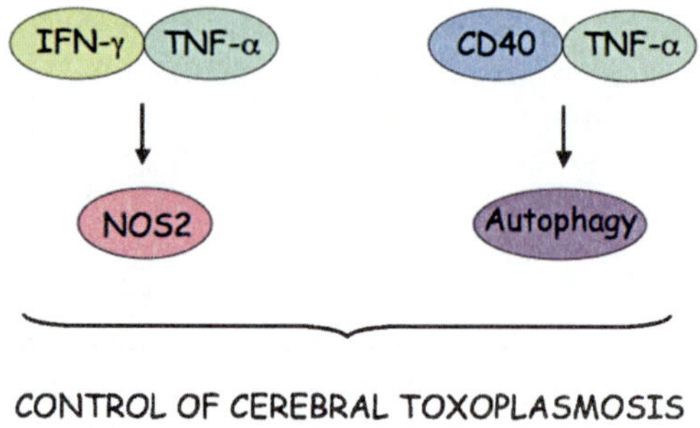

Fig. 4 Both IFN-γ/TNF-α-induced NOS2 and CD40/TNF-α–induced autophagy appear to be required for control of cerebral toxoplasmosis

than in mice (Kroncke et al. 1998). In addition, while mice have 23 IRG genes, among which *Irgm1* (LRG-47), *Irgm3* (IGTP) and *Irga6* (IIGP1) mediate anti-*T. gondii* activity in mouse cells (Collazo et al. 2001, 2002; Martens et al. 2005), IRG genes in humans have been reduced to a truncated gene, *IRGM* (syntenic with the mouse gene *Irgm1*), and *IRGC*, both of which lack an IFN-inducible element (Bekpen et al. 2005). CD40-induced autophagic killing of *T. gondii* may be an important contributor to the control of *T. gondii* in humans because CD40 induces killing of *T. gondii* independently of IFN-γ, STAT1, NOS2 and IRG (Andrade et al. 2005a, 2006; Subauste et al. 2007a; Subauste and Wessendarp 2006). Indeed, defects in the CD40 pathway are relevant to certain groups of patients that develop cerebral and/or ocular toxoplasmosis, including patients with X-linked Hyper IgM syndrome who lack functional CD154 (Subauste et al. 1999), and newborns, since many groups reported impaired expression of CD154 on neonatal $CD4^+$ T cells (Durandy et al. 1995; Han et al. 2004; Julien et al. 2003; Kaur et al. 2007; Nonoyama et al. 1995) and reduced levels of CD40 on dendritic cells (Kaur et al. 2007). Impaired CD154 induction is particularly pronounced in preterm babies (Kaur et al. 2007), a finding relevant to toxoplasmosis in newborns since this is an infection acquired prior to birth. The CD40-dependent pathway of host protection is also relevant to HIV-1-infected patients, because these individuals exhibit a defect in CD154 induction in their $CD4^+$ T cells (Subauste et al. 2001, 2004, 2007b; Zhang et al. 2004).

4 Conclusion

The studies using *T. gondii* have demonstrated that autophagy is not only an innate mechanism of pathogen eradication but that it can be activated by cell-mediated immunity. Given that many pathogens are susceptible to lysosomal degradation, modulation of autophagy may result in novel approaches to enhance host protection. The development of these approaches will require further understanding of the regulation of autophagy, in particular that mediated by the immune system, as well as of the mechanisms used by pathogens to impair autophagy. This knowledge may lead to improved control of toxoplasmosis and other infections, hopefully without adversely affecting the other cellular processes regulated by autophagy.

Acknowledgments The original work by the author was supported by the National Institutes of Health, the American Heart Association (Ohio Valley Affiliate), the Juvenile Diabetes Research Foundation International, the Research to Prevent Blindness Foundation and the Ohio Lions Eye Research Foundation.

References

Andrade RM, Wessendarp M, Subauste CS (2003) CD154 activates macrophage anti-microbial activity in the absence of IFN-γ through a TNF-α-dependent mechanism. J Immunol 171:6750–6756

Andrade RM, Portillo J-AC, Wessendarp M, Subauste CS (2005a) CD40 signaling in macrophages induces anti-microbial activity against an intracellular pathogen independently of IFN-γ and reactive nitrogen intermediates. Infect Immun 73:3115–3123

Andrade RM, Wessendarp M, Portillo J-AC, Yang J-Q, Gomez FJ, Durbin JE, Bishop GA, Subauste CS (2005b) TRAF6 signaling downstream of CD40 primes macrophages to acquire anti-microbial activity in response to TNF-α. J Immunol 175:6014–6021

Andrade RM, Wessendarp M, Gubbels MJ, Striepen B, Subauste CS (2006) CD40 induces macrophage anti-*Toxoplasma gondii* activity by triggering autophagy-dependent fusion of pathogen-containing vacuoles and lysosomes. J Clin Invest 116:2366–2377

Archbarou A, Mercereau-Puijalon O, Sadak A, Fortier B, Leriche MA, Camus D, Dubremetz JF (1991) Differential targeting of dense granule proteins in the parasitophorous vacuole of *Toxoplasma gondii*. Parasitology 103:321–329

Bekpen C, Hunn JP, Rohde C, Parvanova I, Guethlein L, Dunn DM, Glowalla E, Leptin M, Howard JC (2005) The interferon-inducible p47 (IRG) GTPases in vertebrates: loss of the cell autonomous resistance mechanism in the human lineage. Genome Biol 6:R92

Birmingham CL, Smith AC, Bakowski MA, Yoshimori T, Brumell JH (2006) Autophagy controls *Salmonella* infection in response to damage to the Salmonella-containing vacuole. J Biol Chem 281:11374–11383

Bishop GA, Moore CR, Xie P, Stunz LL, Kraus ZJ (2007) TRAF proteins in CD40 signaling. Adv Exp Med Biol 597:131–151

Carruthers VB, Boothroyd JC (2007) Pulling together: an integrated model of *Toxoplasma* cell invasion. Curr Opin Microbiol 10:83–89

Carruthers VB, Sibley D (1997) Sequential protein secretion from three distinct organelles of *Toxoplasma gondii* accompanies invasion of human fibroblasts. Eur J Cell Biol 73:114–123

Cesbron-Delauw MF (1994) Dense-granule organelles of *Toxoplasma gondii*: their role in the host-parasite relationship. Parasitol Today 10:293–296

Checroun C, Wehrly TD, Fischer ER, Hayes SF, Celli J (2006) Autophagy-mediated reentry of *Francisella tularensis* into the endocytic compartment after cytoplasmic replication. Proc Natl Acad Sci USA 103:14578–14583

Collazo CM, Yap GS, Sempowski GD, Lusby KC, Tessarollo L, Vande Woude GF, Sher A, Taylor GA (2001) Inactivation of LRG-47 and IRG-47 reveals a family of interferon γ-inducible genes with essential, pathogen-specific roles in resistance to infection. J Exp Med 194:181–187

Collazo CM, Yap G, Hieny S, Caspar P, Feng CG, Taylor GA, Sher A (2002) The function of gamma interferon-inducible GTP-binding protein IGTP in host resistance to *Toxoplasma gondii* is Stat1 dependent and requires expression in both hematopoietic and nonhematopoietic cellular compartments. Infect Immun 70:6933–6939

Coppens I, Dunn JD, Romano JD, Pypaert M, Zhang H, Boothroyd JC, Joiner KA (2006) *Toxoplasma gondii* sequesters lysosomes from mammalian hosts in the vacuolar space. Cell 125:261–274

Deckert-Schluter M, Bluethmann H, Rang A, Hof H, Schluter D (1998) Crucial role of TNF receptor type 1 (p55), but not of TNF receptor type 2 (p75), in murine toxoplasmosis. J Immunol 160:3427–3436

Djavaheri-Mergny M, Amelotti M, Mathieu J, Besancon F, Bauvy C, Souquere S, Pierron G, Codogno P (2006) NF-kappaB activation represses tumor necrosis factor-alpha-induced autophagy. J Biol Chem 281:30373–30382

Dunn WAJ (1994) Autophagy and related mechanisms of lysosome-mediated protein degradation. Trends Cell Biol 110:1923–1933

Durandy A, de Saint Basile G, Lisowska-Grospierre B, Gauchat JF, Forveille M, Kroczek RA, Bonnefoy JY, Fischer A (1995) Undetectable CD40 ligand expression on T cells and low B cell responses to CD40 binding agonists in human newborns. J Immunol 154:1560–1568

Gavrielescu LC, Butcher BA, del Rio L, Taylor GA, Denkers EY (2004) STAT1 is essential for antimicrobial effector function but dispensable for gamma interferon production during *Toxoplasma gondii* infection. Infect Immun 72:1257–1264

Gazzinelli RT, Hakim FT, Hieny S, Shearer GM, Sher A (1991) Synergistic role of CD4+ and CD8+ T lymphocytes in IFN-γ production and protective immunity induced by an attenuated Toxoplasma gondii vaccine. J Immunol 146:286–292

Gazzinelli RT, Eltoum I, Wynn TA, Sher A (1993) Acute cerebral toxoplasmosis is induced by in vivo neutralization of TNF-α and correlates with the down-regulated expression of inducible nitric oxide synthase and other markers of macrophage activation. J Immunol 151:3672–3681

Gazzinelli RT, Brezin A, Li Q, Nussenblatt RB, Chan CC (1994) *Toxoplasma gondii*: acquired ocular toxoplasmosis in the murine model, protective role of TNF-α and IFN-γ. Exp Parasitol 78:217–229

Gutierrez MG, Master SS, Singh SB, Taylor GA, Colombo MC, Deretic V (2004) Autophagy is defense mechanism inhibiting BCG and *Mycobacterium tuberculosis* survival in infected macrophages. Cell 119:753–766

Hakansson S, Charron AJ, Sibley LD (2001) *Toxoplasma* evacuoles: a two-step process of secretion and fusion forms the parasitophorous vacuole. EMBO J 20:3132–3144

Han P, McDonald T, Hodge G (2004) Potential immaturity of the T-cell and antigen-presenting cell interaction in cord blood with particular emphasis on the CD40-CD40 ligand costimulatory pathway. Immunology 113:26–34

Hayashi S, Chan C-C, Gazzinelli RT, Pham NTH, Cheung MK, Roberge FG (1996a) Protective role of nitric oxide in ocular toxoplasmosis. Br J Ophthalmol 80:644–648

Hayashi S, Chan CC, Gazzinelli R, Roberge FG (1996b) Contribution of nitric oxide to the host parasite equilibrium in toxoplasmosis. J Immunol 156:1476–1481

Ishida TK, Mizushima SI, Azuma S, Kobayashi N, Tojo T, Suzuki K, Aizawa S, Watanabe T, Mosialos G, Kieff E, Yamamoto T, Inoue J (1996) Identification of TRAF6, a novel tumor necrosis factor receptor-associated factor protein that mediates signaling from an amino-terminal domain of the CD40 cytoplasmic region. J Biol Chem 271:28745–28748

Israelski DM, Remington JS (1993) Toxoplasmosis in the non-AIDS immunocompromised host. Curr Clin Top Infect Dis 13:322–356

Janssen R, van Wengen A, Verhard E, de Boer T, Zomerdijk T, Ottenhoff THM, van Dissel JT (2002) Divergent role for TNF-α in IFN-γ-induced killing of *Toxoplasma gondii* and *Salmonella typhimurium* contributes to selective susceptibility of patients with partial IFN-γ receptor 1 deficiency. J Immunol 169:3900–3907

Joiner KA, Fuhrman SA, Mietinnen H, Kasper LH, Mellman I (1990) *Toxoplasma gondii*: fusion competence of parasitophorous vacuoles in Fc receptor transfected fibroblasts. Science 249:641–646

Joiner KA, Roos DS (2002) Secretory traffic in the eukaryotic parasite *Toxoplasma gondii*: less is more. J Cell Biol 157:557–563

Julien P, Cron RQ, Dabbagh K, Cleary A, Chen L, Tran P, Stepick-Biek P, Lewis DB (2003) Decreased CD154 expression by neonatal CD4[+] T cells is due to limitations in both proximal and distal events of T cell activation. Int Immunol 15:1461–1472
Kaur K, Chowdhury S, Greenspan NS, Schreiber JR (2007) Decreased expression of tumor necrosis factor family receptors involved in humoral immune responses in preterm neonates. Blood 110:2948–2954
Kravetz JD, Federman DG (2005) Toxoplasmosis in pregnancy. Am J Med 118:212–216
Kroncke KD, Fehsel K, Kolb-Bachofen V (1998) Inducible nitric oxide synthase in human disease. Clin Exp Immunol 113:147–156
Leiva LE, Junprasert J, Hollenbaugh D, Sorensen RU (1998) Central nervous system toxoplasmosis with an increased proportion of circulating $\gamma\delta$ T cells in a patient with hyper IgM syndrome. J Clin Immunol 18:283–290
Levine B, Klionsky DJ (2004) Development by self-digestion: molecular mechanisms and biological functions of autophagy. Dev Cell 6:463–477
Lieberman LA, Banica M, Reiner SL, Hunter CA (2004) STAT1 plays a critical role in the regulation of antimicrobial effector mechanisms, but not in the development of Th1-type responses during toxoplasmosis. J Immunol 172:457–463
Ling YM, Shaw MH, Ayala C, Coppens I, Taylor GA, Ferguson DJP, Yap G (2006) Vacuolar and plasma membrane stripping and autophagic elimination of *Toxoplasma gondii* in primed effector macrophages. J Exp Med 203:2063–2071
Lingelbach K, Joiner KA (1998) The parasitophorous vacuole membrane surrounding *Plasmodium* and *Toxoplasma*: an unusual compartment in infected cells. J Cell Sci 111:1467–1475
Lu LF, Cook WJ, Lin LL, Noelle RJ (2003) CD40 signaling through a newly identified tumor necrosis factor receptor-associated factor 2 (TRAF2) binding site. J Biol Chem 278:45414–45418
Martens S, Parvanova I, Zerrahn J, Griffiths G, Schell G, Reichmann G, Howard JC (2005) Disruption of *Toxoplasma gondii* parasitophorous vacuoles by the mouse p47-resistance GTPases. PLoS Pathogens 1:187–201
Martin AM, Liu T, Lynn BC, Sinai AP (2007) The *Toxoplasma gondii* parasitophorous vacuole membrane: transactions across the border. J Eukaryot Microbiol 54:25–28
McCannel CA, Holland GN, Helm CJ, et al. (1996) Causes of uveitis in the general practice of ophthalmology. Am J Ophthalmol 121:35–46
Mizushima N, Ohsumi Y, Yoshimori T (2002) Autophagosome formation in mammalian cells. Cell Struct Funct 27:421–429
Montoya JG, Kovacs JA, Remington JS (2005) *Toxoplasma gondii*. In: Mandell GL, Bennett JE, Dolin R (eds) Principles and practice of infectious diseases. Elsevier Churchill Livingstone, Philadelphia, pp 3170–3198
Mordue DG, Sibley LD (1997) Intracellular fate of vacuoles containing *Toxoplasma gondii* is determined at the time of formation and depends on the mechanisms of entry. J Immunol 159:4452–4459
Mukundan L, Bishop GA, Head KZ, Zhang L, Wahl L, Suttles J (2005) TNF receptor-associated factor 6 is an essential mediator of CD40-activated proinflammatory pathways in monocytes and macrophages. J Immunol 174:1081–1090
Nakagawa I, Amano A, Mizushima N, Yamamoto A, Yamaguchi H, Kamimoto T, Nara A, Funao J, Nakata M, Tsuda K, Hamada S, Yoshimori T (2004) Autophagy defends cells against invading Group A *Streptococcus*. Science 306:1037–1040
Navia BA, Petito CK, Gold JW, Cho ES, Jordan BD, Price RW (1986) Cerebral toxoplasmosis complicating the acquired immune deficiency syndrome: clinical and neuropathological findings in 27 patients. Ann Neurol 19:224–238
Nonoyama S, Penix LA, Edwards CP Lewis D, Ito S, Aruffo A, Wilson CB, Ochs HD (1995) Diminished expression of CD40 ligand by activated neonatal T cells. J Clin Invest 95:66–75
Ogawa M, Yoshimori T, Suzuki T, Sagara H, Mizushima N, Sasakawa C (2004) Escape of intracellular *Shigella* from autophagy. Science 307:727–731

Orvedahl A, Alexander D, Talloczy Z, Sun Q, Wei Y, Zhang W, Burns D, Leib DA, Levine B (2007) HSV-1 ICP34.5 confers neurovirulence by targeting the Beclin 1 autophagy protein. Cell Host Microbe 1:23–35

Pullen SS, Miller HG, Everdeen DS, Dang TT, Crute JJ, Kehry MR (1998) CD40-tumor necrosis factor receptor-associated factor (TRAF) interactions: regulation of CD40 signaling through multiple TRAF binding sites and TRAF hetero-oligomerization. Biochemistry 37:11836–11845

Reichmann G, Walker W, Villegas EN, Craig L, Cai G, Alexander J, Hunter CA (2000) The CD40/CD40 ligand interaction is required for resistance to toxoplasmic encephalitis. Infect Immun 68:1312–1318

Rich KA, Burkett C, Webster P (2003) Cytoplasmic bacteria can be targets for autophagy. Cell Microbiol 5:455–468

Robben PM, LaRegina M, Kuziel WA, Sibley LD (2005) Recruitment of Gr-1$^+$ monocytes is essential for control of acute toxoplasmosis. J Exp Med 201:1761–1769

Roberts F, Roberts CW, Ferguson DJP, McLeod R (2000) Inhibition of nitric oxide production exacerbates chronic ocular toxoplasmosis. Parasite Immunol 22:1–5

Scharton-Kersten T, Yap G, Magram J, Sher A (1997) Inducible nitric oxide is essential for host control of persistent but not acute infection with the intracellular pathogen *Toxoplasma gondii*. J Exp Med 185:1261–1273

Schluter D, Lohler J, Deckert M, Hof H, Schwendemann G (1991) Toxoplasma encephalitis of immunocompetent and nude mice: immunohistochemical characterisation of *Toxoplasma* antigen, infiltrates and major histocompatibility complex gene products. J Neuroimmunol 31:185–198

Singh SB, Davis AS, Taylor GA, Deretic V (2006) Human IRGM induces autophagy to eliminate intracellular mycobacteria. Science 313:1438–1441

Subauste CS, Wessendarp M (2000) Human dendritic cells discriminate between viable and killed *Toxoplasma gondii* tachyzoites: Dendritic cell activation after infection with viable parasites results in CD28 and CD40 ligand signaling that controls IL-12-dependent and -independent T cell production of IFN-γ. J Immunol 165:1498–1505

Subauste CS, Wessendarp M (2006) CD40 restrains the in vivo growth of *Toxoplasma gondii* independently of gamma interferon. Infect Immun 74:1573–1579

Subauste CS, Wessendarp M, Sorensen RU, Leiva L (1999) CD40-CD40 ligand interaction is central to cell-mediated immunity against *Toxoplasma gondii*: patients with hyper IgM syndrome have a defective type-1 immune response which can be restored by soluble CD40L trimer. J Immunol 162:6690–6700

Subauste CS, Wessendarp M, Smulian AG, Frame PT (2001) Role of CD40 ligand signaling in defective type-1 cytokine response in HIV infection. J Infect Dis 183:1722–1731

Subauste CS, Wessendarp M, Portillo J-AC, Andrade RM, Hinds LM, Gomez FJ, Smulian AG, Grubbs PA, Haglund LA (2004) Pathogen-specific induction of CD154 is impaired in CD4$^+$ T cells from HIV-infected individuals. J Infect Dis 189:61–70

Subauste CS, Andrade RM, Wessendarp M (2007a) CD40-TRAF6 and autophagy-dependent antimicrobial activity in macrophages. Autophagy 3:245–248

Subauste CS, Subauste A, Wessendarp M (2007b) Role of CD40-dependent down-regulation of CD154 in impaired induction of CD154 in CD4$^+$ T cells from HIV-1-infected patients. J Immunol 178:1645–1653

Suzuki Y, Orellana MA, Schreiber RD, Remington JS (1988) Interferon-γ: the major mediator of resistance against *Toxoplasma gondii*. Science 240:516–518

Suzuki Y, Conley FK, Remington JS (1989) Importance of endogenous IFN-γ for prevention of toxoplasmic encephalitis in mice. J Immunol 143:2045–2050

Talloczy Z, Jiang W, Virgin HW, Leib DA, Scheuner D, Kaufman RJ, Eskelinen EL, Levine B (2002) Regulation of starvation- and virus-induced autophagy by the eIF2α signaling pathway. Proc Natl Acad Sci USA 99:190–195

van Kooten C, Banchereau J (2000) CD40-CD40 ligand. J Leuk Biol 67:2–17

Yano T, Mita S, Ohmori H, Oshima Y, Fujimoto Y, Ueda R, Takada H, Goldman WE, Fukase K, Silverman N, Yoshimori T, Kurata S (2008) Autophagic control of *Listeria* through intracellular innate immune recognition in Drosophila. Nat Immunol 9:908–916

Yap GS, Scharton-Kersten T, Charest H, Sher A (1998) Decreased resistance of TNF receptor p55- and p75-deficient mice to chronic toxoplasmosis despite normal activation of inducible nitric oxide synthase in vivo. J Immunol 160:1340–1345

Zhang R, Fichtenbaum CJ, Hildeman DA, Lifson JD, Chougnet C (2004) CD40 ligand dysregulation in HIV: HIV glycoprotein 120 inhibits signaling cascades upstream of CD40 ligand transcription. J Immunol 172:2678–2686

Zhao Z, Fux B, Goodwin M, Dunay IR, Strong D, Miller BC, Cadwell K, Delgado MA, Ponpuak M, Green KG, Schmidt RE, Mizushima N, Deretic V, Sibley LD, Virgin HW (2008) Autophagosome-independent essential function for the autophagy protein Atg5 in cellular immunity to intracellular pathogens. Cell Host Microbe 4:458–469

Autophagy in Mammalian Antiviral Immunity

Anthony Orvedahl and Beth Levine

Contents

1 Introduction 268
2 Autophagy in Combat with Viruses 268
  2.1 Autophagy Protects Against Neurotropic Viral Infections 270
  2.2 HSV-1 271
  2.3 Potential Protective Mechanisms of Autophagy in Mammalian Viral Diseases 273
3 Viruses Conduct Evasive Maneuvers 278
  3.1 HSV-1 Mounts a Concerted Attack on Autophagy 278
  3.2 Viral Bcl-2 Homologs Antagonize Host Autophagy 280
  3.3 Autophagy Evasion as a Common Virulence Mechanism? 281
4 Conclusion 282
References 282

Abstract Autophagy plays diverse roles in cellular adaptation to stress and promotes vital housekeeping functions by recycling unused or damaged organelles and proteins. As an innate immune defense pathway, autophagy also protects against infection with diverse pathogens, including viruses. Autophagy combats infections with both RNA and DNA viruses, and may function by degrading viral components, by promoting the survival of virally infected cells, and/or by activating innate and adaptive immunity. Viruses have evolved counter-mechanisms to evade host autophagy in order to promote their own survival. This chapter will highlight recent advances and unanswered questions relating to autophagy in mammalian antiviral immunity.

A. Orvedahl
Department of Internal Medicine, University of Texas Southwestern Medical Center, Dallas, TX, 75390, USA; Department of Microbiology, University of Texas Southwestern Medical Center, Dallas, TX, 75390, USA

B. Levine (✉)
Department of Internal Medicine, University of Texas Southwestern Medical Center, Dallas, TX, 75390, USA; Department of Microbiology, University of Texas Southwestern Medical Center, Dallas, TX, 75390, USA; Howard Hughes Medical Institute, University of Texas Southwestern Medical Center, Dallas, TX, 75390, USA
e-mail: beth.levine@utsouthwestern.edu

B. Levine et al. (eds.), *Autophagy in Infection and Immunity,*
Current Topics in Microbiology and Immunology 335,
DOI 10.1007/ 978-3-642-00302-8_13,

1 Introduction

Autophagy involves the envelopment of cytoplasmic contents in double-membraned vesicles that are delivered to the lysosome for degradation (Levine and Kroemer 2008; Xie and Klionsky 2007) (Fig. 1). The basal turnover of long-lived proteins and organelles by autophagy functions in cellular housekeeping and prevents the accumulation of damaged cellular contents. During conditions of stress, autophagic recycling of cellular constituents generates metabolites for essential cellular processes. Autophagy can also deliver viral proteins, nucleic acids, and potentially assembled intact viral particles to endolysosomal compartments, which have important functions in cellular recycling and innate and adaptive immune activation. The autophagic delivery of microbial components for lysosomal degradation has been termed "xenophagy" to distinguish the process from "self-eating" (Levine 2005), and this may function as an important catabolic pathway to prevent accumulation and replication of microbes in the host cytoplasm.

Recent evidence suggests that many immune signaling pathways can regulate autophagy induction (reviewed in the chapter by Tal and Iwasaki in this volume), and autophagy protects against infection with viruses from different families. As a countermeasure, viral antagonism of autophagy may be a common virulence strategy. Additionally, some viruses have evolved to subvert autophagy for their own benefit (reviewed in the chapter by Kirkegaard in this volume). The diverse functions of autophagy in antiviral immunity and the exquisite mechanisms for viral evasion and subversion of the pathway point toward a central role for autophagy in the evolutionary struggle between viruses and their hosts. This chapter will review our current understanding of mammalian autophagy as an innate antiviral immune pathway, will discuss potential mechanisms by which autophagy protects against viral infection and viruses evade host autophagy, and will highlight major unanswered questions in the field.

2 Autophagy in Combat with Viruses

With its seemingly unlimited capacity to sequester cytoplasm for lysosomal degradation, autophagy is also poised to target foreign intracellular microbes. An increasing number of studies have characterized diverse functions of autophagy in immunity and infection, including with viruses (reviewed in Levine and Deretic 2007; Orvedahl and Levine 2009). Autophagy is induced in response to viral infection, and Toll-like receptors (TLRs) that both recognize viral microbial-associated molecular patterns (MAMPs) and induce autophagy include TLR3 (poly IC/dsRNA) and TLR7 (ssRNA) (Delgado et al. 2008; Shi and Kehrl 2008). The delivery of cytosolic ligands to TLR-containing vesicles may also be an important mechanism of innate recognition of viral pathogens, as has been demonstrated for vesicular stomatitis virus and TLR7 (Lee et al. 2007) and recently suggested for B-cell receptor co-activation of TLR9

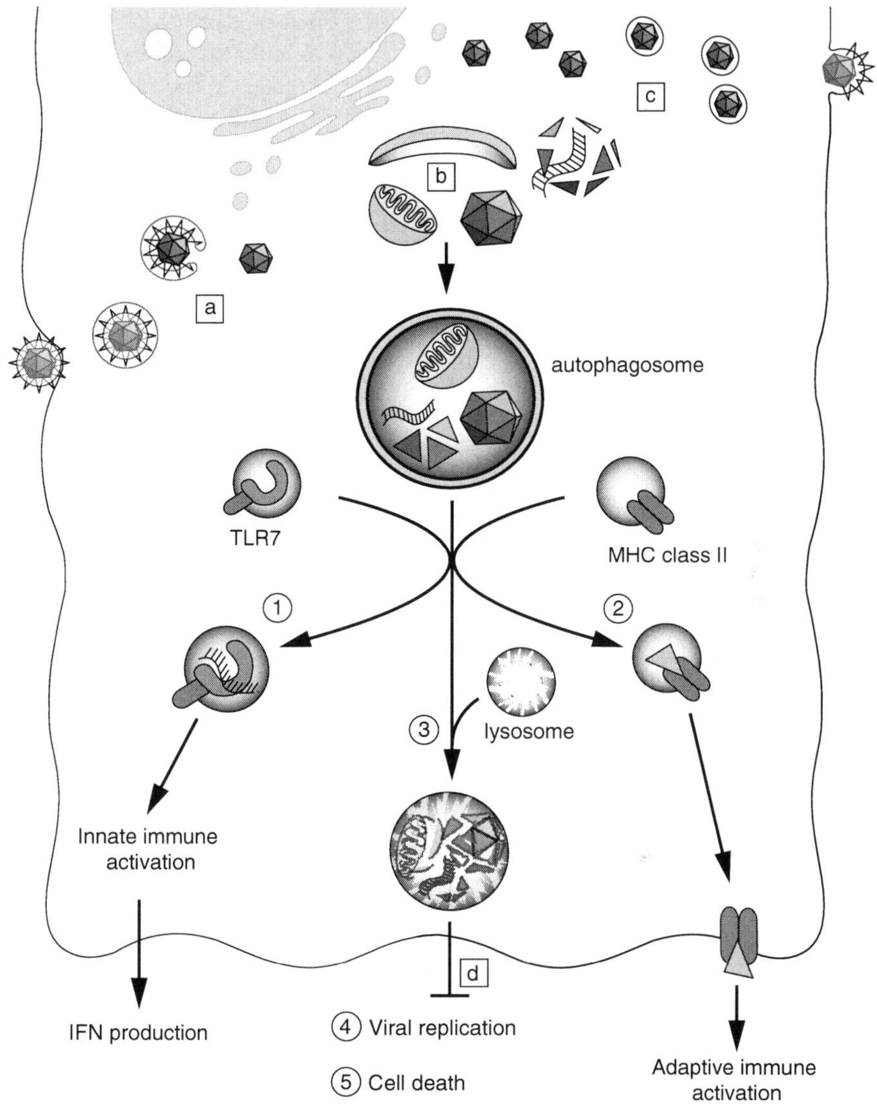

Fig. 1a–d Schematic illustration of potential protective functions of autophagy (**1–5**) during viral infection, and some unanswered questions (**a–d**). Autophagy delivers viral nucleic acids and viral proteins to vesicles containing TLRs (**1**) or MHC class II molecules (**2**) for innate and adaptive immune activation, respectively. Lysosomal degradation (**3**) may rid cells of viral components or generate essential nutrients during infection. Through these effects, and potentially other as-yet undefined mechanisms, autophagy also likely restricts viral replication (**4**) and promotes survival of infected cells (**5**). Unanswered questions include: **a** Does autophagy target viruses during entry and uncoating? **b** Do specific mechanisms exist to selectively target viral vs. cellular components? **c** At which step during viral replication does autophagy target viral proteins and virions for degradation? **d** What are the precise mechanisms by which autophagy restricts viral replication and promotes cellular survival?

(Chaturvedi et al. 2008). The role of TLR-mediated autophagy induction during viral infection in vivo is unknown, and the intersection of innate immune activation of autophagy and activation of innate immunity by autophagy has been reviewed in detail elsewhere (reviewed in the chapter by Tal and Iwasaki in this volume). Evidence for an antiviral role of autophagy has been demonstrated in vivo for different classes of neurotropic viruses in mammals (see below) and in protection against viral infection in plants (Liu et al. 2005; reviewed in the chapter by Seay et al. in this volume). This section will focus on the role of autophagy in mammalian defense against viral infection.

2.1 Autophagy Protects Against Neurotropic Viral Infections

Studies with the neurovirulent alphavirus, Sindbis virus, led to the first identification and characterization of a mammalian autophagy gene, *beclin 1*, the ortholog of yeast *ATG6*, and provided the first indication that autophagy may function as a protective mechanism against viral infection (Liang et al. 1998). Subsequently, we demonstrated that herpes simplex virus type I (HSV-1) targets Beclin 1-mediated autophagy to confer neurovirulence (Orvedahl et al. 2007). To date, evidence for a protective role of autophagy against viral infection in vivo is restricted to studies of neurotropic viruses. As neurons are postmitotic cells that rely on autophagy for routine cellular maintenance, this may indicate that neuronal cells are specialized for engaging viruses through autophagy (Orvedahl and Levine 2008). Alternatively, or in addition, since neurons are postmitotic cells that fail to express Class I MHC molecules and are specialized to avoid cytolytic immune mechanisms for viral clearance, the role of autophagy in antiviral immunity may have increased importance in neurons relative to that in renewable, rapidly dividing cellular targets of viral infection. However, autophagy as an antiviral pathway is likely not limited to the central nervous system (CNS); immune cells activate autophagy through viral-sensing TLRs (as discussed above), and viruses from distinct classes that cause systemic viral diseases antagonize host autophagy (discussed below, see Sect 1.3). Nonetheless, studies with the neurotropic Sindbis and HSV-1 viruses have provided strong evidence for an important role of autophagy in antiviral host defense and accordingly, will be described in more detail below.

2.1.1 Sindbis Virus

Sindbis virus is an arthropod-borne, enveloped single-stranded positive sense RNA virus in the alphavirus family. In humans, Sindbis virus causes a mild rheumatologic disease, but in mice it serves as a model of alphavirus encephalitis (Strauss and Strauss 1994). In mouse models, fatal Sindbis virus encephalitis correlates with the death of infected neurons (Levine 2002). The important role of neuronal survival in protecting against fatal disease has been demonstrated using recombinant Sindbis virus vectors that drive the viral overexpression of proteins that inhibit neuronal apoptosis, including Bcl-2, Bax, cowpox-encoded CrmA, and the peripheral benzodiazepine receptor (Johnston et al. 2001; Levine et al. 1996; Lewis et al. 1999; Nava et al. 1998). In a search for mammalian proteins that interact with Bcl-2 that may play a role promoting survival during Sindbis infection, Liang and colleagues identified Beclin 1 (Liang et al. 1998). The overexpression of Beclin 1 from a double subgenomic promoter in a recombinant Sindbis virus reduced mortality and viral CNS replication in a mouse model of Sindbis encephalitis (Liang et al. 1998). Additionally, the brains of mice infected with Sindbis virus overexpressing Beclin 1 showed significantly less neuronal cell death (Fig. 2a, Liang et al. 1998) and decreased expression of viral RNA (Liang et al. 1998). This study was an important first demonstration that autophagy may confer a protective effect during viral infection, and that enforced expression of an autophagy gene in infected cells was sufficient to reduce viral neuropathogenesis. While the formal possibility remains that Beclin 1 has functions in addition to autophagy that may account for the observed antiviral effects, this is unlikely in light of recent data showing that neuronal-specific deletion of another autophagy gene, *Atg5*, increases lethality during CNS infection with Sindbis virus (MacPherson et al. unpublished data).

2.2 *HSV-1*

HSV-1 is a double-stranded DNA virus in the α-herpesvirus family. In humans, HSV-1 infection results in diseases ranging from mild mucocutaneous disease to fatal encephalitis. HSV-1 encephalitis is devastating; there is a 70% mortality rate in the absence of treatment. Even with treatment, 20–30% of those affected succumb to the disease, and a majority of survivors exhibit permanent neurological dysfunction (Whitley and Roizman 2001). HSV-1 neurovirulence has been mapped to the infected cell protein 34.5 (*ICP34.5*) gene, which is essential for fatal herpes encephalitis (Chou et al. 1990). ICP34.5 contains a GADD homology domain (Chou and Roizman 1994; He et al. 1996) that recruits protein phosphatase-1α to reverse host cell translational shutoff (He et al. 1997). However, *ICP34.5*-deleted mutant viruses in which translational shutoff activity is restored through additional mutations remain avirulent in vivo (Markovitz et al. 1997; Mohr et al. 2001), suggesting that functions of ICP34.5 in addition to reversal of host cell shutoff contribute to neurovirulence.

One clue to an additional function of ICP34.5 came from studies that identified the mammalian autophagy protein Beclin 1 as an interacting partner. ICP34.5 had previously been shown to antagonize PKR signaling-mediated autophagy (Tallóczy et al. 2002). Deletion mapping and interaction studies identified a 20 amino acid region of ICP34.5 (aa68–87) required for binding to Beclin 1 and inhibition of autophagy (Orvedahl et al. 2007). This activity is distinct from the C-terminal GADD34 domain required for eIF2α dephosphorylation and host cell shutoff reversal (Orvedahl et al. 2007). A mutant virus lacking the Beclin 1 interaction domain (34.5Δ68–87) is

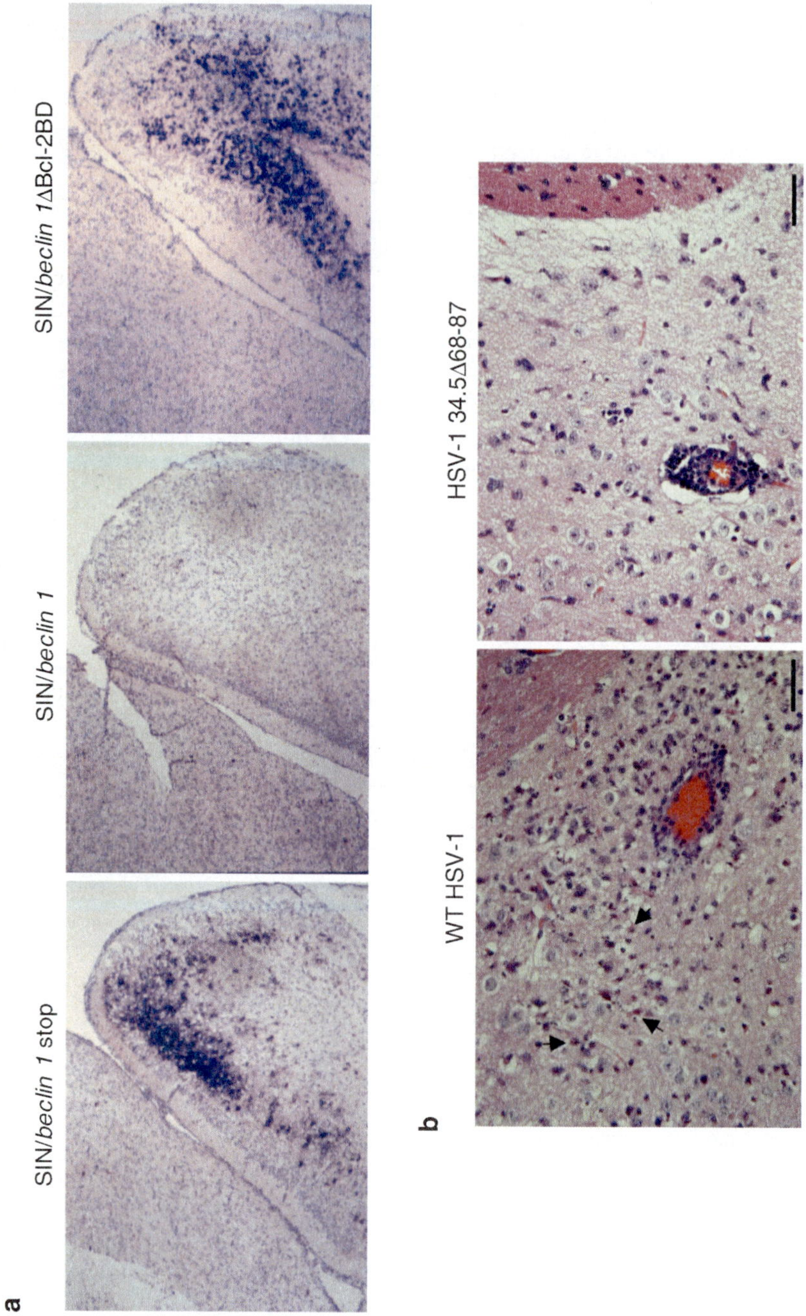
a
SIN/beclin 1 stop
SIN/beclin 1
SIN/beclin 1ΔBcl-2BD
b
WT HSV-1
HSV-1 34.5Δ68-87

unable to inhibit host autophagy and is severely attenuated in its ability to cause fatal herpes encephalitis (Orvedahl et al. 2007) (Fig. 2b). Furthermore, the neurovirulence of the 34.5Δ68–87 mutant virus is restored in *pkr*$^{-/-}$ mice, indicating that PKR signals upstream of Beclin 1 to induce autophagy (Orvedahl et al. 2007). These findings suggest a novel function of ICP34.5 in conferring neurovirulence (i.e., direct antagonism of the host cell autophagic machinery), and further demonstrate an essential role for autophagy in protection against viral disease.

2.3 *Potential Protective Mechanisms of Autophagy in Mammalian Viral Diseases*

The above studies with Sindbis and HSV-1 demonstrate an important role for autophagy in antiviral defense (overexpression of an autophagy protein or abrogation of a viral autophagy evasion strategy, respectively, are sufficient to rescue animals from fatal disease). However, the precise mechanisms through which autophagy functions in host protection against viral disease remain unclear (Fig. 1). Given the role for autophagy in maintaining cellular homeostasis through its recycling functions and metabolite generation during stress, it is possible that autophagy promotes cell survival during viral infection. Additionally, autophagy can serve as a conduit for delivering cytoplasmic material to endolysosomal compartments, which may deliver viral proteins and nucleic acids for immune activation and/or lysosomal degradation.

Indeed, histopathologic studies of mouse brains infected with Sindbis virus and HSV-1 suggest that one important function of autophagy is the prevention of virus-induced cell death (Fig. 2). Compared to mouse brains infected with a virus containing a control insert, mouse brains infected with Sindbis virus overexpressing Beclin 1 exhibit decreased levels of apoptotic neuronal nuclei, while those infected with a virus expressing Beclin 1 with its Bcl-2 interaction domain deleted exhibit increased neuronal death (Liang et al. 1998)(Fig. 2a). Thus, Bcl-2 regulation of Beclin 1 function may be required for autophagy to be neuroprotective rather than neurotoxic. Indeed, Pattingre and colleagues subsequently demonstrated that Bcl-2 regulation of Beclin 1 maintains autophagy within physiological levels that are

Fig. 2a–b Cell death and histopathology of Sindbis virus and HSV-1 infected brains. **a** Mouse brains infected with Sindbis virus expressing different Beclin 1 constructs and labeled with in situ end labeling to detect apoptotic cells (labeled *blue*). Fewer apoptotic cells are observed in brains infected with recombinant Sindbis virus expressing full-length Beclin 1 (SIN/*beclin*) compared to brains infected with virus expressing Beclin 1 with a premature stop codon (SIN/*beclin*stop). Viral expression of Beclin 1 lacking its Bcl-2 binding domain (SIN/*beclin*ΔBcl-2BD) results in increased cell death, at levels higher than with SIN/*beclin*stop. **b** H & E staining of mouse brains infected with the ICP34.5Δ68–87 rescue virus (34.5Δ68–87R) or the 34.5Δ68–87 mutant virus incapable of binding to Beclin 1 and antagonizing autophagy. Increased dying pyknotic neurons (*arrows*) are observed in the brains of mice infected with the rescue virus as compared to the brains of mice infected with the mutant virus. Scale bars: 50 μm. Adapted from Orvedahl et al. (2007) with permission

compatible with cellular survival; expression of Beclin 1 mutants that cannot bind to Bcl-2 results in unregulated autophagy and increased cell death that is inhibitable by knockdown of autophagy genes (Pattingre et al. 2005). Interestingly, some viruses may have captured the autophagy regulatory functions of Bcl-2 and evolved to suppress Beclin 1 autophagy function more effectively than their cellular counterparts (see Sect 1.3.2).

Autophagy may also promote the survival of neurons infected with HSV-1, as fewer dying pyknotic neurons are observed in brains of mice infected with the 34.5Δ68–87 virus compared to wild-type virus expressing full-length ICP34.5 (Orvedahl et al. 2007)(Fig. 2b). The evidence that neuronal apoptosis is a direct cause of fatal encephalitis for HSV-1 is less clear than it is for Sindbis virus, although some studies in mice (Geiger et al. 1997) and in humans (DeBiasi et al. 2002) have shown a direct correlation between HSV-1 infection and apoptosis. While the interplay between autophagy and apoptosis is complex, in some cases autophagy may function as a pro-survival pathway that prevents apoptosis or other forms of cell death (Maiuri et al. 2007). Autophagy promotes the survival of parvovirus B19-infected cells in vitro (Nakashima et al. 2006), and autophagy genes are required to restrict cell death to sites of infection in tobacco mosaic virus-infected plants (Liu et al. 2005). Further studies are needed to determine the relative importance of the pro-cell survival effects of autophagy versus other potential effector functions in protection against CNS, and other types of viral infections.

In both the HSV-1 and Sindbis virus encephalitis models, increased autophagy levels directly correlate with decreased viral titers in infected brains (Liang et al. 1998; Orvedahl et al. 2007) (Fig. 1). In cells infected with HSV-1 lacking *ICP34.5* (Δ34.5), Tallóczy and colleagues found that PKR-mediated autophagy could function to degrade virions and viral proteins (Tallóczy et al. 2006). An increased rate of viral protein degradation was observed in biochemical analyses, and increased autophagosomal capturing of virions was observed in ultrastructural analyses in wild-type neurons infected with HSV-1 Δ34.5 compared to $pkr^{-/-}$ neurons infected with HSV-1 Δ34.5 or in wild-type neurons infected with wild-type virus (Tallóczy et al. 2006)(Fig. 3a). Sindbis virions are also frequently observed in autophagosomes in wild-type but not in $pkr^{-/-}$ neurons, whereas single-membraned viral replication vacuoles are more numerous in $pkr^{-/-}$ neurons, suggesting increased viral replication in the setting of deficient autophagic capturing of the virus (Fig. 3b, unpublished data).

These observations raise a number of questions with respect to xenophagic degradation of viruses (Fig. 1). First, do mechanisms exist to identify and sequester virions and viral proteins selectively in autophagosomes, or does degradation of viral components merely result from bulk cytosolic autophagy (Fig. 1b)? Examples of substrates for selective autophagy in mammalian cells include cytosolic proteins and organelles (Yu et al. 2008), as well as bacteria (Levine and Deretic 2007), and it seems likely that mechanisms may also exist to specifically target viral components. In evolutionary terms, the selective targeting of viral components for degradation would be advantageous since host metabolites that have been parasitized by the virus would be reclaimed, while host proteins and organelles essential for maintaining cellular homeostasis would be spared. Yet, in ultrastructural studies of the two

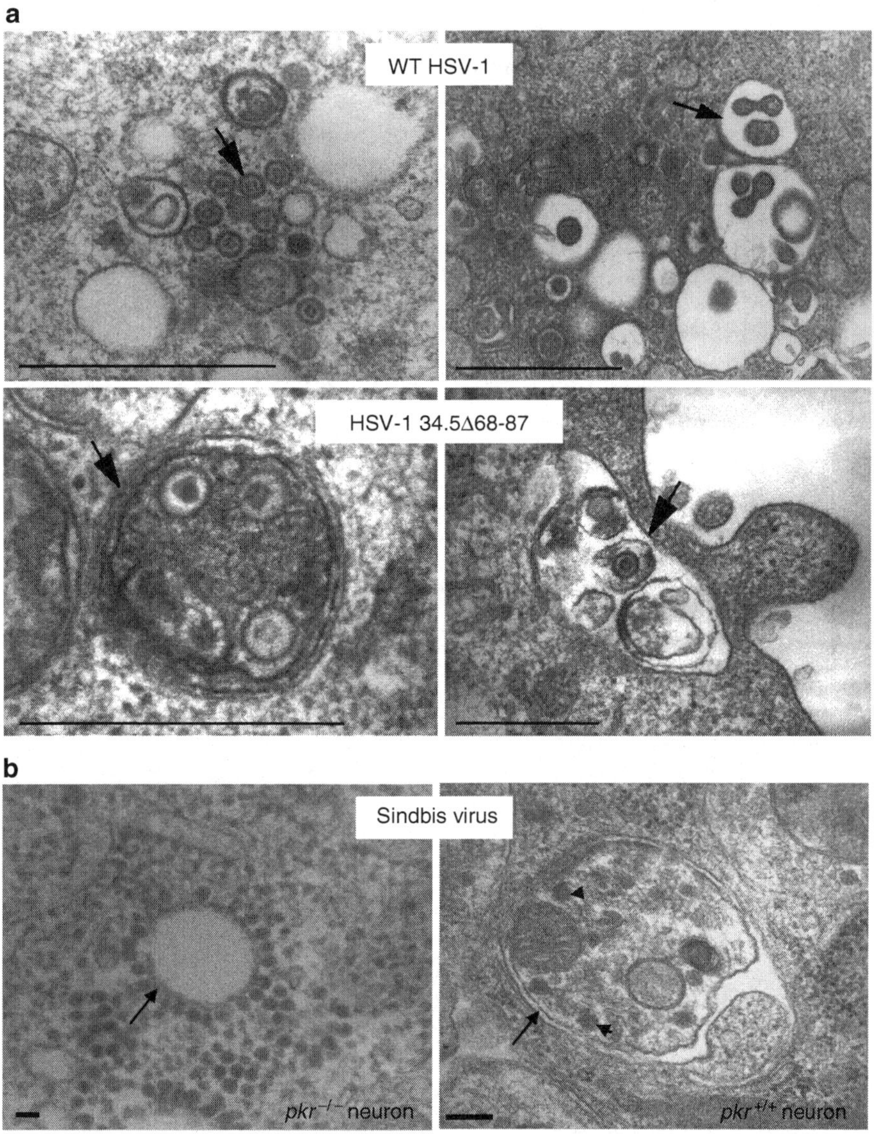

Fig. 3a–b Electron micrographs showing HSV-1 and Sindbis virus in autophagosomes. **a** Wild-type neurons infected with wild-type HSV-1 (*top panels*) or HSV-1 with ICP34.5 (Δ34.5) deleted (*bottom panels*). In neurons infected with HSV-1 that expresses ICP34.5, free cytoplasmic virions (*top left, arrow*) or viral vesicles (*top right, arrow*) are frequently observed. In neurons infected with the Δ34.5 virus, early autophagosomes containing intact virions (*lower left, arrow*) and late autophagosomes with partially degraded viral vesicles are observed (*lower right, arrow*). Scale bars: 1 μm (*top*), 0.5 μm (*bottom*). Adapted from Tallóczy et al. (2006) with permission. **b** *pkr*$^{-/-}$ (*left*) or wild-type (*right*) neurons infected with Sindbis virus. In *pkr*$^{-/-}$ neurons, viral replication vacuoles (*arrow*) are frequently observed, while viral particles (*arrowheads*) are more frequently observed within autophagosomes (*arrow*) in wild-type neurons. Scale bars: 50 nm (*top*), 200 nm (*bottom*)

mammalian viruses for which autophagy has been shown to play a role in host defense in vivo, Sindibis virus and HSV-1, both cytoplasmic contents and viral components are observed in autophagosomes. Thus, it is possible that host cells may not encode factors to specifically target viral proteins and exclude host factors, but host factors may exist that ensure that viruses are included during the selection of cellular substrates such as organelles and/or specific host proteins, by trafficking, tethering, or binding viral components to these host structures. A related question is whether viral proteins or intact virions are targeted by xenophagy. Tallóczy et al. found that the degradation of total levels of HSV-1 proteins was enhanced by PKR-dependent autophagic degradation, and autophagic vacuoles also appeared to degrade intact HSV-1 virions and HSV-1 viral vesicles (Tallóczy et al. 2006) (Fig. 3a). Preliminary ultrastructural studies also suggest that intact Sindbis viral particles can be targeted to autophagosomes (Fig. 3b, unpublished data). Therefore, ultrastructural evidence indicates that intact virions, and by extension viral proteins, can be sequestered and degraded by autophagy, though it remains to be determined whether viral proteins can be targeted before their incorporation into viral particles. In addition, the precise mechanisms for targeting viral proteins and virions remain unclear (Fig. 1).

Second, during which steps in the viral life cycle does autophagy function to target viral components (Fig. 1c)? During Sindbis virus infection, replication complexes are associated with single-membrane vesicles in the cytoplasm, while it is thought that nucleocapsids form freely during genome packaging in the cytosol before budding at the plasma membrane (Strauss and Strauss 1994). Thus, autophagy may potentially target alphavirus replication complexes or naked nucleocapsid virions in the cytoplasm. Although the precise route is controversial, HSV-1 nucleocapsids are thought to enter the cytoplasm after egressing through the nuclear membrane and secondarily acquire their envelope from cytoplasmic vesicles (Mettenleiter et al. 2006). Both cytoplasmic virions and cytoplasmic vesicles containing virions are observed in autophagosomes during HSV-1 infection. Thus it appears that viral proteins, virions, and viral vesicles can be targeted by autophagy, but it is not yet known whether specific mechanisms exist to target viral components during different steps in the replication cycle.

To date, the models in which autophagy has been shown to protect against viral infection demonstrate a protective role relatively late during the viral life cycle. Expression of Beclin 1 from a Sindbis virus double-subgenomic promoter corresponds with viral structural protein synthesis. Additionally, ICP34.5 expression during HSV-1 infection occurs during late gene expression, when structural proteins are expressed. In both models, subgenomic expression of Beclin 1 and ICP34.5 expression occur after viral entry, uncoating, and genome replication, and temporally correspond to viral structural protein synthesis. These temporal associations between antiviral effector functions of autophagy and viral structural protein expression suggest that an important function of autophagy may be to target viral structural proteins or forming virions for degradation. However, an important related but unanswered question is whether autophagy targets viruses during entry and uncoating (Fig. 1a), and this cannot be addressed with the Sindbis virus and HSV-1 models described above. Autophagy targets bacterial pathogens that escape endosomes or phagosomes upon entry into the cytoplasm (reviewed in (Orvedahl and Levine 2009), and xenophagic targeting of uncoating viruses may partially explain the high ratios of viral particles to infectious units that are observed in most viral infections. Also, it should be noted that autophagy in infected cells may contribute to non-cell-autonomous antiviral responses, including activation of innate and adaptive immunity, both of which may also restrict viral replication (see the chapters by Tal and Iwasaki and by Gannagé and Münz in this volume). It is also possible that autophagy restricts viral replication through mechanisms other than xenophagic degradation of viral proteins and/or particles, perhaps through the degradation of essential host factors required for replication. This latter possibility could potentially explain why, unlike with intracellular bacteria, all ultrastructural studies performed to date demonstrate cytoplasmic components in addition to viral components inside autophagolysosomal structures in virally infected cells.

In addition to promoting survival and restricting viral replication in infected cells, autophagy may protect against infection through immune activation (Fig. 1). A role for autophagy in MHC class II antigen presentation was first postulated based on findings that many class II peptides derive from cytosolic proteins. This function has now been supported by a number of studies, which are reviewed elsewhere (see the chapter by Gannagé and Münz in this volume). Although it is not yet known whether this function of autophagy protects against viral infection in vivo, in vitro studies suggest that it is likely to contribute to host antiviral responses. Furthermore, it may represent an attractive strategy for vaccine enhancement, as has been demonstrated by fusion of influenza matrix protein to Atg8/LC3 (Schmid et al. 2007). Neurons, the predominant target cell during Sindbis virus and HSV-1 CNS infection, do not express MHC class II molecules; however, it is possible that autophagy in glial cells in infected brains may play a role in antigen presentation and protection against viral encephalitides. Autophagy may also function to deliver viral replication intermediates to TLR-7-containing vesicles for innate immune activation and interferon production (Lee et al. 2007). Lee and colleagues demonstrated this function in plasmacytoid dendritic cells, but it remains possible that autophagy in neurons or glial cells can also deliver viral MAMPs for TLR recognition and interferon production.

Another recent link between autophagy and innate immune regulation, with potential implications for antiviral functions of autophagy, relates to the production of the proinflammatory cytokine IL-1β. *Atg16L1*- and *Atg7*-deficient macrophages produce aberrantly elevated levels of IL-1β, and *Atg16L1*-deficient chimeric mice are more susceptible to experimental colitis (Saitoh et al. 2008). Autophagy-mediated suppression of IL-1β production may explain the previous observation that IL-1$\beta^{-/-}$ mice are resistant to paralysis and death from Sindbis virus infection (Liang et al. 1999), especially in view of the findings discussed above of a protective role of autophagy in Sindbis virus infection. The role of IL-1β production and other immune functions regulated by autophagy in antiviral host defense is an important area of future investigation. It seems likely that these paracrine as well as cell-autonomous functions of autophagy may contribute to its antiviral effects in Sindbis, HSV-1, or other yet-to-be-characterized viral diseases.

3 Viruses Conduct Evasive Maneuvers

Millions of years of coevolution between microbes and their hosts has resulted in complex and potent immune pathways that ensure host survival, along with the emergence of sophisticated mechanisms by pathogens to evade host immunity. Pathogens, including viruses, have evolved numerous strategies to counteract nearly every aspect of host immunity (Finlay and McFadden 2006). Indeed, the evolution of microbial evasion strategies directed towards a host pathway strongly implies that the pathway is important in host immune defense; in this context, it is noteworthy that mechanisms for viral evasion of host autophagy have recently been identified (reviewed in Orvedahl and Levine 2009) (Fig. 4). Viruses from different classes possess virulence factors that antagonize host autophagy, likely through multiple distinct mechanisms, and at least one virus, HSV-1, targets multiple points within the autophagy pathway. An important question is whether pharmacologic targeting of autophagy evasion factors can lead to effective novel antiviral treatments.

3.1 HSV-1 Mounts a Concerted Attack on Autophagy

The HSV-1 neurovirulence factor ICP34.5 is a potent viral autophagy evasion protein. This essential HSV-1 neurovirulence factor targets the autophagy pathway by at least two mechanisms. The first indication that ICP34.5 functioned in autophagy evasion

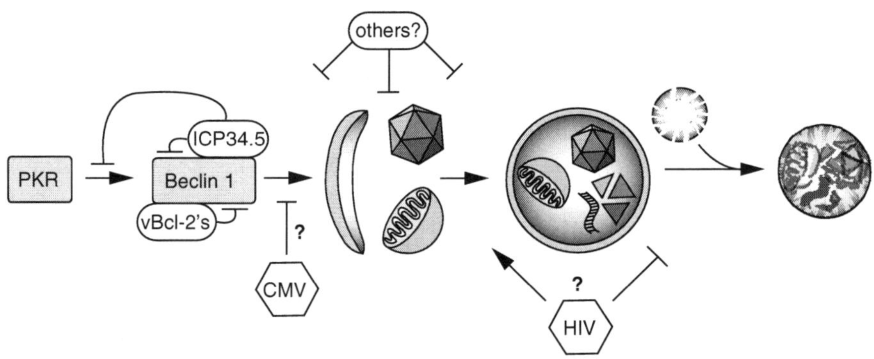

Fig. 4 Viral evasion of autophagy as a common strategy. Schematic illustration of steps in the autophagy pathway targeted by viral virulence factors. Virulence factors from an α-herpesvirus (ICP34.5) and γ-herpesviruses (vBcl-2's) antagonize autophagy by directly targeting Beclin 1 through binding to different regions of Beclin 1. Additionally, ICP34.5 antagonizes upstream signals that induce autophagy by reversing PKR-mediated eIF2α phosphorylation. Viruses that may antagonize autophagy through mechanisms that remain unclear include CMV, which antagonizes autophagy downstream of rapamycin and lithium chloride-regulated steps; and HIV, which may activate or inhibit different steps in the autophagy pathway to promote its replication. Still other as-yet unidentified viral evasion proteins potentially target one or more steps in the autophagy pathway

came from studies of the double-stranded RNA sensing kinase, PKR. Tallóczy et al. found that PKR can function redundantly with the stress-induced kinase GCN2 to induce autophagy in yeast, and that PKR signaling through eIF2α phosphorylation is also required for autophagy induction in mammalian cells in response to HSV-1 infection (Tallóczy et al. 2002). Consistent with the previously described role of ICP34.5 in antagonizing another function of PKR through eIF2α phosphorylation, PKR-mediated host cell shutoff (He et al. 1997), ICP34.5 was also found to antagonize PKR-mediated autophagy induction (Tallóczy et al. 2002). These studies suggested that PKR is important in viral sensing for autophagy induction, and that HSV-1 can antagonize this response. However, as mentioned above, restoration of the host cell shutoff antagonism function in HSV-1 strains with deletion mutations of ICP34.5 did not confer neurovirulence in vivo, suggesting additional functions of ICP34.5.

Yeast two-hybrid studies with ICP34.5 provided a clue to an additional function of ICP34.5 in conferring neurovirulence; ICP34.5 interacts directly with the autophagy protein Beclin 1. Deletion mapping revealed that a 20 amino acid region of ICP34.5 was required for Beclin 1 binding and inhibition of starvation-induced autophagy in mammalian cells (Orvedahl et al. 2007). A mutant virus lacking this interaction domain (34.5Δ68–87) was defective in Beclin 1 binding, unable to inhibit autophagy in primary neurons, and severely attenuated in its ability to cause fatal HSV-1 encephalitis in mice (Orvedahl et al. 2007). Importantly, this mutant virus retained the GADD34 homology region of ICP34.5, and its ability to antagonize host cell shutoff was indistinguishable from the wild-type virus. This study demonstrated that direct antagonism of autophagy by ICP34.5 is important to confer neurovirulence, and these effects are genetically separable from ICP34.5's antagonism of host cell shutoff.

While simultaneously antagonizing PKR signaling through eIF2α dephosphorylation, which is required for autophagy induction, and directly targeting Beclin 1, ICP34.5 has evolved to efficiently block host autophagy. However, while phosphorylation of eIF2α is required for PKR-mediated autophagy induction (Tallóczy et al. 2002), the 34.5Δ68–87 mutant virus which retains its ability to reverse eIF2α phosphorylation remains avirulent in vivo (Orvedahl et al. 2007). This suggests that blockade of eIF2α phosphorylation is insufficient to block host induction of autophagy, and that ICP34.5 evolved an additional mechanism (direct antagonism of Beclin 1) to efficiently evade host autophagy. Importantly, neurovirulence of the 34.5Δ68–87 mutant virus is restored in $pkr^{-/-}$ mice, suggesting that PKR signals upstream of Beclin 1 (Orvedahl et al. 2007). Thus, ICP34.5 appears to have evolved as a multifunctional virulence factor in HSV-1 infection. It reverses host cell shutoff through its GADD34 domain to ensure continued viral protein synthesis, and it mediates autophagy evasion through blockade of PKR signaling and direct antagonism of Beclin 1, which may prevent xenophagic degradation of virions or viral proteins. A recent study indicates that autophagy may not significantly restrict HSV-1 replication in vitro (Alexander et al. 2007), suggesting that cell-type-specific effects or in vivo differences may contribute to apparent differences observed in the requirement of autophagy to affect HSV-1 replication in different studies. Further investigations are needed to assess the relative importance of the cell-autonomous vs. immune activation functions of autophagy in protection against viral diseases and, as a corollary, the effects of HSV-1 ICP34.5 and other viral evasion of these functions in conferring virulence.

3.2 Viral Bcl-2 Homologs Antagonize Host Autophagy

Viral proteins from at least two members of the γ-herpesvirus family evade autophagy by targeting regions of Beclin 1 distinct from the region targeted by ICP34.5. As mentioned, Beclin 1 was identified as an interaction partner with Bcl-2 (Liang et al. 1998), and the Bcl-2/Beclin 1 interaction is important to maintain physiological levels of autophagy (Pattingre et al. 2005). In addition, Pattingre and colleagues first demonstrated Beclin 1 binding and autophagy inhibitory activity for the Kaposi's sarcoma-associated herpesvirus (KHSV) Bcl-2 homolog (Pattingre et al. 2005). This function has also been more recently demonstrated for murine γ-herpesvirus 68 (γ-HV68) (Liang et al. 2006; Sinha et al. 2008). Viral Bcl-2 molecules (vBcl-2's) bind to a BH3 domain present in Beclin 1 (aa 105-128), while yeast two-hybrid studies indicate that residues 1-236 of Beclin 1 are insufficient for HSV-1 ICP34.5 binding (Orvedahl et al. 2007). The observation that multiple different herpesvirus virulence factors target Beclin 1 in distinct regions underscores the likely importance of viral antagonism of autophagy as a virulence strategy (Fig. 4).

There are also unique biochemical aspects of vBcl-2 regulation of Beclin 1 (versus that of cellular Bcl-2 regulation of Beclin 1) which suggest that autophagy antagonism may be particularly important for γ-herpesvirus infections. Levels of Beclin 1 activity are an important determinant of autophagy induction, and the dynamic regulation of Beclin 1 activity is important for cell homeostasis (Levine et al. 2008). Interestingly, some characteristics of vBcl-2's suggest that they have evolved to be super-suppressors of host autophagy. First, vBcl-2's can bind Beclin 1 with higher affinity than their cellular counterparts bind Beclin 1 (Ku et al. 2008; Sinha et al. 2008). Second, vBcl-2's lack a regulatory phosphorylation loop present in cellular Bcl-2 (cBcl-2). During starvation, JNK1-mediated phosphorylation of cBcl-2 results in the release of Beclin 1 and autophagy induction, while vBcl-2 suppression of Beclin 1-mediated autophagy is resistant to this regulation (Wei et al. 2008). Together, these findings suggest that γ-herpesviruses have evolved to constitutively and potently inhibit host autophagy.

These studies also raise some interesting questions regarding the role of autophagy evasion in viral disease. While antagonism of Beclin 1-mediated autophagy by ICP34.5 results in increased neuronal death, increased viral replication, and increased mortality in mice with HSV-1 encephalitis (Orvedahl et al. 2007), the role of vBcl-2's during γ-herpesvirus disease is not yet known. Given that KSHV and γ-HV68 are oncogenic viruses, and Beclin 1 is a tumor suppresser (Levine 2007), it follows that vBcl-2 suppression of autophagy may contribute to the tumorigenic properties of these viruses. Genetic disruption of two autophagy-related genes, *beclin 1* (Qu et al. 2003; Yue et al. 2003) and *bif-1* (Takahashi et al. 2007), leads to malignancies—most prominently lymphomas; cellular Bcl-2 and Bcl-2 family members have long been associated with malignancies and lymphomas (Yip and Reed 2008), and mice infected with γ-HV68 develop lymphoproliferative diseases that resemble human lymphomas (Sunil-Chandra et al. 1994; Tarakanova et al. 2005). Thus, it will be interesting to examine whether the autophagy-inhibitory functions of vBcl-2's play a role in viral-mediated lymphomagenesis.

3.3 Autophagy Evasion as a Common Virulence Mechanism?

Autophagy is emerging as a common target for viral evasion, as recent reports suggest that cytomegalovirus (CMV; a member of a third subfamily of herpesviruses, the β-herpesviruses) and human immunodeficiency virus (HIV), a retrovirus, also antagonize host autophagy. CMV causes infectious mononucleosis in healthy individuals, but may lead to severe disease in neonates and immunocompromised patients (Gandhi and Khanna 2004). Chaumorcel and colleagues found that CMV significantly decreases levels of autophagy in infected cells (Chaumorcel et al. 2008), and that CMV infection activates mTOR (a known inhibitor of autophagy induction). However, treatment of infected cells with two autophagy inducers, either rapamycin (an inhibitor of mTOR) or lithium chloride (which acts in an mTOR-independent fashion to regulate the inositol triphosphate, IP_3, receptor), could not override the autophagy blockade imposed by CMV infection (Chaumorcel et al. 2008). Moreover, Chaumorcel et al. found that suppression of autophagy is restricted to cells infected with CMV, suggesting that a viral factor expressed in infected cells is required for autophagy evasion in a cell-autonomous fashion. Thus, CMV may antagonize autophagy induction downstream of regulation by mTOR and the IP_3 receptor, perhaps at the level of the autophagic machinery, in a manner parallel to members of the α- and γ-herpesvirus families. Intriguingly, analysis of the CMV genome does not reveal any predicted orthologs of Bcl-2 or ICP34.5, indicating that CMV may have evolved a distinct virulence factor involved in autophagy evasion. It will be interesting to determine the identity of this factor(s), if its function overlaps with the known α- and γ-herpesvirus evasion factors that target Beclin 1, or if it has a novel target for autophagy evasion.

A complex picture is emerging with respect to HIV and autophagy. The first study to suggest a link between HIV and autophagy demonstrated that the Env fusion protein expressed on the surface of infected cells induces autophagic death of bystander $CD4^+$ cells (Espert et al. 2006; see the chapter by Espert and Biard-Piechaczyk in this volume). Interestingly, the fusogenic activity of Env is sufficient to induce autophagy in bystander cells (Denizot et al. 2008). Additional evidence has emerged that autophagy may be subverted or inhibited by HIV during infection. Unpublished results suggest that HIV may induce autophagy early during infection but inhibit steps later in the pathway to promote viral production (reported in He and Orvedahl 2007). These findings are supported by the identification of autophagy genes in an siRNA screen for host factors that are required for HIV replication (Brass et al. 2008). Conversely, Zhou and colleagues found that autophagy was inhibited in HIV-infected macrophages (Zhou and Spector 2008). As these studies have been performed in vitro or in cell types that are not the primary target of HIV during disease, it will be important to study HIV in relevant cellular targets to determine the contribution of evasion or subversion of autophagy to HIV pathogenesis. Adding to this complexity, a recent study suggests that soluble factors from HIV-infected glia may inhibit autophagy in neurons, leading to neurodegeneration (Alirezaei et al. 2008). Further studies are needed to examine the role of autophagy in HIV-infected cells and bystander cells in vivo.

4 Conclusion

As a highly conserved pathway for maintaining cellular homeostasis and responding to stressful conditions, it is not surprising that autophagy is emerging as a linchpin in the host antiviral response. Given this linchpin role, it is equally not surprising that viruses have evolved mechanisms to counteract autophagy. While a growing number of studies are defining the role of autophagy as an antiviral pathway, and viral evasion strategies continue to be uncovered and characterized, many questions remain unanswered. For example, the precise functions and relative importance of each effector function of autophagy in protecting against viral disease are still unclear. In addition, the signaling pathways and components of the autophagy machinery that may function uniquely during viral infection remain to be identified. Innate immune signaling pathways that activate autophagy (such as TLRs and PKR) have been identified, but these likely represent the tip of the iceberg, and the precise mechanisms by which these sensors activate autophagy are completely unknown. Future studies are likely to reveal additional protective functions for autophagy, additional viral diseases against which autophagy protects, and additional mechanisms that viruses use to avoid host autophagy.

Acknowledgments We apologize to authors whose work could not be included due to space restrictions. We thank Deborah Shaw for administrative support. The work in the authors' laboratory was supported by NIH grants R01 A10151367 (B.L.) and T32 A1007520 (A.O), and an Ellison Medical Foundation Senior Scholars Award in Infectious Diseases (B.L.).

References

Alexander DE, Ward SL, Mizushima N, Levine B, Leib DA (2007) Analysis of the role of autophagy in replication of herpes simplex virus in cell culture. J Virol 81:12128–12134

Alirezaei M, Kiosses WB, Flynn CT, Brady NR, Fox HS (2008) Disruption of neuronal autophagy by infected microglia results in neurodegeneration. PLoS ONE 3:e2906

Brass AL, Dykxhoorn DM, Benita Y, Yan N, Engelman A, Xavier RJ, Lieberman J, Elledge SJ (2008) Identification of host proteins required for HIV infection through a functional genomic screen. Science 319:921–926

Chaturvedi A, Dorward D, Pierce SK (2008) The B cell receptor governs the subcellular location of Toll-like receptor 9 leading to hyperresponses to DNA-containing antigens. Immunity 28:799–809

Chaumorcel M, Souquère S, Pierron G, Codogno P, Esclatine A (2008) Human cytomegalovirus controls a new autophagy-dependent cellular antiviral defense mechanism. Autophagy 4:46–53

Chou J, Kern ER, Whitley RJ, Roizman B (1990) Mapping of herpes simplex virus-1 neurovirulence to γ134.5, a gene nonessential for growth in culture. Science 250:1262–1266

Chou J, Roizman B (1994) Herpes simplex virus 1 γ134.5 gene function, which blocks the host response to infection, maps in the homologous domain of the genes expressed during growth arrest and DNA damage. Proc Natl Acad Sci USA 91:5247–5251

DeBiasi RL, Kleinschmidt-DeMasters BK, Richardson-Burns S, Tyler KL (2002) Central nervous system apoptosis in human herpes simplex virus and cytomegalovirus encephalitis. J Infect Dis 186:1547–1557

Delgado MA, Elmaoued RA, Davis AS, Kyei G, Deretic V (2008) Toll-like receptors control autophagy. Embo J 27:1110–1121

Denizot M, Varbanov M, Espert L, Robert-Hebmann V, Sagnier S, Garcia E, Curriu M, Mamoun R, Blanco J, Biard-Piechaczyk M (2008) HIV-1 gp41 fusogenic function triggers autophagy in uninfected cells. Autophagy 4:998–1008

Espert L, Denizot M, Grimaldi M, Robert-Hebmann V, Gay B, Varbanov M, Codogno P, Biard-Piechaczyk M (2006) Autophagy is involved in T cell death after binding of HIV-1 envelope proteins to CXCR4. J Clin Invest 116:2161–2172

Finlay BB, McFadden G (2006) Anti-immunology: evasion of the host immune system by bacterial and viral pathogens. Cell 124:767–782

Gandhi MK, Khanna R (2004) Human cytomegalovirus: Clinical aspects, immune regulation, and emerging treatments. Lancet Infect Dis 4:725–738

Geiger KD, Nash TC, Sawyer S, Krahl T, Patstone G, Reed JC, Krajewski S, Dalton D, Buchmeier MJ, Sarvetnick N (1997) Interferon-gamma protects against herpes simplex virus type 1-mediated neuronal death. Virology 238:189–197

He B, Chou J, Liebermann DA, Hoffman B, Roizman B (1996) The carboxyl terminus of the murine *MyD116* gene substitutes for the corresponding domain of the γ*134.5* gene of herpes simplex virus to preclude the premature shutoff of total protein synthesis in infected human cells. J Virol 70:84–90

He B, Gross M, Roizman B (1997) The γ134.5 protein of herpes simplex virus 1 complexes with protein phosphatase 1α to dephosphorylate the α subunit of the eukaryotic translation initiation factor 2 and preclude the shutoff of protein synthesis by double-stranded RNA-activated protein kinase. Proc Natl Acad Sci U SA 94:843–848

He C, Orvedahl A (2007) 2007 keystone symposium on autophagy in health and disease. Autophagy 3:527–536

Johnston C, Jiang W, Chu T, Levine B (2001) Identification of genes involved in the host response to neurovirulent alphavirus infection. J Virol 75:10431–10445

Ku B, Woo JS, Liang C, Lee KH, Hong HS, Kim KS, Jung JU, Oh BH (2008) Structural and biochemical bases for the inhibition of autophagy and apoptosis by viral BCL-2 of murine γ-herpesvirus 68. PLoS Pathog 4:e25

Lee HK, Lund JM, Ramanathan B, Mizushima N, Iwasaki A (2007) Autophagy-dependent viral recognition by plasmacytoid dendritic cells. Science 315:1398–1401

Levine B (2002) Apoptosis in viral infections of neurons: a protective or pathologic host response? Curr Top Microbiol Immunol 265: 95–118

Levine B (2005) Eating oneself and uninvited guests: autophagy-related pathways in cellular defense. Cell 120:159–162

Levine B (2007) Cell biology: autophagy and cancer. Nature 446:745–747

Levine B, Deretic V (2007) Unveiling the roles of autophagy in innate and adaptive immunity. Nat Rev Immunol 7:767–777

Levine B, Goldman JE, Jiang HH, Griffin DE, Hardwick JM (1996) Bc1-2 protects mice against fatal alphavirus encephalitis. Proc Natl Acad Sci USA 93:4810–4815

Levine B, Kroemer G (2008) Autophagy in the pathogenesis of disease. Cell 132:27–42
Levine B, Sinha S, Kroemer G (2008) Bcl-2 family members: dual regulators of apoptosis and autophagy. Autophagy 4:600–606
Lewis J, Oyler GA, Ueno K, Fannjiang YR, Chau BN, Vornov J, Korsmeyer SJ, Zou S, Hardwick JM (1999) Inhibition of virus-induced neuronal apoptosis by Bax. Nat Med 5:832–835
Liang C, Feng P, Ku B, Dotan I, Canaani D, Oh BH, Jung JU (2006) Autophagic and tumour suppressor activity of a novel Beclin1-binding protein UVRAG. Nat Cell Biol 8:688–699
Liang XH, Goldman JE, Jiang HH, Levine B (1999) Resistance of interleukin-1β-deficient mice to fatal Sindbis virus encephalitis. J Virol 73:2563–2567
Liang XH, Kleeman LK, Jiang HH, Gordon G, Goldman J, Berry G, Herman B, Levine B (1998) Protection against fatal Sindbis virus encephalitis by Beclin, a novel Bcl-2-interacting protein. J Virol 72:8586.
Liu Y, Schiff M, Czymmek K, Tallóczy Z, Levine B, Dinesh-Kumar SP (2005) Autophagy regulates programmed cell death during the plant innate immune response. Cell 121:567–577
Maiuri MC, Zalckvar E, Kimchi A, Kroemer G (2007) Self-eating and self-killing: crosstalk between autophagy and apoptosis. Nat Rev Mol Cell Biol 8:741–752
Markovitz NS, Baunoch D, Roizman B (1997) The range and distribution of murine central nervous system cells infected with the $\gamma_1 34.5^-$ mutant of herpes simplex virus 1. J Virol 71:5560–5569
Mettenleiter TC, Klupp BG, Granzow H (2006) Herpesvirus assembly: a tale of two membranes. Curr Opin Microbiol 9:423–429
Mohr I, Sternberg D, Ward S, Leib D, Mulvey M, Gluzman Y (2001) A herpes simplex virus type 1 γ34.5 second-site suppressor mutant that exhibits enhanced growth in cultured glioblastoma cells is severely attenuated in animals. J Virol 75:5189–5196
Nakashima A, Tanaka N, Tamai K, Kyuuma M, Ishikawa Y, Sato H, Yoshimori T, Saito S, Sugamura K (2006) Survival of parvovirus B19-infected cells by cellular autophagy. Virology 349:254–263
Nava VE, Rosen A, Veliuona MA, Clem RJ, Levine B, Hardwick JM (1998) Sindbis virus induces apoptosis through a caspase-dependent, CrmA-sensitive pathway. J Virol 72:452–459
Orvedahl A, Alexander D, Tallóczy Z, Sun Q, Wei Y, Zhang W, Burns D, Leib DA, Levine B (2007) HSV-1 ICP34.5 confers neurovirulence by targeting the Beclin 1 autophagy protein. Cell Host Microbe 1:23–35
Orvedahl A, Levine B (2008) Autophagy and viral neurovirulence. Cell Microbiol 10:1747–1756
Orvedahl A, Levine B (2009) Eating the enemy within: autophagy in infectious diseases. Cell Death Differ 16:57–69
Pattingre S, Tassa A, Qu X, Garuti R, Liang XH, Mizushima N, Packer M, Schneider MD, Levine B (2005) Bcl-2 antiapoptotic proteins inhibit Beclin 1-dependent autophagy. Cell 122:927–939
Qu X, Yu J, Bhagat G, Furuya N, Hibshoosh H, Troxel A, Rosen J, Eskelinen EL, Mizushima N, Ohsumi Y et al (2003) Promotion of tumorigenesis by heterozygous disruption of the *beclin 1* autophagy gene. J Clin Invest 112:1809–1820
Saitoh T, Fujita N, Jang MH, Uematsu S, Yang BG, Satoh T, Omori H, Noda T, Yamamoto N, Komatsu M et al (2008) Loss of the autophagy protein Atg16L1 enhances endotoxin-induced IL-1β production. Nature 456:264–268
Schmid D, Pypaert M, Munz C (2007) Antigen-loading compartments for major histocompatibility complex class II molecules continuously receive input from autophagosomes. Immunity 26:79–92
Shi CS, Kehrl JH (2008) MyD88 and Trif target Beclin 1 to trigger autophagy in macrophages. J Biol Chem 283:33175–33182
Sinha S, Colbert CL, Becker N, Wei Y, Levine B (2008) Molecular basis of the regulation of Beclin 1-dependent autophagy by the γ-herpesvirus 68 Bcl-2 homolog M11. Autophagy 4:989–997
Strauss JH, Strauss EG (1994) The alphaviruses: gene expression, replication, and evolution. Microbiol Rev 58:491–562
Sunil-Chandra NP, Arno J, Fazakerley J, Nash AA (1994) Lymphoproliferative disease in mice infected with murine gammaherpesvirus 68. Am J Pathol 145:818–826

Takahashi Y, Coppola D, Matsushita N, Cualing HD, Sun M, Sato Y, Liang C, Jung JU, Cheng JQ, Mul JJ, et al (2007) Bif-1 interacts with Beclin 1 through UVRAG and regulates autophagy and tumorigenesis. Nat Cell Biol 9:1142–1151

Tallóczy Z, Jiang W, Virgin HW IV, Leib DA, Scheuner D, Kaufman RJ, Eskelinen EL, Levine B (2002) Regulation of starvation- and virus-induced autophagy by the eIF2α kinase signaling pathway. Proc Natl Acad Sci USA 99:190–195

Tallóczy Z, Virgin HW IV, Levine B (2006) PKR-dependent autophagic degradation of herpes simplex virus type 1. Autophagy 2:24–29

Tarakanova VL, Suarez F, Tibbetts SA, Jacoby MA, Weck KE, Hess JL, Speck SH, Virgin HW IV (2005) Murine gammaherpesvirus 68 infection is associated with lymphoproliferative disease and lymphoma in BALB β2 microglobulin-deficient mice. J Virol 79:14668–14679

Wei Y, Pattingre S, Sinha S, Bassik M, Levine B (2008) JNK1-mediated phosphorylation of Bcl-2 regulates starvation-induced autophagy. Mol Cell 30:678–688

Whitley RJ, Roizman B (2001) Herpes simplex virus infections. The Lancet 357:1513–1518

Xie Z, Klionsky DJ (2007) Autophagosome formation: core machinery and adaptations. Nat Cell Biol 9:1102–1109

Yip KW, Reed JC (2008) Bcl-2 family proteins and cancer. Oncogene 27:6398–6406

Yu L, Strandberg L, Lenardo MJ (2008) The selectivity of autophagy and its role in cell death and survival. Autophagy 4:567–573

Yue Z, Jin S, Yang C, Levine AJ, Heintz N (2003) *Beclin 1*, an autophagy gene essential for early embryonic development, is a haploinsufficient tumor suppressor. Proc Natl Acad Sci USA 100:15077–15082

Zhou D, Spector SA (2008) Human immunodeficiency virus type-1 infection inhibits autophagy. AIDS 22:695–699

Something Old, Something New: Plant Innate Immunity and Autophagy

Montrell Seay, Andrew P. Hayward, Jeffrey Tsao, and S.P. Dinesh-Kumar

Contents

1 Introduction 288
2 Autophagy in Plant Development and Senescence 288
3 Autophagy and Plant Innate Immunity 289
4 Does Autophagy Limit Death Signals? A Model to Explain Regulation of HR-PCD 292
5 Autophagy During the Basal Immune Response 295
6 Plant RNA Viruses and the Chloroplast 295
6.1 Physiological Consequences of Viruses in the Chloroplast 296
6.2 Virus-Induced Transport of Chloroplast into the Vacuole: Chlorophagy or Executed Innocent Bystander? 297
7 Plant Viral Clearance in the Vacuole 300
8 Conclusions 302
References 302

Abstract Autophagy performs a variety of established functions during plant growth and development. Recently, autophagy has been further implicated in the regulation of programmed cell death induced during the plant innate immune response. In this chapter we describe specific mechanisms through which autophagy may contribute to a successful defense against pathogen invasion. Accumulating evidence shows that the plant immune system utilizes the chloroplasts as primary sites for the regulation of cell death programs. Viruses also appear to utilize the chloroplast as a site of replication and accumulation, potentially inactivating chloroplast defense signaling in the process. Autophagy-like mechanisms have been observed to target the chloroplast, which we refer to as "chlorophagy," potentially targeting invasive viruses for degradation or regulating chloroplast-based signaling during the immune response. We hypothesize that chlorophagy is significant for the execution of plant immune defenses, during both basal and effector-triggered immunity.

M. Seay, A.P. Hayward, J. Tsao, and S.P. Dinesh-Kumar (✉)
Department of Molecular, Cellular, and Developmental Biology, Yale University, New Haven, CT 06520-8103, USA
e-mail: Savithramma.dinesh-kumar@yale.edu

B. Levine et al. (eds.), *Autophagy in Infection and Immunity,*
Current Topics in Microbiology and Immunology 335,
DOI 10.1007/ 978-3-642-00302-8_14, © Springer-Verlag Berlin Heidelberg 2009

1 Introduction

The increasing recent interest in the machinery of autophagy and its regulation has not gone unnoticed by the plant community. Plant autophagy is now known to play a necessary role in plant senescence and nutrient regulation (Bassham 2007; Bassham et al. 2006; Patel et al. 2006; Thompson and Vierstra 2005). More recently, autophagy has also been shown to limit the spread of programmed cell death (PCD) from infection foci during the plant effector-triggered immune (ETI) response (Liu et al. 2005; Patel and Dinesh-Kumar 2008). In each of these cases, autophagy promotes cell survival, presumably by limitation of pro-death signals. Many lines of evidence point to the plant chloroplast as a generator of pro-death signals, in addition to serving as a niche for virus replication, and a center for defense response initiation. In this review we will discuss recent developments in the field of plant autophagy, with specific emphasis placed on the role of autophagy in plant innate immunity.

2 Autophagy in Plant Development and Senescence

The plant genome contains many of the same basic autophagy machinery found in yeast and mammals (for a review, see Klionsky 2005; Xie and Klionsky 2007). In fact, almost all essential yeast autophagy genes have homologs in *Arabidopsis*, with the notable exception of *Atg2, Atg10*, *Atg14*, *Atg16,* and *Atg17* (reviewed in Seay et al. 2006). Interestingly, while animals have lost *Atg13*, this gene has been retained in plants. In yeast, the coupling of Atg13 to Atg1 is essential in the regulation of the autophagic response. It remains to be determined if Atg13 in plants performs a similar function.

Early studies in plant autophagy have revealed a necessary role for autophagy genes in senescence. Senescence is an age-related genetic response that occurs during late developmental stages in plant leaves, resulting in the liberation of energy stored during vegetative growth to support the plant reproductive phase (Lim et al. 2007). An end result of the ordered breakdown of plant organelles and macromolecules during senescence is PCD. When critical autophagy genes such as *Arabidopsis thaliana* (At) *AtAtg7* (Doelling et al. 2002) or *AtAtg9* (Hanaoka et al. 2002) are disrupted by T-DNA insertions, plant development is unaffected. Upon nutrient starvation, however, plants display enhanced chlorosis and reduced growth rates. Additionally, age-related senescence and leaf detachment-triggered senescence occur more rapidly in these autophagy-compromised plants. Interestingly, chlorosis is one of the most recognizable phenotypes associated with systemic virus infection (Culver 2002). The chlorotic phenotype also occurs in the regions undergoing the hypersensitive response—a pathogen-triggered PCD (Lam 2004; Mur et al. 2008; Soosaar et al. 2005).

3 Autophagy and Plant Innate Immunity

Upon pathogen infection, plants rapidly induce a basal defense response (Bent and Mackey 2007; Zipfel 2008). This response is marked by a characteristic MAPK-based signaling cascade, which leads to the expression of antimicrobial proteins and cuticle cell wall strengthening. In some cases, pathogens are able to circumvent these basal defenses by expressing effector proteins that inhibit the defense signaling of the host cell. In response, plants have evolved resistance (R) genes encoding intracellular immune receptors that recognize the activity of these pathogen-encoded effectors (Caplan et al. 2008b; Chisholm et al. 2006). Plant R proteins are characterized by conserved C-terminal leucine-rich repeats (LRRs) and centrally located nucleotide-binding domains. The N-terminal domain of these R proteins contains either the Toll/interleukin-1 homology (TIR) domain or the coiled-coil domain. R proteins are structurally similar to mammalian immune receptors, including Toll-like receptors (TLRs) and nucleotide-oligomerization domain (NOD) LRR proteins. However, mammalian TLRs and NODs recognize pathogen-associated molecular patterns (PAMPs), while this role is performed by receptor-like kinase proteins in plants (Zipfel 2008).

Recognition of pathogen effectors by cognate R proteins initiates a signaling cascade that culminates in the hypersensitive response (HR), a form of PCD at the site of pathogen infection (Lam 2004; Mur et al. 2008; Soosaar et al. 2005). The signaling cascade involved in effector-triggered immunity (ETI) is remarkably similar to that of basal immunity, requiring a more prolonged influx of Ca^{2+}, MAPK-dependent production of ROS and nitric oxide (NO), salicylic acid (SA), ethylene (ET), and jasmonic acid (JA) hormone signaling, and transcriptional reprogramming (Dangl and Jones 2001; Soosaar et al. 2005). The ultimate induction of HR-PCD, however, is specific to ETI. Interestingly, several downstream markers of HR-PCD are also upregulated during the induction of senescence induced PCD, including *HIN1, ELI3* and *LSC54* (Lim et al. 2007), indicating that both cell death programs may proceed through common components.

The HR-PCD induced during plant innate immunity is thought to provide a physical barrier that prevents the movement of harmful molecules and viral particles from infected cells to their uninfected neighbors. In animals, an apoptotic cell is ultimately packaged into apoptotic bodies, facilitating phagocytosis by neighboring cells or by mobile immune cells (Mur et al. 2008). Such a fate is doubly prohibited in plant cells: first by the cell wall, which prevents phagocytosis, and second by the lack of mobile cells. Instead, damaged or infected plant cells executing HR-PCD will ultimately undergo cellular collapse, with the remaining cellular contents potentially leaking into the apoplast (Lam 2004). How do infected and dying cells prevent the leakage of harmful compounds and pathogen particles into healthy uninfected tissue? One hypothesis is that autophagy may be utilized to degrade components that would otherwise kill healthy tissue prior to membrane collapse. It has been noted that HR-PCD shares physical characteristics with autophagic or Type II cell death, including vacuolization and the appearance of double-membrane vesicles (Mur et al. 2008). Additionally, recent results demonstrate that autophagy plays an important role in plant immune defenses by limiting the activation of PCD to the sites of pathogen infection (Liu et al. 2005; Patel and Dinesh-Kumar 2008).

While a role for autophagy has been demonstrated during basal immune responses in animals (Deretic 2005; Levine and Deretic 2007), Liu et al. (2005) were the first to directly demonstrate a role for autophagy during a plant effector-triggered immune response. In *Nicotiana* plants containing the "N" immune receptor, recognition of *tobacco mosaic virus* (TMV) or the TMV-derived effector p50 culminates in the generation of characteristic macroscopic HR-PCD lesions (Whitham et al. 1994). Using these lesions as a phenotypic marker, a high-throughput virus-induced gene silencing (VIGS) screen demonstrated that *Nicotiana benthamiana* (Nb) *ATG6/BECLIN 1* is required for proper regulation of the HR-PCD during a pathogen-induced immune response. In *NbATG6/BECLIN 1*-deficient plants, HR-PCD spread throughout the infected leaf and to upper leaves of the plant, even though TMV remained restricted to the infection foci. HR-PCD lesions were absent in plants lacking the *N* gene and death remained properly confined in plants infiltrated with the mechanical death-inducer methanol, indicating phenotypic specificity to the effector-triggered immune response. Localized p50 expression further confirmed that TMV movement was not causal to the runaway cell death phenotype. To differentiate between a plant Atg6-specific or autophagy-specific explanation for this phenotype, Liu et al. (2005) demonstrated that the plant orthologs of the autophagy proteins Vps34/PI3K, Atg3 and Atg7 are also required to restrict HR-PCD to infection foci. Autophagy induction in plant tissue during the N-mediated immune response was confirmed by electron microscopy and by LysoTracker staining. Control plants showed characteristic induction of double-membrane autolysosomal structures during TMV infection, while these structures were absent in *NbATG6/BECLIN 1*-silenced plants (Liu et al. 2005). Together, these results confirm a necessary role for autophagy in the restriction of HR-PCD during an effector-triggered immune response.

Recently published results using an *AtATG6*-antisense (*AtATG6-AS*) line further reveal that the role of autophagy during ETI is conserved across plant species (Patel and Dinesh-Kumar 2008). *AtATG6-AS* plants displayed fewer monodansylcadaverine (MDC)- and LysoTracker-stained autolysosomes in response to carbon/nitrogen starvation, confirming that knockdown of *AtATG6* in these lines is sufficient to downregulate autophagy. Mirroring the results discussed above, *AtATG6-AS* plants challenged with the *P. syringae pv tomato* DC3000 bacteria containing the effector protein AvrRPM1 displayed a runaway cell death phenotype, without migration of the pathogen from initial infection foci (Patel and Dinesh-Kumar 2008).

In the *AtATG6-AS* lines (Patel and Dinesh-Kumar 2008) and in *AtATG6* co-supression lines (Qin et al. 2007), plants display stunted growth of leaves, enhanced senescence, and reduced fertility phenotypes. Interestingly, microarray analyses of *AtATG6* T-DNA insertion lines revealed upregulation of the immune response genes *PR1, EDS1, EDS5,* and *SEN1* (Qin et al. 2007). In addition, (Liu et al. 2005) demonstrated that *NbATG6* mRNA is also upregulated after TMV infection in *Nicotiana* plants and during the early phase of virulent and avirulent *Pseudomonas syringae pv tomato* (Pst) DC3000 infection in *Arabidopsis* plants (Patel and Dinesh-Kumar 2008). These results demonstrate that autophagy genes have acquired functions required for plant immune responses.

While the *AtATG6* data described above relied on LysoTracker staining and electron microscopy to monitor autophagy induction, many recently published papers utilize *AtATG8* marker lines. Unfortunately, while *ATG8* is a standard in other systems, the application of *ATG8* in plants has been complicated by the diversification of *ATG8* species. In *Arabidopsis*, there are at least nine known *ATG8* homologs (Seay et al. 2006; Slavikova et al. 2005; Thompson et al. 2005). Analysis of *AtATG8a-i* mRNA transcripts shows increased expression in response to carbon starvation, suggesting that all *AtATG8* species may play a conserved role in starvation-induced autophagy (Thompson et al. 2005). However, microarray analysis has revealed tissue-specific variation of individual *AtATG8* species expression (Thompson et al. 2005). An analysis of *AtATG8* expression profiles under different stress conditions using Genevestigator (Zimmermann et al. 2004) further illustrates a potential diversity of function among *AtATG8* species during defense responses (Fig. 1). *AtATG8* species are predominantly upregulated in response to nutrient starvation and senescence, while showing very little change in response to the basal

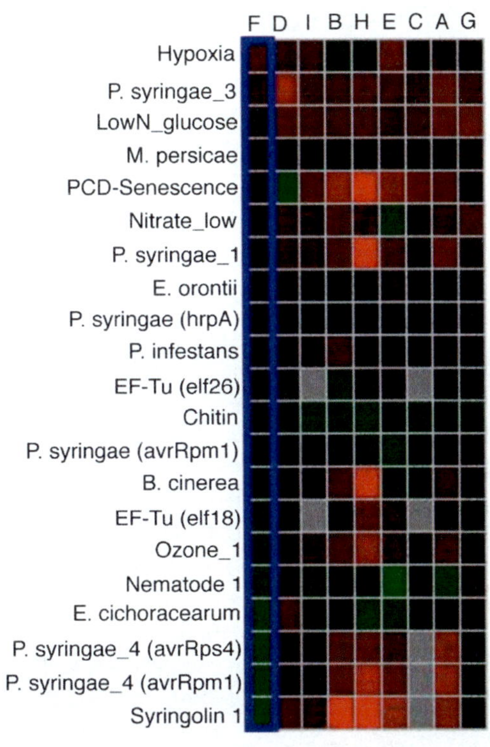

Fig. 1 Genevestigator heat map displaying *Arabidopsis ATG8a-i* mRNA expression levels in response to various stimuli. *Red*, upregulation; *green*, downregulation

immune response inducers EF-Tu and fungal chitin. However, there is notably more specificity of *AtATG8* expression in response to pathogen-encoded effector proteins. For the purpose of monitoring autophagy during plant innate immune responses, an *AtATG8* line must be developed that can be used to visualize autophagy in leaf tissue, the primary site of HR-PCD.

The upregulation of plant *ATG6* and *ATG8* during pathogen infection suggests either that autophagy is required by the pathogen to facilitate pathogenicity or that autophagy is required by the plant for proper induction of defense responses. The runaway cell death phenotype of pathogen-challenged *ATG6*-knockdown plants is notably similar to the accelerated senescence and chlorotic phenotypes of *ATG*-knockout lines. These observations strongly suggest that autophagy is indeed required for plant disease resistance, and also that the regulation of autophagy and cell death are interrelated in plants, as they are in animals.

The deregulation of PCD described in *ATG6*-knockdown plants is particularly interesting in light of the known functions of the mammalian Atg6/Beclin 1 protein. In mammals, Atg6/Beclin 1 appears to act as a bridging protein between the apoptosis and autophagy pathways. Beclin 1 was first identified in a two-hybrid screen for Bcl-2-interacting proteins (Liang et al. 1998), and Bcl-2 was later shown to bind Beclin 1 to disrupt its autophagy function (Pattingre et al. 2005). Prior to these results, Bcl-2 was well known to function endogenously in mammals as an antiapoptotic protein by inhibiting cytochrome c release from the mitochondria (Youle and Strasser 2008). Thus, the binding status of Bcl-2 to Beclin 1 may be a central determinant of whether a cell will undergo Type I apoptotic or Type II autophagic cell death. The existence of such a molecular switch fits well with a large body of work suggesting that autophagy and apoptosis are mutually exclusive, and that the inhibition of one pathway can trigger the other (Maiuri et al. 2007). This data hints at a conserved role for Atg6/Beclin 1 and autophagy across eukaryotes in determining cell survival choices.

4 Does Autophagy Limit Death Signals? A Model to Explain Regulation of HR-PCD

Nitric oxide (NO) and reactive oxygen species (ROS) are highly diffusible secondary messenger molecules produced in abundance during the HR-PCD, and are required for its execution (Fig. 2, Region 1) (for a review, see Heath 2000; Mur et al. 2008). During HR-PCD, an initial extracellular burst of ROS is produced at the membrane-bound NADPH oxidase, encoded by the *Arabidopsis* respirator burst oxidase homolog (rboh) genes *AtrbohD* and *AtrbohF*, among others (Torres et al. 2002). Importantly, both cell-encoded and externally applied antioxidants have been shown to delay or inhibit the HR-PCD (Kiba et al. 2006; Kiraly et al. 2008). NO is also rapidly produced during the initiation phase of HR-PCD by nitric oxide synthase 1 (Zeidler et al. 2004).

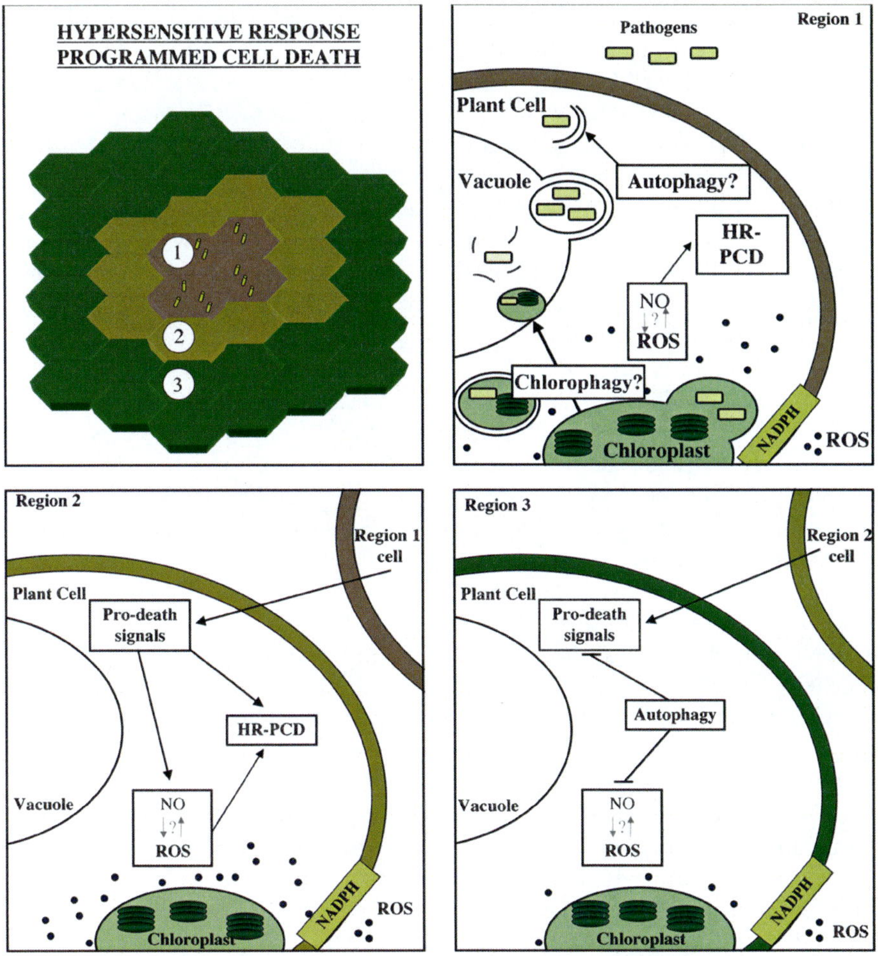

Fig. 2 Model for autophagy activity during plant innate immunity. *Top left panel* shows tissue regions during hypersensitive response programmed cell death (HR-PCD). *Region 1*: At the site of pathogen infection, R proteins detect the presence of pathogen effectors. This triggers a downstream signaling cascade, dependent on signals like reactive oxygen species (ROS) and nitric oxide (NO). It is suspected that autophagy is involved in defense response, as autophagy-compromised plants display enhanced susceptibility to infection. Chloroplasts may be involved in both pathogenesis and plant defense responses. *Region 2*: Regions adjacent to the site of infection perceive pro-death signals and initiate HR-PCD to restrict pathogen and reduce its spread to other tissues. *Region 3*: Uninfected systemic tissues utilize autophagy to downregulate pro-death signals such as ROS

Despite the prominence of the NADPH-mediated ROS burst, recent evidence suggests that NADPH-generated superoxide may have an anti-PCD rather than pro-PCD function. The runaway cell death phenotype of the lesion simulating disease

1 (*lsd1*) lesion mimic in *Arabidopsis* is caused by the failure of these plants to dissipate excess excitation energy generated by the plant chloroplasts, which in turn causes excess cellular ROS in the form of H_2O_2 (Mateo et al. 2004). Therefore, it was predicted that an *lsd1 atrbohD atrbohF* triple mutant would suppress the *lsd1* phenotype (Torres et al. 2005). However, despite showing decreased ROS production, the triple mutant accentuated *lsd1*-associated cell-death. Ectopic superoxide application also activated AtrbohD function to negatively regulate cell death in the *lsd1* mutant (Torres et al. 2005). Temporally downstream of the NADPH-mediated oxidative burst are several other potential intracellular sources of ROS, including the mitochondria, chloroplasts, and peroxisomes. These alternative intracellular sources of ROS may in fact be the primary sources of the "pro-death" ROS signals (Fig. 2, Region 1 and Region 2).

In tissues immediately adjacent to the site of pathogen infection, NO and ROS induce PCD to limit the spread of pathogen (Fig. 2, Region 2). Tissues distal to infection sites normally do not experience PCD, presumably due to the action of autophagy and/or other unknown factors. Ectopic application of NO and ROS induces autophagy in plant tissues, and in the absence of *ATG6*, ectopic application of ROS and NO causes the runaway cell death phenotype of *ATG6*-knockdown plants, as discussed above (Mamillapalli and Dinesh Kumar, unpublished). These results suggest that NO and ROS could function as pro-death signals that induce autophagy during the innate immune response. Induction of HR-PCD in autophagy-deficient plants is not affected in *ATG*-deficient lines, as pathogens are successfully restricted to infection foci. Instead, the runaway cell death phenotype displayed by these plants suggests that autophagy induction may be required to eliminate pro-death signals emanating from the primary infection site (Fig. 2, Region 3) (Liu et al. 2005; Patel and Dinesh-Kumar 2008). In support of this hypothesis, it is noteworthy that "*N*"-containing *Nicotiana* plants showed increased autophagic activity adjacent to infection sites (Liu et al. 2005). Taken with the results from Qin et al. (2007), which show that immune response-related genes are upregulated in *ATG6*-silenced plants, this suggests that *AtATG6* is integral to the repression of defense programming until it is required to defend against invading pathogens or in adjacent uninfected tissues.

As well as targeting these individual pro-death secondary molecules, autophagy may also target proteins necessary for the integration of these signals at neighboring cells, preventing the initiation of the cell-death program and establishing a delineated border between dying and surviving tissue. Xiong et al. recently demonstrated that autophagy can be induced by oxidative stress in *Arabidopsis*, and further that selective autophagic degradation of oxidized proteins occurs during this stress to augment the overburdened 20S proteasome (Xiong et al. 2007b). Interestingly, the disruption of autophagy appears to result in constitutive oxidative stress in *Arabidopsis*, even under otherwise normal growth conditions (Xiong et al. 2007a). These data and other evidence discussed below suggest that autophagy may target ROS-producing organelles for degradation, most notably the chloroplast.

5 Autophagy During the Basal Immune Response

In mammals, autophagy is known to preferentially target and sequester invading bacteria, viruses and parasites, and to promote their degradation, indicating a role for autophagy during a host's defense response to invading pathogens (reviewed in (Levine and Deretic 2007). Plants may also use similar strategies to control viral propagation and spread during non-effector-triggered immune responses. In agreement with this hypothesis, microarray analysis has revealed several genes associated with senescence and vacuolar transport that are upregulated in *Arabidopsis* infected with TMV and in *Vitis vinifera* infected with grapevine leafroll-associated virus (Espinoza et al. 2007). In addition to genes related to defense/stress responses and ROS detoxification, the authors identified the upregulation of two autophagy-related genes, *ATG5* and *ATG8E,* during viral infection and senescence. As further support for a role of autophagy during basal immune responses, autophagy-deficient *Nicotiana* plants infected with GFP-tagged tobacco mosaic virus (TMV-GFP) displayed enhanced virus accumulation and increased TMV-derived RNA at the site of infection (Liu et al. 2005). The virus accumulation could not be attributed to the spread of TMV, because TMV-GFP could not be detected at sites distal to the infection site. Similar results have been reported in *Arabidopsis* plants deficient in the autophagy protein AtAtg6 (Patel and Dinesh-Kumar 2008). Virulent *P. syringae pv tomato* DC3000 bacteria replicated to significantly higher levels shortly after infection of *ATG6*-antisense *A. thaliana* plants compared to wild-type plants. An antiviral and antibacterial role for autophagy in plants may represent an evolutionarily conserved mechanism to limit the replication of intracellular pathogens (Levine and Deretic 2007).

6 Plant RNA Viruses and the Chloroplast

Similar to animal viruses, plant positive-strand RNA viruses replicate in membrane-associated replication complexes (Ahlquist 2006). Different RNA viruses are able to use a variety of intracellular membranes, including endoplasmic reticulum (ER), tonoplast, mitochondria, or chloroplast (reviewed by Ahlquist 2006; Salonen et al. 2005). These intracellular membranes often undergo rearrangements during virus replication leading to the formation of membrane invaginations, formation of multivesicular bodies and other structures.

A wide variety of viruses localize to the chloroplast and presumably utilize it for replication and/or assembly. Like the mitochondria, the chloroplast serves as an ideal manufacturing organelle, since it contains double membranes for proper viral attachment, protection, and synthesis. In addition, like the mitochondria, the chloroplast has the advantage of containing all of the components needed for transcription/translation, including DNA and RNA polymerases and protein synthesis apparatus; up to 50% of all ribosomes in plant leaves are chloroplastic ribosomes

(Ellis and Hartley 1974). The small chloroplast genome coupled with a disproportionate number of ribosomes makes the chloroplast an efficient manufacturing center for viruses.

Esau and Cronshaw (1967) showed some of the first electron micrographs with TMV-like particles localized in the chloroplast of infected cells. Later it was demonstrated that both the TMV virion and coat protein localize in the chloroplast (Reinero and Beachy 1986; Shalla et al. 1975). Recently, Caplan et al. (2008a) demonstrated that during plant innate immunity to TMV, NRIP1, a chloroplastic protein, associates with the p50 helicase domain of TMV and the N receptor during defense, mechanistically connecting the chloroplast to immune responses. These data suggest that TMV may use chloroplastic proteins like NRIP1 for pathogenicity, while the N immune receptor recognizes this association and triggers an immune response.

Similar to TMV, other viruses associate with the chloroplast or target chloroplastic proteins, presumably for pathogenicity. The potato virus Y uses the helper protease HC-Pro to localize to the chloroplast by interacting with chloroplastic proteins (Jin et al. 2007). The bamboo mosaic virus interacts with the chloroplastic phosphoglycerate kinase and knockdown of this kinase results in reduced bamboo mosaic virus coat protein (Lin et al. 2007). Using confocal laser scanning microscopy, it was shown that the triple gene block of movement proteins and the gamma B protein of barley stripe mosaic virus localize to the chloroplast in the presence of viral RNA (Torrance et al. 2006). The cucumber necrosis virus uses a chloroplast transit peptide-like region to transport its coat protein into the chloroplast (Xiang et al. 2006). Interestingly, viruses have also developed mechanisms to localize within the chloroplast. Prod'homme et al. (2003) showed that the chloroplast envelope invaginates during turnip yellow mosaic virus infection. The viral RNA-dependent RNA polymerase then associates with the chloroplast invaginations and other small vesicles formed within the chloroplast.

6.1 Physiological Consequences of Viruses in the Chloroplast

During infection, it has been estimated that 1% of chloroplasts from inoculated leaves and 12% of chloroplasts from systemically infected leaves contain virus-like particles (Granett and Shalla 1970). Accumulation of viral particles within the chloroplast is not without consequences for the plant; several studies have shown that virus accumulation is associated with gross morphological changes within the chloroplast. Although many plant viral infections are asymptomatic, one of the most common phenotypes associated with viral infection is the yellowing of infected leaves, or chlorosis (Culver 2002). Chlorosis is a very obvious phenotype and it is not surprising that references in early literature detailing the chloroplast damage induced by viruses is almost as prevalent as the literature on viruses themselves. Chloroplast damage, disintegration and clearance, increased amino acid accumulation, and the appearance of peculiar double-membrane structures have been noted in most viral infections—especially RNA viruses. Virus-induced chlorosis is caused by a variety of factors that affect the chloroplast. These various effects on the chloroplast can be attributed in part to the propensity of viral proteins and RNAs to localize to the chloroplast; yet the biological relevance of these changes is still unknown.

One hypothesis explaining viral particles in chloroplasts is that they disrupt basal resistance and the ROS-generating machinery. Consistent with this hypothesis, several viruses specifically affect the photosystem of the chloroplast. The plum pox virus (PPV) is associated with the plasmodesmata, the ER, and chloroplast, with replication occurring on the latter two organelles (Martin et al. 1995; Rodriguez-Cerezo et al. 1997). The cylindrical inclusion (CI) protein from PPV interacts with the photosystem protein PSI-K in yeast two-hybrid analysis, and expression of CI causes a decrease in PSI-K accumulation (Jimenez et al. 2006). In support of the hypothesis that viruses target the chloroplast to inhibit basal resistance, knockout of the PSI-K genes in *A. thaliana* and *N. benthamiana* results in higher accumulation of PPV. Additionally, localization of TMV coat protein and movement proteins to the chloroplast results in the depletion of photosystem II, the oxygen-evolving complex 33, and FtsH, all of which are essential components of photosynthesis, ROS production and defense (Lehto et al. 2003; Reinero and Beachy 1989) (Hodgson et al. 1989). These observations strongly suggest that viruses target the chloroplast for replication and may damage the chloroplast as a counter defense against ROS production.

Prolonged viral infection also causes a reduction in the total number of chloroplasts and chloroplast-localized proteins. For example, in bean yellow mosaic virus infection of faba beans, there was a marked reduction in the number of chloroplasts in infected leaves (Radwan et al. 2008). The zucchini yellow mosaic virus also reduces chloroplast numbers (Zechmann et al. 2003). Although interesting, chloroplast clearance and its biological significance in the context of viral infections remains a mystery. Chloroplasts may be targeted for degradation simply due to damage resulting from viral replication. Alternatively, the chloroplasts may be degraded as part of basal defense mechanisms utilizing autophagy to actively target chloroplast associated viral particles for degradation (Fig. 2, Region 1).

6.2 *Virus-Induced Transport of Chloroplast into the Vacuole: Chlorophagy or Executed Innocent Bystander?*

The majority of the literature dedicated to autophagy has focused on the ability of autophagosomes to sequester cytoplasmic material. Most references point to protein turnover and increases in free pools of amino acids as evidence that one of the primary functions of autophagosomes is to provide nutrients during suboptimal conditions. Several excellent reviews have outlined the uptake of the mitochondria (mitophagy) (Mijaljica et al. 2007) and peroxisome (pexophagy) (Dunn et al. 2005); however, there have been fewer discussions explaining the biological relevance of organelle sequestration. Recent observations suggest that damaged chloroplasts become specifically deposited into the vacuole (Niwa et al. 2004). Is this phenomenon purely circumstantial—an unimportant consequence of bulk cytoplasmic uptake—or does chloroplast localization to the vacuole play a vital, unappreciated, and undiscovered role? Further, are chloroplasts actively transported to the vacuole (herein referred to as "chlorophagy"), are mechanisms similar to mitophagy and pexophagy utilized to achieve vacuole localization, and why are the chloroplasts degraded (see Fig. 2, Region 1)?

Two mechanisms have been defined for autophagy-mediated organelle turnover: autophagosome-mediated uptake of peroxisomes or ribosomes and piecemeal autophagy described for the mitochondria and the nucleus (Beau et al. 2008; Dunn et al. 2005; Kim et al. 2007; Roberts et al. 2003). During piecemeal autophagy, mitochondria or nuclei are in juxtaposition with vacuoles until the vacuole invaginates and "pieces" of the organelle are taken into the vacuole. By contrast, during pexophagy, the entire organelle is encapsulated by autophagosomes and transported to the vacuole (Sakai et al. 1998; Tuttle and Dunn, 1995). Organelle dysfunction can trigger autophagic activation and preferential shuttling of defective organelles into the vacuole. In yeast, mitochondria dysfunction triggers autophagy and defective mitochondria are preferentially shuttled to the vacuole, presumably to clear cells of potentially damaging material (Nowikovsky et al. 2007; Priault et al. 2005). Loss of MDM38, an essential component of the mitochondrial K^+/H^+ exchange system, results in a variety of alterations in mitochondrial morphology, loss of membrane potential and eventually autophagy activation (Nowikovsky et al. 2007). Thus a plausible explanation for chloroplast clearing during senescence could be chlorophagy.

There are increasing data to suggest that chloroplasts or chloroplast-like proteins within vesicles are indeed specifically targeted to the vacuole. *Arabidopsis* mutants that lack PPI40, an essential plastid import protein, are defective in chloroplast function and accumulate chloroplast-like vesicles within the vacuole (Figs. 3a and 3b; Niwa et al. 2004). The authors concluded that the size and membrane topology of the organelles within the vacuole most likely represents the remains of chloroplasts. Similar to piecemeal autophagy, there are also instances of portions of the chloroplast transported to the vacuole during autophagy-dependent senescence. Ribulose-1,5-bisphosphate carboxylase/oxygenase (Rubisco) localizes to the stroma and accounts for 12–35% of total plant protein. Interestingly, during senescence, Rubisco is rapidly degraded and it is believed that the nutrients generated are recycled and transported to other organs (Feller et al. 2008; Mae et al. 1983; Makino et al. 1983). Rubisco degradation has been linked to the vacuole (Miller and Huffaker 1981; Thayer and Huffaker 1984; Wittenbach et al. 1982; Yoshida and Minamikawa 1996). Furthermore, the decrease in Rubisco content is much faster than the general decrease in the total number of chloroplasts during senescence, suggesting that parts of the chloroplast may be selectively degraded, similar to piecemeal autophagy instead of autophagosome-mediated whole-organelle encapsulation (Mae et al. 1984; Ono et al. 1995; Wardley et al. 1984). In line with these observations, Chiba et al. (2003) demonstrated the presence of vesicles that bud from the intact chloroplast containing Rubisco and glutamine synthetase but lacking other chloroplastic proteins and features (Figs. 3c and 3d). These vesicles

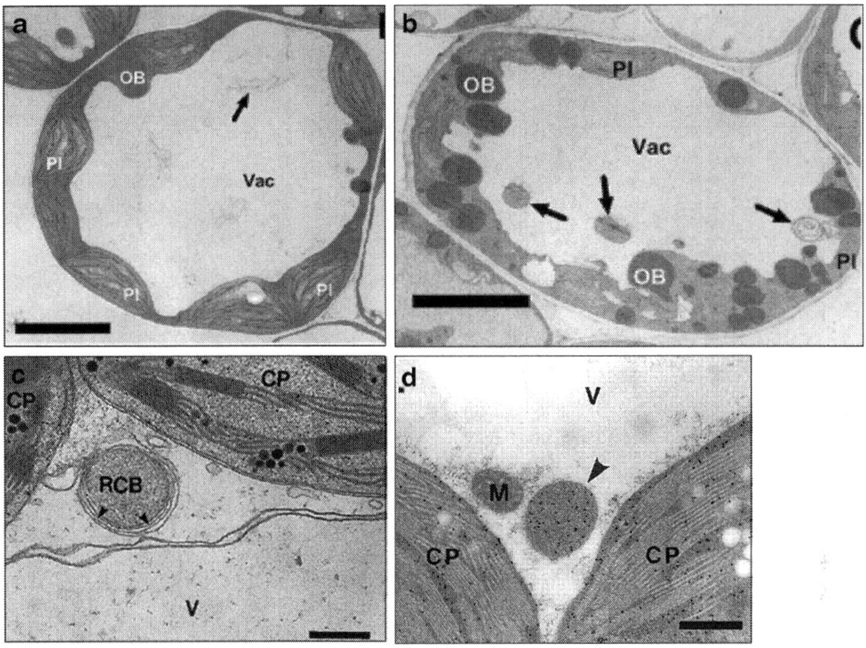

Fig. 3a–d Targeting of chloroplast and plastid proteins to vacuole (**a–b**). Electron micrograph showing cotyledon cells in wild type (*A*) and ppi40 mutant *A. thaliana* plants. Scale bar is 5 μM. *Vac*, central vacuole; *Pl*, plastid; *OB*, oil body. *Arrow* points to vacuolar inclusion. Reproduced from Niwa et al. (2004). Immunolocalization of the large subunit of Rubisco in senescent wheat leaves (**c–d**). *Arrowhead* points to double membrane in *C* and cytoplasmic inclusion in *D* that is labeled by anti-Rubisco immunogold antibodies. Scale bar is 0.5 μM. *RCB*, Rubisco containing bodies; *M*, mitochondrion; *CP*, chloroplast; *V*, vacuole. Reproduced from Chiba et al. (2003) with kind permission from Oxford University Press

then fuse with the vacuole, where the contents are presumably degraded. The observation that these vesicles are only present during autophagy-dependent senescence strongly suggests that the chloroplast in part or in pieces can be delivered to the vacuole via autophagic mechanisms.

There have been fewer reports of whole recognizable chloroplasts in the vacuole. Wittenbach et al. (1982) observed chloroplast loss and chloroplasts positioned around the vacuole and in close proximity to vacuole invaginations during dark-induced senescence in wheat. Eventually the entire chloroplast was deposited into the vacuole. This phenomenon seems specific to senescence-inducing conditions, since prior to induction there are no chloroplasts in or near the vacuole (Minamikawa et al. 2001; Wittenbach et al. 1982). Minamikawa et al. (2001) have shown that autophagy induction in French bean leaves also results in chloroplast localization within the vacuole. In TMV-infected *N. benthamiana* plants, several remnants of chloroplasts can be seen within the vacuole, while chloroplasts remain in the cytoplasm in uninfected plants (Fig. 4). It has already been well established that TMV viral particles reside within the chloroplast and lead to chloroplast malformation and

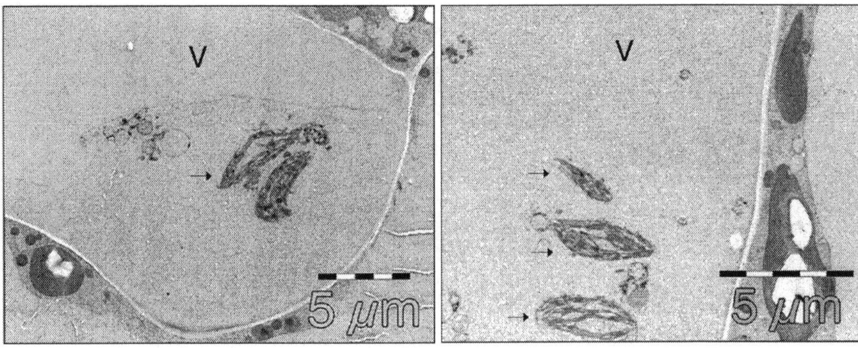

Fig. 4 Targeting of chloroplast to vacuole during tobacco mosaic virus infection in *Nicotiana benthamiana* plants. *Arrowheads* point to damaged chloroplasts in the vacuole (*V*)

dysfunction (Lehto et al. 2003; Seo et al. 2000). One possible explanation for the general reduction in total chloroplasts during prolonged viral infections could be that aberrant or infected chloroplasts trigger transport to the vacuole for degradation, similar to damaged mitochondria and mitophagy.

Although the observations mentioned above are encouraging, genetic evidence for autophagy-mediated chloroplast degradation is less convincing. Preliminary experiments using mutant plants deficient in autophagy have begun to address the role of autophagy in chloroplast functions. Contrary to the hypothesis that autophagy would be required for chloroplast degradation, in some autophagy-deficient plants there is no reduction in Rubisco degradation (Doelling et al. 2002; Hanaoka et al. 2002). One possible explanation for this apparent discrepancy can be found by studying the molecular mechanisms needed for mitophagy. Inhibition of PI3K, an essential component of autophagy, blocks starvation-induced autophagy, but photodamage-induced mitophagy is uncompromised (Kim et al. 2006). Instead, inhibition of PI3K enhances the association of LC3 (an Atg8 homolog) with damaged mitochondria. By contrast, autophagy mutant strains, Δ*atg1*, Δ*atg6*, Δ*atg8* and *atg12*, are defective in mitochondrial degradation (Zhang et al. 2007). Additionally, piecemeal microautophagy of the nucleus is not affected by deletion of *ATG7* (Roberts et al. 2003). Mitophagy and piecemeal autophagy of the nucleus do not use all of the components required for classical autophagy. This may explain why several knockout lines in *Arabidopsis* have no effect on chlorophagy. Electron microscopy and immunocytochemical analysis should be done in several autophagy-deficient lines to determine if chloroplast localization is affected.

7 Plant Viral Clearance in the Vacuole

Although molecular evidence for viral clearance through autophagy is lacking in plants, there are numerous reports indicating the possibility that viruses might be transported to the vacuole for degradation. The tobacco ringspot virus in *Nicotiana tabacum* and southern bean mosaic virus in cowpea cells accumulate in high numbers in the vacuole (Fig. 2, Region 1 and Fig. 5a) (Roberts et al. 1970; Weintraub and Ragetli 1970). Double-membrane vesicles filled with electron-dense material, presumably viral particles, can be seen fusing with a vacuole that already contains viral particles, consistent with an autophagosome type of uptake and transport (Roberts et al. 1970). In grapevine fanleaf virus-infected *Nicotiana clevlandii* plants, accumulation of large groups of viral particles in characteristic crystalline arrays within the vacuole has been observed (Saric and Wrischer 1975). The shapes of the aggregates within the vacuole were very similar to viral aggregates observed in the cytoplasm. Furthermore, a preponderance of small vacuole-like structures in the cytoplasm adjacent to the vacuole was observed. Maratelli et al. demonstrated that, at least for artichoke mottled crinkle virus, it is possible to observe invagination of the tonoplast incorporating virus particles (Fig. 5b, Martelli and Russo 1973). These results demonstrate that the localization of viral particles is most likely a consequence of an active transport system, reminiscent of microautophagy and macroautophagy, as seen in Fig. 5a. With the advancement of microscopy and the availability of a wide range of autophagy-deficient lines, it will be interesting to revisit many of these observations to determine the role(s) of macroautophagy and microautophagy in viral localization/clearance, pathogenicity, and basal immunity.

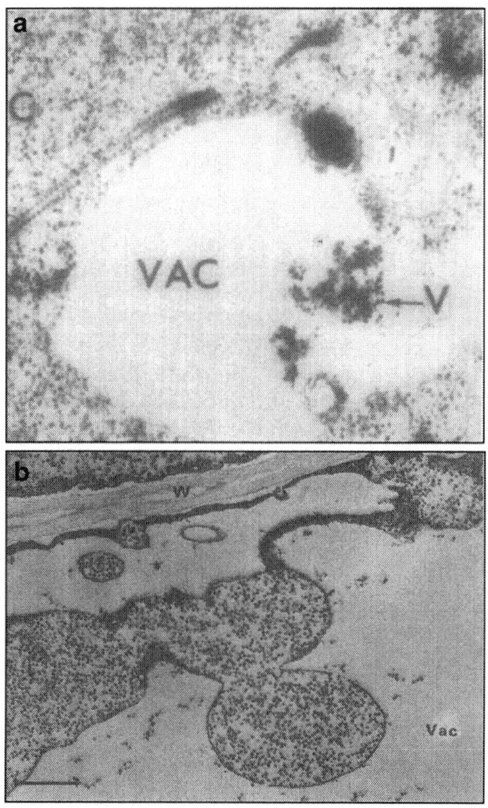

Fig. 5a–b Targeting of viruses to vacuole (**a**). Tobacco ringspot virus particles (*V*) in a double-membrane vesicle (*arrow*) fusing with the central vacuole (*VAC*). *C*, cytoplasm. Reproduced from Roberts et al. (1970) with kind permission from Elsevier Limited. **b** Electron micrograph of artichoke mottled crinkle virus in a vesicle invaginating in the central vacuole (*Vac*). *W*, cell wall. Reproduced from Martelli and Russo (1973) with kind permission from Elsevier Ltd

8 Conclusions

Autophagy-like mechanisms may be important for multiple steps in plant innate immunity. During infections, RNA viruses take advantage of the chloroplast for replication, assembly and the possibly of impairing chloroplastic function in order to reduce ROS generation during the HR-PCD. We propose that chlorophagy functions to eliminate damaged chloroplasts so as to restore cellular homeostasis or to target infected chloroplasts to the vacuole for degradation. Additionally, plant viruses may be directly targeted to the vacuole through micro- and macroautophagy for viral clearance. Autophagy is also used as a pro-survival strategy to limit HR-PCD and viral spread during effector-triggered immunity. The use of fluorescently tagged Atg8 proteins that label autophagosomes, sophisticated microscopy, and autophagy-deficient plants may shed further light on how different classes of viruses are shuttled to the vacuole by autophagy. Taken together, the fields of virology and plant innate immunity will benefit immensely from understanding the various ways that autophagy may control pathogenicity and cell death.

Acknowledgments NIH and NSF grants support autophagy work in the Dinesh-Kumar lab.

References

Ahlquist P (2006) Parallels among positive-strand RNA viruses, reverse-transcribing viruses and double-stranded RNA viruses. Nat Rev Microbiol 4:371–382

Bassham DC (2007) Plant autophagy—more than a starvation response. Curr Opin Plant Biol 10:587–593

Bassham DC, Laporte M, Marty F, Moriyasu Y, Ohsumi Y, Olsen LJ, Yoshimoto K (2006) Autophagy in development and stress responses of plants. Autophagy 2:2–11

Beau I, Esclatine A, Codogno P (2008) Lost to translation: when autophagy targets mature ribosomes. Trends Cell Biol 18(7):311–314

Bent AF, Mackey D (2007) Elicitors, effectors, and R genes: the new paradigm and a lifetime supply of questions. Annu Rev Phytopathol 45:399–436

Caplan J, Mamilapalli P, Burch-Smith TM, Czymmek K, Dinesh-Kumar SP (2008a) Chloroplastic protein NRIP1 mediates innate immune receptor recognition of a viral effector. Cell 132:449–462

Caplan J, Padmanabhan M, Dinesh-Kumar SP (2008b) Plant NB-LRR immune receptors: from recognition to transcriptional reprogramming. Cell Host Microbe 3:126–135

Chiba A, Ishida H, Nishizawa NK, Makino A, Mae T (2003) Exclusion of ribulose-1,5-bisphosphate carboxylase/oxygenase from chloroplasts by specific bodies in naturally senescing leaves of wheat. Plant Cell Physiol 44:914–921

Chisholm ST, Coaker G, Day B, Staskawicz BJ (2006) Host–microbe interactions: shaping the evolution of the plant immune response. Cell 124:803–814

Culver JN (2002) Tobacco mosaic virus assembly and disassembly: determinants in pathogenicity and resistance. Annu Rev Phytopathol 40:287–308

Dangl JL, Jones JDG (2001) Plant pathogens and integrated defence responses to infection. Nature 411:826–833

Deretic V (2005) Autophagy in innate and adaptive immunity. Trends Immunol 26:523–528

Doelling JH, Walker JM, Friedman EM, Thompson AR, Vierstra RD (2002) The APG8/12-activating enzyme APG7 is required for proper nutrient recycling and senescence in *Arabidopsis thaliana*. J Biol Chem 277:33105–33114

Dunn WA, Jr, Cregg JM, Kiel JA, van der Klei IJ, Oku M, Sakai Y, Sibirny AA, Stasyk OV, Veenhuis M (2005) Pexophagy: the selective autophagy of peroxisomes. Autophagy 1:75–83
Ellis RJ, Hartley MR (1974) Nucleic acids of chloroplasts. In: Burton K (ed) Biochemistry of nucleic acids (MTP Int Rev Sci Ser Biochem, vol. 6). Medical and Technical Publishing Co. Ltd, Lancaster
Esau K, Cronshaw J (1967) Relation of tobacco mosaic virus to the host cells. J Cell Biol 33:665–678
Espinoza C, Medina C, Somerville S, Arce-Johnson P (2007) Senescence-associated genes induced during compatible viral interactions with grapevine and *Arabidopsis*. J Exp Bot 58:3197–3212
Feller U, Anders I, Mae T (2008) Rubiscolytics: fate of Rubisco after its enzymatic function in a cell is terminated. J Exp Bot 59:1615–1624
Granett AL, Shalla TA (1970) Discrepancies in the intracellular behavior of three strains of tobacco mosaic virus, two of which are serologically indistinguishable. Phytopathology 60:419–425
Hanaoka H, Noda T, Shirano Y, Kato T, Hayashi H, Shibata D, Tabata S, Ohsumi Y (2002) Leaf senescence and starvation-induced chlorosis are accelerated by the disruption of an *Arabidopsis* autophagy gene. Plant Physiol 129:1181–1193
Heath MC (2000) Hypersensitive response-related death. Plant Mol Biol 44:321–334
Hodgson RA, Beachy RN, Pakrasi HB (1989) Selective inhibition of photosystem II in spinach by tobacco mosaic virus: an effect of the viral coat protein. FEBS Lett 245:267–270
Jimenez I, Lopez L, Alamillo JM, Valli A, Garcia JA (2006) Identification of a plum pox virus CI-interacting protein from chloroplast that has a negative effect in virus infection. Mol Plant Microbe Interact 19:350–358
Jin Y, Ma D, Dong J, Li D, Deng C, Jin J, Wang T (2007) The HC-pro protein of potato virus Y interacts with NtMinD of tobacco. Mol Plant Microbe Interact 20:1505–1511
Kiba A, Takata O, Ohnishi K, Hikichi Y (2006) Comparative analysis of induction pattern of programmed cell death and defense-related responses during hypersensitive cell death and development of bacterial necrotic leaf spots in eggplant. Planta 224:981–994
Kim I, Mizushima N, Lemasters J (2006) Selective removal of damaged mitochondria by autophagy (mitophagy). Hepatology 44(Suppl):188A–286A (241A abstract)
Kim I, Rodriguez-Enriquez S, Lemasters JJ (2007) Selective degradation of mitochondria by mitophagy. Arch Biochem Biophys 462:245–253
Kiraly L, Hafez YM, Fodor J, Kiraly Z (2008) Suppression of tobacco mosaic virus-induced hypersensitive-type necrotization in tobacco at high temperature is associated with downregulation of NADPH oxidase and superoxide and stimulation of dehydroascorbate reductase. J Gen Virol 89:799–808
Klionsky DJ (2005) The molecular machinery of autophagy: unanswered questions. J Cell Sci 118:7–18
Lam E (2004) Controlled cell death, plant survival and development. Nat Rev Mol Cell Biol 5:305–315
Lehto K, Tikkanen M, Hiriart JB, Paakkarinen V, Aro EM (2003) Depletion of the photosystem II core complex in mature tobacco leaves infected by the flavum strain of tobacco mosaic virus. Mol Plant Microbe Interact 16:1135–1144
Levine B, Deretic V (2007) Unveiling the roles of autophagy in innate and adaptive immunity. Nat Rev Immunol 7:767–777
Liang XH, Kleeman LK, Jiang HH, Gordon G, Goldman JE, Berry G, Herman B, Levine B (1998) Protection against fatal Sindbis virus encephalitis by Beclin, a novel Bcl-2-interacting protein. J Virol 72:8586–8596
Lim PO, Kim HJ, Nam HG (2007) Leaf senescence. Annu Rev Plant Biol 58:115–1136
Lin JW, Ding MP, Hsu YH, Tsai CH (2007) Chloroplast phosphoglycerate kinase, a gluconeogenetic enzyme, is required for efficient accumulation of bamboo mosaic virus. Nucleic Acids Res 35:424–432
Liu Y, Schiff M, Czymmek K, Talloczy Z, Levine B, Dinesh-Kumar SP (2005) Autophagy regulates programmed cell death during the plant innate immune response. Cell 121:567–577

Mae T, Kai N, Makino A, Ohira K (1984) Relation between ribulose bisphosphate carboxylase content and chloroplast number in naturally senescing primary leaves of wheat. Plant Cell Physiol 25:333–336

Mae T, Makino A, Ohira K (1983) Changes in the amounts of ribulose bisphosphate carboxylase synthesized and degraded during the life span of rice leaf (*Oryza sativa* L.). Plant Cell Physiol 24:1079–1086

Maiuri MC, Zalckvar E, Kimchi A, Kroemer G (2007) Self-eating and self-killing: crosstalk between autophagy and apoptosis. Nat Rev Mol Cell Biol 8:741–752

Makino A, Mae T, Ohira K (1983) Photosynthesis and ribulose 1,5-bisphosphate carboxylase in rice leaves: changes in photosynthesis and enzymes involved in carbon assimilation from leaf development through senescence. Plant Physiol 73:1002–1007

Martelli GP, Russo M (1973) Electron microscopy of artichoke mottled crinkle virus in leaves of *Chenopodium quinoa* Willd. J Ultrastruct Res 42:93–107

Martin MT, Cervera MT, Garcia JA, Bonay P (1995) Properties of the active plum pox potyvirus RNA polymerase complex in defined glycerol gradient fractions. Virus Res 37:127–137

Mateo A, Muhlenbock P, Rusterucci C, Chang CC, Miszalski Z, Karpinska B, Parker JE, Mullineaux PM, Karpinski S (2004) Lesion simulating disease 1 is required for acclimation to conditions that promote excess excitation energy. Plant Physiol 136:2818–2830

Mijaljica D, Prescott M, Devenish RJ (2007) Different fates of mitochondria: alternative ways for degradation? Autophagy 3:4–9

Miller BL, Huffaker RC (1981) Partial purification and characterization of endoproteinases from senescing barley leaves. Plant Physiol 68:930–936

Minamikawa T, Toyooka K, Okamoto T, Hara-Nishimura I, Nishimura M (2001) Degradation of ribulose-bisphosphate carboxylase by vacuolar enzymes of senescing french bean leaves: immunocytochemical and ultrastructural observations. Protoplasma 218:144–153

Mur LA, Kenton P, Lloyd AJ, Ougham H, Prats E (2008) The hypersensitive response; the centenary is upon us but how much do we know? J Exp Bot 59:501–520

Niwa Y, Kato T, Tabata S, Seki M, Kobayashi M, Shinozaki K, Moriyasu Y (2004) Disposal of chloroplasts with abnormal function into the vacuole in *Arabidopsis thaliana* cotyledon cells. Protoplasma 223:229–232

Nowikovsky K, Reipert S, Devenish RJ, Schweyen RJ (2007) Mdm38 protein depletion causes loss of mitochondrial K^+/H^+ exchange activity, osmotic swelling and mitophagy. Cell Death Differ 14:1647–1656

Ono K, Hashimoto H, Katoh S (1995) Changes in the number and size of chloroplasts during senescence of primary leaves of wheat grown under different conditions. Plant Cell Physiol 36:9–17

Patel S, Caplan J, Dinesh-Kumar SP (2006) Autophagy in the control of programmed cell death. Curr Opin Plant Biol 9:391–396

Patel S, Dinesh-Kumar SP (2008) *Arabidopsis* ATG6 is required to limit the pathogen-associated cell death response. Autophagy 4:20–27

Pattingre S, Tassa A, Qu X, Garuti R, Liang XH, Mizushima N, Packer M, Schneider MD, Levine B (2005) Bcl-2 antiapoptotic proteins inhibit Beclin 1-dependent autophagy. Cell 122: 927–939

Priault M, Salin B, Schaeffer J, Vallette FM, di Rago JP, Martinou JC (2005) Impairing the bioenergetic status and the biogenesis of mitochondria triggers mitophagy in yeast. Cell Death Differ 12:1613–1621

Prod'homme D, Jakubiec A, Tournier V, Drugeon G, Jupin I (2003) Targeting of the turnip yellow mosaic virus 66K replication protein to the chloroplast envelope is mediated by the 140K protein. J Virol 77:9124–9135

Qin G, Ma Z, Zhang L, Xing S, Hou X, Deng J, Liu J, Chen Z, Qu LJ, Gu H (2007) *Arabidopsis* AtBECLIN 1/AtAtg6/AtVps30 is essential for pollen germination and plant development. Cell Res 17:249–263

Radwan DE, Lu G, Fayez KA, Mahmoud SY (2008) Protective action of salicylic acid against bean yellow mosaic virus infection in *Vicia faba* leaves. J Plant Physiol 165:845–857

Reinero A, Beachy RA (1986) Association of TMV coat protein with chloroplast membranes in virus-infected leaves. Plant Mol Biol 6:291–301
Reinero A, Beachy RN (1989) Reduced photosystem II activity and accumulation of viral coat protein in chloroplasts of leaves infected with tobacco mosaic virus. Plant Physiol 89:111–116
Roberts DA, Christie RG, Archer MC, Jr. (1970) Infection of apical initials in tobacco shoot meristems by tobacco ringspot virus. Virology 42:217–220
Roberts P, Moshitch-Moshkovitz S, Kvam E, O'Toole E, Winey M, Goldfarb DS (2003) Piecemeal microautophagy of nucleus in *Saccharomyces cerevisiae*. Mol Biol Cell 14:129–141
Rodriguez-Cerezo E, Findlay K, Shaw JG, Lomonossoff GP, Qiu SG, Linstead P, Shanks M, Risco C (1997) The coat and cylindrical inclusion proteins of a potyvirus are associated with connections between plant cells. Virology 236:296–306
Sakai Y, Koller A, Rangell LK, Keller GA, Subramani S (1998) Peroxisome degradation by microautophagy in *Pichia pastoris*: identification of specific steps and morphological intermediates. J Cell Biol 141:625–636
Salonen A, Ahola T, Kaariainen L (2005) Viral RNA replication in association with cellular membranes. Curr Top Microbiol Immunol 285:139–173
Saric A, Wrischer M (1975) Fine structure changes in different host plants induced by grapevine fanleaf virus. J. Phytopathol. 84:97–104
Seay M, Patel S, Dinesh-Kumar SP (2006) Autophagy and plant innate immunity. Cell Microbiol 8:899–906
Seo S, Okamoto M, Iwai T, Iwano M, Fukui K, Isogai A, Nakajima N, Ohashi Y (2000) Reduced levels of chloroplast FtsH protein in tobacco mosaic virus-infected tobacco leaves accelerate the hypersensitive reaction. Plant Cell 12:917–932
Shalla TA, Petersen LJ, Giunchedi L (1975) Partial characterization of virus-like particles in chloroplasts of plants infected with the U5 strain of TMV. Virology 66:94–105
Slavikova S, Shy G, Yao Y, Glozman R, Levanony H, Pietrokovski S, Elazar Z, Galili G (2005) The autophagy-associated *Atg8* gene family operates both under favourable growth conditions and under starvation stresses in *Arabidopsis* plants. J Exp Bot 56:2839–2849
Soosaar JL, Burch-Smith TM, Dinesh-Kumar SP (2005) Mechanisms of plant resistance to viruses. Nat Rev Microbiol 3:789–798
Thayer SS, Huffaker RC (1984) Vacuolar localization of endoproteinases EP(1) and EP(2) in barley mesophyll cells. Plant Physiol 75:70–73
Thompson AR, Doelling JH, Suttangkakul A, Vierstra RD (2005) Autophagic nutrient recycling in *Arabidopsis* directed by the ATG8 and ATG12 conjugation pathways. Plant Physiol 138:2097–2110
Thompson AR, Vierstra RD (2005) Autophagic recycling: lessons from yeast help define the process in plants. Curr Opin Plant Biol 8:165–173
Torrance L, Cowan GH, Gillespie T, Ziegler A, Lacomme C (2006) Barley stripe mosaic virus-encoded proteins triple-gene block 2 and gammab localize to chloroplasts in virus-infected monocot and dicot plants, revealing hitherto-unknown roles in virus replication. J Gen Virol 87:2403–2411
Torres MA, Dangl JL, Jones JD (2002) *Arabidopsis* gp91phox homologues AtrbohD and AtrbohF are required for accumulation of reactive oxygen intermediates in the plant defense response. Proc Natl Acad Sci USA 99:517–522
Torres MA, Jones JD, Dangl JL (2005) Pathogen-induced, NADPH oxidase-derived reactive oxygen intermediates suppress spread of cell death in Arabidopsis thaliana. Nat Genet 37:1130–1134
Tuttle DL, Dunn WA, Jr. (1995) Divergent modes of autophagy in the methylotrophic yeast *Pichia pastoris*. J Cell Sci 108 (Pt 1):25–35
Wardley TM, Bhalla PL, Dalling MJ (1984) Changes in the number and composition of chloroplasts during senescence of mesophyll cells of attached and detached primary leaves of wheat (*Triticum aestivum* L.). Plant Physiol 75:421–424
Weintraub M, Ragetli HW (1970) Identification of the constituents of southern bean mosaic virus in crystals of infected cells, and their distribution within the virion. Virology 41:729–739

Whitham S, Dinesh-Kumar SP, Choi D, Hehl R, Corr C, Baker B (1994) The product of the tobacco mosaic virus resistance gene N: similarity to Toll and the interleukin-1 receptor. Cell 78:1101–1115

Wittenbach VA, Lin W, Hebert RR (1982) Vacuolar localization of proteases and degradation of chloroplasts in mesophyll protoplasts from senescing primary wheat leaves. Plant Physiol 69:98–102

Xiang Y, Kakani K, Reade R, Hui E, Rochon D (2006) A 38-amino-acid sequence encompassing the arm domain of the cucumber necrosis virus coat protein functions as a chloroplast transit peptide in infected plants. J Virol 80:7952–7964

Xie Z, Klionsky DJ (2007) Autophagosome formation: core machinery and adaptations. Nat Cell Biol 9:1102–1109

Xiong Y, Contento AL, Bassham DC (2007a) Disruption of autophagy results in constitutive oxidative stress in *Arabidopsis*. Autophagy 3:257–258

Xiong Y, Contento AL, Nguyen PQ, Bassham DC (2007b) Degradation of oxidized proteins by autophagy during oxidative stress in *Arabidopsis*. Plant Physiol 143:291–299

Yoshida T, Minamikawa T (1996) Successive amino-terminal proteolysis of the large subunit of ribulose 1,5-biphosphate carboxylase/oxygenase by vacuolar enzymes from French bean leaves. Eur J Biochem 238:317–324

Youle RJ, Strasser A (2008) The BCL-2 protein family: opposing activities that mediate cell death. Nat Rev Mol Cell Biol 9:47–59

Zechmann B, Muller M, Zellnig G (2003) Cytological modifications in zucchini yellow mosaic virus (ZYMV)-infected Styrian pumpkin plants. Arch Virol 148:1119–1133

Zeidler D, Zahringer U, Gerber I, Dubery I, Hartung T, Bors W, Hutzler P, Durner J (2004) Innate immunity in *Arabidopsis thaliana*: lipopolysaccharides activate nitric oxide synthase (NOS) and induce defense genes. Proc Natl Acad Sci USA 101:15811–15816

Zhang Y, Qi H, Taylor R, Xu W, Liu LF, Jin S (2007) The role of autophagy in mitochondria maintenance: characterization of mitochondrial functions in autophagy-deficient *S. cerevisiae* strains. Autophagy 3:337–346

Zimmermann P, Hirsch-Hoffmann M, Hennig L, Gruissem W (2004) Genevestigator. *Arabidopsis* microarray database and analysis toolbox. Plant Physiol 136:2621–2632

Zipfel C (2008) Pattern-recognition receptors in plant innate immunity. Curr Opin Immunol 20:10–6

Autophagy in HIV-Induced T Cell Death

Lucile Espert and Martine Biard-Piechaczyk

Contents

1 Introduction 308
2 HIV Replication Cycle 309
3 Highly Active Antiretroviral Therapy and Autophagy-Inducing Drugs 310
4 Mechanisms Leading to Immunodeficiency 312
5 Role of Autophagy in Env-Mediated CD4 T Cell Death 313
6 Role of Autophagy in HIV-Infected Cells 315
7 Conclusion 315
References 316

Abstract HIV infection leads to progressive CD4 T cell depletion, resulting in the development of AIDS. The mechanisms that trigger T cell death after HIV infection are still not fully understood, but a lot of data indicate that apoptosis of uninfected CD4 lymphocytes plays a major role. HIV directly modulates cell death using various strategies in which several viral proteins, in particular the envelope glycoproteins (Env), play an essential role. Importantly, Env, expressed on infected cells, triggers autophagy in uninfected CD4 T cells, leading to their apoptosis. Furthermore, HIV, like other viruses, has evolved strategies to inhibit this autophagic process in HIV-infected cells. This discovery further increases the level of complexity of the cellular processes involved in HIV-induced pathology. Interestingly, HIV protease inhibitors, currently used in highly active antiretroviral therapy (HAART), are able to induce autophagy in cancer cells, leading to a recent repositioning of these drugs as anticancer agents. This review presents an overview of the relationship between HIV, HAART, and autophagy.

L. Espert and M. Biard-Piechaczyk (✉)
CPBS, CNRS UMR5236, University of Montpellier, Institut de Biologie,
4, Bd Henri IV, CS 69033, 34965 Montpellier Cedex 2, France
e-mail: martine.biard@univ-montp1.fr

B. Levine et al. (eds.), *Autophagy in Infection and Immunity*,
Current Topics in Microbiology and Immunology 335,
DOI 10.1007/ 978-3-642-00302-8_15,

Abbreviations

Env	HIV envelope glycoproteins
HAART	Highly active antiretroviral therapy
KS	Kaposi's sarcoma
PCD	Programmed cell death
PI	HIV protease inhibitor

1 Introduction

Human immunodeficiency virus (HIV) is a retrovirus belonging to the *Lentivirus* genus (Levy 1993). It possesses a complex genome composed of three structural genes (*gag*, *pol*, and *env*), two regulatory genes (*tat* and *rev*), and four accessory genes (*vpu*, *vpr*, *vif*, and *nef*).

HIV can infect a variety of immune cells, such as CD4 T cells, macrophages, and microglial cells, but its cytopathic effects mainly target the CD4 T cell population. Indeed, the macrophage population is preserved in HIV-infected patients, despite progressive T cell depletion (Gendelman et al. 1989). Whether HIV infection within mononuclear phagocytic cells is latent or is subject to active replication has not been proven in vivo, and accurate estimation of the proportions of mononuclear phagocytic cells harboring HIV infection still requires further investigation (Noursadeghi et al. 2006).

There are two viral types, HIV-1 and HIV-2, which bear a strong homology in both structure and sequence (30–60%). They are thought to have entered the human population from Simian immunodeficiency virus (SIV)-infected monkeys and central African chimpanzees for HIV-1 and a West African species of sooty mangabey for HIV-2. In addition, numerous groups (M, N, and O for HIV-1 and A through H for HIV-2), subtypes, subsubtypes, and circulating recombinant forms have emerged (Butler et al. 2007).

HIV-1 is the causative agent of acquired immune deficiency syndrome (AIDS) (Barré-Sinoussi et al. 1983; Gallo et al. 1983; Popovic et al. 1984), defined as a low CD4 T cell count and at least one opportunistic infection (Centers for Disease Control and Prevention, CDC, definition from August 2008). The global pandemic is still expanding, accounting for 2.5 million new infections and 2.1 million deaths in 2007 (December 2007 report, UNAIDS). In the absence of treatment, the vast majority of HIV-1-infected individuals (70–80%) develop profound immunodeficiency within approximately ten years of acquiring HIV-1, correlated with a high viral replication and CD4 T cell depletion (Langford et al. 2007). However, a small proportion of HIV-1-infected individuals (5–15%), called long-term nonprogressors, remain clinically and/or immunologically stable for years, with low to moderate amounts of viremia and minimal rates of CD4 T cell decline. Amongst this population, a small subset, called HIV controllers, maintain viral loads below the limits of detection (<50 copies HIV RNA/mL), present normal CD4 T cell numbers, and fail to develop symptoms for many years after infection (Deeks and Walker 2007).

Furthermore, some individuals who are highly exposed to HIV-1 and yet remain uninfected are likely to be naturally resistant to HIV-1 infection (exposed uninfected individuals, EU).

Unlike HIV-1, HIV-2 infection has remained confined mainly to West Africa and is compatible with a normal lifespan in most infected people (Leligdowicz et al. 2007; Rowland-Jones and Whittle 2007). Thus, the individual risk of acquiring HIV infection and developing AIDS is not the same, and AIDS progression remains a multifactorial event involving both viral and host factors (Jansen et al. 2006). Consequently, understanding the host factors that control disease progression will be fundamental to the development of new therapeutic strategies (Deeks and Walker 2007).

2 HIV Replication Cycle

The first step in HIV-1 infection is mediated by the binding of envelope glycoproteins (Env) to CD4 and two major coreceptors, CCR5 and CXCR4 (Alkhatib et al. 1996; Bleul et al. 1996; Feng et al. 1996), present on target cells. On the virion surface, as well as on the surfaces of HIV-infected cells, Env is arranged in spike-like structures formed from trimeric complexes of gp120, the external envelope glycoprotein, and gp41, the transmembrane subunit. To initiate infection of target cells, gp120 first interacts with CD4, which triggers conformational changes leading to increased exposure of the gp120 regions (including the V3 loop), which are then able to bind to the coreceptor. The gp41 then adopts a trimeric extended pre-hairpin intermediate conformation, which is a target of HIV fusion inhibitors such as T20 and C34 peptides. These events induce the formation of a coiled-coil structure in the gp41 ectodomain that places the hydrophobic, N´-terminal fusion peptide of gp41 in close proximity to the cellular membrane, thereby initiating cell fusion (Weissenhorn et al. 1997). This strategy of interaction provides an effective mechanism for concealing highly conserved neutralization epitopes from the attack of host antibodies.

HIV-1 infection of CD4 T cells is favored by cell-to-cell contact, through the formation of the virological synapse (Piguet and Sattentau 2004). Contact of HIV-infected cells, which express Env, with uninfected CD4 T cells induces a gp41-dependent hemifusion process in which the transfer of lipids from the membrane of Env-expressing cells to the target cells occurs. Complete cell-to-cell fusion can also be triggered, leading eventually to the formation of giant multinucleated cells called syncytia (Munoz-Barroso et al. 1998). Syncytia are not stable over an extended time period (Lifson et al. 1986; Sodroski et al. 1986; Yoffe et al. 1987), and are barely detectable in infected individuals, except in the brain (Navia et al. 1986) and the tonsils (Frankel et al. 1996).

The differential use of CCR5 and/or CXCR4 defines HIV strain variants (R5, X4, and R5X4) that are related to, but distinct from, target cell tropism (Goodenow and Collman 2006). Despite these three phenotypes, R5 viruses are almost exclusively associated with acute infection, irrespective of the route of transmission (sexual, mother-to-child or intravenous) (Margolis and Shattock 2006). R5 strains that use CCR5 in addition to CD4 for entry were previously called M-tropic strains because they preferentially infect macrophages in addition to primary CD4 T cells. One type of natural resistance to HIV infection has been elucidated with the identification of a 32-bp deletion within the coding region of the *CCR5* gene (*CCR5-Δ32*). This introduces a premature stop codon, resulting in a defective receptor that cannot be expressed at the cell surface (Liu et al. 1996; Samson et al. 1996). In homozygotes, this defect confers a high degree of protection against HIV infection, and disease progression is significantly delayed in *CCR5-Δ32* heterozygotes due to reduced CCR5 expression levels. During the progression of the disease, X4 viruses, which use CXCR4 for entry, emerge and are associated with a rapid decline in the number of CD4 T cells. These strains are able to infect CD4 T cell lines in addition to primary CD4 T cells and were thus previously called T-tropic.

After HIV entry into target cells, the viral single-stranded RNA template is reverse transcribed into double-stranded DNA by the viral reverse transcriptase. This process takes place in the cytoplasmic reverse transcription complex, a nucleoprotein complex derived from the core of the infecting virion. The newly synthesized viral DNA remains associated with viral and cellular proteins in a large nucleoprotein complex called the pre-integration complex (PIC), which probably reaches the nuclear envelope by active transport along microtubules. After entry into the nucleus, presumably through the nuclear pore complex, the PIC gains access to chromatin and the viral DNA is integrated into the genome of activated CD4 T cells, an event mediated by the viral integrase (Suzuki and Craigie 2007 for a review). The viral DNA is then replicated along with the cellular DNA during each cycle of cell division. Importantly, HIV productively infects nondividing macrophages and neurons, but cell activation is required for the completion of the viral replication cycle in CD4 T cells (Katz et al. 2005), even if incoming HIV can persist for several weeks in the cytosol of quiescent CD4 T cells as a stable, centrosome-associated, pre-integration intermediate (Zamborlini et al. 2007).

A large number of studies have contributed to our knowledge of the late steps of the HIV-1 life cycle. However, some controversy exists regarding the site of HIV assembly after protein synthesis, depending on the cell types, with two main hypotheses: (1) the plasma membrane is the major site regardless of cell type (Finzi et al. 2007; Welsch et al. 2007), or (2) the site of assembly is the plasma membrane for CD4 T cells and multivesicular bodies (MVB) for macrophages (Pelchen-Matthews et al. 2003; Raposo et al. 2002).

3 Highly Active Antiretroviral Therapy and Autophagy-Inducing Drugs

Although tremendous success has been achieved to date in the treatment of HIV infection, no regimen is curative, and treatments seem to eventually fail in most patients, for reasons related to dose-limiting adverse events, poor patient compliance, evolution of resistance, and persistence of viral reservoirs. Transient rebounds of plasma viremia, "blips," are frequent among patients undergoing highly active antiretroviral therapy (HAART), and are associated with the emergence of drug resistance mutations (Macias et al. 2005). More importantly, a prompt re-emergence of viruses and resumption of CD4 T cell depletion is observed following HAART withdrawal (Chun et al. 2000; Davey et al. 1999).

HAART is usually based on the administration of a combination of two nucleoside reverse-transcriptase inhibitors in association with either a non-nucleoside reverse-transcriptase inhibitor or a protease inhibitor (PI) (Hammer et al. 2006). Drugs with novel mechanisms of action, as well as an HIV vaccine, are under investigation. These comprise agents that inhibit HIV entry by blocking gp41 fusion, such as T20 (Fuzeon®) and C34, or by binding to the coreceptors (CCR5 and CXCR4); agents that target HIV integrase and HIV regulatory gene products; and immunomodulatory agents such as IL-12 and IL-2 (Agrawal et al. 2006). These challenges are coupled with the need for systems that optimize patient access to treatment in resource-limited regions, especially within the developing world (Dunne 2007).

HIV PIs are now an integral part of effective anti-HIV therapy. These drugs block HIV protease and thus prevent proper packaging of virions. It is worth noting that in addition to direct antiviral effects, HIV PIs can also affect caspase and calpain activities, apoptosis-regulatory proteins and mitochondrial transmembrane potential (Hisatomi et al. 2008; Vlahakis et al. 2007). A recent study demonstrates that the most potent HIV protease inhibitor, nelfinavir, induces endoplasmic reticulum (ER) stress, autophagy, and apoptosis in several cancer cell lines in vitro and in xenograft models (Gills et al. 2007a; Gills et al. 2007b). Pyrko and collaborators have also described that several PIs inhibit the proteasome, leading to the accumulation of unfolded and/or misfolded proteins that trigger ER stress and then apoptosis in various malignant glioma cell lines (Pyrko et al. 2007). Initially, PI-mediated autophagy is induced in an attempt to help cancer cells survive, since 3-MA, a compound that inhibits autophagy, increases nelfinavir-mediated cell death. Moreover, treatment with several PIs has reduced Kaposi's sarcoma (KS), indicating that these drugs can be exploited for the therapy of KS and other tumors that occur in both HIV-infected and uninfected individuals (Monini et al. 2004; Sgadari et al. 2002; Sgadari et al. 2003). Nelfinavir is now repositioned as a US Food and Drug Administration (FDA)-approved drug for cancer therapeutics, and is currently being tested in Phase I clinical trials in patients with cancer. However, the activation of ER stress and autophagy by PIs, in addition to their documented blockade of insulin receptor signaling (Ben-Romano et al. 2004; Schutt et al. 2004), could trigger the development of pathological metabolic conditions such as insulin resistance, as has already been described for HAART. This may at least partially explain why the metabolic syndrome, with a substantially increased prevalence of type 2 diabetes, is more common in patients receiving PIs (Gkrania-Klotsas and Klotsas 2007). It is worth noting that the development of dyslipidemia during PI treatment appears to be exposure (dose and time)-related (Nair 2005). However, in HIV-1 infected patients, PI-based therapy results in improved CD4 T cell counts and reduced T cell apoptosis (Phenix et al. 2000), and inhibition of HIV-1 protease by synthetic inhibitors prevents virus-induced T cell death (Ventoso et al. 2005). These contradictory results may be due to the differences in the metabolic rates and the rate of protein synthesis between cancer cells and terminally differentiated T cells, and/or to the therapeutic concentration used.

4 Mechanisms Leading to Immunodeficiency

Just how CD4 T cells are killed during HIV infection is still one of the most controversial issues in AIDS research. It is now accepted that apoptosis of uninfected, bystander, CD4 T cells is a major mechanism responsible for T cell depletion in patients infected with HIV-1 (Ameisen and Capron 1991; Terai et al. 1991). HIV-infected cells also have a shortened lifespan, contributing to the overall CD4 T cell demise (Ho et al. 1995; Perelson et al. 1996; Wei et al. 1995), but they do not undergo apoptosis as readily as uninfected cells (Finkel et al. 1995). This suggests that HIV-1 has evolved multiple mechanisms to ensure survival for a time period sufficient for productive infection. In addition, some activated CD4 T cells that have been infected may survive long enough to revert back to a resting state and become reservoirs, which are a major barrier to HIV eradication from infected individuals (Chun et al. 1995; Chun et al. 1997).

HIV-1 infection results in high activation and turnover of immune cells, and thus accelerates both the production and destruction of CD4 T cells (Fauci 1988; McCune 2001). A strong immune response is a priori beneficial in controlling viral replication, but a chronic, heightened activation of the immune system may contribute in a direct manner to progressive CD4 T cell depletion by destabilizing or progressively changing the homeostatic states of resting T cell populations (Deeks and Walker 2004; Grossman et al. 2002). In addition to the consequences of a general activation of the immune system during HIV-1 infection, several extracellular HIV-1 proteins, such as Env, Tat, Vpr and Nef, are also involved in modulating the death pathways activated in the uninfected CD4 T cells (Varbanov et al. 2006 for a review). The real importance of this process in vivo is not yet fully understood, but cumulative data demonstrate a major role of Env in the cell death of uninfected lymphocytes (Heinkelein et al. 1995; Laurent-Crawford et al. 1993; Ohnimus et al. 1997; Roshal et al. 2001). Indeed, the binding of HIV-1 Env to its receptors (CD4 and a coreceptor) constitutes the primary interface between viruses and target cells, and this event is likely to modulate T cell signaling. However, the mechanisms by which Env triggers bystander CD4 T cell death are very difficult to establish because of the complexity and the multiplicity of the factors involved. These factors include the following: (1) Env activates different cell death mechanisms depending on its presentation (i.e., soluble Env, Env expressed on virions or at the surface of infected cells); (2) Env is composed of two glycoproteins, gp120 and gp41, that are able to trigger different, nonexclusive, cell death pathways; (3) gp120 sequentially binds to CD4 and a coreceptor, both of which are capable of transducing functional responses, such as proliferation, differentiation, chemotaxis, and proinflammatory cytokine secretion (Misse et al. 1999; Weissman et al. 1997) in addition to cell death; (4) gp120-mediated signaling pathways differ depending on whether the coreceptor is CCR5 or CXCR4 (Davis et al. 1997); and (5) only about 15–30% of the CD4 T lymphocytes express detectable levels of CCR5 on the cell surface, in contrast to CXCR4, which is expressed on nearly all of these T cells (Bleul et al. 1997; Grivel and Margolis 1999). This explains, at least in part, that X4 strains exert a profound cytopathic effect on a much wider range of target cells via their particular capacity to induce bystander cell death.

Furthermore, agents interfering with cell-to-cell fusion, such as the peptide T20 which abolishes the correct folding of gp41 following binding of gp120 to its receptor molecules and insertion of the gp41 fusion peptide into the cell membrane (Wild et al. 1994), prevent cell death and T cell depletion (LaBonte et al. 2003; Stocker et al. 2000). Blanco and collaborators have demonstrated that Env-induced cell death of single bystander CD4 T cells requires both gp120 and gp41 functions (Blanco et al. 2003), and recent studies have shown that gp41-induced apoptosis is mediated by caspase-3-dependent mitochondrial depolarization and production of reactive oxygen species (ROS) (Garg and Blumenthal 2006). These results indicate that, in addition to the role of gp120, gp41 could also actively participate in the molecular events leading to Env-induced cell death.

5 Role of Autophagy in Env-Mediated CD4 T Cell Death

Autophagy is a lysosomal degradation pathway essential for cell homeostasis. This process maintains cell survival during nutrient starvation and can rescue cells from apoptosis. However, autophagy is also described as a cell death pathway (type II programmed cell death, PCD). Autophagic cell death has been distinguished from apoptosis (type I PCD) by morphological criteria; i.e., the presence of autophagosomes in dying cells. Cell death with features of autophagy can occur in cells lacking critical apoptosis executioners, indicating that autophagy can compensate for defective apoptosis (Bursch et al. 1996), and has also been described after treatment with chemotherapeutic drugs (Kondo et al. 2005). However, cell death characterized by hallmarks of both type I PCD and type II PCD is frequently observed, making a clear-cut distinction between them difficult (Debnath et al. 2005; Djavaheri-Mergny et al. 2006; Gonzalez-Polo et al. 2005). In addition, autophagy can contribute to cell destruction during necrosis. Indeed, very recently, Samara and collaborators have demonstrated that diverse genetic and environmental insults trigger autophagy in *Caenorhabditis elegans*, and that this process is required for necrotic cell death (Samara et al. 2007). Thus, exclusive definitions of each form of PCD may be artificial due to overlapping signaling pathways shared by these different cell death programs (Broker et al. 2005; Thorburn 2007).

Besides its dual role in controlling life and death of cells, autophagy is a mechanism of both innate and adaptive immunity against intracellular bacteria and viruses (see Deretic 2005; Deretic and Klionsky 2008; Lee and Iwasaki 2008; Levine and Deretic 2007; Schmid et al. 2006; Schmid and Munz 2007; Swanson 2006 for reviews). Autophagy was first proposed as a mechanism that protects against viral infection by degrading the pathogens in autolysosomes. Evidence for an antiviral role for autophagy is strengthened by the fact that interferon (IFN) signaling pathways are involved in its induction.

Autophagy has been implicated in both the delivery of cytosolic antigens to the major histocompatibility complex class II (MHC class II) pathway (Murphy et al. 1993; Paludan et al. 2005) and the digestion of endogenously synthesized viral proteins, allowing their processing for MHC II presentation. It can thus be linked to adaptive immunity (Dengjel et al. 2005). In addition, autophagy allows the sequestration of pathogens inside autophagosomes, leading to their destruction by lysosomes. However, several viruses have evolved strategies to divert autophagy to their own benefit (Espert et al. 2007; Orvedahl and Levine 2008 for a review). Finally, it has been shown that autophagy plays a role in the innate recognition of viral replication products by Toll-like receptors, and is involved in virally induced IFN-α secretion in plasmacytoid dendritic cells (Lee et al. 2007; Seay and Dinesh-Kumar 2007).

Recently, it has been demonstrated in vitro that the autophagic pathway is activated in uninfected CD4 T cells after interaction with Env expressed at the surface of X4 HIV-infected cells (Espert et al. 2006), and that this mechanism leads to apoptosis, since the blockade of autophagy at different steps, by either drugs or siRNAs specific for autophagic genes, totally inhibits apoptosis. The complete autophagic process is needed to trigger apoptosis, suggesting that while autophagic vacuole formation alone does not induce apoptotic cell death, lysosomal degradation of the contents of the autophagic vacuole does. In addition, CD4 T cells still underwent Env-mediated cell death, with autophagic features, when apoptosis was inhibited. The presence of CD4 and CXCR4 on target CD4 T cells is required for this process, but CD4 (Espert et al. 2006) and CXCR4 (Denizot et al. 2008) signaling pathways triggered after gp120 binding are not directly involved. A very recent study demonstrated that the fusogenic function of gp41 is responsible for Env-mediated autophagy of the uninfected CD4 T cells (Denizot et al. 2008). Thus, the entire sequence of events (i.e., gp120 binding to CD4 and CXCR4, and insertion of gp41 into the target membrane) is required to trigger autophagy and may be at least partially responsible for CD4 T cell depletion and immunodeficiency.

Whether Env-induced autophagy is first triggered to rescue the CD4 T cells from cell death or the autophagic process is directly activated to kill the uninfected CD4 T lymphocytes is currently unknown. However, the study described above demonstrates that autophagy plays a crucial role in the Env-mediated CD4 T cell death that characterizes the pathophysiology of AIDS. The results suggest that naive CD4 T cells, which cannot be productively infected by HIV, may take different routes to die after contact with a cell infected by an X4 HIV strain, and that sequential (but not exclusive) cell death pathways are triggered by Env.

In addition to autophagy, Env triggers oxidative stress in the uninfected CD4 T cells with overproduction of both ROS and antioxidant enzymes as counteracting cellular responses. Env also induces an increase in proteins involved in other degradative processes, such as the ubiquitin/proteasome system, and perturbs both global cell metabolism and protein synthesis (Molina et al. 2007).

6 Role of Autophagy in HIV-Infected Cells

Autophagy has been proposed as a mechanism that protects against viral infection by degrading the pathogens into autolysosomes. However, several viruses have evolved strategies to block or divert autophagy to their own benefit (Deretic and Klionsky 2008; Espert et al. 2007; Lee and Iwasaki 2008; Orvedahl and Levine 2008). Very recently it has been demonstrated that, while X4 HIV-1 is able to induce autophagy in uninfected CD4 T cells via Env, it inhibits autophagy in productively infected CD4 T cells (see Zhou and Spector 2008 and our personal data). This suggests that HIV-1 is able to downregulate the autophagic pathway, probably in order to enhance its own replication and to evade the immune system. Beclin 1, which is accumulated in uninfected CD4 T cells after interaction with cells expressing Env (Espert et al. 2006), is markedly decreased in HIV-infected cells (Zhou and Spector 2008). Furthermore, downregulation of autophagy in HIV-infected cells can be overcome by amino acid starvation or rapamycin treatment (Zhou and Spector 2008).

Although the mechanisms by which gp41 induces autophagy in uninfected CD4 T cells and by which HIV-1 controls autophagy in the infected cells are still unknown, these results underscore a potential major role of autophagy in the pathophysiology of HIV-1 strains in triggering the death of the CD4 T cells. However, a very important unsolved question remains concerning the mechanisms by which HIV-infected macrophages, in contrast to CD4 T cells, are able to persist in HIV-1-infected patients.

7 Conclusion

Structural, regulatory and accessory HIV-1 proteins, which are produced throughout the viral life cycle, are major actors in the strategies used by the virus to manipulate different mechanisms of cell death. Their activities depend on their level of expression, localization, biochemical characteristics, presence in the extracellular medium, and the status of the target cells. Overall, while some contradictory results do exist, the data suggest that the same viral proteins are able to trigger both the survival of HIV-1-infected cells for long enough to ensure productive infection as well as the death of bystander CD4 T lymphocytes, which cannot be productively infected by HIV-1.

Furthermore, the observation that different forms of PCD are strongly interconnected, and can substitute for each other, suggests that apoptosis is not the only PCD pathway targeted by HIV-1. Initial evidence of this new concept came from the demonstration that Env triggers autophagy, leading to apoptosis, in bystander CD4 T cells. The rapidly accumulating data concerning the tight relationship between viral infection and autophagy emphasize that the latter could be a fundamental and general process governing the pathogenesis of viral infections. Indeed, this process is involved in both adaptive and innate immunity, contributing to the clearance of intracellular pathogens, and in cell survival or cell death. Moreover, viruses are able to evolve strategies to counteract autophagy, or even to exploit it for their own benefit. Indeed, several viruses can block the triggering of autophagy to avoid their destruction in autolysosomes. In contrast, other viruses use autophagosomes for their own replication. Unraveling the specific role of autophagy in different infections would be a crucial advance in our understanding of viral pathogenesis.

The discovery that Env induces autophagy in uninfected CD4 T cells, leading to their apoptosis, was a major milestone, since these results indicate that autophagy could play a major role in the induction of immunodeficiency during HIV infection. A high research priority is to gain a better understanding of the mechanisms by which autophagy is triggered and regulated in CD4 T cells. Moreover, the development of new therapeutics based on the modulation of autophagy to complement drugs that target viral replication and entry may hold a promising future.

Acknowledgments Institutional funds from the Centre National de la Recherche Scientifique (CNRS) and the University of Montpellier, as well as grants from SIDACTION and the Agence Nationale de Recherches sur Le SIDA (ANRS) supported this work. We thank I. Robbins for helpful scientific discussions and careful critical reading of the manuscript.

References

Agrawal L, Lu X, Jin Q, Alkhatib G (2006) Anti-HIV therapy: current and future directions.Curr Pharm Des 12:2031–55

Alkhatib G, Combadiere C, Broder CC, Feng Y, Kennedy PE, Murphy PM, Berger EA (1996) CC-CKR5 : a RANTES, MIP-1a, MIP-1b receptor as a fusion cofactor for macrophage-tropic HIV-1. Science 272:1955–1958

Ameisen JC, Capron A (1991) Cell dysfunction and depletion in AIDS: the programmed cell death hypotheses. Immunol Today 12:102–105

Barré-Sinoussi F, Chermann J-C, Rey F, Nugeyre M-T, Chamaret S, Gruest J, Dauget C, Axler-Blin C, Brun-Vezinet F, Rouzioux C, Rozembaum W, Montagnier L (1983) Isolation of a T-lymphotropic retrovirus from patient at risk for acquired immunodeficiency syndrome (AIDS). Science 220:868–870

Ben-Romano R, Rudich A, Tirosh A, Potashnik R, Sasaoka T, Riesenberg K, Schlaeffer F, Bashan N (2004) Nelfinavir-induced insulin resistance is associated with impaired plasma membrane recruitment of the PI 3-kinase effectors Akt/PKB and PKC-zeta. Diabetologia 47:1107–17

Blanco J, Barretina J, Ferri KF, Jacotot E, Gutierrez A, Armand-Ugon M, Cabrera C, Kroemer G, Clotet B, Este JA (2003) Cell-surface-expressed HIV-1 envelope induces the death of CD4 T cells during GP41-mediated hemifusion-like events. Virology 305:318–29

Bleul C, Wu LHJA, Springer TA, MacKay CR (1997) The HIV coreceptors CXCR4 and CCR5 are differencially expressed and regulated on human T lymphocytes. Proc Natl Acad Sci USA 94:1925–1930

Bleul CC, Farzan M, Choe H, Parolin C, Clark-Lewis I, Sodroski J, Springer TA (1996) The lymphocyte chemoattractant SDF-1 is a ligand for LESTRE/fusin and blocks HIV-1 entry. Nature 382:829–833

Broker LE, Kruyt FA, Giaccone G (2005) Cell death independent of caspases: a review. Clin Cancer Res 11(9):3155–3162

Bursch W, Ellinger A, Kienzl H, Torok L, Pandey S, Sikorska M, Walker R, Hermann RS (1996) Active cell death induced by the anti-estrogens tamoxifen and ICI 164 384 in human mammary carcinoma cells (MCF-7) in culture: the role of autophagy. Carcinogenesis 17(8):1595–1607

Butler IF, Pandrea I, Marx PA, Apetrei C (2007) HIV genetic diversity: biological and public health consequences. Curr HIV Res 5(1):23–45

Chun TW, Davey RT Jr, Ostrowski M, Shawn Justement J, Engel D, Mullins JI, Fauci AS (2000) Relationship between pre-existing viral reservoirs and the re-emergence of plasma viremia after discontinuation of highly active anti-retroviral therapy. Nat Med 6(7):757–761

Chun TW, Finzi D, Margolick J, Chadwick K, Schwartz D, Siliciano RF (1995) In vivo fate of HIV-1-infected T cells: quantitative analysis of the transition to stable latency. Nat Med 1(12):1284–1290

Chun TW, Stuyver L, Mizell SB, Ehler LA, Mican JA, Baseler M, Lloyd AL, Nowak MA, Fauci AS (1997) Presence of an inducible HIV-1 latent reservoir during highly active antiretroviral therapy. Proc Natl Acad Sci U S A 94(24):13193–13197

Davey RT Jr, Bhat N, Yoder C, Chun TW, Metcalf JA, Dewar R, Natarajan V, Lempicki RA, Adelsberger JW, Miller KD, Kovacs JA, Polis MA, Walker RE, Falloon J, Masur H, Gee D, Baseler M, Dimitrov DS, Fauci AS, Lane HC (1999) HIV-1 and T cell dynamics after interruption of highly active antiretroviral therapy (HAART) in patients with a history of sustained viral suppression. Proc Natl Acad Sci USA 96(26):15109–15114

Davis CB, Dikic I, Unutmaz D, Hill CM, Arthos J, Siani MA, Thompson DA, Schlessinger J, Littman DR (1997) Signal transduction due to HIV-1 envelope interactions with chemokine receptors CXCR4 or CCR5. J Exp Med 186(10):1793–1798

Debnath J, Baehrecke EH, Kroemer G (2005) Does autophagy contribute to cell death? Autophagy 1(2):e10–e18

Deeks SG, Walker BD (2004) The immune response to AIDS virus infection: good, bad, or both? J Clin Invest 113(6):808–810

Deeks SG, Walker BD (2007) Human immunodeficiency virus controllers: mechanisms of durable virus control in the absence of antiretroviral therapy. Immunity 27(3):406–416

Delgado M, Singh S, De Haro S, Master S, Ponpuak M, Dinkins C, Ornatowski W, Vergne I, Deretic V (2009) Autophagy and pattern recognition receptors in innate immunity. Immunol Rev 227(1):189–202

Dengjel J, Schoor O, Fischer R, Reich M, Kraus M, Muller M, Kreymborg K, Altenberend F, Brandenburg J, Kalbacher H, Brock R, Driessen C, Rammensee HG, Stevanovic S (2005) Autophagy promotes MHC class II presentation of peptides from intracellular source proteins. Proc Natl Acad Sci USA 102(22):7922–7927

Denizot M, Varbanov M, Espert L, Robert-Hebmann V, Sagnier S, Garcia E, Curriu M, Mamoun R, Blanco J, Biard-Piechaczyk M (2008) HIV-1 gp41 fusogenic function triggers autophagy in uninfected cells. Autophagy 4(8):998–1008

Deretic V (2005) Autophagy in innate and adaptive immunity. Trends Immunol 26(10):523–528

Djavaheri-Mergny M, Amelotti M, Mathieu J, Besancon F, Bauvy C, Souquere S, Pierron G, Codogno P (2006) NF-kappaB activation represses tumor necrosis factor-alpha-induced autophagy. J Biol Chem 281(41):30373–30382

Dunne M (2007) Antiretroviral drug development: the challenge of cost and access. Aids 21(Suppl 4):S73–79

Espert L, Codogno P, Biard-Piechaczyk M (2007) Involvement of autophagy in viral infections: antiviral function and subversion by viruses. J Mol Med 85(8):811–823

Espert L, Denizot M, Grimaldi M, Robert-Hebmann V, Gay B, Varbanov M, Codogno P, Biard-Piechaczyk M (2006) Autophagy is involved in T cell death after binding of HIV-1 envelope proteins to CXCR4. J Clin Invest 116(8):2161–2172

Fauci AS (1988) The human immunodeficiency virus: infectivity and mechanisms of pathogenesis. Science 239(4840):617–622

Feng Y, Broder CC, Kennedy PE, Berger EA (1996) HIV-1 entry cofactor : functional cDNA cloning of a seven-transmembrane, G protein-coupled receptor. Science 272:872–877

Finkel TH, Tudor-Williams G, Banda NK, Cotton MF, Curiel T, Monks C, Baba TW, Ruprecht RM, Kupfer A (1995) Apoptosis occurs predominantly in bystander cells and not in productively infected cells of HIV- and SIV-infected lymph nodes. Nat Med 1(2):129–134

Finzi A, Orthwein A, Mercier J, Cohen EA (2007) Productive human immunodeficiency virus type 1 assembly takes place at the plasma membrane. J Virol 81(14):7476–7490

Frankel SS, Wenig BM, Burke AP, Mannan P, Thompson LD, Abbondanzo SL, Nelson AM, Pope M, Steinman RM (1996) Replication of HIV-1 in dendritic cell-derived syncytia at the mucosal surface of the adenoid. Science 272(5258):115–117

Gallo RC, Sarin PS, Gelmann EP, Robert-Guroff M, Richardson E, Kalyanaraman VS, Mann D, Sidhu GD, Stahl RE, Zolla-Pazner S, Leibowitch J, Popovic M (1983) Isolation of human T-cell leukemia virus in acquired immune deficiency syndrome (AIDS). Science 220(4599):865–867

Garg H, Blumenthal R (2006) HIV gp41-induced apoptosis is mediated by caspase-3-dependent mitochondrial depolarization, which is inhibited by HIV protease inhibitor nelfinavir. J Leukoc Biol 79(2):351–362

Gendelman HE, Orenstein JM, Baca LM, Weiser B, Burger H, Kalter DC, Meltzer MS (1989) The macrophage in the persistence and pathogenesis of HIV infection. Aids 3(8):475–495

Gills JJ, Lopiccolo J, Dennis PA (2008a) Nelfinavir, a new anti-cancer drug with pleiotropic effects and many paths to autophagy. Autophagy 4(1):107–9

Gills JJ, Lopiccolo J, Tsurutani J, Shoemaker RH, Best CJ, Abu-Asab MS, Borojerdi J, Warfel NA, Gardner ER, Danish M, Hollander MC, Kawabata S, Tsokos M, Figg WD, Steeg PS, Dennis PA (2007b) Nelfinavir, A lead HIV protease inhibitor, is a broad-spectrum, anticancer agent that induces endoplasmic reticulum stress, autophagy, and apoptosis in vitro and in vivo. Clin Cancer Res 13(17):5183–5194

Gkrania-Klotsas E, Klotsas AE (2007) HIV and HIV treatment: effects on fats, glucose and lipids. Br Med Bull 84:49–68

Gonzalez-Polo RA, Boya P, Pauleau AL, Jalil A, Larochette N, Souquere S, Eskelinen EL, Pierron G, Saftig P, Kroemer G (2005) The apoptosis/autophagy paradox: autophagic vacuolization before apoptotic death. J Cell Sci 118(Pt 14):3091–102

Goodenow MM, Collman RG (2006) HIV-1 coreceptor preference is distinct from target cell tropism: a dual-parameter nomenclature to define viral phenotypes. J Leukoc Biol 80(5):965–972

Grivel JC, Margolis LB (1999) CCR5- and CXCR4-tropic HIV-1 are equally cytopathic for their T-cell targets in human lymphoid tissue. Nat Med 5(3):344–346

Grossman Z, Meier-Schellersheim M, Sousa AE, Victorino RM, Paul WE (2002) $CD4^+$ T-cell depletion in HIV infection: are we closer to understanding the cause? Nat Med 8(4):319–323

Hammer SM, Saag MS, Schechter M, Montaner JS, Schooley RT, Jacobsen DM, Thompson MA, Carpenter CC, Fischl MA, Gazzard BG, Gatell JM, Hirsch MS, Katzenstein DA, Richman DD, Vella S, Yeni PG, Volberding PA (2006) Treatment for adult HIV infection: 2006 recommendations of the International AIDS Society—USA panel. Top HIV Med 14(3):827–843

Heinkelein M, Sopper S, Jassoy C (1995) Contact of human immunodeficiency virus type 1-infected and uninfected $CD4^+$ T lymphocytes is highly cytolytic for both cells. J Virol 69(11):6925–6931

Hisatomi T, Nakazawa T, Noda K, Almulki L, Miyahara S, Nakao S, Ito Y, She H, Kohno R, Michaud N, Ishibashi T, Hafezi-Moghadam A, Badley AD, Kroemer G, Miller JW (2008) HIV protease inhibitors provide neuroprotection through inhibition of mitochondrial apoptosis in mice. J Clin Invest 118(6):2025–2038

Ho DD, Neumann AU, Perelson AS, Chen W, Leonard JM, Markowitz M (1995) Rapid turnover of plasma virions and CD4 lymphocytes in HIV-1 infection. Nature 373:123–126

Jansen CA, van Baarle D, Miedema F (2006) HIV-specific $CD4^+$ T cells and viremia: who's in control? Trends Immunol 27(3):119–124

Katz RA, Greger JG, Skalka AM (2005) Effects of cell cycle status on early events in retroviral replication. J Cell Biochem 94(5):880–889

Kondo Y, Kanzawa T, Sawaya R, Kondo S (2005) The role of autophagy in cancer development and response to therapy. Nat Rev Cancer 5(9):726–734
LaBonte J, Lebbos J, Kirkpatrick P (2003) Enfuvirtide. Nat Rev Drug Discov 2(5):345–346
Langford SE, Ananworanich J, Cooper DA (2007) Predictors of disease progression in HIV infection: a review. AIDS Res Ther 4:11
Laurent-Crawford AG, Krust B, Riviere Y, Desgranges C, Muller S, Kieny MP, Dauguet C, Hovanessian AG (1993) Membrane expression of HIV envelope glycoproteins triggers apoptosis in CD4 cells. AIDS Res Hum Retroviruses 9(8):761–773
Lee HK, Iwasaki A (2008) Autophagy and antiviral immunity. Curr Opin Immunol 20(1):23–29
Lee HK, Lund JM, Ramanathan B, Mizushima N, Iwasaki A (2007) Autophagy-dependent viral recognition by plasmacytoid dendritic cells. Science 315(5817):1398–1401
Leligdowicz A, Yindom LM, Onyango C, Sarge-Njie R, Alabi A, Cotten M, Vincent T, da Costa C, Aaby P, Jaye A, Dong T, McMichael A, Whittle H, Rowland-Jones S (2007) Robust Gag-specific T cell responses characterize viremia control in HIV-2 infection. J Clin Invest 117(10):3067–3074
Levine B, Deretic V (2007) Unveiling the roles of autophagy in innate and adaptive immunity. Nat Rev Immunol 7(10):767–777
Levy JA (1993) Pathogenesis of human immunodeficiency virus infection. Microbiol Rev 57(1):183–289
Lifson JD, Feinberg MB, Reyes GR, Rabin L, Banapour B, Chakrabarti S, Moss B, Wong-Staal F, Steimer KS, Engleman EG (1986) Induction of CD4-dependent cell fusion by the HTLV-III/LAV envelope glycoprotein. Nature 323(6090):725–728
Liu R, Paxon WA, Choe S, Ceradini D, Martin SR, Horuk R, MacDonald ME, Stuhlmann H, Koup RA, Landau NR (1996) Homozygous defect in HIV-1 coreceptor accounts for resistance of some multiply-exposed individuals to HIV-1 infection. Cell 86:367–377
Macias J, Palomares JC, Mira JA, Torres MJ, Garcia-Garcia JA, Rodriquez JM, Vergera S, Pineda JA (2005) Transient rebounds of HIV plasma viremia are associated with the emergence of drug resistance mutations in patients on highly active antiretroviral therapy. J Infect 51(3):195–200
Margolis L, Shattock R (2006) Selective transmission of CCR5-utilizing HIV-1: the 'gatekeeper' problem resolved? Nat Rev Microbiol 4(4):312–317
McCune JM (2001) The dynamics of CD4+ T-cell depletion in HIV disease. Nature 410(6831):974–979
Misse D, Cerutti M, Noraz N, Jourdan P, Favero J, Devauchelle G, Yssel H, Taylor N, Veas F (1999) A CD4-independent interaction of human immunodeficiency virus-1 gp120 with CXCR4 induces their cointernalization, cell signaling, and T-cell chemotaxis. Blood 93(8):2454–2462
Molina L, Grimaldi M, Robert-Hebmann V, Espert L, Varbanov M, Devaux C, Granier C, Biard-Piechaczyk M (2007) Proteomic analysis of the cellular responses induced in uninfected immune cells by cell-expressed X4 HIV-1 envelope. Proteomics 7(17):3116–3130
Monini P, Sgadari C, Toschi E, Barillari G, Ensoli B (2004) Antitumour effects of antiretroviral therapy. Nat Rev Cancer 4(11):861–875
Munoz-Barroso I, Durell S, Sakaguchi K, Appella E, Blumenthal R (1998) Dilation of the human immunodeficiency virus-1 envelope glycoprotein fusion pore revealed by the inhibitory action of a synthetic peptide from gp41. J Cell Biol 140(2):315–323
Murphy KM, Sweet MJ, Ross IL, Hume DA (1993) Effects of the tat and nef gene products of human immunodeficiency virus type 1 (HIV-1) on transcription controlled by the HIV-1 long terminal repeat and on cell growth in macrophages. J Virol 67(12):6956–6964
Nair D (2005) Antiretroviral therapy-induced hyperlipidaemia. Int J STD AIDS 16(Suppl 1):2–10; discussion 10–13, 41–13
Navia BA, Cho ES, Petito CK, Price RW (1986) The AIDS dementia complex: II. Neuropathology. Ann Neurol 19(6):525–535
Noursadeghi M, Katz DR, Miller RF (2006) HIV-1 infection of mononuclear phagocytic cells: the case for bacterial innate immune deficiency in AIDS. Lancet Infect Dis 6(12):794–804

Ohnimus H, Heinkelein M, Jassoy C (1997) Apoptotic cell death upon contact of CD4+ T lymphocytes with HIV glycoprotein-expressing cells is mediated by caspases but bypasses CD95 (Fas/Apo-1) and TNF receptor 1. J Immunol 159(11):5246–5252

Orvedahl A, Levine B (2008) Eating the enemy within: autophagy in infectious diseases. Cell Death Differ 16(1):57–69

Paludan C, Schmid D, Landthaler M, Vockerodt M, Kube D, Tuschl T, Munz C (2005) Endogenous MHC class II processing of a viral nuclear antigen after autophagy. Science 307(5709):593–596

Pelchen-Matthews A, Kramer B, Marsh M (2003) Infectious HIV-1 assembles in late endosomes in primary macrophages. J Cell Biol 162(3):443–455

Perelson AS, Neumann AU, Markowitz M, Leonard JM, Ho DD (1996) HIV-1 dynamics in vivo: virion clearance rate, infected cell life-span, and viral generation time. Science 271(5255):1582–1586

Phenix BN, Angel JB, Mandy F, Kravcik S, Parato K, Chambers KA, Gallicano K, Hawley-Foss N, Cassol S, Cameron DW, Badley AD (2000) Decreased HIV-associated T cell apoptosis by HIV protease inhibitors. AIDS Res Hum Retroviruses 16(6):559–567

Piguet V, Sattentau Q (2004) Dangerous liaisons at the virological synapse. J Clin Invest 114(5):605–610

Popovic M, Sarngadharan MG, Read E, Gallo RC (1984) Detection, isolation, and continuous production of cytopathic retroviruses (HTLV-III) from patients with AIDS and pre-AIDS. Science 224(4648):497–500

Pyrko P, Kardosh A, Wang W, Xiong W, Schonthal AH, Chen TC (2007) HIV-1 protease inhibitors nelfinavir and atazanavir induce malignant glioma death by triggering endoplasmic reticulum stress. Cancer Res 67(22):10920–10928

Raposo G, Moore M, Innes D, Leijendekker R, Leigh-Brown A, Benaroch P, Geuze H (2002) Human macrophages accumulate HIV-1 particles in MHC II compartments. Traffic 3(10):718–729

Roshal M, Zhu Y, Planelles V (2001) Apoptosis in AIDS. Apoptosis 6(1-2):103–116

Rowland-Jones SL, Whittle HC (2007) Out of Africa: what can we learn from HIV-2 about protective immunity to HIV-1? Nat Immunol 8(4):329–331

Samara C, Syntichaki P, Tavernarakis N (2007) Autophagy is required for necrotic cell death in *Caenorhabditis elegans*. Cell Death Differ 15(1):105–12

Samson M, Libert F, Doran BJ, Rucker J, Liesnard C, Farber CM, Saragosti S, Lapoumeroulie C, Cognaux J, Forceille C, Muyldermans G, Verhofstede C, Bourtonboy G, Georges M, Imai T, Rana S, Yi Y, Smyth RJ, Collman RG, Doms RW, Vassart G, Parmentier M (1996) Resistance to HIV-1 infection in caucasien individuals bearing mutant alleles of the CCR-5 chemokine receptor gene. Nature 382:722–725

Schmid D, Dengjel J, Schoor O, Stevanovic S, Munz C (2006) Autophagy in innate and adaptive immunity against intracellular pathogens. J Mol Med 84(3):194–202

Schmid D, Munz C (2007) Innate and adaptive immunity through autophagy. Immunity 27(1):11–21

Schutt M, Zhou J, Meier M, Klein HH (2004) Long-term effects of HIV-1 protease inhibitors on insulin secretion and insulin signaling in INS-1 beta cells. J Endocrinol 183(3):445–454

Seay MD, Dinesh-Kumar SP (2007) Autophagy takes its TOLL on innate immunity. Cell Host Microbe 2(2):69–70

Sgadari C, Barillari G, Toschi E, Carlei D, Bacigalupo I, Baccarini S, Palladino C, Leone P, Bugarini R, Malavasi L, Cafaro A, Falchi M, Valdembri D, Rezza G, Bussolino F, Monini P, Ensoli B (2002) HIV protease inhibitors are potent anti-angiogenic molecules and promote regression of Kaposi sarcoma. Nat Med 8(3):225–232

Sgadari C, Monini P, Barillari G, Ensoli B (2003) Use of HIV protease inhibitors to block Kaposi's sarcoma and tumour growth. Lancet Oncol 4(9):537–547

Sodroski J, Goh WC, Rosen C, Campbell K, Haseltine WA (1986) Role of the HTLV-III/LAV envelope in syncytium formation and cytopathicity. Nature 322(6078):470–474

Stocker H, Scheller C, Jassoy C (2000) Destruction of primary CD4(+) T cells by cell-cell interaction in human immunodeficiency virus type 1 infection in vitro. J Gen Virol 81(Pt 8):1907–1911
Suzuki Y, Craigie R (2007) The road to chromatin—nuclear entry of retroviruses. Nat Rev Microbiol 5(3):187–196
Swanson MS (2006) Autophagy: eating for good health. J Immunol 177(8):4945–4951
Terai C, Kornbluth RS, Pauza CD, Richman DD, Carson DA (1991) Apoptosis as a mechanism of cell death in cultured T lymphoblasts acutely infected with HIV-1. J Clin Invest 87:1710–1715
Thorburn A (2007) Apoptosis and autophagy: regulatory connections between two supposedly different processes. Apoptosis 13(1):1–9
Varbanov M, Espert L, Biard-Piechaczyk M (2006) Mechanisms of CD4 T-cell depletion triggered by HIV-1 viral proteins. AIDS Rev 8(4):221–236
Ventoso I, Navarro J, Munoz MA, Carrasco L (2005) Involvement of HIV-1 protease in virus-induced cell killing. Antiviral Res 66(1):47–55
Vlahakis SR, Bennett SA, Whitehead SN, Badley AD (2007) HIV protease inhibitors modulate apoptosis signaling in vitro and in vivo. Apoptosis 12(5):969–977
Wei X, Ghosh SK, Taylor ME, Johnson VA, Emini EA, Deutsch P, Lifson JD, Bonhoeffer S, Nowak MA, Hahn BH, Saag MS, Shaw GM (1995) Viral dynamics in human immunodeficiency virus type 1 infection. Nature 373:117–122
Weissenhorn W, Dessen A, Harrison SC, Skehel JJ, Wiley DC (1997) Atomic structure of the ectodomain from HIV-1 gp41 [see comments]. Nature 387(6631):426–430
Weissman D, Rabin RL, Arthos J, Rubbert A, Dybul M, Swofford R, Venkatesan S, Farber JM, Fauci AS (1997) Macrophage-tropic HIV and SIV envelope proteins induce a signal through the CCR5 chemokine receptor. Nature 389(6654):981–985
Welsch S, Keppler OT, Habermann A, Allespach I, Krijnse-Locker J, Krausslich HG (2007) HIV-1 buds predominantly at the plasma membrane of primary human macrophages. PLoS Pathog 3(3):e36
Wild CT, Shugars DC, Greenwell TK, McDanal CB, Matthews TJ (1994) Peptides corresponding to a predictive alpha-helical domain of human immunodeficiency virus type 1 gp41 are potent inhibitors of virus infection. Proc Natl Acad Sci USA 91(21):9770–9774
Yoffe B, Lewis DE, Petrie BL, Noonan CA, Melnick JL, Hollinger FB (1987) Fusion as a mediator of cytolysis in mixtures of uninfected $CD4^+$ lymphocytes and cells infected by human immunodeficiency virus. Proc Natl Acad Sci USA 84:1429–1433
Zamborlini A, Lehmann-Che J, Clave E, Giron ML, Tobaly-Tapiero J, Roingeard P, Emiliani S, Toubert A, de The H, Saib A (2007) Centrosomal pre-integration latency of HIV-1 in quiescent cells. Retrovirology 4:63
Zhou D, Spector SA (2008) Human immunodeficiency virus type-1 infection inhibits autophagy. AIDS 22(6):695–699

Subversion of the Cellular Autophagy Pathway by Viruses

Karla Kirkegaard

Contents

1 Introduction 323
2 Viruses for Which the Autophagy Pathway is Advantageous 325
3 What Can Autophagy Do for a Virus? 327
4 Conclusion 331
References 332

Abstract Autophagy is a cellular process that creates double-membraned vesicles, engulfs and degrades cytoplasmic material, and generates and recycles nutrients. A recognized participant in the innate immune response to microbial infection, a functional autophagic response can help to control the replication of many viruses. However, for several viruses, there is functional and mechanistic evidence that components of the autophagy pathway act as host factors in viral replicative cycles, viral dissemination, or both. Investigating the mechanisms by which viruses subvert or imitate autophagy, as well as the mechanisms by which they inhibit autophagy, will reveal cell biological tools and processes that will be useful for understanding the many functional ramifications of the double-membraned vesicle formation and cytosolic entrapment unique to the autophagy pathway.

1 Introduction

Viruses, especially RNA viruses, have the capacity to evolve quickly in response to changes in host cells. As a consequence, it is difficult to think of any antiviral response for which there is not an example of a viral mechanism to avoid the

---

K. Kirkegaard
Department of Microbiology and Immunology, Stanford University School of Medicine, Stanford, CA 94305
e-mail: karlak@stanford.edu

B. Levine et al. (eds.), *Autophagy in Infection and Immunity*,
Current Topics in Microbiology and Immunology 335,
DOI 10.1007/ 978-3-642-00302-8_16,

response or, even more beneficially for the virus, to subvert the response to provide some advantage for the virus. A good example of the wide range and mechanisms of viral reactions to antiviral mechanisms is that of viral strategies to contend with cell death by apoptosis, an effective defense against many catastrophes, including infection by microorganisms. Most viruses, especially those whose success has not elevated them to fame as pathogens, are undoubtedly cleared by cellular apoptosis before they have a chance to spread. Almost every virus that is known to thrive in animals or in tissue culture cells has, of necessity, developed a mechanism to inhibit cellular apoptosis. Indeed, one of the first antiapoptotic proteins identified, p35, was discovered by Lois Miller and her laboratory through their studies of baculoviruses, a class of insect-infecting DNA viruses (Clem et al. 1991). Inhibitors of nearly every step in the extrinsic and intrinsic apoptotic pathways have been identified from viruses, and the actions of these range from derailing signal transduction to blocking mitochondrial channel formation to inhibiting the effector caspases themselves (reviewed in Best 2008; Galluzzi et al. 2008). We could conclude that the viral repertoire is exceedingly thorough, but this would be a bit unfair, because often (as was the case for p35) it was the identification of the viral inhibitor that defined the step in the first place.

It is perhaps not surprising that, given the sequence and cell biological space that viruses can explore, some viruses would have evolved to exploit the apoptotic pathway for their own benefit, rather than just inhibiting it. How could the induction of apoptosis facilitate viral replication? The eventual lysis of infected cells is the mechanism of spread of many viruses; it is likely that the ultimate failure of antiapoptotic mechanisms in many viral infections results in a delayed apoptosis that facilitates cell spread. Another, more surprising mechanism has also been demonstrated: for mink cell focus-forming virus (Best et al. 2003), human astrovirus (Mendez et al. 2004) and feline calicivirus (Al-Molawi et al. 2003), caspases induced during apoptosis are required for viral capsid processing (reviewed in Best 2008).

The cellular autophagy pathway, as discussed throughout this volume, was originally discovered as a response to starvation, and provides a route for cytoplasmic constituents to be targeted to the lysosomal machinery for degradation (reviewed in Mizushima et al. 2002, 2008). Cytosolic components become enwrapped by a double-membraned structure, termed the "immature autophagosome." The resulting double-membraned structures contain the marker LC3, which becomes covalently linked to phosphatidylethanolamine and thereby membrane-associated upon the induction of autophagy. LC3 was originally identified as MAP-LC3, a microtubule-associated protein (Kuznetsov and Gelfand 1987), and may be part of the mechanism by which autophagosomes interact with the microtubule network, on which they traffic by an anterograde route toward the microtubule organizing center (Fass et al. 2006; Jahreiss et al. 2008; Kimura et al. 2008; Koechl et al. 2006). Immature autophagosomes mature by fusion with vesicles from the endosomal pathway, then with lysosomes, in experimentally separate steps. At the "mature autophagosome" stage, the inner membranes become degraded, accomplishing the topological transformation of cytosol into lumen, something that happens rarely within cells. Eventually, condensed "autolysosomes" release amino acids, lipids and nucleotides for further cellular use. A possibility considered here is that autophagosomes and autolysosomes have the capacity to release cytosolic contents to the extracellular milieu by a process similar to the known pathways of lysosomal exocytosis (reviewed in Stinchcombe et al. 2004).

From first principles, how might the autophagy pathway be related to the replicative cycles of viruses? A clear role in the degradation of viruses and viral factories can be readily envisaged, given the known ability of the autophagy pathway to degrade cellular components. Indeed—as discussed in the chapters by Tal and Iwasaki, Orvedahl and Levine, and Seay et al. in this volume—autophagy has been shown to participate in the control of viral infections as diverse as human herpes and plant RNA viruses, and has been justifiably hailed as a branch of the innate immune response. In such cases, reducing the function of the autophagy pathway removes this control and allows increased proliferation of the virus

2 Viruses for Which the Autophagy Pathway is Advantageous

In contrast, the opposite situation, in which reduction in the function of some component of the autophagy pathway reduces viral replication, has been demonstrated for a handful of viruses. As shown in Fig. 1, reduction of the intracellular concentration of either LC3 or Atg12 using small interfering RNAs was shown to reduce both the intracellular and extracellular yields of poliovirus, a small, nonenveloped positive-strand RNA virus, in single-cycle infections (Jackson et al. 2005). For coxsackievirus B3, a closely related picornavirus, reduction in the intracellular concentration of Atg7 substantially reduced the amount of viral capsid protein synthesized in a single-cycle infection (Fig. 1c) (Wong et al. 2008). For dengue virus, an enveloped positive-strand RNA virus, experiments to compare single-cycle infections of murine embryonic fibroblasts derived from autophagy-proficient and autophagy-deficient mice showed clear reductions in the yield of extracellular virus in the absence of a functional autophagy pathway (Lee et al. 2008). Recently, a genomic screen was performed to identify those protein-coding human genes whose targeting by specific small interfering RNAs reduced the yield of human immunodeficiency virus. This screen identified 36 known host factors, including coreceptors, and more than 100 others, which included autophagy genes *ATG7*, *GABARAPL2* (a homolog of *LC3*), *ATG12*, and *ATG16* (Brass et al. 2008). As is the case for the positive-strand RNA viruses mentioned, the mechanism of any advantage conferred to the replicative cycle, and whether the effects are direct or indirect, is not yet known.

What mechanistic inferences can be drawn from data such as those in Fig. 1? First, the effects are not absolute: in no case is the yield of intracellular virus or virus product reduced more than threefold, and, for extracellular virus, the largest observed reduction was 20-fold (Fig. 1a). For the RNAi experiments, it is likely that the reductions in autophagy protein concentration were not complete, so residual autophagy function remained. However, for genetic knockouts such as that

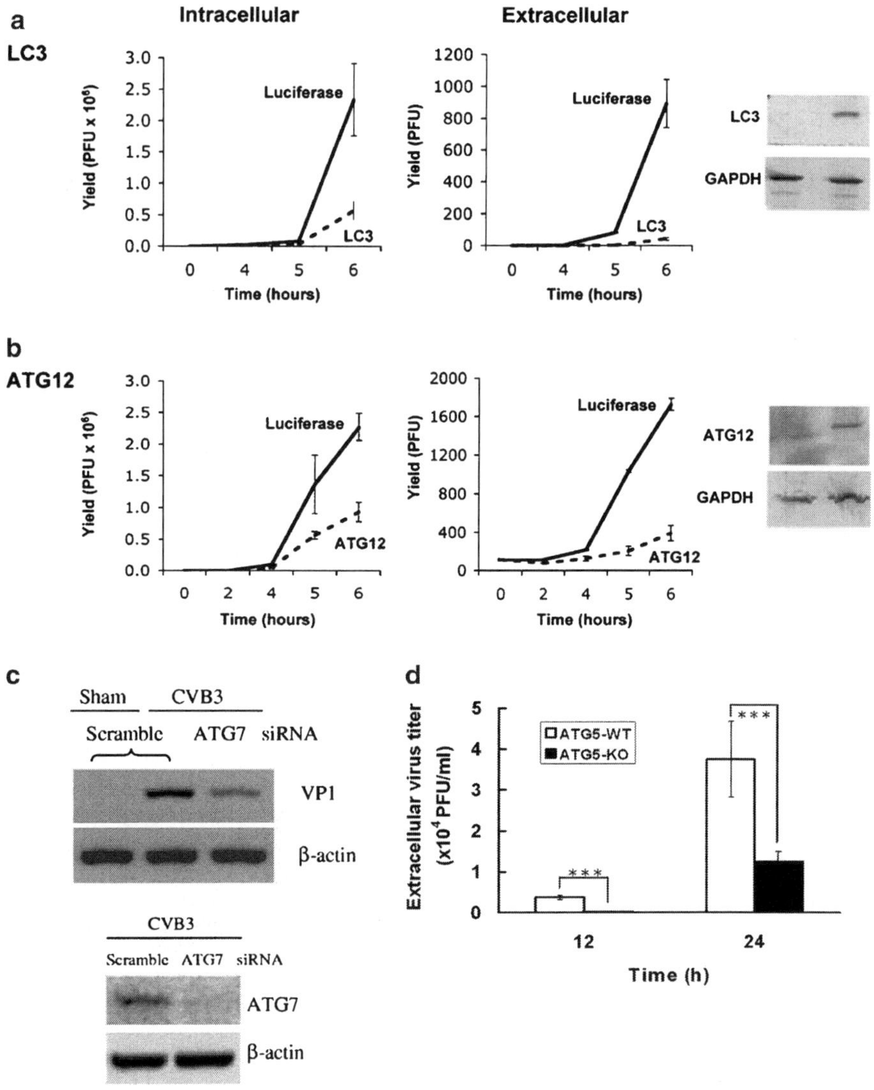

Fig. 1 Functional evidence that autophagy pathway function can have a positive effect on viral replicative cycles in tissue culture. The intracellular and extracellular yields of poliovirus were quantified in HeLa cells after treatment with (**a**) 12.5 pm each of eight different RNA duplexes targeted to the *LC3A* and *LC3B* mRNAs or with 100 pm of an RNA duplex targeted to firefly luciferase, or (**b**) with 25 pm each of four RNA duplexes targeted to *ATG12* mRNA or with 100 pm of an RNA duplex targeted to firefly luciferase, for 24 h at 37°C. For both panels, *insets* show the relative abundance of the targeted protein upon treatment. Taken from Jackson et al. (2005) with permission. (**c**) Effect of reducing Atg7 protein abundance on intracellular accumulation of VP1 during infection with coxsackievirus B3 (Wong et al. 2008). (**d**) Differences in the yield of dengue virus after infection of murine embryonic fibroblasts from $Atg5^{+/+}$ and $Atg5^{-/-}$ mice at two different times postinfection. Taken from Lee et al. (2008) with permission

shown for dengue virus in *Atg5*$^{-/-}$ cells in Fig. 1d, the conclusion is inescapable that the autophagy pathway or its components are helpful to the formation or release of extracellular dengue virus, but not essential.

3 What Can Autophagy Do for a Virus?

A few possibilities for positive roles of the autophagy pathway or autophagosome formation in viral infections have been suggested by data in the literature thus far. In 1965, electron microscopic observations of poliovirus-infected cells revealed the presence of large numbers of double-membraned vesicles (Dales et al. 1965), which were recognized by these authors as "autophagolysomes." Then, led by the work of Kurt Bienz and Denise Eggers (Bienz et al. 1983), it became increasingly appreciated that all positive-strand RNA viruses replicate their RNA genomes on the cytosolic faces of cytoplasmic membranes. In the case of poliovirus, these cytosolic-facing surfaces were shown to be those of the hundreds of membranous vesicles that accumulated in the cytoplasm of infected cells. Might these membranous vesicles be related to the double-membraned, autophagosome-like structures originally observed by the laboratory of George Palade? Over the last several years, my laboratory has shown that such double-membraned structures are readily visualized in both poliovirus-infected cells (Fig. 2a, b) and human rhinovirus-infected cells by electron microscopy following fixation by high-pressure

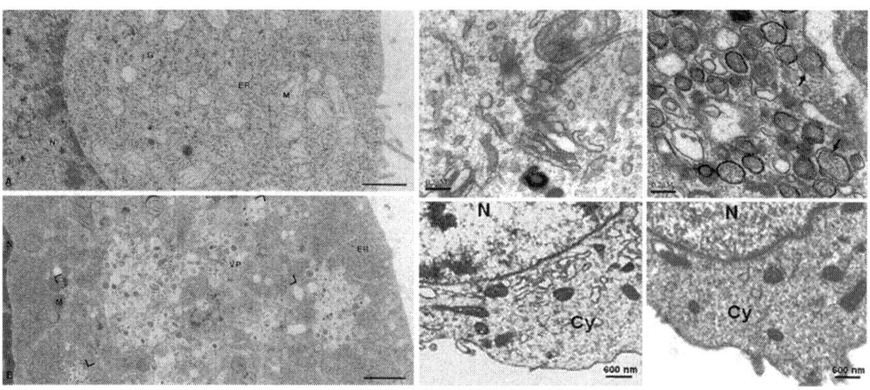

Fig. 2 Ultrastructural changes in cells infected with viruses that are suspected of subverting the cellular autophagy pathway. Transmission electron microscopy of uninfected (**a**) and poliovirus-infected (**b**) Hela cells preserved by high-pressure freezing and freeze substitution in the presence of osmium tetroxide. Size bar: 1000 nm. Taken from Schlegel et al. (1996) with permission. Uninfected (**c**) and coxsackievirus 3-infected (**d**) HEK293 cells visualized following glutaraldehyde fixation and osmium tetroxide staining. Size bar in inset: 200 nm. Taken from Wong et al. (2008) with permission. Dengue virus-infected murine embryonic fibrobasts from either *Atg5*$^{+/+}$ (**e**) or *Atg5*$^{-/-}$ mice (**f**) after glutaraldehyde fixation and osmium tetroxide staining. Taken from Lee et al. (2008) with permission

freezing and freeze substitution (Schlegel et al. 1996; Dodd et al. 2001; Suhy et al. 2000; Jackson et al. 2005). Immunoelectron microscopy has shown that virally encoded proteins that are known to be found in the RNA replication complex localize to the cytoplasmic surface of these double-membraned vesicles, as do LC3 and LAMP-1 (Schlegel et al. 1996; Suhy et al. 2000; Jackson et al. 2005). Mechanistic insight into the assembly of double-membraned vesicles in cells may be provided by the finding that two viral proteins, termed 2BC and 3A, are sufficient to induce double-membraned vesicles that contain LC3 and LAMP-1 (Jackson et al. 2005; Suhy et al. 2000), with 2BC alone being sufficient to cause the accumulation of lipidated LC3 (Taylor and Kirkegaard, 2007). The observation that RNA replication complexes can assemble on the cytoplasmic surfaces of double-membraned vesicles that resemble autophagosomes in many respects has led to the hypothesis that the autophagy pathway is subverted by poliovirus, and possibly other positive-strand viruses, to generate such membranous surfaces. As can be seen in Fig. 2, double-membraned vesicles have recently been observed in cells infected with coxsackievirus B3 and Dengue virus as well, both of which show some functional dependence on the autophagy pathway (Fig. 1) (Lee et al. 2008; Wong et al. 2008).

But why would viruses use double-membraned vesicles and autophagosome-related proteins? While the hypothesis that the membranes and membrane constituents of the autophagy pathway contribute to the platforms on which viral RNA replication complexes can be built is attractive, it is clear that even poliovirus, coxsackievirus B3 and dengue viruses must be able to utilize other membrane surfaces as well, given that the reduced availability of autophagic proteins resulted in only small reductions in virus production (Fig. 1). For example, in *Atg5*$^{-/-}$ cells infected with dengue virus (Fig. 1f), no double-membraned vesicles were observed (Lee et al. 2008), yet the virus must have assembled its RNA replication complex somewhere. And what is to be made of related viruses, such as human rhinovirus and murine hepatitis virus, which induce the accumulation of double-membraned vesicles in infected cells, but for which single replicative cycles show no apparent dependence on the presence or absence of a functional autophagic pathway substitution (Brabec-Zaruba et al. 2007; Zhao et al. 2007)? It is interesting, in this context, to consider the case of flock house virus, a positive-strand RNA virus that normally infects insects, and whose RNA replicative cycle can be observed in a variety of cell types including *S. cerevisiae* In all cell types examined, the normal site of assembly of the RNA replication complexes of flock house virus is the outer mitochondria membrane. However, clever genetic engineering experiments allowed the assembly of these RNA replication complexes on the ER of yeast cells; at this retargeted location, the rate of viral RNA replication increased sixfold (Miller et al. 2003). Thus, either nature has made a mistake by targeting flock house virus RNA replication to the outer mitochondrial membrane, or it is targeted there for some reason other than that which can be observed as an increased rate of a mechanistic process in a single RNA replicative cycle.

One such potentially alternative function for autophagosome-like membranes in virally infected cells was suggested to us by the data in Fig. 1. At 6 h after infection with poliovirus, no apparent cell lysis is usually observed. Yet, 1,000 plaque-forming units of extracellular virus accumulated in control infected cells, and 20-fold less when the abundance of LC3 was reduced. This effect was much more pronounced than the threefold reduction in intracellular virus titer, and it led us to hypothesize that, since it is known that the double-membraned vesicles that form late in poliovirus infection can contain cytoplasmic contents that include infectious virions and other cytoplasmic components, these membranes can provide a route of nonlytic viral escape from infected cells. We term this proposed pathway of cell exit "autophagic exit without lysis" or AWOL (Taylor et al., 2009), and suspect that it may be a component of viral pathogenesis.

There is ample precedent in the literature for the nonlytic spread of nonenveloped cytoplasmic viruses. Elegant experiments with Theiler's virus, a murine virus related to poliovirus, have argued that the transfer of Theiler's virus from optic nerves to the oligodendrocytes that myelinate them is an obligate step in the establishment of persistent infections, and can occur without neuronal lysis (Roussarie et al. 2007). In another potential example from picornaviruses, hepatitis A virus is known to spread through liver tissue in the absence of any apparent cell lysis; no mechanism for this observation has been demonstrated (reviewed in Hollinger and Ticehurst 1990).

Even at early times postinfection, it is difficult to distinguish whether a few cells lyse or many cells leak when so few submicroscopic viruses are present in the extracellular milieu (Fig. 1). Similarly, although the sporadic, apparently nonlytic spread of toxic aggregated proteins between cells has been observed, it is difficult to exclude stochastic events like occasional cell lysis (Ren et al. 2009). However, the demonstrably nonlytic release of *Cryptococcus neoformans*, a pathogenic fungus, from infected macrophages, which has been shown to correlate with association with complex membranes that contain LAMP-1 (Alvarez and Casadevall 2006), may be a spectacular example of AWOL by an intracellular pathogen.

The integrity of contents delivered by a pathway such as AWOL should depend critically on the extent of maturation of the autophagosome-like compartment. As illustrated in Fig. 3, the fusion of an immature autophagosome with the plasma membrane would secrete a lipid-rich, single-membraned vesicle, similar to an exosome, into the extracellular milieu. On the other hand, a mature autolysosome would, upon fusion with the plasma membrane, directly release degraded material and any cytosolic constituents that could withstand the environment of the lysosome. Clearly, intermediates between these stages are likely to exist as well.

Recent experiments have shown directly that autophagosomal-like membranes within virally infected cells can be less degradative than bona fide autophagosomes. As shown in Fig. 4a, decreasing the concentration of LAMP-2 was shown to increase the intracellular accumulation of coxsackievirus B3 VP1 protein, presumably by changing the functionality of the virus-induced membranes. During hepatitis C infection, autophagy is known to be induced, although any benefit of this induction for viral replication is not yet known (Ait-Goughoulte et al. 2008; Sir et al. 2008). In any case, as shown in Fig. 4b, The GFP-LC3-containing structures induced during HCV infection do not colocalize with a marker of acidified lysosomes, unlike the GFP-LC3 in starved cells. It remains to be explored whether these viruses affect the maturation of autophagosomes to inhibit an antiviral response in the host cell, or for their own direct benefit.

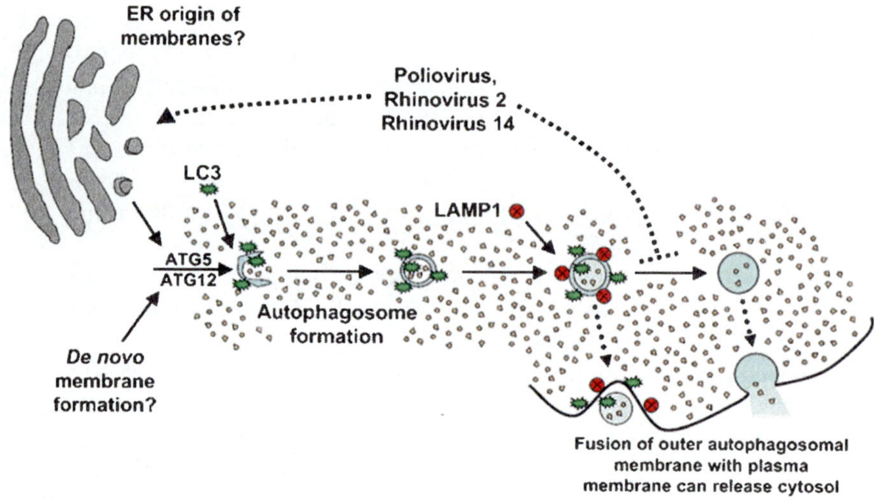

Fig. 3 Model for AWOL: autophagosome-mediated exit without lysis. Cytosolic constitutents can be surrounded by double-membraned vesicles derived from or resembling membranes of the autophagy pathway. Partial or complete degradation of the inner membrane can effectively mix cytosol and luminal components. Then, a process akin to the known lysosomal secretion pathway could facilitate the extracellular release of the autophagosomal contents. Modified from Jackson et al. (2005) with permission

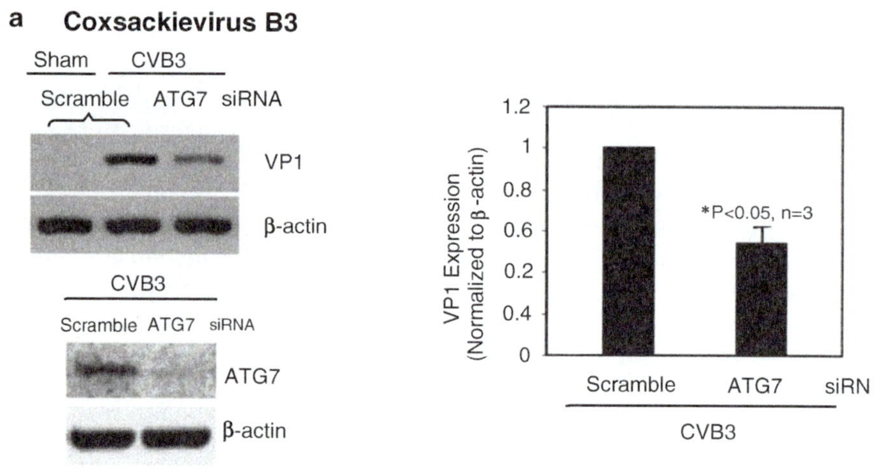

Fig. 4 Potential blockage to maturation of autophagosome-like compartments in virally infected cells. (**a**) Immunoblot analysis of extracts from mock-infected and coxsackievirus B3-infected HeLa cells in the absence and presence of reduced intracellular amounts of LAMP-1. Taken from Wong et al. (2008) with permission. (**b**) Fluorescence microscopy to determine colocalization of Lysotracker and GFP-LC3 in Huh-7 cells that were either starved to induce autophagy or infected with the JFH-1 strain of hepatitis C virus. Taken from Sir et al. with permission of John Wiley & Sons, Inc

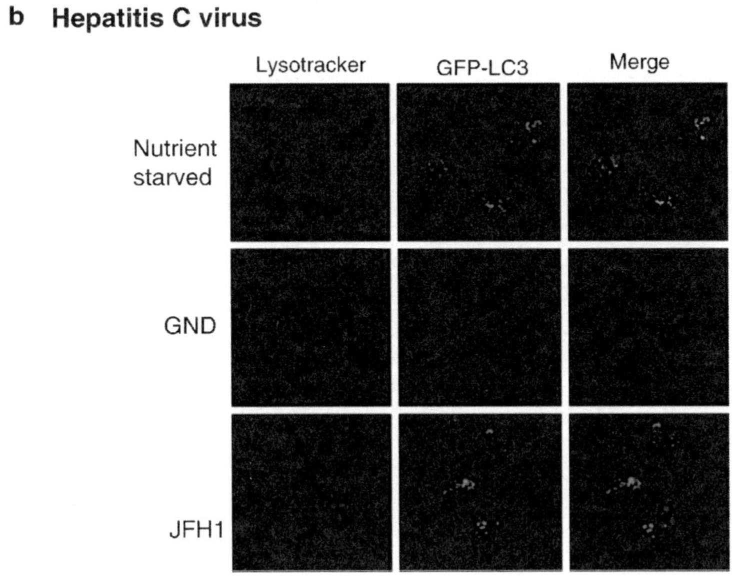

Fig. 4 (continued)

4 Conclusion

In summary, for most viruses and other intracellular pathogens, cellular defenses such as apoptosis and autophagy are induced upon infection and participate actively in clearance. Any viruses famous enough to be studied, however, are likely to encode one or multiple inhibitors of both these processes. Several viruses have been shown to induce autophagosome-like structures and to benefit from their formation. These benefits, so far, are likely to be the proliferation of membranes for RNA replication complexes and the generation of unique topologies that allow the escape of cytoplasmic constituents without cell lysis, which may be critical for viral spread within infected tissues.

Acknowledgments I would especially like to thank Thomas H. Giddings, Jr., whose electron micrographs allowed us to rediscover the double-membraned character of poliovirus-induced vesicles, and Andrew Staehelin, who first taught me about the autophagy pathway. I am grateful to the members of the laboratory who have worked on this topic over several years: Andreas Schlegel, David Suhy, William T. Jackson and Matthew Taylor. I thank Michel Brahic for stimulating discussions of non-lytic viral release and Arturo Casadevall for communicating results prior to publication. This work has been funded by the National Institutes of Health.

References

Ait-Goughoulte M, Kanda T, Meyer K, Ryerse JS, Ray RB, Ray R (2008) Hepatitis C virus genotype 1a growth and induction of autophagy. J Virol 82:2241–2249

Al-Molawi N, Beardmore VA, Carter MJ, Kass GE, Roberts LO (2003) Caspase-mediated cleavage of the feline calicivirus capsid protein. J Gen Virol 84:1237–1244

Alvarez M, Casadevall A (2006) Phagosome extrusion and host-cell survival after *Cryptococcus neoformans* phagocytosis by macrophages. Curr Biol 16:2161–2165

Best SM (2008) Viral subversion of apoptotic enzymes: escape from death row. Annu Rev Microbiol 62:171–192

Best SM, Shelton JF, Pompey JM, Wolfinbarger JB, Bloom ME (2003) Caspase cleavage of the nonstructural protein NS1 mediates replication of Aleutian mink disease parvovirus. J Virol 77:5305–5312

Bienz K, Egger D, Rasser Y, Bossart W (1983) Intracellular distribution of poliovirus proteins and the induction of virus-specific cytoplasmic structures. Virology 131:39–48

Brabec-Zaruba M, Berka U, Blaas D, Fuchs R (2007) Induction of autophagy does not affect human rhinovirus type 2 production. J Virol 81:10815–10817

Brass AL, Dykxhoorn DM, Benita Y, Yan N, Engelman A, Xavier RJ, Lieberman J, Elledge SJ (2008) Identification of host proteins requried for HIV infection through a functional genomic screen. Science 319:921–926

Clem RJ, Fechheimer M, Miller LK (1991) Prevention of apoptosis by a baculovirus gene during infection of insect cells. Science 254:1388–1390

Dales S, Eggers HJ, Tamm I, Palade GE. (1965) Electron microscopic study of the formation of poliovirus. Virology 26:379–389

Dodd DA, Giddings TH, Jr., Kirkegaard K. (2001) Poliovirus 3A protein limits interleukin-6 (IL-6), IL-8, and beta interferon secretion during viral infection. J Virol 75:8158–8165

Fass E, Shvets E, Degani I, Hirschberg K, Elazar Z (2006) Microtubules support production of starvation-induced autophagosomes but not their targeting and fusion with lysosomes. J Biol Chem 281:36303–36316

Galluzzi L, Brenner C, Morselli E, Touat Z, Kroemer G (2008) Viral control of mitochondrial apoptosis. PLoS Pathog 4:e1000018

Hollinger FB, Ticehurst J (1990) Hepatitis A virus. In: Fields BN et al. (eds) Fields virology, 2nd edn. Raven, New York, pp 631–667

Jackson WT, Giddings TH Jr, Taylor MP, Mulinyawe S, Rabinovitch M, Kopito RR, Kirkegaard K (2005) Subversion of cellular autophagosomal machinery by RNA viruses. PLoS Biol 3:e156

Jahreiss L, Menzies FM, Rubinsztein DC (2008) The itinerary of autophagosomes: from peripheral formation to kiss-and-run fusion with lysosomes. Traffic 9:574–587

Kimura S, Noda T, Yoshimori T (2008) Dynein-dependent movement of autophagosomes mediates efficient encounters with lysosomes. Cell Struct Funct 33:109–122

Kirkegaard K, Jackson WT (2005) Topology of double-membraned vesicles and the opportunity for non-lytic release of cytoplasm. Autophagy 1:182–184

Koechl R, Hu XW, Chan EYW, Tooze S (2006) Microtubules faciliate autophagosome formation and fusion of autophagosomes with endosomes. Traffic 7:129–145

Kuznetsov SA, Gelfand VI (1987) 18 kDa microtubule-associated protein: identification as a new light chain (LC-3) of microtubule-associated protein 1 (MAP-1). FEBS Lett 212:145–148

Lee YR, Lei HY, Liu MT, Wang JR, Chen SH, Jiang-Shieh YF, Lin YS, Yeh TM, Liu CC, Liu HS (2008) Autophagic machinery activated by dengue virus enhances virus replication. Virology 374:240–248

Mendez E, Salas-Ocampo E, Arias CF (2004) Caspases mediate processing of the capsid precursor and cell release of human astroviruses. J Virol 78:8601–8608

Miller DJ, Schwartz MD, Dye BT, Ahlquist P (2003) Engineered retargeting of viral RNA replication complexes to an alternative intracellular membrane. J Virol 77:12193–12202

Mizushima N, Ohsumi Y, Yoshimori T (2002) Autophagosome formation in mammalian cells. Cell Struct Funct 27:421–429
Mizushima N, Levine B, Cuervo AM, Klionsky DJ (2008) Autophagy fights disease through cellular self-digestion. Nature 451:1069–1075
Ren PH, Lauckner JE, Kachirskaia I, Heusor JE, Melki R, Kopita RR (2009) Cytoplasmic penetration and persistent infection of mammalian cells by polyglutamine. Nat Cell Biol 11(2): 219–225 (epub 2009)
Roussarie JP, Ruffie C, Edgar JM, Griffiths I, Brahic M (2007) Axon myelin transfer of a non-enveloped virus. PLoS ONE 2:e1331
Schlegel A, Giddings TH, Jr., Ladinsky MS, Kirkegaard K. (1996) Cellular origin and ultrastructure of membranes induced during poliovirus infection. J Virol 70:6576–6588
Sir D, Chen WL, Choi J, Wakita T, Yen TS, Ou JH. (2008) Induction of incomplete autophagic response by hepatitis C virus via the unfolded protein response. Hepatology 48:1054–1061
Stinchcombe J, Bossi G, Griffiths GM. (2004) Linking albinism and immunity: the secrets of secretory lysosomes. Science 305:55–59
Suhy DA, Giddings TH, Jr., Kirkegaard K (2000) Remodeling the endoplasmic reticulum by poliovirus infection and by individual viral proteins: an autophagy-like origin for virus-induced vesicles. J Virol 74:8953–8965
Taylor MP, Kirkegaard K (2007) Modification of cellular autophagy protein LC3 by poliovirus. J Virol 81:12543–12553
Taylor MP, Burgon TB, Kirkegaard K, Jackson WT (2009) Role of microtubules in extracellular release of poliovirus. J Viral 83(13):6599–6609 (epub 2009)
Wong J, Zhang J, Si X, Gao G, Mao I, McManus BM, Luo H (2008) Autophagosome supports coxsackievirus B3 replication in host cells. J Virol 82:9143–9153
Zhao Z, Thackray L, Miller B, Lynn T, Becker M, Ward E, Mizushima N, Denison M, Virgin HT (2007) Coronavirus replication does not require the autophagy gene *ATG5*. Autophagy 3:581–585

Index

A

Accessory gene regulator (*agr*) system, 224–225
Acetylation, 50–52
Acquired immune deficiency syndrome (AIDS), 172, 308, 309, 312, 314
Activating molecule in beclin 1-regulated autophagy (AMBRA1), 50
Ag85B antigen, 130, 131
Aging, 78, 79
Akt, 175, 176
Amino acid effluxer, 24
AMP-activated protein kinase (AMPK), 7, 10, 38–41, 58
Anaplasma phagocytophilum, 241–242
  infection process, 241, 242
  intracellular trafficking, 241, 242
Apoptosis. *See* Cell death
Arabidopsis, 288, 291, 292, 294, 295, 298, 300
Atg9 cycling, 22
ATPases, 54–55
Autoimmunity, 136
Autolysosome, 220, 222, 224
Autophagic body, 2, 3, 23, 24
Autophagic or type II cell death, 290, 292
Autophagosome, 2–4, 8, 10, 11, 13, 18, 19, 21–24, 90, 91, 93, 94, 96–98, 219–222, 224–225, 297, 298, 301, 302
Autophagosome exit without lysis (AWOL), 329, 330
Autophagy (AT6) genes, 2, 3, 25, 274, 281
  Atg1, 4–9, 11, 13, 15, 19–23, 39, 41, 46–47, 52, 59
  Atg4, 222
  Atg5, 42, 46, 50–52, 90, 92, 93, 95, 97, 98, 150, 152–157, 161, 220–223, 271
  Atg7, 92–95, 97
  Atg8, 12, 13, 15, 17, 18, 22, 23, 125, 126, 130
  Atg8/LC3, 151, 277
  Atg11, 7, 11–13, 15, 19–21, 23
  Atg12, 42, 46, 50, 51, 150
  Atg16L, 51
  beclin 1, 270, 271, 273, 274, 276, 278–280

B

Bacteria, 142, 145–148, 150, 152, 154, 155, 160, 161
Bax-interacting factor 1 (Bif-1), 49, 50
Bcl-2, 89, 90, 93–95, 97
Beclin 1, 35, 45–50, 52, 55, 59, 90, 95, 97, 98, 181, 256, 258, 259
BH3-mimetic, 59
B lymphocytes, 86, 97–98
Brucella abortus, 242–245
  endoplasmic reticulum (ER), 243
  intracellular niche, 242–245

C

CD40, 175, 253–260
CD154, 253, 254, 256, 258–260
$CD4^+$ T cells, 124, 126–136, 172
$CD8^+$ T cells, 124, 130–132, 134, 173
Cell death, 77, 271, 273, 274, 311–316, 324, 331
Chaperone-mediated autophagy, 75
Chlorophagy, 297–300, 302
Chloroplast, 288, 293–300, 302
Chlorosis, 288, 296, 297
Cholesterol, 219, 221
Class III PI3K, 255, 258
Complement C5, 128, 130
Conjugation systems, 15–18

Coxiella burnetii, 228–232
antigenic variation, 228–229
Beclin 1, 231
biphasic cycle, 229
parasitophorous vacuole (PV), 229–232
Coxsackievirus B3, 325, 326, 328, 330
Crohn's disease, 135, 174
Cryptococcus neoformans, 329
Cvt vesicle, 3, 4, 11–13, 18, 23
CXCR4, 91
Cyclins, 125
Cytokines, 86–88, 112, 117, 145, 146, 151, 153–155, 159, 160
Cytoplasm-to-vacuole trafficking (Cvt) pathway, 3, 5–7, 10–15, 18, 19, 25
Cytoskeleton, 38, 46, 52–53
Cytosolic and nuclear proteins, 125, 132

D
Damage-regulated autophagy modulator (DRAM), 41, 52, 56, 59
Death-associated protein kinase (DAP kinase), 45
Dendritic cells (DCs), 133, 181
Dengue virus, 325–328
Development, 76–77
Dextran sodium sulfate (DSS), 144, 145, 153–155
Double-membrane vesicles, 290, 301
Double-stranded RNA (dsRNA), 181

E
Early endosome antigen 1 (EEA1), 256
E64d, 221
Effector-triggered immune response, 288, 290, 295
eIF2α kinases, 42–43
Endocytic pathway, 219
Endoplasmic reticulum (ER), 146, 156
Endosomal sorting complex required for transport (ESCRT), 55–56
Endosome, 218–222
Epstein–Barr virus (EBV), 126, 132
Exocytosis, 325
Extrinsic death pathway, 88

F
Fas-associated death domain (FADD), 88, 95, 97
Flock house virus, 328
Fusion proteins, 130

G
GAS-containing autophagosome-like vacuoles (GcAVs), 219–222, 224, 225
Gcn4, 9, 10
GFP-LC3, 219–220
Glutamate decarboxylase 65 (GAD65), 128
Glyceraldehyde-3-phosphate dehydrogenase (GAPDH), 126
Graft *versus* host disease (GVHD), 136
Group A *Streptococcus* (*Streptococcus pyogenes;* GAS), 217–225

H
Hepatitis A virus, 329
Hepatitis C virus, 330, 331
Heterotrimeric G proteins, 45
Highly active antiretroviral therapy (HAART), 310–312
HIV, 89, 91, 95, 172, 173, 181
Homeostatic niche, 87
Homeostatic survival, 86, 89, 92
Hyper IgM syndrome, 253, 260
Hypersensitive response (HR), 288, 289
Hypersensitive response programmed cell death (HR-PCD), 288–290, 292–294, 302

I
IFN-β promoter stimulator 1 (IPS-1), 179, 182
IL-4, 171, 175, 176
IL-13, 171, 175, 176
IL-4Rα, 176
IL-13Rα1, 176
IL-13Rα2, 1476
Immunity-related GTPase family M member (IRGM), 171, 175, 177, 178
Immunity-related GTPases (IRGs), 171, 176–178, 254, 258–260
Induction, 4–10, 13, 18
Inflammatory bowel disease, 205–208
ATG16L1, 206–207
Crohn's disease, 205
IRGM, 206
NOD2, 206
paneth cells, 206–207
Influenza matrix protein 1, 126, 128
Innate immunity, 268, 269
IL-1β, 277
PKR, 273–276, 278, 279, 282
toll-like receptors (TLR's 3, 7 and 9), 109–116, 170, 171, 173–183, 208, 268–270, 277, 282

Interferon gamma (IFN-γ), 171, 172, 175–178, 253, 254, 256, 258–260
Intracellular antigen processing, 127–128
Intrinsic death pathway, 88
Invasins, 218
Irga6, 178
Irgm1, 142, 144, 148–149, 151–152, 171, 175–178, 181
Irgm2, 178
Irgm3, 177, 178
Isolation membrane, 219–222

L
Latent membrane protein 1 (LMP1), 129, 132
LC3, 39, 40, 46, 50–52, 90, 91, 150–152, 180–181, 256, 257
  -I, 220
  -II, 220
Legionella pneumophila, 232–234
  dot/icm gene complex, 233
  lipid rafts, 234
  replicative compartment, 232–234
Leupeptin, 221
Lipase, 24
Listeria, 148, 171, 181–183
Listeria monocytogenes, 145, 191, 196–204, 208
  ActA, 198–201
  listeriolysin O (LLO), 199, 219
  phospholipases C (PLCs), 199
  PrfA, 197, 200
  spacious listeria-containing phagosome, 201–202
Lithium chloride, 41, 58
LRG-47, 175–178
Lymphopenia, 86, 88
Lysosomal-associated membrane protein-1 (LAMP-1), 221, 254–256
Lysosomal-associated membrane protein-2 (LAMP-2), 56
Lysosome, 324, 325, 330

M
Macroautophagy, 72, 79
Macrophage, 171–173, 176, 180, 181
Major histocompatability complex (MHC) class I molecules, 124, 126, 172
Maturation step, 53–58
Melanoma differentiation-associated gene 5 (MDA-5), 181
3-Methyladenine (3-MA), 59, 231, 234, 239, 242, 243
MHC class II molecules, 124–129, 131, 133, 134, 136, 172, 179
MHC/peptide, 86–88
Microautophagy, 72
Microtubules, 52, 57, 59
Mitochondria, 72, 76, 78, 89, 93, 94, 182, 183
Mitogen-activated kinases (MAPKs), 43–44, 57
Mitophagy, 93, 94, 297, 298, 300
Mouse embryonic fibroblasts (MEFs), 220–221
mTORC1, 37–41, 43, 47, 58
mTORC2, 37–41, 58
mTOR-independent pathways, 41–42
Mycobacterium tuberculosis, 131
Myeloid differentiation primary response gene (88) (MyD88), 180, 181

N
Natural peptide ligands, 125, 126
Nelfinavir, 311
Neomycin phosphotransferase II (NeoR), 128–130
Neurodegeneration, 74, 75
Nitric oxide (NO), 289, 292–294
NOD2, 144, 145, 149, 153, 154, 159, 161
NOS2, 253, 254, 259, 260
Nuclear antigen 1 of EBV (EBNA1), 126, 128, 129, 132
Nuclear factor κB (NF-κB), 42, 45, 180
Nucleotide-binding and oligomerization domain (Nod), 179
Nutrient starvation, 288, 290, 291
NY-ESO-1, 130

P
p62, 13, 72, 75, 76
Paneth cells, 142, 145, 146, 151, 153, 155–161
Parasitophorous vacuole, 252–258
Pathogen associated molecular patterns (PAMP), 171, 179
Pathogen recognition receptors (PRR), 109, 113, 170, 171, 179–183
Peptidoglycan-recognition protein (PGRP), 200
Pexophagy, 3, 4, 11, 25
Phagocytosis, 113–117
  phagosome, 191, 201
Pho85, 7, 10
Phosphoinositide-3-kinase (PI3K), 255, 258
Piecemeal autophagy, 298, 300
Plasmacytoid dendritic cells, 110–112
Poliovirus, 325–329, 331

Porphyromonas gingivalis, 238–240
dissemination, 240
gingipains, 240
phagosomal compartment, 239
Preautophagosomal structure (PAS), 3, 4, 11–23
Pro-death signals, 288, 293, 294
Programmed cell death (PCD), 288–290, 292–294
Protease inhibitor (PI), 311
Protein kinase A (PKA), 7, 9
Protein kinase C (PKC), 44
Pseudodimer model, 133
PtdIns(3), 14, 15, 20
PtdIns3K, 7, 8, 13–15, 19, 20, 22
PtdIns3K complex, 7, 14, 15, 19, 20, 22
P2X7 receptor, 173
Pyroptosis, 194, 205, 208

Q
Quality control, 74–75, 78

R
Rab5, 256
Rab7, 254, 255, 257, 258
Rab proteins, 54
Rapamycin, 35, 37–40, 42, 43, 47, 58, 231, 234, 237, 238, 242
Reactive oxygen species (ROS), 289, 292–295, 297, 302
Recycling, 3–5, 24–25
Retinoic acid-inducible gene I (RIG-I), 179, 182
Ribulose-1,5-bisphosphate carboxylase/ oxygenase (Rubisco), 298–300
Rickettsia, 182, 183
Rickettsia conorii, 203
Rig I-like receptors (RLR), 109, 179, 181, 182
RNA viruses, 295–300, 302
R proteins, 289, 293
Runaway cell death, 290, 292–294

S
Salmonella, 148, 151
Salmonella enterica serovar Typhimurium, 191–196
effectors, 194
Salmonella-containing vacuole (SCV), 192, 194–196, 208
Salmonella-induced filaments (Sifs), 192, 193
Salmonella pathogenicity island (SPI), 193–195
type three secretion system (T3SS), 192–195, 204
Sch9, 7, 9
Selective types of autophagy, 10–13, 25
Senescence, 288–292, 295, 298, 299
Shigella, 170, 171, 182, 183
Shigella flexneri, 191, 203–205
IcsB, 204, 205
Ipaf, 205, 208
VirG, 204, 205
Single nucleotide polymorphisms (SNPs), 149
Single-stranded RNA (ssRNA), 180
Soluble NSF attachment protein receptors (SNAREs), 54
Spacious *Listeria*–containing phagosome (SLAP), 221–222
Specificity of AtATG8 expression, 292
Sphingolipids, 42, 46
Staphylococcus aureus, 224–225, 235–238
agr and sar system, 235–236
apoptosis, 236, 238
internalization, 236
methicillin-resistant (MRSA), 224
secreted enzymes and toxins, 236, 237
Starvation, 73, 74, 76, 78
STAT-1, 175
STAT-6, 175, 176
Streptolysin O (SLO), 219
Stromal cell population, lymph node cortex, 135

T
Tachyzoite, 252, 254, 255, 258
Tannic acid, 220, 221
Target of rapamycin (Tor), 176
T cells, 172, 173
proliferation, 95–97
Th1, 170, 171, 174–176, 182, 183
Th2, 170, 171, 174–176, 182, 183
Theiler's virus, 329
Thymic epithelial cells, 134
TLR2, 180, 181
TLR3, 180, 181
TLR4, 180
TLR7, 179–181
TLR8, 180, 181
TLR9, 180, 181
T lymphocyte, 147, 152
TNF-α, 143, 146, 148, 172, 175, 253, 257–260
TNF receptor-associated factors (TRAFs), 257
Tolerance, 133–136

Toll-like receptor (TLR), 109–116, 170, 171, 173, 179, 181–183, 208
TORC1, 5–10
TOR-dependent signaling pathways, 37–41
Toxoplasma, 175, 177, 182, 183
Toxoplasma gondii, 251–261
TRAF6, 257, 258
T regulatory cells, 136
Tuberculosis, 169–183
Tumor, 76–78
Tumor necrosis factor (TNF), 88, 89
Type I interferon, 180, 182
Type IV secretion system (T4SS), 229, 233

U
Ubiquitin-like protein, 15–18
Ulk1, 8, 13, 21
UVRAG, 49, 50

V
Vacuolar hydrolases, 3, 12
Vacuole membrane protein 1 (VMP1), 49, 50
Vesicle breakdown, 4, 5, 24–25
Vesicle fusion, 4, 23
Viral proteins, 268, 273, 280
  degradation, 269, 274, 276, 277, 279
  HSV-1 ICP34.5, 271, 273–276, 278–280
  influenza virus matrix protein, 277
  viral Bcl-2 (vBcl-2), 278, 280–281
Viruses, 268–282, 288, 290, 295–302
  cytomegalovirus (CMV), 278, 281
  herpes simplex virus type I (HSV-1), 270–280
  human immunodeficiency virus (HIV), 278, 281, 282
  Kaposi's sarcoma-associated herpesvirus (KHSV), 280
  murine γ-herpesvirus 68 (γ-HV68), 280, 281
  parvovirus B19, 274
  Sindbis virus, 270–271, 273–277
  tobacco mosaic virus, 274
Virus-induced chlorosis, 297

W
Wortmannin (WM), 231, 238, 239, 243

X
Xenophagy, 268, 276

Y
Yeast, 72, 73, 76–78